Lecture Notes in Networks and Systems 1456

The series "Lecture Notes in Networks and Systems" publishes the latest developments in Networks and Systems—quickly, informally and with high quality. Original research reported in proceedings and post-proceedings represents the core of LNNS.

Volumes published in LNNS embrace all aspects and subfields of, as well as new challenges in, Networks and Systems.

The series contains proceedings and edited volumes in systems and networks, spanning the areas of Cyber-Physical Systems, Autonomous Systems, Sensor Networks, Control Systems, Energy Systems, Automotive Systems, Biological Systems, Vehicular Networking and Connected Vehicles, Aerospace Systems, Automation, Manufacturing, Smart Grids, Nonlinear Systems, Power Systems, Robotics, Social Systems, Economic Systems and other. Of particular value to both the contributors and the readership are the short publication timeframe and the world-wide distribution and exposure which enable both a wide and rapid dissemination of research output.

The series covers the theory, applications, and perspectives on the state of the art and future developments relevant to systems and networks, decision making, control, complex processes and related areas, as embedded in the fields of interdisciplinary and applied sciences, engineering, computer science, physics, economics, social, and life sciences, as well as the paradigms and methodologies behind them.

Indexed by SCOPUS, EI Compendex, INSPEC, WTI Frankfurt eG, zbMATH, SCImago.

All books published in the series are submitted for consideration in Web of Science.

For proposals from Asia please contact Aninda Bose (aninda.bose@springer.com).

Maria Lapina · Zahid Raza · Mikhail Babenko ·
Mohammad Sajid · Vyacheslav Zolotarev
Editors

AISMA-2025: International Workshop on Advanced Information Security Management and Applications

Springer

Editors
Maria Lapina
North-Caucasus Federal University
Stavropol, Russia

Mikhail Babenko
North-Caucasus Federal University
Stavropol, Russia

Vyacheslav Zolotarev
Reshetnev Siberian State University
of Science and Technology
Krasnoyarsk, Russia

Zahid Raza
Jawaharlal Nehru University
New Delhi, Delhi, India

Mohammad Sajid
Aligarh Muslim University
Aligarh, Uttar Pradesh, India

ISSN 2367-3370 ISSN 2367-3389 (electronic)
Lecture Notes in Networks and Systems
ISBN 978-3-032-07274-0 ISBN 978-3-032-07275-7 (eBook)
https://doi.org/10.1007/978-3-032-07275-7

This Springer imprint is published by the registered company Springer Nature Switzerland AG
The registered company address is: Gewerbestrasse 11, 6330 Cham, Switzerland

About Book

AISMA-2025: International Workshop on Advanced in Information Security Management and Applications

This book is based on the best papers accepted for presentation during the AISMA-2025: International Workshop on Advanced in Information Security Management and Applications on May 11–15, 2025, in Stavropol, Russia, and online.

The book includes research on information security problems and solutions in the field of security awareness, blockchain and cryptography, data analysis, authentication, key distribution, and security incidents. The scope of research in information security management includes original studies, mathematical models, and software implementations related to the following topics: description of security incidents, blockchain technology, machine learning–based approaches in wireless sensor networks, phishing attack response scenarios, biometric authentication information security audit procedures, and depersonalization processes.

In addition, some papers focus on dynamics risks infrastructural genesis at critical information infrastructure facilities. Finally, the book gives insights into some problems in forecasting the development of information security events. The book intends for readership specializing in the field of information security management and applications, information security methods and features.

Acknowledgements

We would like to thank all the authors for their submissions, presentation of the results of scientific research, and our Program Committee members and additional reviewers for their precious work.

Organization

Co-chairs

Maria Lapina	North-Caucasus Federal University, Stavropol, Russia
Zahid Raza	Jawaharlal Nehru University, New Delhi, India
Mikhail Babenko	North-Caucasus Federal University, Stavropol, Russia
Mohammad Sajid	Aligarh Muslim University, Aligarh (UP), India
Vyacheslav Zolotarev	Reshetnev Siberian State University of Science and Technology, Krasnoyarsk, Russia

Program Committee members

Shafiqul Abidin	Aligarh Muslim University, Aligarh (UP), India
Safwat C. Al-Galda	University of Misan, Misan, Iraq
Morteza Amini	Sharif University of Technology, Iran
Vijay Anant Athavale	Walchand Institute of Technology (Autonomous), Solapur, India
Arutyun Avetisyan	Academician of RAS, Director of Ivannikov Institute for System Programming RAS, Moscow, Russia
Iskandar Azhmukhamedov	Astrakhan State University named after V.N. Tatishchev, Astrakhan, Russia
Ivan Azarov	North Caucasus Federal University, Stavropol, Russia
Faisal Anwer	Aligarh Muslim University, Aligarh (UP), India
Anastasia Arkhipova	Novosibirsk State Technical University, Novosibirsk, Russia
Ahmad Yahya Bahaa	Alistiqlal University, Jericho, Palestine
Deepanraj Balakrishnan	Prince Mohammad Bin Fahd University, Saudi Arabia
Elena Basan	Southern Federal University, Taganrog, Russia
Radhakrishna Bhat	Manipal Academy of Higher Education, Manipal, Karnataka, India
Vinodha D.	Christ University, Bengaluru, India
Maxim Deryabin	Samsung Advanced Institute of Technology, Suwon, Republic of Korea

Mary Anita E. A.	Christ University, Bengaluru, India
Tatiana Ermakova	University of Applied Sciences for Engineering and Economics, Berlin, Germany
Deepak Gupta	National Institute of Technology, Arunachal Pradesh, India
Harisingh Gour	Vishwavidaylaya Sagar (M.P.), India
Prakasha G. S.	Christ University, Bengaluru, India
Yevgenia Ishchukova	Southern Federal University, Taganrog, Russia
Saleema J. S.	Christ University, Bengaluru, India
Evgeny Kostyuchenko	Tomsk State University of Control Systems and Radioelectronics, Tomsk, Russia
Sandeep Kumar	Christ University, Bengaluru, India
Elena Maksimova	Russian Technological University, Moscow, Russia
Umar Muhammad Modibbo	Modibbo Adama University, Yola, Nigeria
Ekaterina Maro	Southern Federal University, Taganrog, Russia
Mohammed Hasan Najajra	Alistiqlal University, Jericho, Palestine
Neelakandan S.	Smart Neurocognitive Knowledge and Data Intelligence Research Lab - SNKDIR, R.M.K. Engineering College, Chennai, India
Sumbwanyambe Mbuyu	University of South Africa, Johannesburg, South Africa
Dmitrii Orel	North-Caucasus Federal University, Stavropol, Russia
Ekaterina Pakulova	Southern Federal University, Rostov-on-Don, Russia
Nikolai Parotkin	Reshetnev Siberian State University of Science and Technology, Krasnoyarsk, Russia
Aleksander Petrenko	Head of Software Engineering Department, Ivannikov Institute for System Programming RAS, Moscow, Russia
Mukesh Prasad	University of Technology Sydney, Sydney, Australia
Alexey Pavlychov	Far Eastern Federal University, Vladivostok, Russia
Ranjit Rajak	Vishwavidaylaya Sagar (M.P.), India
Mohammad Saleh	Alistiqlal University, Jericho, Palestine
Tamnna Siddiqui	Aligarh Muslim University, Aligarh (UP), India
Jagendra Singh	Bennett University, Greater Noida (UP), India
Mohammad Shahid	Aligarh Muslim University, Aligarh, India
Nurlan Tursynkhanov	Samarkand State University, Uzbekistan
Nguyen Viet Hung	LeQuyDon Technical University, Hanoi, Vietnam

Marina Zhukova	Reshetnev Siberian State University of Science and Technology, Krasnoyarsk, Russia

Contents

Development of Key Exchange Application Based on Homomorphic Encryption in IP Telephony ... 1
Olga Abramkina, Muborak Yakubova, and Tetiana Babenko

Automated Braille Page Orientation Assessment Using Flexible Constraints ... 11
Akmal Rustamovich Akhatov, Shokh Abbos Bakhodir ugli Ulugmurodov, Munis Xamidov, and Kumar Singla Manish

Blockchain Driven Generative AI: Ensuring Data Provenance and Model Integrity ... 19
Mahfooz Alam, Deepak, Divya Mishra, Mohammad Shahid, and Maria Lapina

The Theoretical Rationale for Using Quantum Computers to Crack Cryptosystems ... 31
Yulia Andrusenko, Dmitry Lukyanov, Alexey Dmitrienko, Alisa Kushchenko, and Diana Fedash

Methodology of Decision-Making in Information Security Specialist Training Management Systems ... 40
Anastasiya Arkhipova

Designing an Intelligent SOC Framework with Azure Sentinel for Enhanced Security Incident Management ... 51
Dipali Awasekar, Vijay Anant Athavale, L. M. R. J. Lobo, and Vitalii Lapin

Optimisation of the Genetic Circuit Alignment Algorithm on Low-Power Devices ... 63
Ekaterina Bezuglova, Vladislav Lutsenko, and Vijay Anant Athavale

Developing an Integrated Approach to Creating a Safe and Effective Collaborative Learning Environment ... 72
Vasiliy Bogadurov, Svetlana Govorova, Sergey Melnikov, and Egor Govorov

Big Data Protection Methods Using Neural Networks ... 83
Petr Chetyrbok, Anatoliy Kazak, Yelena Lukyanova, Angela Mayorova, Anna Dorofeeva, and Daria Syedneva

AI-Powered Botnet Detection Systems: A Critical Review of Current Approaches and Challenges ... 93
D. Vinodha, E. A. Mary Anita, and Maria Lapina

Budget Aware Load Balancing Algorithm with Strict Deadline Constraints for Efficient Resource Utilization in Fog and Edge Environments ... 105
Deepak, Manoj Kumar Upadhyay, and Mahfooz Alam

Enhancing Real Estate Asset Transfers with Blockchain: A Decentralized Protocol Approach ... 120
Rashmi Kamlakar Dixit, Vijay Anant Athavale, L. M. R. J. Lobo, and Vitalii Lapin

Interobject Anthropomorphism as a Threat to Information Security of Critical Information Infrastructure ... 131
Sergey Dolzhenkov and Elena Maksimova

An Enhanced Secure Message Authentication Protocol for Internet of Vehicles ... 145
E. A. Mary Anita, S. Sajini, J. Jenefa, and Vitalii Lapin

Adaptive Data Storage, Transmission and Processing System in Fog Computing Using Residue Number System and Artificial Neural Networks ... 156
Dmitriy Gorlacev, Nikita Mirny, Aisanat Geryugova, Vladislav Lutsenko, and Mikhail Zgonnikov

Investigation of Machine Learning Models for Detecting Network Anomalies ... 168
Svetlana Govorova, Sergey Melnikov, Egor Govorov, and Mohammad Shahid

On the General Approach and Criteria for Optimizing the Selection of Protective Measures in Preparation for the Procedure for Assessing Compliance with Information Security Requirements for Financial Organizations ... 177
Anastasia Iakovleva and Marina Zhukova

Assessment of the Impact of Security Measures on the Acceptable Password Trust Level in the Authentication System ... 186
Andrey Ivanov and Dmitry Nechaev

Increasing the Reliability of Text Information Based on PR-Classifiers of Linguistic Variables ... 200
Isroil Jumanov, Khusan Karshiev, Munis Xamidov, and Tharewal Sumegh

Increasing Reliability of Information on the Basis of Information Proximity of Document Elements 212
Isroil Jumanov, Khusan Karshiev, Munis Xamidov, and Tharewal Sumegh

Automate Dynamic Response to Phishing Attacks 224
Nadezhda E. Karpova, Vyacheslav V. Zolotarev, and Victor O. Goncharov

Algorithm for Improving the Reliability of Data Processing and Transmission in Fog Environment 234
Nikolay Kucherov, Egor Shiriaev, Andrey Gladkov, and E. A. Mary Anita

Random Time-Slotted Multiple Access Protocol with Load Balancing for Direct-to-Satellite IoT 243
Nikita Kustov

Machine Learning Methods for Cyber Attacks Detection and Classification 255
Vitalii Lapin, Walaa H. Elashmawi, Veronika Abakumova, and Marina Tokmakova

Detecting Information Security Threats Using Machine Learning: A Survey of Algorithms and Application Problems 268
Vitalii Lapin, Marina Tokmakova, Alexey Dmitrienko, and Mohammad Sajid

Data Integrity in Transactions and Document Verification Based on Blockchain Technologies 277
Maria Lapina, Alina Bagautdinova, Umar Muhammad Modibbo, and Elena Maksimova

Detection of Malware Attacks Based on Machine Learning Technologies 286
Maria Lapina, Evgeny Yurdanov, Zahid Raza, and Irina Ivanova

Development of an Algorithm for Finding Errors in Modular Codes 298
Artem B. Lidzhiev, Victor S. Peleshenko, Tatyana A. Peleshenko, Alexander S. Gish, and Vijay Anant Athavale

The Concept of a Methodology for Calculating the Deviation of an Employee's Psychological Profile from the Actual Average Norm of Tolerant Ethics of Behavior for Information Protection in Small Businesses 310
Igor Mandritsa, Vyacheslav Petrenko, Igor Makarov, Maksim Sokolovskii, and Olga Mandritsa

A Comparative Analysis of Machine Learning Models for Financial Fraud Detection 323
Shreya Shivkumar Mathpati, Pratibha C. Kaladeep Yalagi, Vijay Anant Athavale, and Ivan Azarov

Enhancing Scheduling with Q-Learning and Firefly Algorithm in IaaS Cloud Environment: A Security-Aware Approach 335
Zeya Mustafa, Mohammad Qasim, and Mohammad Sajid

Algorithms for Detecting Suspicious Transactions in Payment Systems Based on Artificial Intelligence 352
Fayzullo Nazarov, Sherzodjon Yarmatov, Munis Xamidov, and Asliddin Sayidkulov

Post-quantum Cryptography in Practice: A Survey of Algorithms, Applications, and Deployment Challenges 362
Sharlene Anna Pereira, Melsa Magdalene Fernandes, E. A. Mary Anita, and Tatyana Grobova

A Performance-Driven Firefly Algorithm for Scheduling Workflow in Cloud Computing 370
Mohammad Qasim, Mohammad Sajid, and Mohammad Shahid

Content Analysis Methods and Development of a System for Profile Assessing for Information Object of Social Influence 384
Marina Rudenko, Milada Krapivina, Kuhta Dmitrii, and Andrei Rudenko

System for Integrating Social Network Analysis and Information Security in the Context of Countering Radicalism 397
Marina Rudenko, Milada Krapivina, Dmitrii Kuhta, Andrei Rudenko, and Yulia Andrusenko

Designing an Architecture for a Confidential Data Analysis System in a Fog Computing Network for Low-Power Devices 413
Egor Shiriaev, Nikolay Kucherov, Ekaterina Bezuglova, and Andrey Gladkov

UK-IDS-Machine Learning Based Intrusion Detection System for Unknown Attack Detection 425
T. Sowmya, E. A. Mary Anita, and Maria Lapina

Development of an Information Security Risk Assessment Model Through the Calculation of the Vulnerability Danger Factor 434
Maria Utenkova, Andrey Komkov, Elena Maksimova, and Oksana Mezentseva

Solving High-Performance Computing Problems Using Distributed Neural Networks with Numerical Methods ... 442
Nikolay Vershkov, Mikhail Babenko, Vladislav Lutsenko, and Nataliya Kuchukova

Modern Methods for Anomaly Detection in Enterprise System Logs: Algorithms, Implementations, and Practical Case Studies ... 451
Sergey Yarushev and Aleksandr Anurov

Algebraic Generalization of Wilson's Theorem and Its Applications in Cryptographic and Intelligent Systems ... 462
Vladimir N. Yevstifeyev, Yelizaveta S. Vitulyova, Dina B. Shaltykova, and Ibragim E. Suleimenov

Modeling Intruder Behavior in Information Systems: Methodology Using Graphs with Non-standard Reachability ... 471
Boris Zakhir and Anna Oleynikova

SFOA-Driven Defense: Leveraging SFOA for Optimal Feature Selection in Network Intrusion Detection Systems ... 482
Mohammad Zayeem and Mohammad Sajid

Provision of Security of Wireless Information Transmission Systems in View of Correlation Properties of Code Sequences ... 497
Alexander Zhuk, Kirill Stogniy, and Elsi Midaev

Evaluation of Approaches to Modeling Information Risk of Telecommunication Systems Based on Hypergraphs ... 510
Vyacheslav Zolotarev, Elena Gorodetskaya, and Alexey Kuznetsov

Author Index ... 523

Editors of AISMA-2025

Dr. Maria Lapina is Associate Professor of the Department of Computational Mathematics and Cybernetics at North Caucasus Federal University, Stavropol, Russia. She has more than 20 years of teaching experience. She received PhD in mathematics from Stavropol State University in 2005. She defended her master's thesis in Information security at the North Caucasus Federal University with honors in 2015. Her research interests include Data Science, Machine Learning, Big Data, Cybersecurity, Cryptography, Trusted Artificial Intelligence, Edu-Tech, Teacher Education, & Higher Ed. She has published in peer-reviewed journals indexed in Scopus and WoS. She has more than 80 research papers. She is Active Editorial Board Member for Springer Nature, IGI Global. She is Active Reviewer of MDPI publishing journals. She is Organizer, Co-chair and Editor of the proceedings of international scientific events: AISMA-2021, 2022, 2023, 2024: International Workshop on Advanced in Information Security Management and Applications, DMAI-2021: International Workshop on Data Mining and Al. She has participated in international internships at Sapienza University of Rome (Italy) in 2018, the University of Ljubljana (Slovenia) in 2018, the Technical University of Dresden (Germany) in 2019 and the 2nd ASEF Innovation Laboratory of Higher Education (ASEFInnoLab) at Fudan University in Shanghai, China, in 2020. She received a grant from the Vladimir Potanin Foundation and an award as the Best Lecturer of Russia. She is a finalist of the All-Russian management competition "Leaders of Russia".

Prof. Zahid Raza is currently working as Dean, School of Computer and Systems Sciences, Jawaharlal Nehru University, India. Prof. Raza has a M.Sc. degree in Electronics in which he was the Gold Medalist, M.Tech. degree in Computer Science and Ph.D. in Computer Science. Prior to joining Jawaharlal Nehru University, he served as Lecturer in Banasthali Vidyapith University, Rajasthan, India. His research interest is in the area of Parallel and Distributed Systems. He has proposed various resource provisioning models for job scheduling for Computational Grid, Cloud and Parallel Systems. Prof. Raza has published more than 30 research papers in various peer reviewed International Journals and Transactions and more than 30 research papers in proceedings of various peer-reviewed conferences in India and abroad along with contributing towards chapter in an edited book. Prof. Raza co-authored a book in Springer Briefs in Computer Science entitled "Auction based Resource Provisioning in Cloud Computing". Various invited talks have been delivered by him throughout his academic career. Prof. Raza has also served in various committees for academic, administrative and evaluation purposes.

Prof. Mikhail Babenko is Head of the Department of Computational Mathematics and Cybernetics North-Caucasus Federal University, Russia. He is Head of the Department of Theoretical and Numerical Systems of the Regional Scientific and Educational Mathematical Centre of the North Caucasus Centre for Mathematical Research. He is

Member of the scientific school named after Professor Chervyakov "Neuromathematics, modular neurocomputers and high-performance computing". He received PhD in Mathematical modeling, numerical methods, and software. Mikhail took internships in CICESE Research Center in México; Le Quy Don Technical University, Hanoi, Vietnam in 2015; University of Lorraine, Nancy, France in 2017. Research area: Security, Cryptography, Residue Number System, Neurocomputer Technologies, Digital Signal Processing, Big Data, Internet of Things, Fog-Edge-Cloud Computing, Trust, Uncertainty, Scalable and Reliable Systems, Elliptic Curve, Cryptographic Methods of Information Protection, and Cybersecurity.

Dr. Mohammad Sajid is Assistant Professor in the Department of Computer Science at Aligarh Muslim University, India. He has completed a Ph.D., M. Tech. and M.C.A. degree from the School of Computer and Systems Sciences, Jawaharlal Nehru University (JNU), New Delhi. His research interests include Evolutionary and swarm Intelligence Algorithms, Machine Learning, Combinatorial Optimization problems, Cloud Computing, Applications of Scheduling and Routing, and Intelligent Transportation Systems. He has published many research papers in international conferences and journals of repute (including IEEE, ACM, Elsevier, Springer, Wiley, Taylor & Francis, IGI Global, InderScience, etc.). He is Reviewer of several international journals of repute (including ACM, IEEE, Taylor and Francis, Elsevier, Springer, Wiley, etc.)

Dr. Vyacheslav Zolotarev is working as Head of the Department of Information Security at Siberian State University of Science and Technology, Krasnoyarsk, Russia. He has extensive experience in scientific and practical projects in the field of information security, an information security management specialist. Works both with research teams and in the educational field. Has good reviews in the field of teaching information security management. Teaches courses on information security management, ethical hacking (in terms of social engineering) and investigation of computer incidents. He has worked as Head of an innovative enterprise and has the skills to manage project teams, modern development techniques.

Development of Key Exchange Application Based on Homomorphic Encryption in IP Telephony

Olga Abramkina[1,2](✉), Muborak Yakubova[1], and Tetiana Babenko[2]

[1] Almaty University of Power Engineering and Telecommunications named after Gumarbek Daukeyev, 050013 Almaty, Kazakhstan
olga.manank@gmail.com

[2] International IT University, 050000 Almaty, Kazakhstan

Abstract. The paper is devoted to the development and implementation of a key exchange application using homomorphic encryption to enhance security in IP telephony. With the increasing dependence on Voice over IP (VoIP) technologies, protecting sensitive data has become a critical issue. Traditional key exchange protocols, despite their effectiveness, are often vulnerable to various cyber threats, requiring the search for more secure alternatives. An analysis of existing key exchange protocols such as ZRTP and TLS-SRTP is performed, assessing their strengths and weaknesses in terms of security, performance, and usability. Then, the principles and advantages of homomorphic encryption, which allows computing over encrypted data without decrypting it, are explored. A prototype application is developed that integrates homomorphic encryption into the key exchange process, ensuring secure communication while maintaining operational efficiency. The application is thoroughly tested under various conditions to evaluate its effectiveness in real-world scenarios, paying special attention to key performance metrics such as latency, computational load, and bandwidth requirements.

Keywords: Secure Communication · Cryptography · Security Protocol · Privacy · VoIP Security

1 Introduction

In the context of modern IP-telephony systems, ensuring secure key exchange remains a critical challenge, especially in real-time communication environments where both performance and confidentiality are essential [1–3]. Traditional cryptographic protocols-such as Diffie-Hellman and its variants, as well as TLS-SRTP-offer a foundational level of protection but are increasingly susceptible to well-documented threats, including man-in-the-middle (MITM) attacks, replay attacks, and quantum-based cryptographic attacks [4–6]. These vulnerabilities have intensified the need for more resilient cryptographic methods, particularly in systems where maintaining low latency is crucial.

Homomorphic encryption (HE) has emerged as a promising cryptographic technique for enhancing the privacy and integrity of data in communication systems [7].

M. Lapina et al. (Eds.): AISMA 2025, LNNS 1456, pp. 1–10, 2026.
https://doi.org/10.1007/978-3-032-07275-7_1

Unlike conventional encryption, HE enables computations to be performed directly on encrypted data, which allows sensitive information to remain confidential throughout processing. A comprehensive classification of homomorphic schemes includes partially homomorphic encryption (PHE), somewhat homomorphic encryption (SHE), and fully homomorphic encryption (FHE), each offering different levels of arithmetic capability and computational complexity [8, 9].

The application of homomorphic encryption to voice-over-IP (VoIP) systems poses practical challenges. While FHE provides strong theoretical guarantees, its high computational requirements and processing delays limit its direct use in time-sensitive environments [10, 11]. In response, researchers have proposed optimization strategies such as reducing key sizes, refining encryption parameters, and improving arithmetic operations to support faster performance in communication systems [12–14].

Recent work has also focused on the deployment of homomorphic schemes in multi-user systems, where trade-offs between computational efficiency and confidentiality must be considered. Solutions based on PHE or hybrid cryptographic approaches have demonstrated potential in scenarios that involve real-time secure communication [15, 16]. Furthermore, advanced schemes have explored applications in encrypted key exchange, secure computation, and identity-based models of encryption [17–19].

Some studies have investigated the feasibility of deploying homomorphic encryption on constrained hardware platforms, including microcontrollers. These implementations, although limited in functionality, represent important steps toward expanding the applicability of HE in lightweight environments [20, 21]. Despite these advances, limitations remain regarding support for complex arithmetic operations and overall system performance [22–24].

In addition to standalone cryptographic designs, several studies have examined the integration of homomorphic encryption with existing communication protocols such as TLS-SRTP and ZRTP. These hybrid approaches seek to combine the strengths of both classical and modern encryption techniques to ensure secure voice and video transmission [25, 26]. Other contributions have explored floating-point implementations of he and privacy-preserving methods for domain-specific applications [27].

This paper examines the applicability of homomorphic encryption in securing key exchange for IP-telephony systems. Particular attention is paid to computational efficiency, architectural constraints, and integration with existing VoIP frameworks. The goal is to assess the extent to which homomorphic encryption can provide meaningful improvements to security without compromising the performance requirements of real-time communication.

2 Methodology

The study implemented a secure key exchange using homomorphic encryption in the Asterisk environment and the SIP protocol. The main work is focused on protecting the encryption keys that are used to establish a secure VoIP connection. Homomorphic encryption performs calculations on encrypted data without decrypting it. Figure 1 shows a structural diagram of the system and its components.

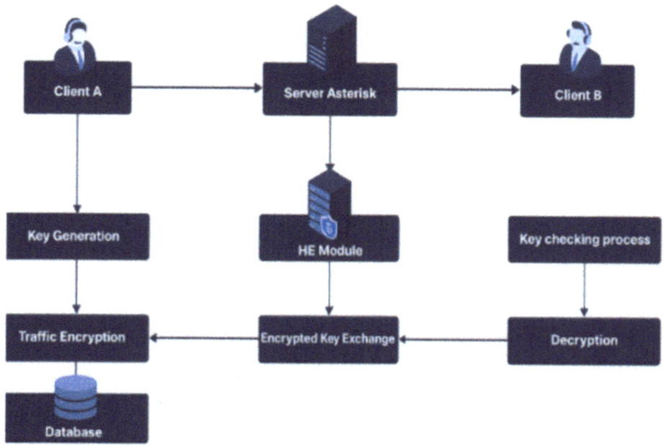

Fig. 1. System architecture diagram.

The server platform is Asterisk PBX (20.6.0) an open-source framework designed for building IP telephony systems. It was able to manage SIP calls and user registration and also manage AGI scripts. Ubuntu 24.04.02 LTS was selected as the essential operating system because it is the most reliable, has the largest number of packages, and is the easiest to deploy with the Asterisk PBX. Each of the system's building blocks such as Asterisk, Python scripts, and MariaDB server were set up and conFig.d in this environment. Linphone was the SIP client that was used to place and receive calls. The configuration of Linphone was not changed, and all the keys were controlled via the external scripts developed. Python 3.12 was chosen for its simplicity and ease of use, its large collection of libraries and good support for cryptographic and machine learning tasks.

It runs on the SIP protocol, which is one of the most common call control protocols in VoIP networks. The client part of the project is implemented through the Linphone application. The application helps to implement the interface through which the user will perform authorization, registration and free calls. The study also uses a database for storing service information about users, an encryption module for implementing the functions of generating and storing keys. In the diagram, the Asterisk server is the core of the system, its job is to handle user registration, call routing, and authentication for security. The Linphone client service is the user interface that logs into the system and can initiate a SIP call, as well as handle all call control and data transfer issues. Key management is a pair of Paillier encryption keys during the key exchange process. The database stores user accounts, call records, and other configurations, retrieved from the servers when needed. The arrows in the diagram indicate the data flows between the components. All of these components work together to create a secure and efficient IP telephony system.

The Asterisk server does the following:

1. Users log in using SIP credentials validated in the database;
2. Handles call setup, media transfer, and call termination;
3. Generates and securely shares keys to protect data;
4. Uses homomorphic encryption + ECDH for secure SRTP keys;

5. Asterisk AGI runs custom scripts for additional features such as call logging or encryption.

The Linphone client application allows users to interact with the server. The SIP-based VoIP client works with the following actions:

1. Connects to Asterisk using SIP credentials;
2. Makes and receives calls over a secure channel;
3. Generates ECDH keys, encrypts the public key using Paillier and sends it to Asterisk.

Linphone and Asterisk interact as follows. After connecting to the server, Linphone sends data in real time. At this time, an encryption key is generated at the client level, which is negotiated with the server. Encryption keys work only during an active call. After the call, the key information is permanently deleted. Also worth mentioning is an important part of the project, this is homomorphic encryption, the workflow is described below:

1. Generates, stores (during calls) and securely exchanges keys;
2. Uses Paillier homomorphic encryption to protect ECDH public keys;
3. ECDH to create a shared SRTP key for secure VoIP calls.

The client pre-registers on the server, and the program is authenticated before installation. If authentication is successful, the server creates a secure channel. A symmetric ECDH session key is created between the participants. And for more reliable encryption, Paillier protects the exchange of public keys. During the call, all data is encrypted on the client side using SRTP. The server does not decrypt or process this data, which guarantees its confidentiality. After the session is completed, the server deletes temporary data and erases the keys. This is necessary to prevent reuse.

3 Result and Discussion

The system was tested and Table 1 describes the tests performed and their results.

Table 1. System testing result

Test scenario	Expected result	Actual result	Status
Registering users in SIP	User successfully authenticated	Successful	+
Generating ECDH keys	A key pair (private + public) was created	Successful	+
Encrypting Paillier keys	Key successfully encrypted	Successful	+
Transferring keys via Asterisk	Encrypted key was transferred to the second client	Successful	+
Decrypting the key	Client decrypts the key	Successful	+

The system was tested and the tests performed and their results are described in Table 1. Based on the results of the tests that were performed to verify different usage scenarios from SIP registration to key deletion, it can be seen that from SIP registration to key deletion, the processes were successful, which confirmed the correctness of the implemented logic and the reliability of the encryption and decryption mechanisms in the IP telephony system. In the last scenario, a pre-trained machine learning model was used to check the quality of service (QoS) criteria for the call. Using such data, the model could find unknown problems in the network and advise how to solve them, for example, by reducing the codec bitrate. This indicated that the system could work to perform QoS and fault management in real time, thereby increasing its intelligence and ability to be resilient to errors. We can see that the users have successfully logged into their SIP accounts. As shown in Fig. 2, the «pjsip show contacts» command showed a list of users and their details.

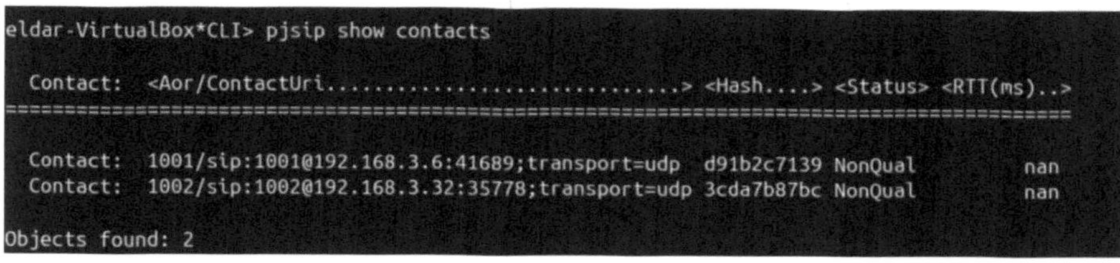

Fig. 2. User successfully authorized the server.

Figure 3 shows that our Python scripts are running, and that the key creation and encryption was successful.

Fig. 3. Key generation and encryption log.

Figure 4 shows the transmission of a signaling message from the first user to the second user. The syslog verifies that the SIP INVITE or the matching message has been dealt with properly and that the device shows that the message arrived at the final destination smoothly with no problems.

Fig. 4. Message successfully sent to second user.

Figures 5 and 6 show the results of session key authentication during call setup, where shows that the keys were correctly decrypted and delivered to each other, and the message "Keys match, continue call" appeared in the system log. As a result, the call was not blocked. In Fig. 7, however, the system logs the message "Error: Keys do not match" and hence the call cannot be initiated. Thus, the mechanism only secures matched calls.

Fig. 5. Keys match and call initiation.

Figure 7 that shows the machine learning scripts being run on the terminal. More particularly, after hang up with a customer, the system will load the predict_action.py script, and the algorithm conFig.d will predict the best QoS using previous data. The log states that the script has begun, the data has been processed, and the work is done without any errors, which indicates that the QoS feedback function has worked seamlessly.

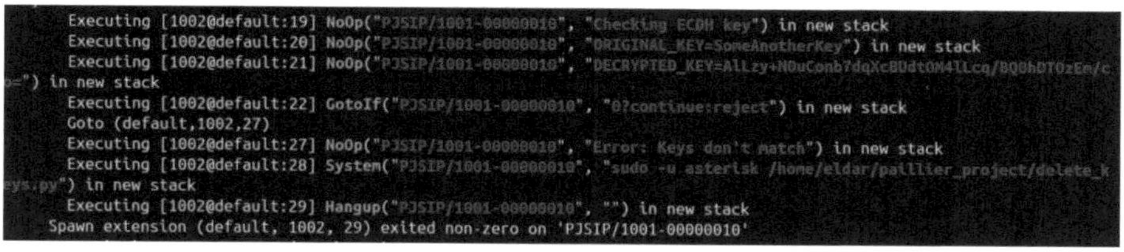

Fig. 6. Call not started because keys don't match.

Fig. 7. Logs of successful ML scripts finishing.

In system testing carried out, it was confirmed, that the IP telephony system developed not only is correct, stable, and reliable, but is also greatly secure, extensible, and can be implemented in real-time environments where service quality and confidentiality are critical factors. The IP telephony system has proved its resolvability, stability, and dependability during the conducted testing. All critical elements were tested, and the results show that the employed architecture is robust and the given requirements are satisfied. The test cases that were completed successfully for the normal and the error situations confirm that the system's architecture is strong and the conditions of the initial design are met. All in all, the testing phase is a full demonstration that the new system is not just operational but also secure, extensible.

To verify the performance and resilience of our system, a simulated man-in-the-middle (MITM) attack was performed. In this simulation, an attacker attempted to intercept and subsequently decrypt an encrypted message transmitted between a client and a server over the network. This process is illustrated in Fig. 8, which demonstrates the security evaluation setup and the attempted attack scenario.

During the experiment, a public key structure based on the assumed parameter after that, an attempt to call the decryption function using only the public key resulted in an error (Attempting to decrypt). This is because the library does not specifically provide the ability to decrypt data without a private key. The system works in such a way that it preserves the fundamental principle of cryptography as protecting the confidentiality of data. This is due to the fact that it is impossible to decrypt without the private key, even in the presence of the public key and the ciphertext itself.

The results showed that this confirms the resistance of the applied encryption algorithm to attacks. These attacks that are completed by attackers related to the interception of data and attempts to decrypt them. This indicates the correctness of the implementation of Paillier in the selected library that corresponds to the cryptographic scheme basic security requirements. As a result, the encryption scheme protects information during its transmission over the network preventing the recovery of the original data without access to the private key.

Another check was related to the resistance to bruteforce attacks in the hybrid cryptographic scheme ECDH + Paillier. The thesis focuses on the study of the robustness

Fig. 8. Wireshark result with Encrypted key in SIP Message Body.

of a hybrid cryptographic scheme combining Elliptic Curve Diffie-Hellman (ECDH) and Paillier homomorphic encryption against one of the potentially dangerous types of bruteforce attacks. The topic we are studying is related to the protection of transmitted data in an open and vulnerable environment as in Internet protocols.

The ECDH method, unlike the classical Diffie-Hellman algorithm, uses elliptic curves like secp256r1. It provides cryptographic strength at 256 bits and has a compact key size. To put it simply, an attacker deciding to try all possible variants of the private key would need to make relatively 2^{256} attempts. This is equivalent to the number 1.16×10^{77}. Even given the hypothetical presence of powerful supercomputers, this remains beyond the realm of reality. Additionally, the smaller key size makes it harder to work on traffic analysis and dictionary attacks.

This level of protection may not be enough if an attacker can intercept or spoof the ECDH public key. To further protect the data Paillier scheme was used. Homomorphic encryption is based on the complexity of factorization of large numbers as in RSA. Robustness is maintained for modulus lengths of 2048 bits or more. In terms of operation, this means that if an attacker obtains the encrypted public key, he cannot decrypt it without knowing the Paillier secret key.

4 Conclusion

Based on the results of the work, it can be said that the system using homomorphic encryption ensures a secure exchange of session keys when establishing voice connections, eliminating the possibility of their compromise even in the event of interference in the communication channel. During the study, a number of limitations were identified that the required level of security imposes on the possibilities of implementing homomorphic encryption.

ECDH was used as a method, which made it possible to quickly reach a common secret agreement between two endpoints. Paille homomorphic encryption was also used to further improve the security of test public keys over the communication channel. A prototype was implemented on the Asterisk platform using Linphone SIP clients, with which a number of tests were carried out, which included an assessment of the functionality and performance of the system, checking the integrity of the key generation, exchange and deletion functions. As a result, it was concluded that the solution is reliable and its value will be sufficient for modern communication systems with strict security requirements.

The scientific value of the work lies in the successful practical implementation of a hybrid cryptographic scheme in the open VoIP platform Asterisk. In modern commercial systems, such solutions are rarely used, but the relevance of the topic increases as the number of cyberattacks on corporate communication systems increases. Additionally, incorporating algebraic methods for logical operation optimization [28] could enhance key exchange efficiency. Another promising avenue is the integration of homomorphic encryption with hardware solutions, such as optical neural networks [29], to reduce computational overhead in real-time VoIP systems.

Funding. This research is funded by the Committee of Science of the Ministry of Science and Higher Education of the Republic of Kazakhstan (Grant No. AP25796395).

References

1. Yakubova, M., Manankova, O., Mukasheva, A., Baikenov, A., Serikov, T.: The development of a secure internet protocol (IP) network based on asterisk private branch exchange (PBX). Appl. Sci. Switzerland. **13**(19), 10712 (2023). https://doi.org/10.3390/app131910712
2. Abdalla, M., Bellare, M., Rogaway, P.: DHAES: an encryption scheme based on the Diffie-Hellman problem. IEEE. **P1363a** (1998)
3. Babenko, T., Kolesnikova, K., Panchenko, M., Meish, Y., Mazurchuk, P.: Risk assessment of cryptojacking attacks on endpoint systems: threats to sustainable digital agriculture. Sustainability Switzerland. **17**(12), 5426 (2025). https://doi.org/10.3390/su17125426
4. Manankova, O.A., Yakubov, B.M., Serikov, T.G., Yakubova, M.Z., Mukasheva, A.K.: Analysis and research of the security of a wireless telecommunications network based on the IP PBX asterisk in an opnet environment. J. Theor. Appl. Inf. Technol. **99**(14), 3617–3630 (2021)
5. Yakubova, M., Abramkina, O., Baikenov, A.: Cryptanalysis of the SHA-256 hash function using rainbow tables. Indones. J. Electr. Eng. Inform. **10**(4), 930–944 (2022). https://doi.org/10.52549/ijeei.v10i4.4247
6. Manankova, O.A., Yakubova, M.Z., Rakhmatullaev, M.A., Baikenov, A.S.: Simulation of the rainbow attack on the SHA-256 hash function. J. Theor. Appl. Inf. Technol. **101**(4), 1594–1603 (2023)
7. Acar, A., Aksu, H., Uluagac, A.S., Conti, M.: A survey on homomorphic encryption schemes: theory and implementation. ACM Comput. Surv. **51**(4), 1–35 (2018)
8. Varnovsky, N.P., Shokurov, A.V.: Homomorphic encryption. In: Proceedings of ISP RAS. Available: https://cyberleninka.ru/article/n/gomomorfnoe-shifrovanie. Accessed 22 Feb 2025
9. Chaudhary, P., Gupta, R., Singh, A., Majumder, P.: Analysis and comparison of various fully homomorphic encryption techniques. In: 2019 International Conference on Computing Power and Communication Technologies (GUCON), pp. 58–62. New Delhi, India (2019)
10. Wang, D., Guo, B., Shen, Y., Cheng, S.-J., Lin, Y.-H.: A faster fully homomorphic encryption scheme in big data. In: 2017 IEEE 2nd International Conference on Big Data Analysis (ICBDA), pp. 345–349. Beijing, China (2017)

11. Rohloff, K., Cousins, D.B., Sumorok, D.: Scalable, practical VoIP teleconferencing with end-to-end homomorphic encryption. IEEE Trans. Inf. Forensics Secur. **12**(5), 1031–1041 (2017)
12. Lee, S., Shin, D.-J.: Overflow-detectable floating-point fully homomorphic encryption. IEEE Access. **12**, 6160–6180 (2024)
13. Alanwar, A., etc.: Privacy-preserving set-based estimation using partially homomorphic encryption. Eur. J. Control. **68**(1), 100786 (2023)
14. Tu, G., Liu, W., Zhou, T., Yang, X., Zhang, F.: Concise and efficient multi-identity fully homomorphic encryption scheme. IEEE Access. **12**, 49640–49652 (2024)
15. Yi, X., Paulet, R., Bertino, E.: Homomorphic Encryption and Applications SpringerBriefs in Computer Science. Springer, Cham (2014)
16. Jiefeng, L., Sheng, L.: An efficient and more secure searchable encryption algorithm based on fully homomorphic encryption. In: 2023 International Conference on Automation, Control and Electronics Engineering (CACEE), pp. 83–92. Chongqing, China (2023)
17. Kim, J., Yun, A.: Secure fully homomorphic authenticated encryption. IEEE Access. **9**, 107279–107297 (2021)
18. Han, J.-L., Yang, M., Wang, C.-L., Xu, S.-S.: The Implementation and Application of Fully Homomorphic Encryption Scheme. In: 2012 IMCCC, pp. 714–717. Harbin, China (2012)
19. Temirbekova, Z., Pyrkova, A., Abdiakhmetova, Z., Berdaly, A.: Library of fully homomorphic encryption on a microcontroller. In: 2022 International Conference on Smart Information Systems and Technologies (SIST), pp. 1–5. Nur-Sultan, Kazakhstan (2022)
20. Mohammed, S.J., Taha, D.B.: Performance evaluation of RSA, ElGamal, and Paillier partial homomorphic encryption algorithms. In: 2022 International Conference on Computer Science and Software Engineering (CSASE), pp. 89–94. Duhok, Iraq (2022)
21. Ermukhambetova, B.B., Mun, G.A., Kabdushev, S.B., Vitulyova, Y.S., Suleimenov, I.E.: New approaches to the development of information security systems for unmanned vehicles. Indones. J. Electr. Eng. Comput. Sci. **31**(2), 810–819 (2023). https://doi.org/10.11591/ijeecs.v31.i2.pp810-819
22. Mittal, S., Singh, S.: Retrospective study on classical homomorphic encryption algorithms. In: 2023 International Conference on IoT, Communication and Automation Technology (ICICAT), pp. 1–6. Gorakhpur, India (2023)
23. Grechko, V., Babenko, T., Myrutenko, L.: Secure Software Developing Recommendations. In: 2019 IEEE International Scientific Practical Conference Problems of Infocommunications Science and Technology Pic S and T 2019 Proceedings, pp. 45–50 (2019). https://doi.org/10.1109/PICST47496.2019.9061529
24. Babenko, T., Hnatiienko, H., Bigdan, A.: Model for determining the protection level of a complex system. In: CEUR Workshop Proceedings 3132, pp. 156–165 (2022)
25. Boyu, M., Hong, C.: Adaptive secure homomorphic encryption scheme. In: 2024 IEEE 2nd ICCECT, pp. 1–5. Jilin, China (2024)
26. Abramkina, O., Yakubova, M., Serikov, T., Begimbayeva, Y., Yakubov, B.: Implementation of lattice theory into the TLS to ensure secure traffic transmission in IP networks based on IP PBX asterisk. Int. J. Adv. Comput. Sci. Appl. **15**(10), 749–756 (2024)
27. Stobbe, P., Keijzer, T., Ferrari, R.M.G.: A fully homomorphic encryption scheme for real-time safe control. In: 2022 IEEE 61st Conference on Decision and Control (CDC), pp. 2911–2916. Cancun, Mexico (2022)
28. Suleimenov, I.E., Vitulyova, Y.S., Kabdushev, S.B., Bakirov, A.S.: Improving the efficiency of using multivalued logic tools: application of algebraic rings. Sci. Rep. **13**, 22021 (2023). https://doi.org/10.1038/s41598-023-49593-1
29. Kalimoldayev, M.N., et al.: To the question of physical implementation of optical neural networks. News Natl. Acad. Sci. Repub. Kazakhstan, Ser. Geol. Tech. Sci. **2**, 217–224 (2019). https://doi.org/10.32014/2019.2518-170X.57

Automated Braille Page Orientation Assessment Using Flexible Constraints

Akmal Rustamovich Akhatov[1], Shokh Abbos Bakhodir ugli Ulugmurodov[2], Munis Xamidov[1(✉)], and Kumar Singla Manish[3]

[1] Samarkand State University, 140100 Samarkand, Uzbekistan
samsungmunis@gmail.com
[2] Jizzakh branch of the National University of Uzbekistan, 130100, Jizzakh, Uzbekistan
[3] Applied Science Private University, Amman 11931, Jordan

Abstract. This article delves into the Optical Character Recognition (OCR) techniques used for recognizing the Latin Braille alphabet in Uzbek, the script designed for individuals with visual impairments. The text recognition process was achieved by applying a combination of various image processing and analysis methods. In particular, significant attention was given to adaptive thresholding techniques and the determination of text tilt angles.

Keywords: Adaptive constraints · tilt angle · optical character recognition · grid structure

1 Introduction

The purpose of this article is to develop an OCR that recognizes optical signs for Braille code (a type of writing used by the blind). In doing this work, we considered the process of identifying Braille characters using the following 5 main steps:

- Input image quality management;
- Processing scanned images to determine the location of the points that make up the text written in Braille.
- Processing the points found to obtain line and column measurement values in the text. This method is known as Braille mesh detection.
- Using the Latin Braille alphabet in the Uzbek language to transform the combination of identified points into an expression in the ASCII code;
- User interface.

We have the results related to the above five parts. We will work with the results presented in this article [1].

Incoming image quality management. In this article, we chose the Twen standard [2]. This standard has been promoted by many industrial firms (HP, Logitech, Caere, and Aldus). The purpose of this standard is to improve the quality of the incoming image and reduce various noise values. This standard allows for the acquisition of images from multiple sources, not just scanners. This is the priority task of getting the characters in the image.

M. Lapina et al. (Eds.): AISMA 2025, LNNS 1456, pp. 11–18, 2026.
https://doi.org/10.1007/978-3-032-07275-7_2

2 Localization of Points

This is the first stage of the logical process, and the input image should have low noise parameters. In the image written in Braille, the following two types of speech can be seen: convex points and concave points. The blind person only identifies points in the text and performs the reading process. Concave points are convex points printed on the page on the back of this text (Fig. 1). However, in this case, we suggest that you also separate the characters on both sides by scanning the image once [3].

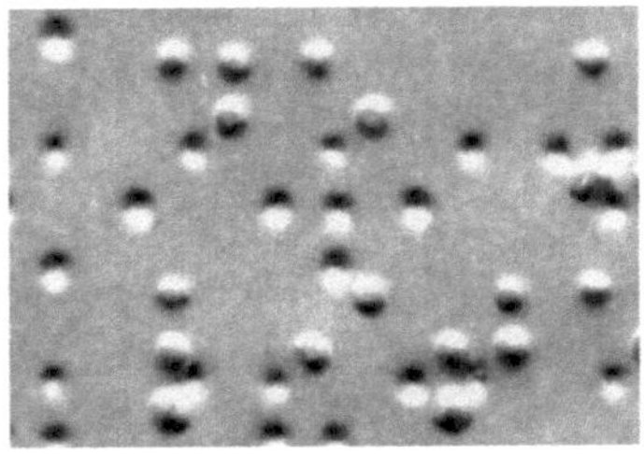

Fig. 1. Double-sided text in Braille.

The starting point in Braille is the observed model of the existing pattern. During the experiment, we found a bright crescent-shaped field above the protruding point on the scanned images, and a dark field of the same shape at the bottom. The concave points had an inverse pattern (dark at the top, bright at the bottom) (see Fig. 2). We used two methods of adaptive splitting when dividing the given areas into white and black spots. Boundary brightness levels are defined as points on the brightness histogram [4] that divide its total area into two areas, so the given proportion of distribution data is above/below the corresponding threshold. The percentage was experimentally set at 5%.

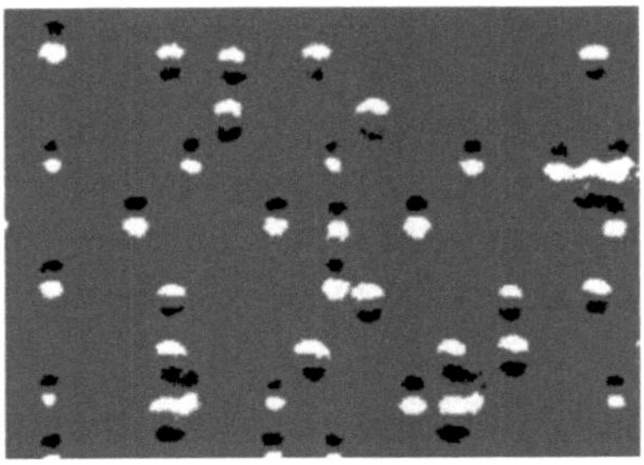

Fig. 2. Image after the limiting process.

Next, we define the centers of the black and white fields and take them as the black and white dot coordinates. Next, we will identify neighboring pairs, which are the white dot above and the black dot below. These pairs are later designated as "swelling points" and "concave points" [5].

This method is characterized by ease of use and high execution speed. Nevertheless, this method has two significant drawbacks [6, 7]. Firstly, the disappearance of some points is observed, and secondly, the appearance of false points.

The errors are due to the presence of closely spaced spots that combine after the boundary change process (this occurs especially when the swelling and settling points are very close; See the dots in the last column in Fig. 3. The problem can be solved by transferring all intelligent processing to the next stage - network identification [8]. The lattice construction algorithm we developed effectively discards points located in an incorrect order and identifies missing point fields belonging to the source.

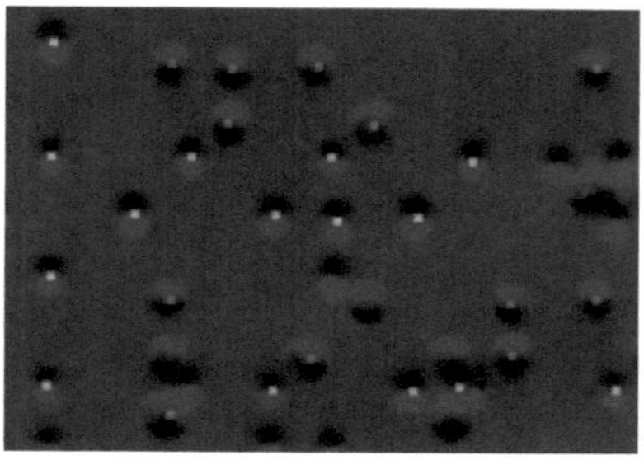

Fig. 3. Localized image points

This algorithm naturally discards points that need not be identified. In addition to the intuitive approach to working with dots, we have identified alternative efficiency methods for identifying patterns. You can review these methods.

3 Network Identification

At this stage, we need to determine the location and direction of the lines and columns of the text. These positions make up what we call Braille mesh. In Braille, we have a grid of 6 points. Each point may or may not be active (marked as a raised Braille point in the text). Each combination represents a separate character. If we have N characters (text), the distances between them are normalized [9]. Next, we have a grid consisting of 6xN points (Braille grid). The previous algorithm identifies a portion of the points, but their number is sufficient to construct a grid. When a complete lattice exists, it can be studied in detail to check for the presence of points close to each potential location (each point has several points located above and below, but sometimes their location is decentralized) [10].

Using this algorithm, we restore the lost points. Also, it is necessary to mention false points - points that do not comply with the distance rules. The algorithm allows them to be discarded as a simple exception.

The first step of the algorithm is to determine the angle of inclination, which is based on the following method [11, 12].

After that, the algorithm selects the starting point and performs a sequential shift of standard distances to build the entire network. Our main choice is not to choose the wrong position. To solve this problem, the system analyzes 10 field points closest to the field of certain corresponding points. For each of these points, the system attempts to establish a relationship based on a given distance from the original corresponding entry point [13].

In each attempt, the system calculates the number of points found in the standard positions by a certain method. The maximum value of these 10 points is used to select the starting point (the action is only performed vertically, and the points are sought along horizontal lines, which ensures independence from the relative position of the point within the symbol).

Another important task is to determine the relative position of the starting point within the Braille symbol under consideration. As you can see in Fig. 4, there are 6 positions [14, 15]. We simply approximate all possible positions and then use the method described above to find the maximum point. First, we determine which column (from 2) the point belongs to, then which row (from 3).

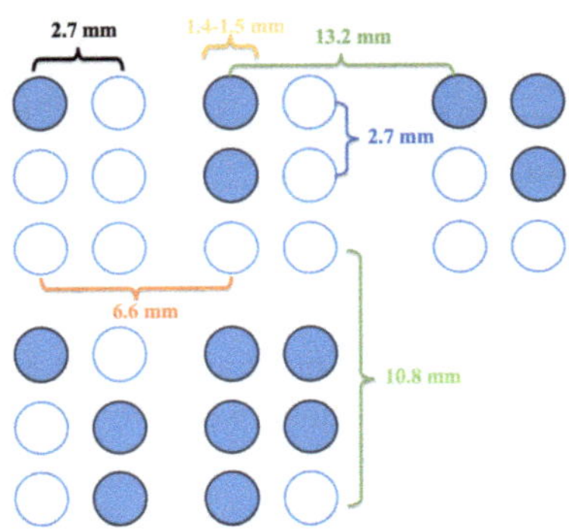

Fig. 4. Braille mesh.

The final stage is considered to be the process from the starting point to the construction of the entire network. This action is performed flexibly. The essence of this method lies in shifting a 6-dimensional lattice for each subsequent point. Once the next character is found, we check the points near the lattice nodes (Fig. 5). Next, we align the lattice position based on these points. We continue this process to the right edge of the image [16].

If the boundaries are set correctly and clearly, the points in the wrong position will not be considered in this process.

3.1 ASCII Code Translation

After stage 3, we have Braille text represented by six-dot characters (six bits, which can be equal to 1 or 0). Now we need to convert it to ASCII code. In other words, we use Braille (Fig. 6).

The proposed method takes the text (T) and dictionary (D) and uses D as a reference to correct possible grammatical errors in the words in T. The algorithm uses the comparison

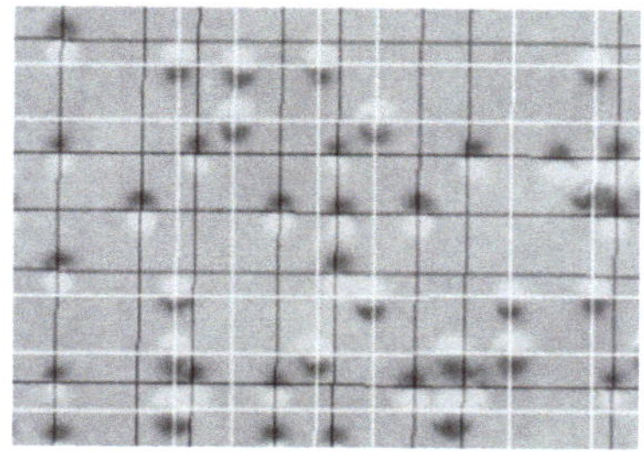

Fig. 5. Braille grid calculated for two pages of the image.

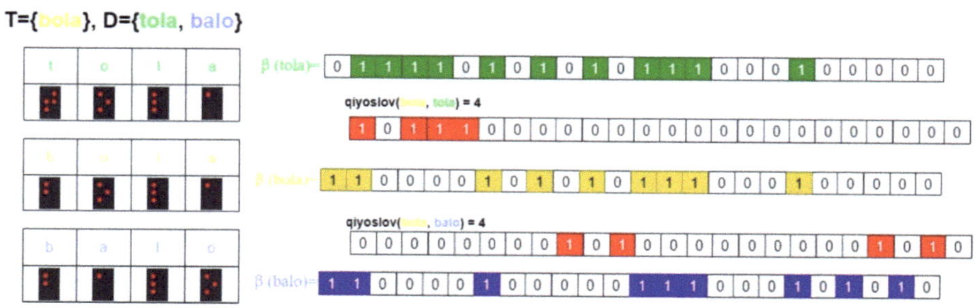

Fig. 6. Example of executing the proposed algorithm when |T| and |D|.

qiyoslov(Pt, pd) to compare the $pt \in T$ word with the word $pd \in D$ (where $|pt| = |pd|$). Equation 1 represents the value returned by the comparison function [17].

$$qiyoslov(p_t, p_d) = \sum_{i=1}^{|p_t|\cdot 6} | \beta(p_t)[i] - \beta(p_d)[i] | \quad (1)$$

4 User Interface

The user interface was created using the Streamlit framework, allowing access from any device [19]. The program code is written in Python and C++ programming languages for the user interface, and the image processing procedures are written in C (DLL libraries are built to connect to the user interface (Fig. 7).

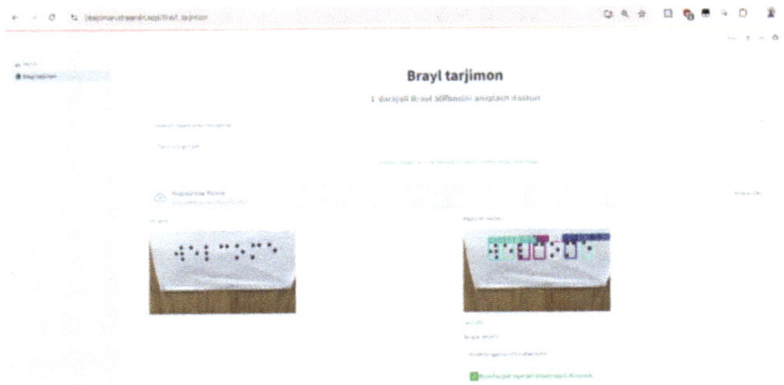

Fig. 7. Overview and operating principle of the program.

5 Required Recommendations, Achievements and Discreptions in Using the Program

Today, we can highlight the following things we still need to do:

- Enhance algorithm performance. We recognize the entire page in about 2.8 s (Intel i5-1035G1 (8) @ 3.600GHz; GPU: Intel Iris Plus Graphics G1 Memory: 7715MiB). This is a fairly fast process.

In the current version of the program, the user must manually set the distance between Braille dots. This is due to the ability to work with non-standard Braille. The system, of course, uses default distances by default, but automatic distance detection is very useful. It is advisable to use artificial intelligence technologies to detect automatic measurements [19].

- Introduction of computerized (8-dot) Braille alphabet and sequences of special symbols (e.g. mathematical symbols).
- In this program, the error rate for determining text written in Braille is 3%.

This app contains some features that are of great interest to users. Although these features are not related to image processing algorithms, we will mention their additional capabilities below:

This OCR is used to identify Braille characters on pages larger than A4. In this case, the scanned image is loaded in two parts, which are reduced to a single page. Usually, educational literature for the blind is also available on pages larger than A4 [20].

Users can create projects. The project is designed to scan corresponding sheets (e.g., books) [21]. The program and scanner parameters are stored with the project

6 Conclusion

We have developed and implemented an Optical Character Recognition (OCR) system for Braille text. The final application is a professional application for all types of devices. It works similarly to the standard text OCR and is a very useful tool for blind people.

References

1. Twain, Reference Manual. HP Corporation
2. Whal, F.M.: Digital Image Signal Processing. Artech House Inc. (1987)
3. Fernández, X., Martín, F., Corbacho, A.: Algoritmos de Reconocimiento de Patrones: Aplicación a O.C.R. Braille. In: Proceedings of URSI-94. Gran Canaria, Spain (September, 1994)
4. O.N.C.E. Normas de Escritura Braille (Spanish Edition of Braille Standards)
5. Farrow, G.S.D., Ireton, M.A., Xydeas, C.S.: Detecting the skew angle in document images. Signal Process. Image Commun. **6**, 101–114 (1994)
6. Axatov Akmal Rustamovich, Ulug'Murodov Shoh Abbos Baxodir O'G'Li: "Deep learning" yordamida brayl yozuvidagi matnlarni tanib olishning usullari. Илм-фан ва инновацион ривожланиш / Наука и инновационное развитие. **6**(1), 32–41 (2023). https://doi.org/10.36522/2181-9637-2023-1-5
7. Akhatov, A.R., Ulugmurodov, S.A.B.: Braille classification algorithms using neural networks. In: Artificial Intelligence, Blockchain, Computing and Security, vol. 2, pp. 654–659. CRC Press (2023). https://doi.org/10.1201/9781032684994-106
8. Akhatov, A., Ulugmurodov, S.A.: Training data selection and labeling for machine learning braille recognition models. Int. J. Contemp. Sci. Tech. Res. (Special Issue), 15–21 (2023) Retrieved from https://journal.jbnuu.uz/index.php/ijcstr/article/view/326
9. Axatov Akmal Rustamovich, Ulug murodov Shoh Abbos Baxodir ogli: INKLUZIV TA'LIMDA INNOVATSION SENSORLI O'QITISH TEXNOLOGIYASI. Int. J. Contemp. Sci. Tech. Res. **1**(2), 213–216 (2022) Retrieved from https://journal.jbnuu.uz/index.php/ijcstr/article/view/160
10. Akhatov, A., Ulugmurodov, A.: Methods and algorithms for separation of text written in braille into classes using neural network technologies. Евразийский журнал математической теории и компьютерных наук. **2**(11), 4–8 (2022) извлечено от https://www.in-academy.uz/index.php/EJMTCS/article/view/4248
11. Xuramov, L.Y., Xafizova, S., Mustaffaqulov, M.: Calculating singular integrals with Cauchy Kernels in digital processing of gastroenterological medical signals. In: 2025 International Russian Smart Industry Conference (SmartIndustryCon), Sochi, Russian Federation, pp. 97–103 (2025). https://doi.org/10.1109/SmartIndustryCon65166.2025.10985974
12. Xuramov, L.Y., Baxromo, A.B., Sanayev, M.E.: Advanced noise-resistant electrogastroenterological classification employing convolutional neural networks and hybrid wavelet transform denoising. In: 2025 International Russian Smart Industry Conference (SmartIndustryCon), Sochi, Russian Federation, pp. 89–96 (2025). https://doi.org/10.1109/SmartIndustryCon65166.2025.10986224
13. Hodjaeva, D., Xuramov, L.: Evaluation of signals before and after filtration in the process of filling and heating water in the smart home system. In: 2025 International Conference on Industrial Engineering, Applications and Manufacturing (ICIEAM), Sochi, Russian Federation, pp. 701–706 (2025). https://doi.org/10.1109/ICIEAM65163.2025.11028172
14. Rana, S., Rai, H.M., Khuramov, L., Mardonov, D.: Shaping the future with quantum computing: an exploration of its emerging field and revolutionary potential. Procedia Comput. Sci. **259**, 844–853., ISSN 1877-0509 (2025). https://doi.org/10.1016/j.procs.2025.04.036
15. Shiriaev, E., Kucherov, N., Movzalevskaya, V., Khamidov, M.: Modification and adaptation of methods and algorithms of the active security concept for fog systems. In: Lapina, M., Raza, Z., Tchernykh, A., Sajid, M., Zolotarev, V., Babenko, M. (eds.) AISMA-2024: International Workshop on Advanced Information Security Management and Applications AISMA 2024 Lecture Notes in Networks and Systems, vol. 863. Springer, Cham (2024). https://doi.org/10.1007/978-3-031-72171-7_28

16. Nazarov, F.M., Xamidov, M.M.: Eye state classification method for detecting physiological deviations in drivers based on CNN algorithm. In: 2024 International Russian Automation Conference (RusAutoCon), Sochi, Russian Federation, pp. 802–807 (2024). https://doi.org/10.1109/RusAutoCon61949.2024.10694309
17. Nurmamatov, M.Q., Sariyev, S.N., Eshonqulov, B.S.: Application of evolutionary algorithms to enhance the efficiency of neural networks and machine learning algorithms. In: 2025 International Russian Smart Industry Conference (SmartIndustryCon) | 979-8-3315-1124-1/25/$31.00 ©2025, pp. 533–537. IEEE. https://doi.org/10.1109/smartindustrycon65166.2025.10986257
18. Makhmadiyarovich, N.F., Sherzodjon, Y.: Development of algorithms for predictive evaluation of investment projects based on machine learning. In: Artificial Intelligence, Blockchain, Computing and Security, vol. 2. eBook ISBN: 9781032684994, pp. 681–685 (2023)
19. Rai, H., Pal, A., Khamidov, M., Bobokhonov, A., Ugli, R.: Computational intelligence transforming healthcare 4.0: innovations in medical image analysis through AI and IoT integration. In: Data-Driven Decision Support System in Intelligent HealthCare (2025). https://doi.org/10.1201/9781003507505-3
20. Akhatov, A., Nurmamatov, M., Nazarov, F.: Intelligent modeling and optimization of processes in the labour market. In: Artificial Intelligence Blockchain Computing and Security Proceedings of the International Conference on Artificial Intelligence Blockchain Computing and Security Icabcs 2023, vol. 2, pp. 694–699 (2024)
21. Basan, E., et al.: UAV security analysis framework. In: Kumar, S., B., K., Kim, J.H., Bansal, J.C. (eds.) Fourth Congress on Intelligent Systems. CIS 2023 Lecture Notes in Networks and Systems, vol. 865, pp. 379–391. Springer, Singapore (2024). https://doi.org/10.1007/978-981-99-9043-6_31

Blockchain Driven Generative AI: Ensuring Data Provenance and Model Integrity

Mahfooz Alam[1], Deepak[2](✉), Divya Mishra[3], Mohammad Shahid[4], and Maria Lapina[5]

[1] Department of MCA, G. L. Bajaj Institute of Technology and Management, Greater Noida 201306,, India
mahfoozalam.amu@gmail.com
[2] Department of Computer Science, Institute of Engineering and Technology, Dr. B.R. Ambedkar University, Khandari, Agra, India
dsingh.cs22@gmail.com
[3] Department of CSE (AIML), Noida Institute of Engineering and Technology, Greater Noida, UP, India
[4] Department of Commerce, Aligarh Muslim University, Aligarh 202002,, India
[5] North-Caucasus Federal University, Stavropol, Russia

Abstract. Integrating blockchain technology with generative artificial intelligence (GAI) provides a transformative method for ensuring data provenance, model integrity, and trust in AI-generated content. Blockchain's decentralized and immutable qualities improve transparency, security, and accountability in AI workflows, addressing issues related to data authenticity, model tampering, and ethical standards. This paper examines the intersection of blockchain and GAI, discussing major challenges such as computational overhead, scalability limits, regulatory restrictions, and energy efficiency concerns. It also offers potential solutions, including hybrid blockchain models, Layer 2 scaling techniques, privacy-preserving AI frameworks, and energy-efficient consensus mechanisms. By proposing a structured framework for blockchain-enabled GAI, this research emphasizes its potential to develop resilient and trustworthy AI ecosystems while tackling important technical and ethical issues.

Keywords: Blockchain · Artificial intelligence · Generative AI · Model integrity

1 Introduction

Generative Artificial Intelligence (GAI) has emerged as a transformational force in content production, pushing the bounds of creativity in text, photos, video, and synthetic media. Using advanced deep learning architectures such as transformer models, Generative Adversarial Networks (GANs), and diffusion frameworks, GAI systems can now produce outputs that are indistinguishable from human-generated content, enabling unprecedented applications in art, entertainment, education, and scientific research.

M. Lapina et al. (Eds.): AISMA 2025, LNNS 1456, pp. 19–30, 2026.
https://doi.org/10.1007/978-3-032-07275-7_3

However, the rapid improvements of today have surpassed the innovations for mechanisms to keep accountability, transparency, and ethical supervision. Some of the significant risks include trust-destroying hyper-realistic deepfake technologies, AI-generated content's ownership disputes, as well as the amplification of biases within algorithms and attacks targeting the model's integrity. The concealment of the origins of training data, the absence of model evolution auditability, unrestricted propagation of AI-created content, and misinformation add to the risks to society and governance.

The technology of blockchain addresses the problems outlined above due to its mechanisms of reaching consensus without centralization, cryptographically secured immutability, and transparent ledgers. Stakeholders can use GAI workflows from data lineage through model training, content creation, and distribution to anchor the blockchain infrastructure and create chains of custody attesting to audit trail verification and frameworks for attribution that are irreversible and resistant to forgery. For example, timestamping and hashing training datasets can facilitate authentication, while compliance with licensing contracts or ethical guidelines could be fulfilled through the use of self-executing contracts. In the same manner, decentralized identifiers (DIDs) along with non-fungible tokens (NFTs) will become important new tools for certifying the ownership of AI content and tracing their lifecycle [1, 2]. This paper tries to examine ways of employing blockchain and GAI because they have the promise of strengthening systems that are generative to have trust, accountability, and security; furthermore, the manuscript proposes multidisciplinary approaches on zero-knowledge proof, decentralized oracle networks, and consensus governance models to preserve fairness algorithmically. We analyze the ethical implications of AI innovations through various case studies, including synthetic media, healthcare data synthesis, and automated journalism, to show the potential of blockchain technology beyond merely mitigating risks.

The analysis also examines other, more directly technological issues like the constraints of scalability in the network of blockchain for high-frequency AI operations, the interoperability between heterogeneous AI and blockchain protocols, and the energy efficiency of decentralized systems. Additionally, we examine evolving regulatory landscapes, including the EU AI Act and the U.S. NIST AI Risk Management Framework, to align technical solutions with global compliance standards. By bridging the gap between decentralized systems and generative AI, this work provides an advanced roadmap for building resilient, transparent, and socially responsible AI ecosystems in the Web3 era.

The rest of this paper is organized as follows: Sect. 2 provides the background and related work connected to this study. Section 3 covers blockchain solutions for generative AI. Section 4 discusses challenges and open issues, the most important concerns, and unresolved questions of using blockchain with generative AI in terms of data provenance and model integrity. Section 5 presents some potential solutions and describes future research. Finally, Sect. 6 offers the conclusions of the paper.

2 Background and Related Work

In recent years, Generative Artificial Intelligence (GAI) has given rise to the ability to create sophisticated content such as text, images, and video. These advancements, with their transformative potential, raise important issues of trust, authenticity, and ethical governance. The lack of creative frameworks regarding the data used to train these models. The increasing possibilities of high-tech AI creating real-false data on computers, and unwanted interference in the functioning of the models point to the more important issue of ensuring accountability, confirming trust, and transparency.

The blockchain approach is different from any other proposal because it is a decentralized, immutable digital registry, which helps to improve the quality of traceability of the data source and the integrity of the model. Through the integration of blockchain technology into GAI work processes, it helps in the establishment of evidence trails, verifications of content concerning training data, and tracking for the AI-generated content. Autonomously, smart contracts support compliance with laws regarding intellectual property, while DIDs alongside cryptographic hashing authenticate claims of AI-generated outputs. Hence, these functionalities assist in solving critical problems for the responsible deployment of AI technology. Regardless of the potential, the integration of blockchain with GAI still encounters issues of a technical and operational nature. High-volume AI processes' blockchain scalability, optimization across different AI and blockchain systems, and the financial expenditure on cryptographic validation are each distinguishable challenges.

We will focus on the combination of blockchain and GAI with the aim of evaluating how blockchain-based models impact security, transparency, ethical responsibility, and generative AI systems in their application. This analysis, along with other initiatives through the opportunities and challenges, is aimed at crafting a resilient and reliable approach towards the AI ecosystem.

2.1 Related Work

The implementation of GAI alongside blockchain is still emerging, with some studies looking into the proprietary data guarantees, model trust, and reliability of AI content as a proof of concept. Existing work is limited to the use of blockchain for automating machine learning workflow security, explainability, adversarial risk management, and other factors in GAI outputs.

In [3], Villegas-Ch et al. proposed a novel method for boosting security and improving processes within IoT networks by merging blockchain with AI. In [4], Ghani et al. integrate blockchain privacy into facial recognition systems utilizing GANs with an innovation framework. In [5], the authors look into integrating GAI in blockchain systems to solve problems of scalability, security, privacy, and even interoperability.

Shinde et al. [6] present a Systematic Literature Review (SLR) to examine the significance of Blockchain technology in enhancing the security of AI-based healthcare applications. In [7], the authors proposed an innovative framework that integrates the Internet of Medical Things (IoMT), Federated Learning (FL), and blockchain technology. The authors' proposed framework has the ability to strengthen the credibility and validity of public media's dissemination of COVID-19 information. Devarapu [8] explores the integration of blockchain technology with AI. Using this integration author tries to enhance medical imaging and diagnostic processes. In this paper, the author discusses how combining these technologies can address challenges related to data security, privacy, and interoperability in healthcare systems. In this study, we combine GAI with blockchain technology to ensure data provenance and model integrity. In this section, we also discussed the GAI and its challenges as well as the use of blockchain technology in the field of AI.

2.2 Generative AI and Its Challenges

Generative AI is defined as a computation technique for generating new, relevant content from text, images, and audio from training data [9]. Generative AI models like GANs [10] and Variational Auto Encoders (VAEs) [11], as well as transformer-based architectures like GPT-4 [12], DALL-E [13], and Stable Diffusion [14], are becoming increasingly popular for creative tasks. Generative AI systems can not only be artistic; they can generate new texts imitating writers or new images imitating illustrators, but they can also assist mankind as an intelligent question-answering system [9]. Despite their powerful capabilities, such models tend to raise enormous amounts of issues:

Data Authenticity and Bias. These GAI models are trained on a lot of unverified heterogeneous data collected from public sources, which raises issues of data provenance, infringement of copyrights, and biases inherent within the data. For example, language models like GPT-3 have been known to magnify societal biases in their training texts, hence maintaining stereotypes [15]. Similar racial and gender biases are perpetuated in the results of image generators like Stable Diffusion due to slanting the sources of training data [16]. The matter is dubious and needs fair and trusted data by providing means of auditing data lineage and representativeness [17].

Misinformation and Fake Content. The fact that more people can use tools to make hyper-realistic deepfakes and AI-generated text has made it easier for bad people to spread false information on a large scale. For example, AI-generated videos of famous people have been used as weapons in political manipulation [18], and chatbots have made phishing content that looks real [19]. These threats make people less trusting and call for systems to verify the sources of content.

Model Security and Tampering. GAI systems are vulnerable to adversarial attacks, where imperceptible perturbations to input data lead to erroneous outputs [20], cybersecurity challenges in the IoT [21] world, and "model poisoning," where adversaries inject malicious data during training to corrupt behavior [22]. In some recent studies, authors demonstrate that models like BERT can be compromised via backdoor attacks [23], so this highlights the need for tamper-proof model versioning.

2.3 Blockchain Technology in AI

Blockchain is a decentralized ledger technology which provides cryptographic immutability and consensus-based transparency. It has been identified as a strong potential facilitator for trustworthy AI systems. Its fundamental attributes include immutable record-keeping, decentralized verification, and smart contracts. It fills significant gaps in GAI governance, such as:

Immutable Provenance Tracking. Blockchain allows training datasets, model parameters, and produced content to be anchored to a tamper-proof ledger [24]. Platforms such as ocean protocol use blockchain to tokenise data access, guaranteeing licensable compliance with license terms [25].

Decentralized Model Auditing. By distributing model updates across a blockchain network, stakeholders can collaboratively validate changes, preventing unauthorized tampering. In [26], the authors proposed a federated learning (FL) framework that integrates blockchain technology and enables distributed addressing of privacy and security challenges in collaborative AI-driven COVID-19 diagnosis using chest CT scans and confidential model training in healthcare without requiring the sharing of raw patient data. As previously noted, AI service fairness audits within the Singularity NET framework are also governed in a decentralized manner [27].

Smart Contracts for Automated Compliance. Self-enforcing smart contracts may impose ethical criteria, i.e., prevent GAI from creating toxic content. For instance, a smart contract may automatically terminate a model's permission if its output violates pre-set policies [28]. Projects like Fetch.ai use blockchain and AI to make data marketplace trades transparent, such as royalty payments for AI-created content [29]. The use of blockchain in AI protection, data privacy, and decentralized AI marketplaces has been the focus of recent research; however, scalability and integration problems remain to be solved.

3 Blockchain-Enabled Solutions for Generative AI

Blockchain technology presents a secure platform to leverage the security, transparency, and trustworthiness of generative AI by utilizing data provenance, model integrity, and decentralized AI marketplaces. Blockchain applications in generative AI can be structured into a framework, as shown in Fig. 1. The framework comprises three interconnected layers: The Data Provenance Layer, the Model Integrity Layer, and the Decentralized AI Marketplace Layer. The subsequent section presents a detailed description of the operation of each module within this framework.

3.1 Data Provenance Layer

This layer is responsible for maintaining the trustworthiness, transparency, and traceability of data used in AI training. The following ways are to secure data provenance using blockchain such as:

- **Immutable Data Records:** It stores dataset metadata such as source, timestamps, ownership data, and usage permissions using the blockchain's immutable ledger.
- **Smart Contract-Based Verification.** It ensures that the verification process is automated, and hence, only verified and authentic data is used to train generative AI models.
- **Decentralized Identity Management System.** It ensures data integrity, controls proper attributions, ownership, and prevents unapproved changes.
- **Access Control**. Specific zones of data can be restricted, enhanced, or utilized based on their modification, therefore using access control.

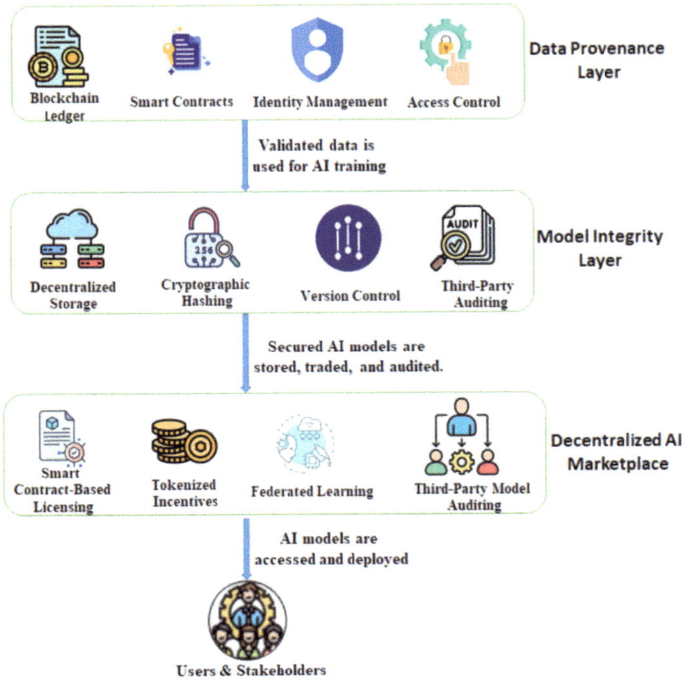

Fig. 1. Blockchain integrated GAI framework.

3.2 Model Integrity Layer

The main goal of the model integrity layer is providing security to AI models. This security prevents AI models from tampering, unauthorized modification, and untraceable updates. Blockchain technology enhances the model integrity, ensuring AI model reliability, and prevents it from using:

- **Decentralized storage solutions:** After training of AI model using datasets with the help of a provenance layer, it is stored in decentralized storage solutions such as IPFS (Inter Planetary File System) or blockchain-based repositories.
- **Cryptographic Hashing:** This technique prevents of unauthorized changes. For this, each AI model is hashed cryptographically, meaning that any alteration to the model will be immediately detectable.
- **Version Control and Traceability:** The technique is useful to track model updates and to revert them to older versions in case an issue or security vulnerability is discovered.
- **Third-party auditing:** Third-party auditing is incorporated into audit mechanisms for fairness, transparency, and compliance checks to ensure ethical AI practices before the deployment of the model. With this security implemented, the model integrity layer guarantees the consistency, robustness, and immunity of AI outputs against adversarial threats.

3.3 Decentralized AI Marketplace Layer

This layer ensures secure commercial use, licensing, and distribution of AI models while offering full transparency and fair compensation to all participants. AI models undergo integrity verification before they are uploaded to the marketplace. Their primary traits are as follows:

- **Smart Contract-Based Licensing:** It enables automated transactions using digital rights of the assets without the need for middlemen.
- **Tokenized Incentives:** Contributors receive tokens on the blockchain as payment for AI development, data provision, and all other services offered. Tokenized payment systems are fair and transparent.
- **AI Training Federated with Blockchain:** These strategies allow for the collaborative training of AI models at decentralized nodes without the need to expose raw data, which ensures privacy and improves the models' accuracy.
- **Model Auditing:** Models are scrutinized and verified by external experts, which provides trust and assurance to the ecosystem, and is therefore fundamental for deployment.

4 Challenges and Open Issues

While the integration of blockchain and generative AI could transform data provenance and model integrity, many challenges restrict this technology's fulfillment. Some of them include computation overhead, scalability constraints, regulatory and compliance issues, energy consumption, and complexity in integration.

4.1 Computational Overhead

Confirmation of transactions and integrity in a blockchain network is done with consensus mechanisms such as Proof of Work (PoW) and Proof of Stake (PoS) [30]. These protocols use resources, due to which it takes a lot of time and energy to complete the work of these protocols. As many GAI models as there are, they are not able to make decisions in real time due to fulfilling the demand of these protocols, along with their processing requirements. Due to which throughput decreases, and latency also increases.

4.2 Scalability Limitations

Bitcoin and Ethereum are Traditional blockchain systems which, unfortunately, only handle a few transactions per second [31]. This is a tremendous bottleneck for artificial intelligence, which needs massive amounts of data to be processed quickly. Additional information and processing weight from AI functions could increase scaling restrictions because the blockchain now has to accommodate both basic transactions and AI activities. Sharding systems, layer-2 scalability, and new blockchains optimised for AI-backed high-throughput applications are possible solutions.

4.3 Regulatory and Compliance Issues

Blockchain technology has an immutable nature, in which information cannot be removed or modified as easily. This immutable nature raises the question of how it can be assured to comply with data protection legislation such as the General Data Protection Regulation (GDPR). GDPR provides individuals with the "right to be forgotten," where they can ask for their personal data to be erased. This right is at odds with the technical nature of blockchain, and a problem of legal and technical magnitude exists. Regulatory obligations can be met by taking innovative measures, i.e., by keeping some data off the blockchain but the reference or the hashes of the data on the blockchain. Additionally, advanced cryptographic methods need to be used as those methods make information unavailable without erasing it.

4.4 Energy Consumption Concerns

PoW consensus mechanism is very energy-intensive [30]; these processes are involved in maintaining blockchain networks. The computational effort required for mining and validating transactions consumes significant amounts of electricity, raising environmental and sustainability concerns. With the integration of AI, the combined energy requirements can become substantial. Processes also demand considerable computational resources for AI training and inference. To fix this problem, we should look into more energy-efficient consensus mechanisms like PoS or Delegated PoS (DPoS), and we should also try to make AI algorithms use less energy.

4.5 Integration Complexity

Combining blockchain and A.I. networks is difficult precisely because the two are such different forms of technology. AI systems remain data-hungry and require centralization in order to work, be efficient, and perform well. In contrast, blockchain operates on secure, immutable, decentralized, and distributed architectures [32]. In order to achieve better integration, we will have to build special design frameworks and intermediary services to bridge the AI and blockchain ecosystems. They have to consider the constraints of the data integration, the system interoperability, and ensuring the two systems have scalable performance, security, and trust.

These problems have to be resolved to make blockchain work with generative AI. There is significant R&D activity aimed at busting these barriers, working towards the development of resilient, efficient, and compliant AI systems empowered by blockchain technology.

5 Potential Solutions and Future Directions

Throughout this section, We have looked at some possible solutions and discussed other research areas that can be explored to resolve the difficulties with implementing blockchain technology in generative AI.

5.1 Hybrid Blockchain Models

The traditional blockchain systems often encounter challenges regarding scalability and efficiency because of their on-chain data storage and processing. To address these issues, some proposed hybrid blockchain models were based on cross-sectional data with off-chain storage. This model tends to keep sensitive information, which requires immutability and transparency on the blockchain, and the rest off the chain. This reduces strain on the entire blockchain network and thus less costly transactions and system expenses, while improving performance [33]. For example, in healthcare systems, a hybrid approach could manage patient data by off-chain storing the more sensitive information while on-chain storing the more important information that needs to be kept intact for retrieval and form verification [34].

5.2 Layer 2 Scaling Solutions

Combining the large number of available applications alongside AI into one leads to significant concern about scalability issues in blockchain networks. As a solution, Layer 2 scaling methods such as sidechains, rollups, and state channels were designed to improve transaction throughput and reduce network congestion [35]. These solutions work above the existing blockchain infrastructures and can allow faster and more efficient transactions while keeping intact the security of the main chain.

5.3 Privacy-Preserving AI Techniques

The tightening of the AI blockchain, coupled with data security mandates, still needs compliance with the GDPR. It presents a few challenges but easily obtainable solutions, such as privacy-preserving techniques, such as zero-knowledge proofs, federated learning, and differential privacy. Zero-knowledge proofs allow for the verification of the authenticity of data without actually disclosing that data, so the data is confidential [36, 37]. Federated learning creates an infrastructure for training AI models across decentralized devices with sample-held local data, since private data is retained within local devices and not transferred. Differential privacy works by injecting statistical noise into a dataset and thus shielding individual data points while keeping the utility of the dataset overall. Overall, these methods enable the creation of AI models that respect user privacy and comply with regulatory standards.

5.4 Energy-Efficient Consensus Mechanisms

The sustainability of energy-guzzling consensus protocols like PoW raises concerns in blockchain technologies. Other consensus models have been studied to solve the problem, such as PoS and Directed Acyclic Graph (DAG) models. PoS is less energy-intensive because it allows validators to put their tokens as collateral, thus not requiring computations. Being truly non-linear, DAG models process different transactions independently, thus enhancing scale and efficiency. The acceptance of such energy-efficient consensus models will help in lowering the carbon footprints of blockchain networks, in line with worldwide sustainability efforts.

5.5 Decentralized AI Governance

Today, establishing governance structures for ensuring ethics and accountability, and responsibility in AI systems becomes relevant. These decentralized AI governance models have advocated for bottom-up community validation as a means of development and use in AI systems by all stakeholders, either directly or indirectly concerned. This way, AI technologies will be in compliance with social and ethical norms. Such decentralized platforms will enable the stakeholders to vote on various proposals regarding the development of the AI systems, monitor behaviors of the AI, or even enforce compliance with ethical guidelines. Such frameworks, such as smart access to government agriculture scheme [38], promote open systems, inclusive and trusting in AI, thereby addressing bias and fairness concerns about those systems, in addition to accountability.

6 Conclusion

Generative AI using blockchain solutions makes it secure, trustworthy, and transparent in AI content. The proof of data provenance is ensured through the immutable ledger, smart contracts, and cryptographic mechanisms of blockchain, thus preventing tampering of the model and ensuring moral governance toward AI practices. The approval indeed has to face many challenges regarding computational overhead, scalability, and regulatory compliance. The critical research priority would be to optimize the blockchain architecture concerning AI workloads, address privacy-enhancing AI principles, and develop energy-efficient blockchain solutions to handle energy consumption in developing networks using a multi-stakeholder approach. As it evolves, however, multicultural and interdisciplinary approaches will likely define an accountable AI ecosystem within the global ethical and regulatory framework.

References

1. Yoon, D., Mun, S., Park, K., Noh, S.: Decentralized identifiers and NFT-driven Blockchain solution for personal data trading. In: 2023 14th International Conference on Information and Communication Technology Convergence (ICTC), pp. 1254–1256. IEEE (2023, October)
2. Battah, A., et al.: Blockchain and NFTs for trusted ownership, trading, and access of AI models. IEEE Access. **10**, 112230–112249 (2022)
3. Villegas-Ch, W., Govea, J., Gurierrez, R., Mera-Navarrete, A.: Optimizing security in IoT ecosystems using hybrid artificial intelligence and Blockchain models: a scalable and efficient approach for threat detection. IEEE Access. **13**, 16933–16958 (2025)
4. Ghani, M.A.N.U., et al.: Securing synthetic faces: a GAN-blockchain approach to privacy-enhanced facial recognition. J. King Saud University-Comput. Inform. Sci. **36**(4), 102036 (2024)
5. Nguyen, C.T., et al.: Generative AI-enabled blockchain networks: fundamentals, applications, and case study. IEEE Netw. **39**, 232–241 (2024)
6. Shinde, R., et al.: Securing AI-based healthcare systems using blockchain technology: a state-of-the-art systematic literature review and future research directions. Trans. Emerg. Telecommun. Technol. **35**(1), e4884 (2024)
7. Samuel, O., et al.: IoMT: a COVID-19 healthcare system driven by federated learning and blockchain. IEEE J. Biomed. Health Inform. **27**(2), 823–834 (2022)

8. Devarapu, K.: Blockchain-driven AI solutions for medical imaging and diagnosis in healthcare. Technol. Manage. Rev. **5**(1), 80–91 (2020)
9. Feuerriegel, S., Hartmann, J., Janiesch, C., Zschech, P.: Generative AI. Bus. Inf. Syst. Eng. **66**(1), 111–126 (2024)
10. Goodfellow, I., et al.: Generative adversarial nets. Adv. Neural Inf. Proces. Syst. **27** (2014)
11. Kingma, D.P.: Auto-encoding variational bayes. arXiv preprint arXiv:1312.6114 (2013)
12. Achiam, J., et al.: Gpt-4 technical report. arXiv preprint arXiv:2303.08774 (2023)
13. Ramesh, A., et al.: Zero-shot text-to-image generation. In: International Conference on Machine Learning, pp. 8821–8831. Pmlr (2021, July)
14. Rombach, R., Blattmann, A., Lorenz, D., Esser, P., Ommer, B.: High-resolution image synthesis with latent diffusion models. In: Proceedings of the IEEE/CVF conference on computer vision and pattern recognition, pp. 10684–10695 (2022)
15. Bender, E.M., Gebru, T., McMillan-Major, A., Shmitchell, S.: On the dangers of stochastic parrots: can language models be too big? In: Proceedings of the 2021 ACM conference on fairness, accountability, and transparency, pp. 610–623 (2021, March)
16. Friedrich, F., et al.: Fair diffusion: instructing text-to-image generation models on fairness. arXiv preprint arXiv:2302.10893 (2023)
17. Mitchell, M., et al.: Model cards for model reporting. In: Proceedings of the conference on fairness, accountability, and transparency, pp. 220–229 (2019, January)
18. Malik, A., Kuribayashi, M., Abdullahi, S.M., Khan, A.N.: DeepFake detection for human face images and videos: a survey. IEEE Access. **10**, 18757–18775 (2022)
19. Fakhouri, H.N., et al.: AI-driven solutions for social engineering attacks: detection, prevention, and response. In: 2024 2nd International Conference on Cyber Resilience (ICCR), pp. 1–8. IEEE (2024, February)
20. Szegedy, C.: Intriguing properties of neural networks. arXiv preprint arXiv:1312.6199 (2013)
21. Lone, A.N., Mustajab, S., Alam, M.: A comprehensive study on cybersecurity challenges and opportunities in the IoT world. Secur. Priv. **6**(6), e318 (2023)
22. Gu, T., Dolan-Gavitt, B., Garg, S.: Badnets: identifying vulnerabilities in the machine learning model supply chain. arXiv preprint arXiv:1708.06733 (2017)
23. Li, L., et al.: Backdoor attacks on pre-trained models by layerwise weight poisoning. arXiv preprint arXiv:2108.13888 (2021)
24. Lo, S.K., et al.: Toward trustworthy AI: Blockchain-based architecture design for accountability and fairness of federated learning systems. IEEE Internet Things J. **10**(4), 3276–3284 (2022)
25. McConaghy, T.: Ocean protocol: tools for the Web3 data economy. In: Handbook on Blockchain, pp. 505–539. Springer International Publishing, Cham (2022)
26. Kumar, R., et al.: Blockchain-federated learning and deep learning models for COVID-19 detection using CT imaging. IEEE Sensors J. **21**(14), 16301–16314 (2021)
27. Goertzel, B., Giacomelli, S., Hanson, D., Pennachin, C., Argentieri, M.: SingularityNET: a decentralized, open market and inter-network for AIs. Thoughts, Theories Stud. Artif. Intell. Res (2017)
28. Kosba, A., Miller, A., Shi, E., Wen, Z., Papamanthou, C.: Hawk: the blockchain model of cryptography and privacy-preserving smart contracts. In: 2016 IEEE Symposium on Security and Privacy (SP), pp. 839–858. IEEE (2016, May)
29. Weeks, M.: The evolution and design of digital economies (2018). https://fetch.ai/uploads/Fetch. AI-Economics-white-paper.pdf
30. Pineda, M., Jabba, D., Nieto-Bernal, W., Pérez, A.: Sustainable consensus algorithms applied to Blockchain: a systematic literature review. Sustainability. **16**(23), 10552 (2024)
31. Lathamani, B., et al.: Enhancing the scalability of Blockchain networks using a data partitioning technique. Eng. Technol. Appl. Sci. Res. **14**(6), 17711–17716 (2024)

32. Choi, N., Kim, H.: Technological convergence of Blockchain and artificial intelligence: a review and challenges. Electronics. **14**(1), 84 (2025)
33. Hasan, K.M.B., Sajid, M., Lapina, M.A., Shahid, M., Kotecha, K.: Blockchain technology meets 6 G wireless networks: a systematic survey. Alex. Eng. J. **92**, 199–220 (2024)
34. Ali, A., et al.: Blockchain-powered healthcare systems: enhancing scalability and security with hybrid deep learning. Sensors. **23**(18), 7740 (2023)
35. Sguanci, C., Spatafora, R., Vergani, A.M.: Layer 2 blockchain scaling: a survey. arXiv preprint arXiv:2107.10881 (2021)
36. Garcia, R.D., et al.: A survey of Blockchain-based privacy applications: an analysis of consent management and self-sovereign identity approaches. arXiv preprint arXiv:2411.16404 (2024)
37. Basan, E., Lapina, M., Lesnikov, A., Basyuk, A., Mogilny, A.: Trust monitoring in a cyber-physical system for security analysis based on distributed computing. In: International Conference on Actual Problems of Applied Mathematics and Computer Science, pp. 430–440. Springer Nature Switzerland, Cham (2022, October)
38. Vinodha, D., et al.: Framework based on IoT, AI, and Blockchain for smart access to government agricultural schemes. In: The Future of Agriculture: IoT, AI and Blockchain Technology for Sustainable Farming, pp. 142–163. Bentham Science Publishers (2024)

The Theoretical Rationale for Using Quantum Computers to Crack Cryptosystems

Yulia Andrusenko, Dmitry Lukyanov(✉), Alexey Dmitrienko, Alisa Kushchenko, and Diana Fedash

North Caucasus Federal University, Stavropol, Russia
ilia.alekseev.kenig@gmail.com

Abstract. Currently, the use of quantum computers to crack cryptosystems is becoming increasingly publicized. Many governments are making huge investments in the development of quantum computers to prepare for possible threats. This article discusses the potential threats of hacking cryptosystems using quantum computers. As a result of the research, ways to protect data and information from quantum attacks will be considered. This article will be useful for information security specialists and researchers involved in the study of data encryption methods that are resistant to quantum attacks.

Keywords: Quantum Computers · Cryptography · Quantum Cryptography · Cryptosystems · Information Security · Hacking

1 Introduction

In today's world, technology plays a key role in human life. Every day, scientists improve the devices and devices that help us in everyday tasks. Computers, which have become an integral part of our lives, are capable of analyzing a huge amount of data. But often the computing power of ordinary computers is not enough to solve any problems. To solve these problems, quantum computers have been invented, which use the principles of quantum physics to operate.

Quantum computers are devices that use the phenomena of quantum physics (the principle of superposition, entanglement and tunneling) to perform various mathematical tasks. Unlike conventional computers that use a system of 0 and 1 (bits), quantum computers use a qubit [1], bits that are in a superposition of several states at once.

Although quantum computers are the solution to many of humanity's problems, they are also sources of threat to information security. They have a huge potential for hacking cryptosystems. Using quantum algorithms to solve mathematical problems that are used in cryptography. An example of an algorithm used to crack cryptosystems by quantum computers is the Shor algorithm. It is precisely such algorithms that pose a threat to the security of everyone's personal data.

The relevance of this topic is due to the fact that many states invest in the development of quantum technologies, which stimulates their rapid growth and development. You

M. Lapina et al. (Eds.): AISMA 2025, LNNS 1456, pp. 31–39, 2026.
https://doi.org/10.1007/978-3-032-07275-7_4

need to know in advance what threats are taking place and how to protect your data from quantum intruders. Thus, this article is aimed at informing readers about the potential threats posed by quantum computers.

The work of quantum computers is based on the principles and laws of quantum physics. While traditional computers use bits to solve mathematical problems, quantum computers use qubits. The most important difference between qubits and ordinary bits is that qubits use the principles of quantum physics. For example, the principle of superposition [2] bodies, allows qubits to be simultaneously in two states 1 and 0. Bits in traditional computers can take the value either 1 or 0, which reduces the speed of their mathematical calculations by several times, compared with quantum computers. Due to the principle of superposition and the ability of qubits to be in two states simultaneously, quantum computers are able to solve mathematical problems that would take traditional computers hours or minutes to solve.

Using the principles of quantum physics allows quantum computers to solve mathematical problems. Figure 1 shows the differences between traditional computers and quantum computers.

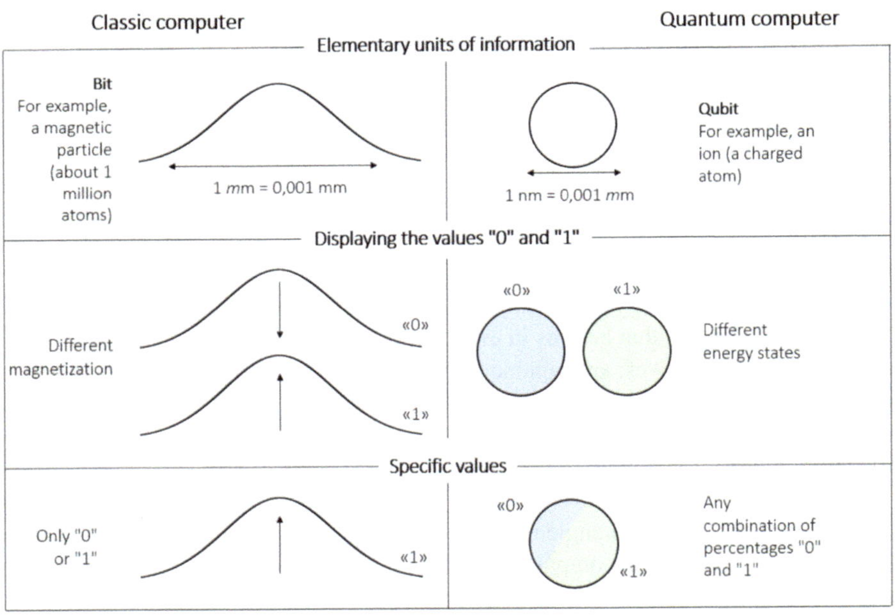

Fig. 1. Comparison of the operation of ordinary bits and qubits.

Quantum computers have a huge potential to solve a large number of complex problems that traditional computers cannot solve. One of the most important tasks that quantum computers can solve is the optimization of power transmission lines, transportation routes, and resource allocation routes. Due to the phenomenon of quantum tunneling, a quantum computer searches for the most optimal and fastest way to solve a problem.

2 The History of the Quantum Computer

In the 1980s, other people began to theoretically explore the concept of a quantum computer and its applications. Paul Benioff proposed the idea of a quantum Turing machine, an idea that eventually led to the definition of a universal quantum computer [15]. This was followed by several theoretically interesting but largely impractical quantum algorithms, including Deutsch's Quantum Oracle algorithm [16], Bernstein-Vazirani algorithm [17], and Simon's algorithm [18].

Peter Shore's discovery in 1994 of a polynomial-time quantum algorithm for factoring integers and calculating discrete logarithms [19] was a breakthrough that stimulated great interest in quantum computing [20]. It was the first quantum algorithm to demonstrate superpolynomial acceleration compared to classical algorithms capable of solving the same problem.

Shor's algorithm was followed by Grover's unstructured search algorithm in 1996 [21], followed by a 1981 proof of Feynman's hypothesis showing that quantum computers can simulate quantum systems without the exponential overhead incurred by classical computers [22].

Shortly after this discovery, the theory of fault-tolerant quantum computing was developed [24]. Of particular importance was the discovery of the threshold theorem, which established that if the error rate underlying a quantum computer could be reduced low enough, then error correction codes could be used to exponentially suppress errors with polynomial costs [24–27].

Taken together, Shor's algorithm and the discovery of quantum error correction have further fueled interest in quantum computer science. This is clearly evident from the explosive growth of articles on quantum information and quantum computing, which began around 1994. Importantly, this rapid growth also encompassed advances in the use and control of quantum mechanical systems to make them perform calculations, with the first quantum gate in 1995 [28] based on a concept presented earlier that year [29], followed by the first demonstration of Shor's algorithm in a nuclear power system. magnetic resonance in liquid state 6 quantum error correction [23].

Shor's algorithm, along with the proof of Feynman's 1981 hypothesis and the discovery of quantum error correction, implied that a quantum computer, if successfully created, would have an extraordinary impact on society and humanity. As a result of these initial efforts, numerous groups began exploring the possibilities of creating quantum hardware based on various basic types of qubits.

Simultaneously, but at a slower pace of progress, scientists have developed and improved their understanding of quantum complexity classes and initiated the complex process of discovering other possible classes of algorithms that benefit from the existence of a quantum computer.

For most of the next two decades, research into quantum hardware, algorithms, and software was largely independent. It wasn't until the mid-2010s that attempts to create full-stack quantum computing appeared.

3 Literature Review

Table 1 contains the basic principles of operation and systems that are associated with quantum computers.

Table 1. Basic principles of operation and systems related to quantum computers.

Aspect	Source
Basic Principles of Quantum Physics for a Quantum Computer	[3–6]
Principles of operation of a quantum computer	[1–7]
Mathematical mechanisms	[8]
Shor's algorithm	[9]
Grover's algorithm	[10]
Classification of cryptosystems	[13]
Building Cryptosystems	[13]
Security algorithms	[11, 12]
Security methods	[13, 14]

Along with the principle of superposition, an equally important phenomenon is used in the work of quantum systems – the phenomenon of quantum entanglement [3]. This phenomenon is based on the interconnection of two or more qubits, regardless of the distance between them. Discovered in 1935 by Erwin Schrodinger and described by Einstein as "Ghostly Long-range" [3] quantum entanglement allows qubits to exchange information across vast distances. This state is maintained between qubits, even if they are separated beyond the boundaries of any interactions, which allows you to receive information from any point in the shortest time.

Special attention should be focused on the algorithms of Shor and Grover, which show particular effectiveness in quantum computing.

The results that are expected after the study:

- Systematization of information about the principles of operation of quantum computers.
- Identification of the influence of quantum systems on cryptography.
- Identification of vulnerability of cryptographic systems to attacks by quantum computers.
- Formulation of recommendations for protecting your data.

In addition to the principle of superposition and quantum entanglement, qubits are characterized by another unique quantum phenomenon – quantum tunneling [4, 5]. The principle of this phenomenon is that a particle is able to overcome barriers that exceed it in energy. This phenomenon relates exclusively to quantum physics and contradicts the law of traditional mechanics. The use of quantum tunneling in quantum computers is to find the fastest possible solution to a problem [6].

Another area of application of quantum computers is the modeling and analysis of complex systems. Thanks to the superposition, qubits are able to process a huge amount of data, which makes it possible, which is useful when analyzing data or modeling complex chemical chains.

But the main use of quantum computers is cryptography. Quantum systems are capable of solving the problems of key selection and factorization of huge numbers, which form the basis of cryptography. Algorithms such as Shor's algorithm [8] make it possible to crack cryptographic systems using public keys. Thus, the improvement of quantum technologies and systems opens up opportunities for both scientific research and practical applications for hacking cryptographic systems.

Currently, scientists are actively studying the potential threats of quantum attacks on encryption systems. Peter Shore's work [8] has become key to understanding the vulnerabilities of cryptographic systems to quantum attacks. In his work, Peter Schoen outputs an algorithm that allows solving the problems of factoring numbers in polynomial time on a quantum computer.

Another important job belongs to Lov Grover. He developed an algorithm [9] that allows us to find solutions for equations where the function is Boolean in n variables.

At the moment, quantum computers are not yet sophisticated enough to pose a threat to standard encryption systems. But the study [10] shows that it is necessary to prepare for the era of quantum calculus right now.

At the moment, quantum computers are not powerful enough to crack, they cannot perform complex mathematical calculations for long enough. To crack the standard encryption algorithm for banks, VPN services, and HTTPS (RSA-2048), a quantum computer using about 20 million stable qubits is needed. At the moment, the most powerful quantum computer has about 1121 qubits, which is not enough to crack modern systems [11].

Currently, quantum computers are used in large technology companies (Google, Microsoft, IBM). These organizations collect information and use quantum computers for routine analysis and even cryptanalysis.

At the moment, hackers can collect user data and decrypt it in the future, when quantum computers become powerful and stable enough. This is especially true for state secrets, military information, medical research, and finance.

If quantum computers become powerful enough to crack structures soon, then the following will be the number one target:

- RSA (2048, 4096 bits), Diffie-Hellman. The protection of these structures is based on factorization (a method of sorting through possible divisors of numbers) of numbers and calculating the discrete logarithm. This structure was the first to be able to be used for encryption and electronic signature creation [12].
- AES (Symmetric encryption) and hash functions. These structures are protected for quantum hacking, but a special method (Grover's method [9]) helps to select the access key n times faster. AES-256 will remain protected as quantum computers improve, but AES-128 will become vulnerable.

Another way to encrypt data using a method that is resistant to quantum attacks is hybrid encryption – public key encryption. This method includes two ways to encrypt data at once: symmetric and asymmetric encryption. Symmetric encryption has higher speed and shorter keys, but cannot send them encrypted, while the asymmetric method sends an encoded key, but has low performance. In hybrid encryption, a symmetric key is used to encrypt data, and an asymmetric key is used to encrypt a symmetric key, which provides high reliability and resistance to quantum attacks.

4 Potential Threats

Our attention to fault-tolerant implementations is due to the fact that to date, no approach based on variational quantum algorithms has been discovered for which there would be a provable (or even plausible) threat to existing public-key cryptosystems. The variational approaches considered to date consist of 2 parts. First, compare the task of attacking the cryptosystem with the task of combinatorial optimization (using a certain amount of classical preprocessing to simplify it). Secondly, solve the resulting optimization problem using variational algorithms.

The main threat that quantum computers pose to existing cryptosystems is an attack on asymmetric cryptography. For cryptosystems with a symmetric key, a suitable quantum attack is the Grover search algorithm [21], which would speed up the search for the key using the brute force method. However, this attack can be eliminated by doubling the key length for the affected algorithms. In fact, there are mitigating factors suggesting that Grover's algorithm will not speed up the brute-force key search as dramatically as one might assume: in 1997, it was proved that to obtain a complete quadratic acceleration, all steps of Grover's algorithm must be performed sequentially.

Although there are other attacks on symmetric-key cryptosystems, the full consequences of these attacks are currently unknown. Today, various public key algorithms are widely used, including RSA, Diffie-Hellman key Agreement, digital Signature Algorithm (DSA), and elliptic curve cryptography (ECC) algorithms. The mathematics underlying these cryptosystems are different, which matters for how quantum computers can crack them (Table 2).

Table 2. Cryptosystems and their mathematical mechanisms.

Cryptosystem	The principle of operation
RSA	Is based on arithmetic operations with integers, and its security is related to the hardness of factoring large numbers by their
Diffie-Hellman; DSA	Are examples of finite field cryptography (FFC), which is based on modular finite field arithmetic
ECC	Is based on discrete logarithm problems over an elliptic curve and includes self-discrete logarithm

The way a quantum computer would crack these cryptosystems is to use Shor's algorithm [9]. This algorithm has several variants: a version that can be used to crack the ECC and FCC cryptosystems, and a version that can be used to crack RSA. As we will see below, these factoring options require various quantum computing resources to crack the cryptosystem they are attacking. If a large-scale fault-tolerant quantum computer is built, all FFC, RSA, and ECC algorithms will be vulnerable to attacks from quantum computers.

5 Potential Threats

Even today, it is necessary to think about protecting data from the increasing pace of development of quantum decryption technologies. To protect against future quantum cyber attacks, it is recommended to implement post-quantum algorithms [13] for data encryption using methods resistant to quantum attacks. Cryptography based on error correction codes [14] represents a cryptosystem with public keys. It uses key generation using random parameters during creation. The disadvantage of this method is the increase in the size of the encrypted message compared to the initial one.

Thus, the threat of quantum hacking remains theoretical, but requires immediate data protection solutions right now. The use of post-quantum algorithms and hybrid encryption is becoming the main step towards protecting your data from quantum hackers. Modern quantum computers are not yet capable of hacking standard encryption methods, but attackers can already collect user data and crack it in the future. That is why the transition to methods capable of protecting information from quantum attacks is necessary right now. Thus, the introduction of technologies to protect data from quantum hacking will guarantee the security and safety of data for many years to come.

References

1. Ladd, T.D., et al.: Quantum computers. Nature. **464**(7285), 45–53 (2010)
2. Khrennikov, A.: Roots of quantum computing supremacy: superposition, entanglement, or complementarity? Eur. Phys. J. Spec. Top. **230**(4), 1053–1057 (2021)
3. Steane, A.: Quantum computing. Rep. Prog. Phys. **61**(2), 117 (1998)
4. Abel, S., Chancellor, N., Spannowsky, M.: Quantum computing for quantum tunneling. Phys. Rev. D. **103**(1), 016008 (2021)
5. Singh, J., Singh, M.: Evolution in quantum computing. In: 2016 International Conference System Modeling & Advancement in Research Trends (SMART), pp. 267–270. IEEE (2016)
6. Quantum tunneling: how the microcosm helps to escape from reality. https://dzen.ru/a/ZuvTjfKy9wkeFZrT
7. Solovyov, V.: Quantum computers and quantum algorithms. Part 1. Quantum computers. Mathematics. Mechanics. Computer science. №4 (2015)
8. Monz, T., et al.: Realization of a scalable Shor algorithm. Science. **351**(6277), 1068–1070 (2016)
9. Zhang, K., Korepin, V.E.: Depth optimization of quantum search algorithms beyond Grover's algorithm. Phys. Rev. A. **101**(3), 032346 (2020)
10. Lukashev, A., Shabunya, V., Sarafannikov, V., Bilan, V., Ivanov, D.: Trends in increasing vulnerability of modern information systems from quantum computers. In: Izvestiya TulSU Technical sciences, №11 (2023)
11. Sharma, M., et al.: Leveraging the power of quantum computing for breaking RSA encryption. Cyber Phys. Syst. **7**(2), 73–92 (2021)

12. Sadikin, M.A., Wardhani, R.W.: Implementation of RSA 2048-bit and AES 256-bit with digital signature for secure electronic health record application. In: 2016 International Seminar on Intelligent Technology and Its Applications (ISITIA), pp. 387–392. IEEE (2016)
13. Dam, D.T., et al.: A survey of post-quantum cryptography: Start of a new race. Cryptography. **7**(3), 40 (2023)
14. Minder L.: Cryptography based on error correcting codes: Verlag nicht ermittelbar (2007)
15. Benioff, P.: The computer as a physical system: A microscopic quantum mechanical Hamiltonian model of computers as represented by Turing machines. J. Stat. Phys. **22**, 563–591 (1980). https://doi.org/10.1007/BF01011339
16. DavidDeutsch: Quantumtheory, the Church–Turing principle and the universal quantum computer. In: Proc. Roy. Soc. Lond. A, vol. 400, pp. 97–117 (1985). https://doi.org/10.1098/rspa.1985.0070
17. Bernstein, E., Vazarani, U.: Quantum complexity theory. SIAM J. Comput. **26**, 1411–1473 (1997). https://doi.org/10.1137/S0097539796300921
18. Simon, D.R.: On the power of quantum computation. In: Proceedings 35th Annual Symposium on Foundations of Computer Science, pp. 116–123. IEEE (1994). https://doi.org/10.1109/SFCS.1994.365701
19. Shor, P.W.: Algorithms for quantum computation: discrete logarithms and factoring. In: Proceedings 35th Annual Symposium on Foundations of Computer Science, pp. 124–134. IEEE (1994). https://doi.org/10.1109/SFCS.1994.365700
20. von Burg, V., et al.: Quantum computing enhanced computational catalysis. Phys. Rev. Res. **3**(3) (2021). https://doi.org/10.1103/physrevresearch.3.033055
21. Grover, L.K.: A fast quantum mechanical algorithm for database search. In: Proceedings of the twenty-eighth annual ACM symposium on theory of computing, pp. 212–219 (1996). https://doi.org/10.1145/237814.237866
22. Lloyd, S.: Universal quantum simulators. Science. **273**(5278), 1073–1078 (1996). https://doi.org/10.1126/science.273.5278.1073
23. Shor, P.W.: Scheme for reducing decoherence in quantum computer memory. Phys. Rev. A. **52**(4), R2493 (1995)
24. Aharonov, D., Ben-Or, M.: Fault-tolerant quantum computation with constant error. In: Proceedings of the twenty-ninth annual ACM symposium on Theory of computing, pp. 176–188 (1997)
25. Shor, P.W.: Fault-tolerant quantum computation. In: Proceedings of 37th Conference on Foundations of Computer Science, pp. 56–65 (1996). https://doi.org/10.1109/SFCS.1996.548464
26. Knill, E., Laflamme, R., Zurek, W.H.: Resilient quantum computation. Science. **279**(5349), 342–345 (1998)
27. Knill, E., Laflamme, R., Zurek, W.H.: Resilient quantum computation: error models and thresholds. In: Proceedings of the Royal Society of London. Series A: Mathematical, Physical and Engineering Sciences 454.1969, pp. 365–384 (1998)
28. Monroe, C., et al.: Demonstration of a fundamental quantum logic gate. Phys. Rev. Lett. **75**(25), 4714 (1995). https://doi.org/10.1103/PhysRevLett.75.4714
29. Cirac, J.I., Zoller, P.: Quantum computations with cold trapped ions. Phys. Rev. Lett. **74**(20), 4091 (1995). https://doi.org/10.1103/PhysRevLett.74.4091
30. Lapina, M., Neelakandan, S., Donchenko, M., Ardeev, D., Boykov, N.: Comparative analysis of basic operations of Helib and microsoft seal libraries for cloud computing. In: Nature-Inspired Optimization Algorithms for Cyber-Physical Systems, pp. 341–345 (2025) https://www.igi-global.com/chapter/comparative-analysis-of-basic-operations-of-helib-and-microsoft-seal-libraries-for-cloud-computing/364786

31. Shadab, M., et al.: Generating highly nonlinear S-boxes using a hybrid approach with particle swarm optimization. In: Nature-Inspired Optimization Algorithms for Cyber-Physical Systems, pp. 1–30 (2025) https://www.igi-global.com/chapter/generating-highly-nonlinear-s-boxes-using-a-hybrid-approach-with-particle-swarm-optimization/364776
32. Vinodha, D., et al.: Framework based on IoT, AI, and blockchain for smart access to government agricultural schemes. In: The future of agriculture: IoT, AI and Blockchain technology for sustainable farming, pp. 142–163 (2024). https://doi.org/10.2174/9789815274349124010011

Methodology of Decision-Making in Information Security Specialist Training Management Systems

Anastasiya Arkhipova(✉)

Novosibirsk State Technical University, 20 Prospekt K. Marksa, Novosibirsk 630073, Russia
arhipova_ab@mail.ru

Abstract. Ensuring the security of networks, devices and software, timely detection of threats and response to incidents is a modern urgent task of information security at the state level. In the article provided a comprehensive methodology that integrates decision-making frameworks to enhance the effectiveness of training programs in the modern labor market. This methodology encompasses various components, including needs assessment, curriculum design, instructional strategies, and evaluation metrics. By employing a systematic approach, the article provides to optimize training outcomes and ensure that information security specialists are equipped with the necessary skills and knowledge to address contemporary security challenges. In the article implemented that the technology of ontological control model building process in information security. It consists of five continuous stages such as generation of knowledge area of educational and career trajectories, formation of a fuzzy set of educational and career trajectories and calculation of the indicator of compliance of the educational system for the training of an information security specialist. Author submitted that the target functionality will provide an extreme value of indicator of compliance and information security requirements. Experimental part consists of applying methodology of decision-making in information security specialist training management systems. It reveals a varied landscape of readiness across different roles within the field of information security.

Keywords: Decision-Making Process · Indicator of Compliance · Fuzzy Logic · Membership function · Competencies · Information Security · Cybersecurity Learning · Ontological control model

1 Introduction

Widespread digitalization has increased the role of information technology and data volumes, but at the same time spawned new cyber threats, which are now a central issue of information security. Information security vendor reports for 2024 contain the following information about key figures and trends [11]:

1. The number of incidents increased by 13% compared to the same period in last year.
2. Malicious software remains the main tool of attackers.

M. Lapina et al. (Eds.): AISMA 2025, LNNS 1456, pp. 40–50, 2026.
https://doi.org/10.1007/978-3-032-07275-7_5

3. There is an increase in interest in spyware in attacks on organizations - indicators in the 2024 quarter are 4% higher than in 2023.
4. In 53% of attacks on organizations, confidential information was disclosed, disruption of activities was observed in 32% of incidents.

Thus, ensuring the security of networks, devices and software, timely detection of threats and response to incidents is a modern urgent task of information security at the state level.

Currently, there are conflicting opinions regarding information security education, both officially, in various forums, and unofficially, including privately. In particular, as a polar one, an opinion is expressed about the inability of regional universities (or, in a more general form, all universities) to train specialists capable of at least at the minimum level to meet the needs of state, municipal authorities, specialized companies. Indeed, a modern employer imposes high requirements on job seekers in the form of knowledge and skills that meet the requirements of the labor market. Unlike IT specialties, the field of information security is characterized by the peculiarities of the processes of applying organizational and technical measures, the implementation of which is described in various federal laws, government decrees, decrees of the Pres-ident of the Russian Federation, orders of federal executive bodies [7]. Legislation in the field of information protection is characterized by a clear identification of objects of protection, which include: personal data information systems, state information systems, critical information infrastructure, automated control systems, state secrets, and other confidential information [3]. On the other hand, there are a large number of regulators: the Federal Security Service, the Federal Service for Technical and Ex-port Control, the Federal Service for Supervision of Communications, Information Technology and Mass Media, responsible for various aspects of information security, issuing their own departmental acts [4]. In addition, the number of software and firmware vulnerabilities is steadily increasing, which necessitates the use of a wide range of protection measures, which also need to be regulated by law [7].

In the field of education, direct regulation takes place by the Ministry of Science and Higher Education, the Ministry of Education in cooperation with the Ministry of Labor of the Russian Federation [3]. The issue of introducing federal educational standards is regulated by the Federal Educational and Methodological Association in the field of information security.

In aggregate, the above factors create a situation in which education in the field of information security has a multi-level structure and a complex formalized large amount of educational base in the form of regulatory legal acts approved at different times by different departments, a wide range of studied software and hardware protection systems, etc. [3]. This fact makes it difficult to study the field of knowledge and makes it necessary to systematize this information for compliance with labor market requirements.

One possible solution to this problem may be to build a management decision-making system using ontological modeling [5].

This methodological approach is used to create formal models of knowledge in certain studied subject areas. An ontology in this context is a structure or schema that describes the basic entities of a subject area, their properties, and the relationships between them.

2 Ontological Control Model Building Process in Information Security

Information security education is classified as a poorly structured subject area that has the potential to form a knowledge base with which to organize their effective management [6]. This requires not only the integration of knowledge, but also the correct formalization that satisfies information search methods [6]. In addition, modern processes in education are characterized by rapid changes, taking into account various factors of influence - speed of implementation, waiting time in personnel matters, degree of importance, changes in the labor market, etc. [7]. This is characteristic of information processing and analysis tasks, especially those that are perceived in several contexts at the same time, as well as in conditions of fierce competition, where effective knowledge management is a reliable tool for the qualitative identification, creation and dissemination of knowledge with minimal cost and response time to changing business conditions [10].

Ontological models provide users with access to large amounts of complex structured information and are one way to exchange knowledge.

When designing complex systems and developing methods for structuring information, a hierarchical approach was used as a methodological technique for dividing a formally described system into levels. The technology involves a consistent solution at all stages of which it consists (from top to bottom) [7].

Stage 1 "Generation of knowledge areas of the educational trajectory" is laborious, since it requires the collection and processing of information. The first stage implies the analysis of curricula, work programs of disciplines, control materials, educational and methodological complexes. Thus, four types of tasks of professional activities within the framework of information security training for a specific curriculum have been previously identified: operational, design and technological, experimental research and organizational and managerial. We denote their binary representation of the influence for the i-th discipline, respectively, in the form of a_j^i, j = 1,..., 4.

Taking into account the above, the formal statement of the task "Generation of knowledge areas of the educational trajectory" is as follows.

As part of the study, we will assume that the curriculum of the specialty in the field of information security represents a tuple of the form [7]:

$$F = < X, Y, Z >,$$

where $X = \{x_1^{t1}, x_2^{t2}, \ldots, x_n^{tn}\}$ - set of the work programs of professional disciplines of training $\{1, \ldots, N\}$ for training tasks $t_i = \{a_1^i, a_2^i, a_3^i, a_4^i\}$;
$Y = \{y_1^{t1}, y_2^{t2}, \ldots, y_n^{tn}\}$ - set of controlling materials of professional training disciplines {1,..., N} for training tasks $t_i = \{a_1^i, a_2^i, a_3^i, a_4^i\}$;
$Z = \{z_1^{t1}, z_2^{t2}, \ldots, z_n^{tn}\}$ - a set of teaching materials for professional training disciplines $\{1, \ldots, N\}$ for training tasks $t_i = \{a_1^i, a_2^i, a_3^i, a_4^i\}$.

Stage 2 "Generation of career trajectory knowledge areas." At this stage, a scheme of career tracks in the field of information security was formed based on labor market proposals (analysis in the field of HR technologies).

In total, seven areas of development of information security specialists were identified:

1. information protection in critical information infrastructure (CIE);
2. safety study;
3. administration of information security tools;
4. Vulnerability management
5. compliance management;
6. Secure application development.
7. Security operation center (SOC).

As part of the study, we will assume that the direction of development of a specialist in the field of information security is a tuple of the form:

$$R = < L, U, E >,$$

where $L = \{l_1, l_2, \ldots, l_n\}$ – set of work tasks within the framework of the professional activities of an information security specialist {1,..., N} [2–4];
$U = \{u_1, u_2, \ldots, u_k\}$ - set of hard-skills as part of the professional activities of an information security specialist [2–4];
$E = \{e_1, e_2, \ldots, e_m\}$ - set of soft-skills as part of the professional activities of an information security specialist [7].

Stage 3 "Formation of a fuzzy set of educational trajectories" is aimed at fuzzification, clustering and determination of fuzzy tendencies of the set of ontologies formed at the first stage depending on the seven identified areas of development of information security specialists [7].

Stage 4 "Formation of a fuzzy set of a career trajectory" is aimed at fuzzification, clustering and determination of fuzzy tendencies of a set of ontologies formed at stage 2 according to enlarged groups of directions/specialties of information security specialists training.

Stage 5 "Calculation of the indicator of compliance of the educational system for the training of an information security specialist" in the form of a fuzzy distance of indicators of Stages 3 and 4.

3 Methodology of Decision-Making in Information Security Specialist Training Management Systems

In general, the optimal control problem is as follows. For the educational system S (x) under the control of *L*, we find a control that will provide an extreme value to the target functionality *F* at the set of indicators *I*, i.e.

$$F = \max_{i \in I, l \in L} F(x, l, i).$$

The indicator of compliance of the educational system for training an information security specialist is a control parameter that displays the level of correlation between the education sector and the labor market. The resulting indicator can be interpreted as the level of compliance of the quality of educational training of the educational institution with the career tracks of potential employers.

Management properties reflect the most significant connections and relations of the management parties (labor market and education system):

- compatibility of systems and controls;
- unity and subordination of performance criteria;
- unity of organizational and methodological bases of profession at all levels of management;
- presence of feedback during interaction of system components;
- single focus on the purpose of the control function.

Thus, the indicator of compliance of the educational system for training an information security specialist with the requirements of the labor market will identify problem segments of training modules in the form of qualitative and quantitative assessments [3].

Structurally, the decision-making methodology in information security specialist training management systems contains the following main sections [3, 7]:

1. General provisions (Purpose, tasks, models used, evaluation criteria).
2. Labor market monitoring procedure.
3. Procedure for monitoring the educational program.
4. Composition of means for assessing the parameters of qualitative and quantitative indicators.
5. Essence of the technology for forming indicators of the compliance of the educational system for training an information security specialist with labor market requirements.
6. Presentation and analysis of the obtained results [3, 7].

This decision-making methodology in information security specialist training management systems is intended for software and computational assessment of readiness of information security specialists [4].

4 Experimental Part

The developed model for calculating the indicator of compliance of the educational system for training an information security specialist with the requirements of the labor market was a constructive basis for planning and conducting experimental work, which was carried out in order to test the formulated hypothesis of the study [3].

In accordance with the set objectives of the study, experimental work was carried out by us in three stages.

At the first stage, a research base was formed in the form of curricula for the specialty 10.05.03 “Information Security of Automated Systems” of Novosibirsk State Technical University in 2020–2024 years. The tool for creating OWL Protégé ontologies was used to visualize and form an ontological graph. The description of the ontology structure was based on specialization, educational level, many disciplines and competencies. Subclasses of ontology were information security tasks according to the requirements of federal educational standards [5, 9, 10].

The analysis showed a deepening of the structure of the last year of student enrollment in the context of the inclusion of professional spectrum disciplines according to the career tracks of an information security specialist. Subclasses of ontology were the corresponding information security tasks (see Fig. 1).

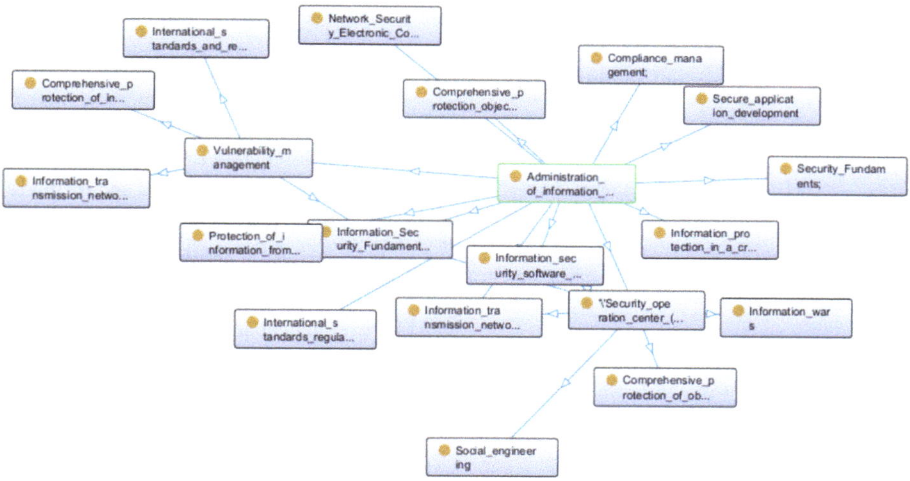

Fig. 1. Ontological graph of education tracks of information security

The second stage of the study included an analysis of labor market requirements for open source information security specialists. Based on the requirements of more than 500 vacancies, the main seven career tracks (directions of development) were identified. Each of the directions of development involves the fulfillment of a certain range of tasks and requires specific competencies [3]. The description of the ontology structure was based on the directions of development of career tracks. Subclasses of ontology were the corresponding information security tasks (see Fig. 2).

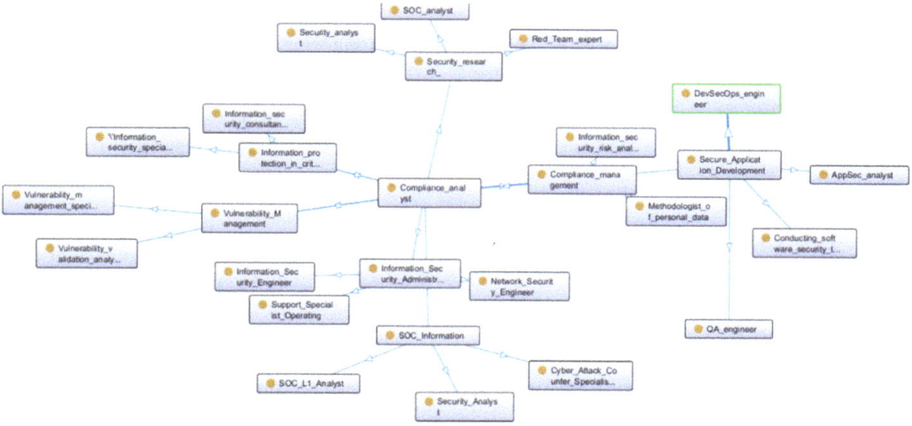

Fig. 2. Ontological graph of career tracks of information security

Each of these areas of development involves the fulfillment of a certain range of tasks and requires specific competencies and indicators (see Table 1).

Table 1. Tasks and set of indicators of roles in the main areas of information security

№	Area of information Security / Role	Set of tasks / indicators L_j
1	Information protection in critical information infrastructure [3]	L_1^1- Assessment of the security of significant objects in their operating conditions; L_2^1 - Development of programs and methods for certification tests of significant objects; L_3^1 - Performance of certification tests with the use of information security controls; L_4^1 - Development and execution of reporting materials based on the results of certification tests.
2	Security research *Role*: Security analyst, Red Team expert, SOC analyst	L_1^2—Conducting penetration tests; L_2^2—research work on the search for vulnerabilities of the application and network level; L_3^2—Assessment of the security of systems of various purposes and composition; development of reports and documentation; L_4^2—Work with security analysis tools; L_5^2—development of tools and techniques for security analysis; L_6^2—Development of scenarios for implementation of security threats.
...	...	...
7	SOC Information *Role*: Security Analyst, SOC L1 Analyst, Cyber Attack Counter Specialist	L_1^7—Monitoring security events received using SIEM alerts or other security tools; L_2^7—Processing of incidents received from users via email and applications; L_3^7—Analysis and response to information security incidents; L_1^2—Preparation of reporting uploads on information security status.

An analysis of information security vacancies also revealed a list of the most mentioned soft skills needed by an information security specialist. Figure 3 displays skills and how many times they are repeated among announcements of information security work in 2025 year:

Based on the analysis of labor market requirements, a basic term set of work tasks (L) has been formed as part of the professional activities of an information security

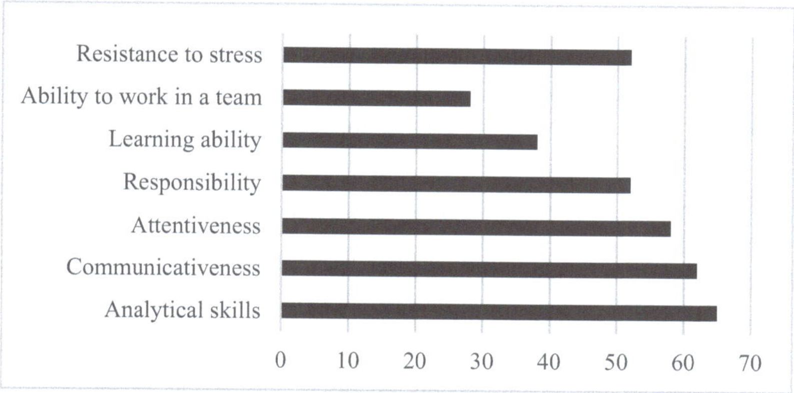

Fig. 3. Soft-skills of the information security vacancies revealed

specialist of the form [3, 8]:

$$< L, T(L), M >,$$

where *T(L)* is the term set of the corresponding *L*-problem [7];

M is a semantic rule that maps a fuzzy variable from *T(L)* to a fuzzy subset of the universal term set.

Under the conditions of the problem, we define the base term set with five fuzzy terms:

$$T = \{T_1, T_2, T_3, T_4, T_5\},$$

where T_1- "low," T_2- "below average," T_3- "medium," T_4- "above average," T_5- "high."

Then reference fuzzy ones are constructed using one of the membership function $M = \{m^2, m^{1/2},\ldots\}$ (Fig. 4) [7]:

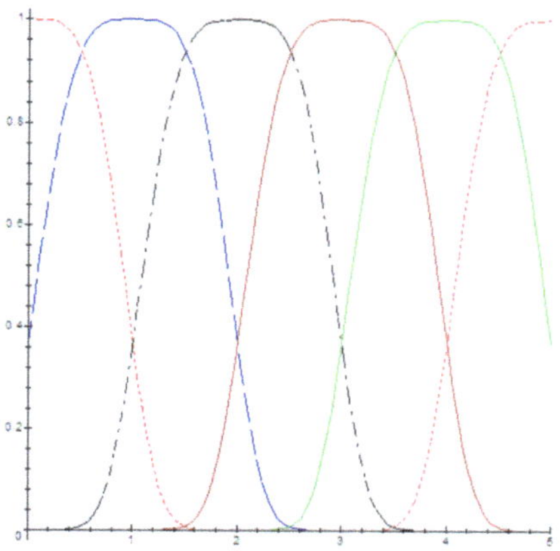

Fig. 4. Membership function in the term set of the corresponding *L*-problem

Based on the selected fuzzy models, an indicator of compliance of the educational system for training an information security specialist is determined. The resulting indicator is interpreted through the correspondence of the basic term set of career tracks of the labor market to the term set of the curriculum of the specialty (Table 2).

Table 2. Tasks and set of indicators of roles in the main areas of information security

№	Role of information Security	Fuzzy Indicator of compliance of the educational system for training an information security specialist	Indicator of compliance of the educational system for training an information security specialist, %
1	Information security consultant, information security specialist, compliance analyst	Above average	76
2	Security analyst, Red Team expert, SOC analyst	Medium	70
3	Information Security Engineer, Network Security Engineer, Support Specialist Operating	Above Average	78
4	Vulnerability management specialist, vulnerability validation analyst, compliance analyst	Medium	70
5	Information security risk analyst, compliance analyst, methodologist in the direction of personal data	Below Average	62
6	AppSec analyst, DevSecOps engineer, QA engineer Conducting software security testing	Below Average	64
7	Security Analyst, SOC L1 Analyst, Cyber Attack Counter Specialist	Medium	72

The output is presented in both linguistic and numerical form as fuzzy numbers and levels. Based on the proposed synthesis methodology, it is possible to build both software and real-time software systems designed to effectively assess the level of training of a specialist in accordance with the requirements of the modern labor market [3, 4, 7].

5 Conclusion

The article represents the methodology of decision-making in information security specialist training management systems.

The article discusses that field of information security is characterized by the peculiarities of the processes of applying organizational and technical measures, the implementation of which is described in various federal laws, government decrees, decrees of the President of the Russian Federation, orders of federal executive bodies [7]. This fact makes it difficult to study the field of knowledge and makes it necessary to systematize this information for compliance with labor market requirements [2, 3]. One possible solution to this problem may be to build a management decision-making system using ontological modeling. The construction of an ontological graph for the domain of information security in critical information infrastructure provides a structured representation of various roles and responsibilities within this field [4]. By categorizing positions such as consultants, specialists, analysts, and engineers into distinct areas researchers gain a clearer understanding of the complex landscape of information security.

In the article we implemented that the technology of ontological control model building process in information security. It consists of five continuous stages from generation of knowledge area of educational and career trajectories to "Calculation of the indicator of compliance of the educational system for the training of an information security specialist".

The analysis of the "Fuzzy Indicator of compliance of the educational system for training an information security specialist" reveals a varied landscape of readiness across different roles within the field. The data indicates that some roles of information security specialists (Information Security Engineer, Network Security Engineer, Support Specialist Operating, Information security consultant, information security specialist) are well-supported by the university. This suggests that the training programs are effectively equipping students with the necessary skills and knowledge for these positions. Overall, while certain areas of the educational system demonstrate strong alignment with industry needs, there remains significant place for improvement in preparing specialists for specific roles within information security. Addressing these gaps will be essential to ensure that future professionals are adequately prepared to meet the evolving challenges in the field of information security. Continuous evaluation and adaptation of training programs will be crucial in fostering a skilled workforce capable of safeguarding against emerging threats.

References

1. Khaleefah, A.D., Al-Mashhadi, H.M.: Methodologies, requirements, and challenges of cybersecurity frameworks: a review. Iraqi J. Sci. **65**(1), 468–486 (2024). https://doi.org/10.24996/ijs.2024.65.1.38
2. Arkhipova, A.B.: Information security specialist readiness indicator as a part of the trusted digital environment. In: CEUR Workshop Proceedings. - 2021. - vol. 3057: Distance Learning Technologies (DLT 2021): 6 intern. scientific and practical conf., Yalta, 20–22 Sept. 2021. - 2021. pp. 36–45. https://www.scopus.com/record/display.uri?eid=2-s2.0-85122819371&origin=resultslist&sort=plf-f

3. Arkhipova, A.B.: Multisociometrical readiness characteristics in information security management. In: CEUR Workshop Proceedings. - 2022. - vol. 3094: Advanced in Information Security Management and Applications: proc. of the intern. workshop on advanced in information security management and applications (AISMA 2021), Stavropol–Krasnoyarsk, 1 Oct. 2021. - 2022. pp. 25–34. http://ceur-ws.org/Vol-3094/paper_2.pdf
4. Arkhipova, A.: General model for assessing the quality of socially significant activities as a part of information security management. https://doi.org/10.1007/978-3-031-77229-0_1. direct. Lecture Notes in Networks and Systems. - 2024. - vol. 1207: International Workshop on Advanced Information Security Management and Applications (AISMA–2023) : proc., Stavropol, 28 Sept. - 1 Oct. 2023. pp. 1–8
5. Zaoui, F., Souissi, N.: Onto-digital: an ontology-based model for digital transformation's knowledge. Int. J. Inform. Technol. Compute. Sci. **10**(12), 1–12 (2018). https://doi.org/10.5815/ijitcs.2018.12.01
6. Gusarova, N.F., Sysoeva, V.V.: Ontological modeling of a semi-structured subject area using fuzzy logic. Sci. Tech. Bull. Inform. Technol. Mech. Opt. **17**(4), 711–718 (2017). https://doi.org/10.17586/2226-1494-2017-17-4-711-718
7. Krokhaleva A.B.: Mathematical and the software of support of decision-making in the system of training of specialists (on the example of the information security direction). Master's thesis. [Management in social and economic systems. Master's thesis]. Novosibirsk, 2015. 201 p. (in Russian)
8. Zadeh, L.A.: Fuzzy sets. Inf. Control. **8**(3), 338–353., ISSN 0019-9958 (1965). https://doi.org/10.1016/S0019-9958(65)90241-X
9. Meriah, I., Ben Arfa Rabai, L., Khedri, R.: Building a comprehensive and multi-dimensional information security ontology: elicitation process and OWL implementation. Knowl. Inf. Syst. **67**, 167–195 (2025). https://doi.org/10.1007/s10115-024-02308-y
10. Afanasyev, T.V., Moshkin, V.S., Namestnikov, A.M., Timin, I.A., Yarushkin, N.G.: Ontological and fuzzy analysis of semi-structured information resources. Under scientific. Yarushkina, N.G. (ed.). UlSTU, Ulyanovsk, p. 130 (2016)
11. Rodríguez-Correa, P.A., Valencia-Arias, A., Martínez Rojas, E., et al.: Information security education: a thematic trend analysis. F1000Res. **14**, 5 (2025). https://doi.org/10.12688/f1000research.159828.2
12. Fernando, C.A., Thomas, R., Mary Anita, E.A., Lapina, M.: A feature selection study on the bot-IoT dataset using ensemble classification techniques. In: 2024 IEEE International Conference on Contemporary Computing and Communications (InC4), Bangalore, India, pp. 1–9 (2024). https://doi.org/10.1109/InC460750.2024.10649035
13. Qasim, M., Sajid, M., Lapina, M., Babenko, M.: A generalized Bi-objective scheduling algorithm for batch-of-tasks on heterogeneous computing systems. Adv. Syst. Sci. Appl. (2024). https://doi.org/10.25728/assa.2024.2024.02.1430. EID: 2-s2.0–85199979370, Part of ISSN: 10786236, https://ijassa.ipu.ru/index.php/ijassa/article/view/1430
14. Chincheti, A., Dixit, R.K., Athavale, V.A., Lobo, L.M.R.J., Lapina, M.: A safe and secure online system for bidding using Blockchain technology. In: Lapina, M., Raza, Z., Tchernykh, A., Sajid, M., Zolotarev, V., Babenko, M. (eds.) AISMA-2024: International Workshop on Advanced Information Security Management and Applications. AISMA 2024 Lecture Notes in Networks and Systems, vol. 863, pp. 43–54. Springer, Cham (2024). https://doi.org/10.1007/978-3-031-72171-7_5

Designing an Intelligent SOC Framework with Azure Sentinel for Enhanced Security Incident Management

Dipali Awasekar[1], Vijay Anant Athavale[2(✉)], L. M. R. J. Lobo[1], and Vitalii Lapin[3]

[1] Department of Information Technology, Walchand Institute of Technology, Solapur, India
[2] Department of Computer Science and Engineering, Walchand Institute of Technology, Solapur, India
vijay.athavale@gmail.com
[3] North Caucasus Federal University, Stavropol, Russia

Abstract. The number of breaches in security has risen to a great extent and this has become a question of concern in companies around the world. An analysis done lately illustrates the urgent need for robust cyber-security measures. Normally breaching of data has worldwide reached a cost of approximately $4.45 million as reported in 2023 which is 15% greater than the previous three years. The urgent identified need requirement for special systems for detecting threats and getting responses to incidents is seeded in this emerging pattern. Azure Sentinel which is Microsoft's cloud-native Security Information and Event Management (SIEM), and Security Orchestration, Automation, and Response (SOAR) are used in this study to build and execute an intelligent Security Operations Center (SOC) framework. The suggested SOC architecture serves as a centralized location for real-time security incident monitoring, detection, analysis, and response. Data ingestion from various sources across on-premises and cloud environments via data connectors, alert-triggering analytic rule configuration, and notification mechanism integration through channels like Gmail and Microsoft Teams are all part of the solution. In order to decrease the mean time to respond (MTTR), the study also uses Azure Logic Apps (Sentinel Playbooks) to automate incident response. By integrating threat intelligence, the framework improves threat detection capabilities, empowering companies to proactively detect and neutralize new attacks. The results of the study demonstrate that adding Azure Sentinel to SOC procedures improves incident detection, expedites response times, and strengthens overall cyber defences.

Keywords: Azure Sentinel · Cyber Defence · Incident Response · Threat Detection · and Security Operations Centre (SOC)

1 Introduction

The quick development of digital technology in today's linked world has completely changed how businesses function and do business [1]. Greater flexibility, scalability, and efficiency have been made possible by the transition from traditional on-premise

M. Lapina et al. (Eds.): AISMA 2025, LNNS 1456, pp. 51–62, 2026.
https://doi.org/10.1007/978-3-032-07275-7_6

infrastructures to cloud-based settings. But enterprises are now more vulnerable to a wider range of cyber threats as a result of this digital transition [2–4]. By standardizing digital transactions and remote deployment, the COVID-19 pandemic accelerated this tendency and gave hackers more options to take advantage of. According to The Sift Trust & Safety Index a 20.7% increase in internet traffic during 2020 triggered more cyberattacks. The number of global cyberattacks climbed 38% between 2022 and 2023 and India experienced 2,138 weekly attacks per firm in 2023 which was 15% higher than in 2022. The FBI's Internet Crime Complaint Centre (IC3) documented 859,532 cybercrime cases in the United States for 2023 which shows a 16% increase. In 2018 organizations spent 196 days locating breaches followed by 69 days to address them. To defend against constantly evolving hacker tactics organizations, need to enforce a proactive cyber-security strategy which focuses on continuous monitoring and swift threat detection along with immediate incident reaction [5]. A centralized Security Operations Canter (SOC) is established to enable monitoring to be done by organizations to detect violations in security and respond to them in real time after analysis [6]. Such a canter represents a caution against cyberattacks. A committed group works with tools and procedures required to protect the digital assets of an organization. In the end, the SOC lowers cyber-security risks and decreases impact by ensuring 24-h surveillance, early threat detection, and prompt incident response [7]. The main advantage of working with a SOC includes its capabilities that support continuity in surveillance, consolidated options supporting visibility, reduced cost in context of cybersecurity, better organizational links, and protection inbuilt that works proactively concerning essential assets.

The fundamental technology powering modern Security Operations Centres (SOCs) is Security Information and Event Management (SIEM). It enables collecting security data from multiple sources. These sources include firewalls and antivirus software which give complete picture of an organization's security at different time intervals. These systems allow analysts to monitor events at time and chain them with other data thus performing identification quickly and further resolving them. A strong cloud-native SIEM and Security Orchestration, Automation, and Response (SOAR) solution in this regard is Microsoft's Azure Sentinel [8]. Azure Sentinel's real-time threat detection, automated incident response, and powerful analytics enable enterprises to create and manage an intelligent SOC. This article focuses on creating and deploying a SOC framework using Azure Sentinel to improve cyber-security posture, reduce risks, and protect organizations from changing threats.

The present study was carried out to exhibit the role of Azure Sentinel in building of SOC with inherent features of intelligence and scalability. This system tracks eventually cyber-security problems and gives organizations solutions to resistance from cyberattacks.

2 Literature Review

One of the most powerful tools for enhancing Security Operation Centers is Azure Sentinel. It supports threat detection, facilitates incident responses and continuous monitoring. A number of articles highlighting its theoretical basis, real world applications and its integration with modern frameworks of security have been examined [9].

DiCola and Copeland (2020) focused on the best use of Azure Sentinel in their article.

This work provides a detailed set of steps to set up, configure. and use Sentinel to protect organizations. They focus on connecting other Azure services to form a strong architecture providing security [10]. To practically work upon this concept, Miller (2010) examined some core components of Security Information and Event Management (SIEM) systems of support managing logs, detecting threats and generating responses for incidents. This expertise in SIEM actual implementation generates the foundation for efficient use of Azure Sentinel [11]. Concentrating on risk management and its implementation on the cloud Mathew et al. (2009) went deeper into the theoretical stand of cloud principles. In their study they emphasized on the actual usefulness of Azure Sentinels and like available solutions for dealing with problems like SOC operating using a cloud [12]. A major principle of Zero Trust security introduced by Gilman and Barth (2017), supports constrained access restrictions and exhibits a continuous verification process, thus a support of the capabilities of Azura Sentinel enhance security and continuous monitoring [13].

It was Lowe et al. (2016) that used the integration of operations related to security to be introduced into design of data Centre's of modern times. He focused on parameters like automation, flexibility and scalability of the system which prove to be supportive for SOC operations to run effectively when Azure Sentinel is used [14]. It is reported by industry that Azure Sentinel has provided valuable results in financial and operational fronts. Forrester Consulting's (2021) the results provided w.r.t financial benefits were better than traditionally achieved SIEM solutions. Sentinel offered additional operational effectiveness, real time threat detection and lowered the cost [15]. It was outlined by Microsoft Learn (2022) through recommendations that deployment of Azure Sentinel covered aspects like secure data intake, management of logs and supported by altering operations to exhibit best performances of corporate [16]. Azure Sentinel is now associated with various real time applications like visibility of enhanced threats and faster responses for incidents across many industries [17]. As stated in the Microsoft Tech Community Blog (2021) developing queries for finding and analyzing security threats is a complex procedure. This may be like a hunting carried our using Kusto Query Language (KQL). Therefore, great importance has to be given to continuous surveillance and conduction of threat hunting proactively for SOC procedures [18]. It is however seen that new threats create the need for tools like Azure Sentinel that solve such problems involved. Referring to the Microsoft Security Blog (2023) a deep diving of cyber-security threats is mentioned [19]. The Microsoft Whitepaper (2022) further addresses problems faced by Small and Medium-sized Business (SMBs) for protecting their IT based infrastructures recommending the use of Azure Sentinel to provide solutions to monitoring of security and responses [20]. Articles in The Journal of Network and Computer applications contribute to various scenarios and provided solutions for Intrusion Detection Systems (IDS) under the Cloud environment. These articles offer strong foundations to understand the role of Sentinel in cloud-based security and highlight the use of SIEM technologies like Azure Sentinel boost SOC capabilities [21, 22].

3 Methodology

With the aid of an Agile procedure and use of Azure Sentinel the deployment of Security Operation Centre (SOC) was carried out. This was eventually a systematic and flexible approach for implementation. A modified architecture was developed to satisfy basic security purposes. This largely emphasised the use of Agile on an iterative scenario, teamwork and timely responses being an integrated part. The three-phase procedure was followed in the development process involving gathering of requirements, design and implementation.

In the phase of requirement gathering, analysis was conducted using Azure Sentinel on the existing SOC which was to be implemented. Here priorities are assigned to goals, potential data sources and concentrating on functional and non-functional needs. Here data feeding methods were defined, appropriate mechanisms were ensured and tenancy arrangements were handled for management of multi-environment deployments.

In the design stage, the requirements collected in the requirement gathering phase are transformed into architectural drawings with system specifications highlighted. There is a development here of workflow frameworks, models of data and detailed diagrams to ensure a design alignment of SOC with respective organization set goals for security. Here there is a focus on developing automated systems supporting responses, alerting and logic detection. Finally in the implementation phase there is actual configuring of Azure Sentinel and integration with the systems in the organization.

3.1 System Requirements

To have a smooth functioning of Security Operations Centre (SOC) deployment and operation, a stronger infrastructure was required by Azure Sentinel incorporating hardware, software and associated elements. The hardware included a dual core processor, 8GB RAM and 256 GB storage running on an appropriate platform of Linux or Windows. A reliable connection with a minimum speed of 10 Mbps is necessary for continuity in data ingestion and communication between systems.

With respect to software, web browsers like Firefox, Google Chrome and Microsoft edge are used. Additionally, SSH clients like PuTTY or OpenSSH were utilized to operate Linux virtual machines; Azure CLI was optional for more effective command-line operations. Furthermore, both a current Azure subscription and an Azure current Directory (AAD) tenancy were necessary for identity and access management.

3.2 System Architecture

As shown in Fig. 1, the system combines multiple components that work in symphony to absorb, assess, and respond to security events in diverse situations in order to guarantee comprehensive coverage and effective operations. Azure Sentinel, the main Security Information and Event Management (SIEM) platform, forms the basis of the design. Azure Sentinel provides a unified solution for researching, evaluating, and addressing security issues by combining security data from many sources. The architecture consists of a number of essential elements, all of which are required to improve cyber-security posture:

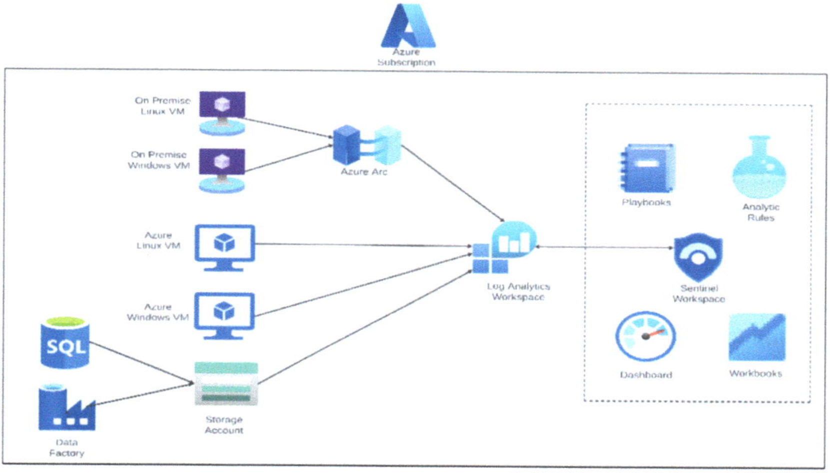

Fig. 1. System Architecture of SOC Implementation using Azure Sentinel.

Data Connectors:
These interfaces make it easier to import data from a variety of sources, including as third-party apps, cloud environments, and on-premises infrastructure. Thus, esurience of a collection of comprehensive data that allows a personnel appointed to check the security can be taken care of from a single defined window. The main storehouse of for telemetry data and security events is the Log Analytics work space. With the use of Kusto Query Language (KQL), it is now possible for an Analyst to query this data for purposes like analysis of forensic, detecting anomaly and advanced threats.

Workbooks:
These provide dashboards that are interactive and help to get visualization of security in real time. Workbooks assist stakeholders to watch critical matrices and give responses to eventual events with incites that are data driven. Thus, they enhance situation awareness. A robust SOC framework is hence enabled with a systematic methodology to monitor continuously, detect rapidly and mitigate efficient security threats.

Analytics Rules:
By these standard procedures suspicious activities are identified, alarms are generated when anomalies are encountered. Automation of threat identification helps get excellent timely responses.

Playbooks:
Developed using Azure Logic Apps, playbooks automate pre-programmed reaction actions, like identifying compromised endpoints, alerting users, or initiating procedures for additional research. This automation reduces response times and ensures that security events are handled consistently. These components work together to create a comprehensive security architecture that helps security experts more effectively monitor, detect, and eliminate threats. These are as shown in Fig. 1.

4 Experimental Setup and Results

4.1 Azure Sentinel Configuration

Setting up the Log Analytics Workspace, which is the foundation for log analysis, is the first stage in putting Azure Sentinel into practice (Figs. 2, 3, 4, 5, 6, 7, 8, 9, 10, 11, and 12). The first step to be followed is to construct the Workspace for Log Analytics by looking for and choosing it in the Azure interface. After entering the workspace creation process, the user is required to provide information about the workspace, including the resource group, the workspace name, the area in which it will be deployed, and the subscription under which it will be established.

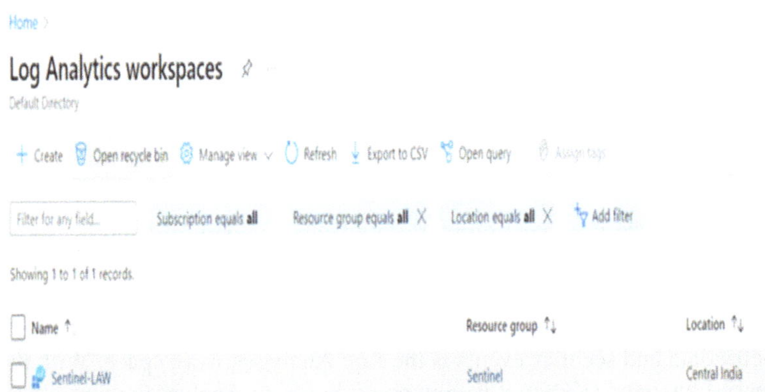

Fig. 2. Creation of Log Analytics Workspace in the Azure Portal.

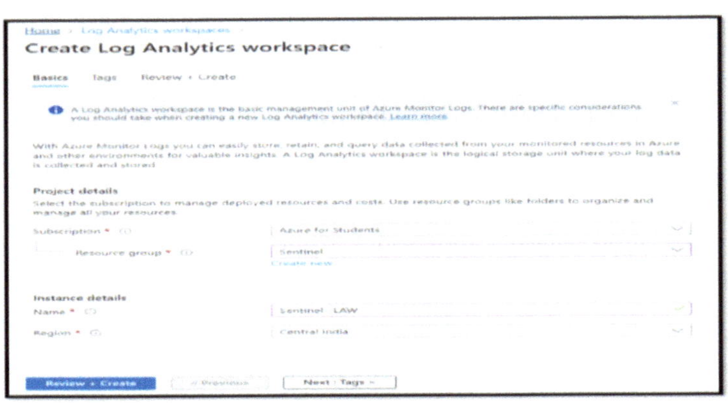

Fig. 3. Microsoft Sentinel dashboard after adding it to Log Analytics Workspace.

Setting up Microsoft Sentinel is the next step. This entails finding and choosing Microsoft Sentinel via the Azure interface. After that, the user creates a Sentinel

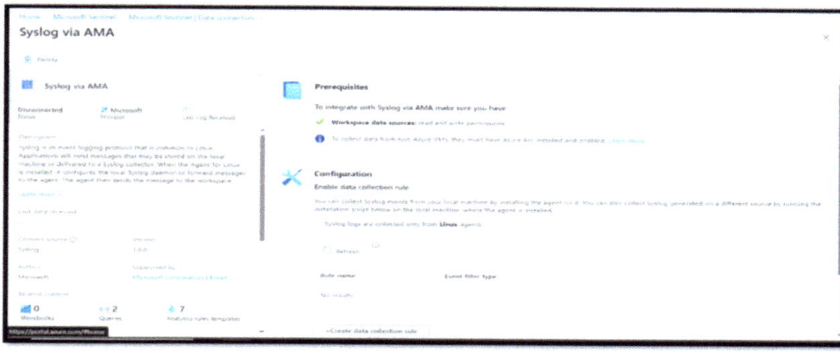

Fig. 4. Installing the Syslog extension from Content Hub for log ingestion.

Fig. 5. Data Connectors tab displaying available connectors for log ingestion.

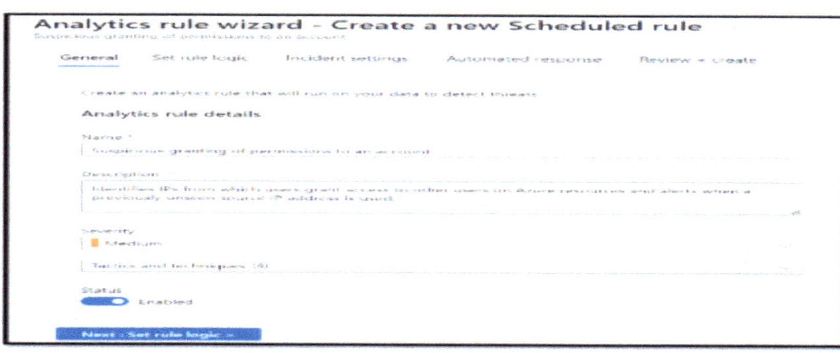

Fig. 6. Configuration page for creating Data Connection Rules in Sentinel.

workspace, which needs to be connected to the Log Analytics Work-space that was already created. For further configurations, the Sentinel workspace dashboard is now available.

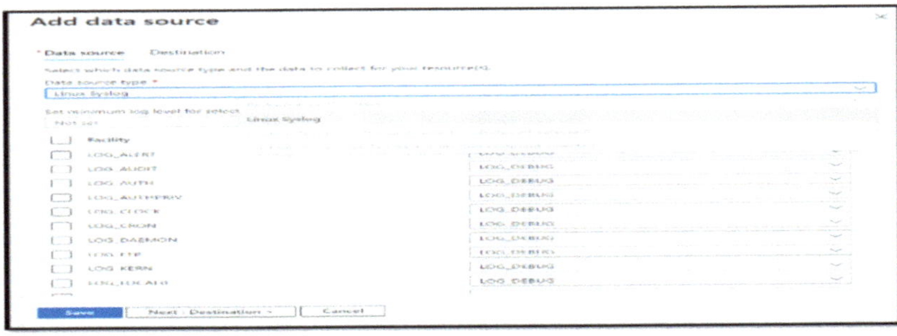

Fig. 7. Selection of resources and log types for data ingestion from Azure Linux VM.

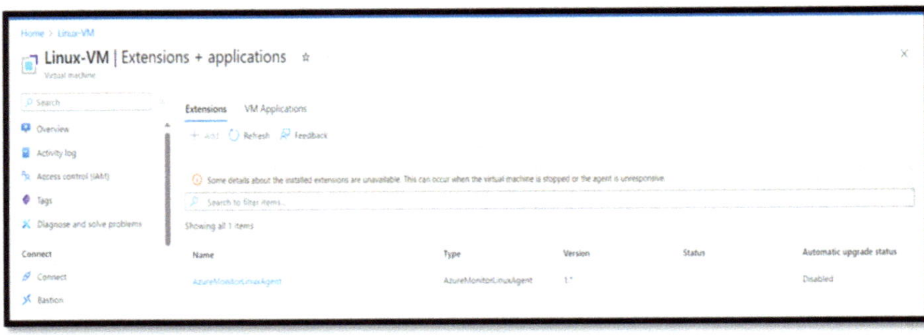

Fig. 8. Extensions and Applications tab showing installed Syslog extension on the Linux VM.

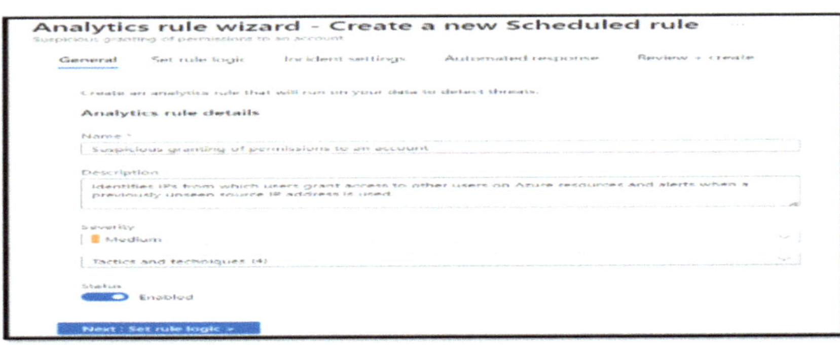

Fig. 9. Rule templates available for monitoring threats from Analytics tab

4.2 Data Ingestion

Azure Data Factory, Azure SQL Database, virtual machines on premises (Windows and Linux), and virtual machines of Azure (Windows and Linux) are some of the sources from which data can be ingested into Azure Sentinel. The ingestion of logs from an Azure Linux Virtual Machine (VM) is the main emphasis of this implementation. Installing the syslog extension from the Content Hub is required before you can start consuming

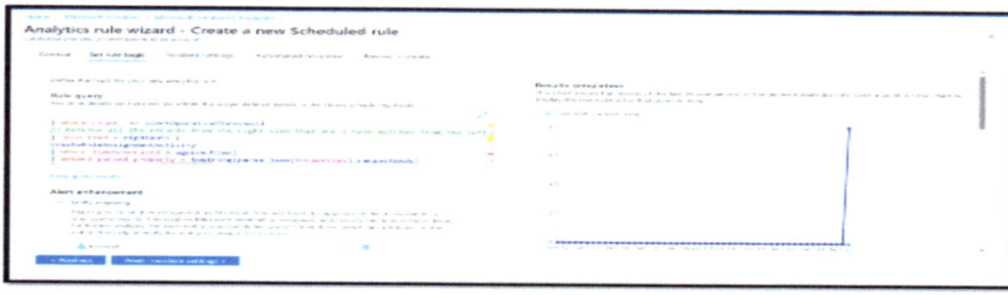

Fig. 10. Defined analytics rules, alert thresholds, and execution from Configuration page.

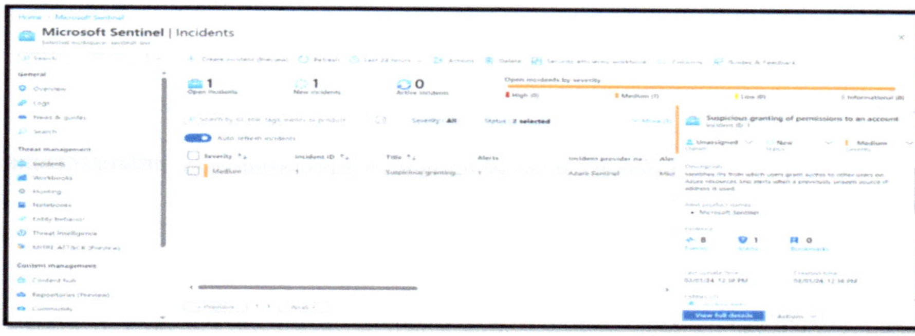

Fig. 11. Generated incidents with crucial risk levels and timestamps displays in Incident tab.

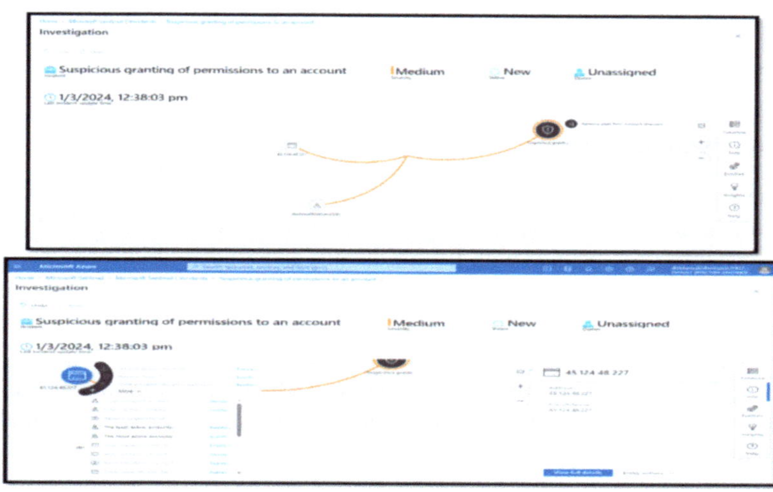

Fig. 12. Investigation graph displaying incident details with connected entities.

logs from an Azure Linux virtual machine. To accomplish this, search for Syslog, pick the relevant extension, and click Install.

It is now need to configure. The Syslog via AMA data connector. To continue with configuration, you must navigate to the Data Connectors menu, choose Syslog via AMA, and then open the connector page.

It is crucial to confirm that all prerequisites are satisfied before moving further with the data connection rules. The user chooses Create Data Connection Rules after confirmation. The required information, such as the rule's name and description, forms the required information is supplied on the basic tab.

On the Resources tab, the user chooses which Azure Linux virtual machine (VM) to connect to and what kinds of logs, like Linux Syslog and Performance Counters, should be gathered. The user goes to the Linux VM resource, chooses the Extensions and Applications Tab, and then makes that the installed syslog extension is present in order to ensure that it has been installed appropriately.

4.3 Analytical Rule Creation

After setting the intake of data we proceed for applying procedures require for tracking and identifying risks associated with security. Be follow steps like navigating through the analytical tab of Azure Sentinel dashboard. This step is followed with choosing complete for rules for identifying those that are necessary. The user then clicks a create rule button.

Query logic and the time window during which the query will execute is entered as information by the user. To guarantee that an alert is triggered when the query yields a predetermined number of results, an alert threshold is also set. Additionally, the user chooses how alerts are organized, either by creating an alert for each event or by combining all events into a single alert. The analytical rule is saved and made ready to run after all configurations are finished.

4.4 Resolving an Incident

During the creation of an incident the Azure Sentinel's Incidents tab is displayed. Now a user examines a specific incident giving desired input. This offers related information of entities, addresses associated with IP, timestamps and crucial levels of risk.

The Investigate option enables visual investigation of the incident through a graph representation for a more thorough analysis. The user can obtain more information and possible security risks by enlarging the nodes in the research graph.

A strong security monitoring solution is provided by this organized deployment of Azure Sentinel, which guarantees smooth configuration, data intake, rule formulation, and incident management.

5 Future Enhancements and Conclusion

5.1 Future Enhancements

Improving detection, automation, and reaction capabilities will be the main goals of the Security Operations Center's (SOC) development. Proactive threat finding can be made possible by integrating threat intelligence platforms such as TAXII 2.0, and automated incident resolution can be streamlined with Azure Sentinel playbooks driven by Azure Functions and Logic Apps.

In order to better anticipate and reduce risks, future enhancements will also incorporate cloud-native security and AI-driven threat detection. Further reducing false positives and increasing response efficiency can be achieved by refining analytics queries and detection procedures. The SOC will become a proactive, flexible security system as a result of these developments. Block chain can also be used to provide advanced security [23].

5.2 Conclusion

Threat detection, analysis, and reaction have greatly improved since Azure Sentinel was implemented. Its cloud-native architecture improves its cyber-security posture by guaranteeing scalability, flexibility, and cost effectiveness. This approach positions the SOC as a robust and flexible cyber-security framework, laying the groundwork for future developments such as AI-driven security improvements and increased automation.

References

1. Athavale, V.A., Sultana, A., Bilquees, A., Fathima, M., Basheer, M., Khatana, K.: IoT security using machine learning: applications and difficulties based on forthcoming directions. In: 2023 2nd International Conference on Automation, Computing and Renewable Systems (ICACRS), pp. 316–322 (2023). https://doi.org/10.1109/ICACRS58579.2023.10404364
2. Lobo, L.M.R.J., Athavale, V.A.: System for providing security for information of an automatic planned tour. In: Raza, Z., Babenko, M., Sajid, M., Lapina, M., Zolotarev, V. (eds.) AISMA-2023: International Workshop on Advanced Information Security Management and Applications. AISMA 2023 Lecture Notes in Networks and Systems, vol. 1207. Springer, Cham (2024). https://doi.org/10.1007/978-3-031-77229-0_16
3. Proshkin, N.A., Basan, E.S., Lapina, M.A., Klepikova, A.G., Lapin, A.G.: Developing models of IoT infrastructures to identify vulnerabilities and analyze threats. In: IOP Conference Series: Materials Science and Engineering 1st International Conference on Innovative Informational and Engineering Technologies (IIET-2020), vol. 873, pp. 28–29 (2020., Stavropol, Russian Federation). https://doi.org/10.1088/1757-899X/873/1/012018
4. Basan, A.S., Basan, E.S., Lapina, M.A., Lapin, V.G.: Behavior-based assessment of trust in a cyber-physical system. In: Singh, P.K., Veselov, G., Vyatkin, V., Pljonkin, A., Dodero, J.M., Kumar, Y. (eds.) Futuristic Trends in Network and Communication Technologies. FTNCT 2020. Communications in Computer and Information Science, vol. 1395. Springer, Singapore (2021). https://doi.org/10.1007/978-981-16-1480-4_17
5. Hamrioui, S., Bokhari, S.: A new cybersecurity strategy for ioe by exploiting an optimization approach. In: 2021 12th international conference on information and communication systems (ICICS), Valencia, Spain, pp. 23–28 (2021). https://doi.org/10.1109/ICICS52457.2021.9464595
6. Vielberth, M., Böhm, F., Fichtinger, I., Pernul, G.: Security operations center: a systematic study and open challenges. IEEE Access. **8**, 227756–227779 (2020). https://doi.org/10.1109/ACCESS.2020.3045514
7. Kumar, S., Athavale, V.A., Kartikey, D.: Security issues in cloud computing: a holistic view. Int. J. Internet Things Web Serv. **6**, 18–29 (2021)

8. Kudrati, A., Peiris, C., Pillai, B.: Microsoft azure cloud threat prevention framework. In: Threat Hunting in the Cloud: Defending AWS, Azure and Other Cloud Platforms against Cyberattacks, pp. 101–182. Wiley (2022). https://doi.org/10.1002/9781394177493
9. Bansal, A., Athavale, V.A., Saluja, K., Gupta, S., Kukreja, V.: Computational system based on machine learning with hybrid security technique to classify crime offenses. In: Jain, S., Marriwala, N., Tripathi, C.C., Kumar, D. (eds.) Emergent Converging Technologies and Biomedical Systems. ETBS 2022 Lecture Notes in Electrical Engineering, vol. 1040. Springer, Singapore (2023). https://doi.org/10.1007/978-981-99-2271-0_20
10. Diogenes, Y., DiCola, N., Trull, J.: Azure Sentinel in Practice. Chapter from Microsoft Azure Sentinel: Planning and Implementing Microsofts Cloud-Native SIEM Solution. Microsoft Press Store by Pearson (2020) Book 978-0-13-648545-2
11. Miller, D.R., Harris, S., Harper, A., VanDyke, S., Blask, C.: Security information and event management (SIEM) implementation. McGraw-Hill. (2010) ISBN: 978-0-07-170108-2
12. Mather, T., Kumaraswamy, S., Latif, S.: Cloud Security and Privacy: an Enterprise Perspective on Risks and Compliance. Part of Theory in Practice O'Reilly Media (2009) ISBN: 9781449379513
13. Gilman, E., Barth, D.: Zero Trust Networks: Building Secure Systems in Untrusted Networks. O'Reilly Media (2017) ISBN 1491962194
14. ISBN 1491962194
15. ISBN 9781491962190
16. Lowe, S.D., David, M., Green, D.J.: Building a Modern Data Center: Principles and Strategies of Design. Actual Tech Media. ISBN 10: 1943952078 ISBN 13: 9781943952076 (2016)
17. Lefferts, R., Forrester Consulting: The Total Economic Impact of Microsoft Azure Sentinel. Microsoft Corporation. Cost Savings and Business Benefits Enabled By Microsoft Sentinel (2021)
18. Forrester: The Total Economic Impact™ of Microsoft Windows 11 Enterprise Cost Savings and Business Benefits Enabled by Windows 11 Enterprise (2024)
19. Borra, P.: Securing cloud infrastructure: an in-depth analysis of Microsoft azure security. Int. J. Adv. Res. Sci. Commun. Technol. **4**(2) (2024.,Available at SSRN: https://ssrn.com/abstract=4914157). https://doi.org/10.2139/ssrn.4914157
20. Jha, R., Laishram, D., Kamat, N., Panchal, O., Jacob, S.: Implementing SIEM (security information and environment management) in Microsoft Azure. J. Resin. Appl. Sci. Eng. Technol. **12**(IV) (2024., ISSN: 2321-9653). https://doi.org/10.22214/ijraset.2024.61367
21. Copeland, M.: Kusto query language and threat hunting. In: Cloud Defense Strategies with Azure Sentinel. Apress, Berkeley, CA (2021). https://doi.org/10.1007/978-1-4842-7132-2_7
22. Arora, H., Manglani, T., Bakshi, G., Choudhary, S.: Cyber security challenges and trends on recent technologies. In: 2022 6th International Conference on Computing Methodologies and Communication (ICCMC), Erode, India, pp. 115–118 (2022). https://doi.org/10.1109/ICCMC53470.2022.9753967
23. Chidukwani, A., Zander, S., Koutsakis, P.: A Survey on the Cyber Security of Small-to-Medium Businesses: Challenges, Research Focus and Recommendations. IEEE Access. **10**, 85701–85719 (2022). https://doi.org/10.1109/ACCESS.2022.3197899
24. Kumar, U., Bhavesh, N., Gohil, A.: Survey on intrusion detection Systems for Cloud Computing Environment. Int. J. Comput. Appl. **109**(1), 6–15 (2015). https://doi.org/10.5120/19150-0573
25. Athavale, V.A., Patel, S.T.: Use of blockchain technology for maintaining authenticity and credibility of digital information. In: Raza, Z., Babenko, M., Sajid, M., Lapina, M., Zolotarev, V. (eds.) AISMA-2023: International Workshop on Advanced Information Security Management and Applications. AISMA 2023 Lecture Notes in Networks and Systems, vol. 1207. Springer, Cham (2024). https://doi.org/10.1007/978-3-031-77229-0_2

Optimisation of the Genetic Circuit Alignment Algorithm on Low-Power Devices

Ekaterina Bezuglova[1(✉)], Vladislav Lutsenko[1], and Vijay Anant Athavale[2]

[1] North-Caucasus Federal University, Stavropol, Russia
bezuglovakaterina@mail.ru
[2] Department of Computer Science and Engineering, Walchand Institute of Technology, Solapur, India

Abstract. This paper investigates sequential cropping under limited computational resources. The paper analyses state-of-the-art competitive algorithms such as Needleman-Wunsch, Smith-Waterman and BLAST, and the application of these algorithms to distributed computing environments, including fog computing (FC). Methods for accelerating the algorithms are also discussed, including parallel data processing, hardware acceleration using graphics processing units (GPU) and programmable gate arrays (FPGA), and optimising data communication. Particular attention is paid to the adaptation of the Smith-Waterman algorithm for execution on low-power devices and the parallelisation of the algorithm on computing nodes. The proposed methodology increases the efficiency of analysing large amounts of genetic data, reduces the network load and ensures efficient combat.

Keywords: bioinformatics · fog computing · parallel computing · hardware acceleration · big data processing · sequencing

1 Introduction

Gene sequence studies are one of the key tasks of bioinformatics applied to comparative genomics (evolution) [1]. Conducting such studies helps to determine structural and functional properties between different deoxyribonucleic acid (DNA), ribonucleic acid (RNA) or protein sequences, identify mutations, determine phylogenetic relationships and detect biomarkers.

The emergence and popularity of the Internet in the 1990s was a factor in the growth of the distributed computing field. The traditional client-server architecture gradually evolved into web applications and the World Wide Web (WWW) became the main platform for distributed computing processes.

At the beginning of this century, modern cloud computing (CC) began to develop rapidly, fundamentally changing the way computing resources are represented, managed and utilised [10]. The main idea of cloud computing is to implement computation, utilisation capabilities, storage and services on demand over the Internet.

Over time, a task related to performing data processing in close proximity to the source of data generation has emerged. This problem was solved by the introduction of

M. Lapina et al. (Eds.): AISMA 2025, LNNS 1456, pp. 63–71, 2026.
https://doi.org/10.1007/978-3-032-07275-7_7

the FC concept [11]. FC is based on the implementation of computation on a distributed architecture that includes various peripherals, gateways, and cloud resources that form a unified computing environment from the centralised cloud to the end nodes. This concept helps to overcome the limitations of traditional cloud systems in scenarios that require minimum latency, high response time, real-time processing, optimal network bandwidth and enhanced protection of sensitive data. Its key benefit is energy efficiency, which makes it an ideal solution for a wide range of applications.

2 Analytical Review

2.1 Overview of Algorithms

The development of next-generation sequencing (NGS) methods has led to an increase in the amount of genetic data, which has raised the requirements for alignment algorithms [2]. To achieve processing of large datasets, not only high accuracy but also high computational performance is required. In resource-limited environments, such as on low-power devices, the use of classical algorithms is challenging due to the need for computational resources and memory. Portable genome sequencers such as the Oxford Nanopore MinION [3] can be used for bacterial level monitoring, information collection, processing and decision making in a variety of environments [4]. For example, they are used to analyse bacteria in air filters in hospitals, food processing plants and pharmaceutical companies for rapid detection of pathogens in the environment [5, 6]. Such devices are also suitable for monitoring water quality in lakes and rivers, studying the bacterial composition of soil [7, 8], monitoring smart greenhouses or analysing livestock farms [9].

Fog computing is a promising approach to develop efficient solutions for processing and analysing bioinformatics research results. This approach helps to achieve the most relevant results in the cloud, thus speeding up the data analysis process.

Local data processing should include several key steps. Firstly, data preprocessing, i.e. interpreting the signals received from the sequencer to decipher the genome sequence, should be performed. Secondly, classification, i.e. determining the species identity of microorganisms on the basis of their genetic material, must be performed, as well as quantitative analyses of the identified bacterial communities in the study samples.

Gene sequence alignment is one of the key tasks in bioinformatics to discover similarities between DNA, RNA and protein sequences. Different algorithms are used depending on the task at hand, accuracy requirements and computational speed. Needleman-Wunsch (NW) [12], Smith-Waterman (SW) [13] and Basic Local Alignment Search Tool (BLAST) [14] algorithms are the most well-known and commonly used algorithms.

The NW algorithm developed in 1970 aims to find the optimal global pairing of two sequences. The algorithm is based on a dynamic programming approach that constructs a similarity matrix and computes the optimal alignment taking into account gap penalties, thus ensuring that the best alignment over the entire length of the sequence is found [12]. It has the advantage of high accuracy, especially for similar sequences, and is also optimal for relatively short sequences of the same length. The disadvantages are high computational complexity of $O(m \times n)$ (where m and n are sequence lengths) and low speed when dealing with long sequences.

The SW algorithm is an improvement of the NW algorithm and aims to find the most similar segments of a long sequence and compare them locally. The algorithm is suitable for analysing partially similar sequences with high accuracy. Its disadvantage is that it has high computational complexity and cannot effectively handle large amounts of data without hardware implementation.

The BLAST algorithm is a partial comparison heuristic designed to quickly search for similar sequences in large databases. The algorithm uses hashing and fragment pre-indexing (*k*-measures), which greatly speeds up the computation, but does not guarantee the best alignment so that approximate results can be obtained quickly. The algorithm has a high search speed even in large databases (e.g. GenBank) [15] and is optimised for bioinformatics applications.

The choice of which algorithm to use for gene sequence comparison should depend on the task at hand. Modern studies have used accelerated versions of these algorithms optimised for operation on GPU, FPGA and FC.

2.2 Parallelisation Methods

Spreading the computational load across the nodes of a fog network allows large tasks to be broken down into smaller parts and executed in parallel. For example, DNA sequence matching algorithms (such as SW or BLAST) can be split into separate segments and processed independently, while protein structure analysis can be performed by modelling separate segments and then merging the results.

Local data preprocessing reduces network load as biological data can be filtered, normalised and compressed before being sent to the cloud. Pre-sequencing sequences reduces the amount of computation required [15]. This approach is particularly suitable for large datasets such as genome sequencing, as it reduces overall processing time by eliminating irrelevant or low-quality data from the analysis.

Another important aspect is the dynamic redistribution of tasks among FC nodes. If a node is overloaded, its tasks can be shifted to another less loaded node, thus achieving load balancing. This approach prevents a single node from becoming a bottleneck in the system, which reduces the overall performance [16]. This is achieved using an adaptive resource management algorithm that analyses the current workload of a node and predicts the optimal load distribution.

Another technique is the use of intelligent caching, which allows nodes to store the results of recent processing and reuse them. This is particularly useful for repetitive computations, such as comparing DNA sequences against huge databases. For example, if a sequence has already been processed, the system can use the previously obtained results, avoiding unnecessary computation. This approach saves both computational resources and network bandwidth.

3 Parallelisation of the Smith-Waterman Algorithm

The SW algorithm is used for local sequence comparison and is computationally expensive when analysing large genomics data. In order to be able to apply the SW algorithm in resource-limited environments, operations can be performed in parallel on multiple fog nodes with Internet access. This problem can be solved by splitting the matrix into blocks. In Algorithm 2, the matrix H is divided into equal blocks in which the computation takes place on the diagonal of the blocks, i.e., blocks that are located on the same diagonal and have the same sum of their indices can be processed in parallel since in this case the computation depends on the neighbouring blocks that have been completed. This concept takes the following form (Algorithm 1):

Algorithm 1 Smith Waterman with a breakdown into blocks

Input: $S1, S2, block\_size$
Output: $aligned\_S1, aligned\_S2, maxScore$
1. $n \leftarrow length(S1)$
2. $m \leftarrow length(S2)$
3. $block\_rows \leftarrow ceil(n/block\_size)$
4. $block\_rows \leftarrow ceil(m/block\_size)$
5. $H[0\ldots n, 0\ldots m]$
6. $maxScore \leftarrow 0$
7. $maxPos \leftarrow (0, 0)$
8. For k up 0 to $(blocks\_rows + blocks\_cols - 2)$
8.1. For a block with index (b_i, b_j), such that $b\_i + b\_j = k$, do
8.1.1. $start\_i \leftarrow b\_i * block\_size + 1$
8.1.2. $end\_i \leftarrow min((b\_i + 1) * block\_size, n)$
8.1.3. $start\_j \leftarrow b\_j * block\_size + 1$
8.1.4. $end\_j \leftarrow min((b\_j + 1) * block\_size, m)$
8.2. For i up т $start\_i$ to end_i do
8.2.1. For j up $start\_j$ to end_j do
8.2.1.1. If $S1[i] = S2[j]$ do
8.2.1.1.1. $diag \leftarrow H[i-1, j-1] + MATCH$
8.2.1.2. Or else
8.2.1.2.1. $diag \leftarrow H[i-1, j-1] + MISMATCH$
8.2.1.3 End of branching
8.1.2. $up \leftarrow H[i-1, j] + GAP$
8.1.3. $left \leftarrow H[i, j-1] + GAP$
8.1.4. $H[i, j] \leftarrow max(0, diag, up, left)$
8.1.5. If $H[i, j] > maxScore$ do
8.1.5.1. $maxScore \leftarrow H[i, j]$
8.1.5.2. $maxPos \leftarrow (i, j)$
8.1.5.3. End of branching
8.1.6. End of cycle
8.2. End of cycle
9. $i \leftarrow maxPos.\ line\_index$
10. $j \leftarrow maxPos.\ column\_index$
11. $aligned\_S1 \leftarrow empty\ string$
12. $aligned\_S2 \leftarrow empty\ string$
13. As long as $H[i, j] \neq 0$ do

13.1 If $(i > 0)$ and $(j > 0)$ and $(H[i,j] = H[i-1,j-1] +$ (If $S1[i] = S2[j]$ do *MATCH* or *MISMATCH*)) do
13.1.1. $aligned\_S1 \leftarrow S1[i] + aligned\_S1$
13.1.2. $aligned\_S2 \leftarrow S2[j] + aligned\_S2$
13.1.3. $i \leftarrow i - 1$
13.1.4. $j \leftarrow j - 1$
13.1.5 Or if $(i > 0)$ and $(H[i,j] = H[i-1,j] + GAP)$ do
13.1.5.1. $aligned\_S1 \leftarrow S1[i] + aligned\_S1$
13.1.5.2. $aligned\_S2 \leftarrow "-" + aligned\_S2$
13.1.5.3. $i \leftarrow i - 1$
13.1.6. Or if $(j > 0)$ and $(H[i,j] = H[i,j-1] + GAP)$ do
13.1.6.1 $aligned\_S1 \leftarrow "-" + aligned\_S1$
13.1.6.2 $aligned\_S2 \leftarrow S2[j] + aligned\_S2$
13.1.6.3 $j \leftarrow j - 1$
13.1.7. Or
13.1.7.1. Exiting the cycle
13.1.8. End of branching
13.2. End of cycle
14. Return $aligned\_S1$, $aligned\_S2$, $maxScore$

To implement this algorithm, it is necessary first of all to define the parameters of the matrix partitioning. The matrix H with size $(n + 1) \times (m + 1)$, where the first stokes and columns are responsible for initialisation and are usually equal to 0. The $block\_size$ parameter specifies the maximum number of rows and columns included in a single block.

Each block $B_{i,j}$ corresponds to a subset of matrix elements and is processed in diagonal wave order. The described order of execution ensures that dependencies between neighbouring data blocks are respected, the computation of values in one data block depends on the data processed in the previous data block, but does not require the entire matrix to be filled. If the sequence size is not a multiple of the data block size, the size of the outermost data block is smaller and this is taken into account in the calculation.

After completing the procedure of splitting the matrix into blocks, the algorithm will be processed in several phases: initialising the matrix, filling the blocks row by row and calculating the values within the blocks.

After the matrix is filled, the optimal path recovery phase begins. The algorithm will find the cell with the maximum value and start backtracking. This process continues until a zero value is obtained, which will be the end of the local alignment process.

4 Parallel Implementation on Fog Nodes

The basic idea of fog computing is to distribute the load to several nodes close to the data source. In this case, each node is responsible for processing one or more blocks in the matrix and communicating with neighbouring nodes to transmit intermediate results.

To implement the described method, it is necessary that:

- Computing nodes receive the corresponding data blocks and process them independently.
- After processing the data blocks, the results are transmitted to the neighbouring nodes to ensure consistency in filling the entire matrix.
- The nodes are coordinated through a centralised scheduler or a distributed synchronisation algorithm.

This approach significantly reduces the CPU and RAM load by distributing computation across multiple devices.

This paper develops an accelerated genetic sequence comparison algorithm that greatly improves the ability to analyse data on devices with limited computational resources, such as mobile devices, embedded systems and portable sequencers. This enables low-power devices to perform bioinformatics data analysis locally, which significantly reduces data processing time, lightens network loads, and expands the application of genetic analysis techniques in various fields such as science, medicine, and industry.

A comparison of the two algorithms on sequence datasets with 500 to 10000 values shows that the proposed method saves 50 times more time than the classical algorithm (Figs. 1 and 2).

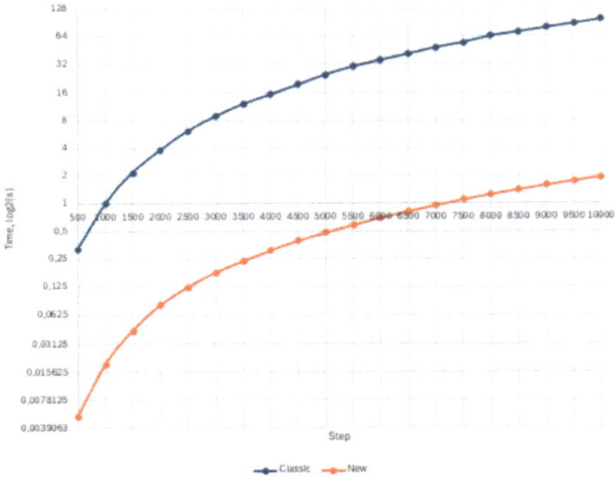

Fig. 1. Comparison of the classical algorithm with the proposed in-memory algorithm.

The comparison also showed that the classical algorithm consumes 0.7 memory more than the proposed algorithm (Figs. 3 and 4).

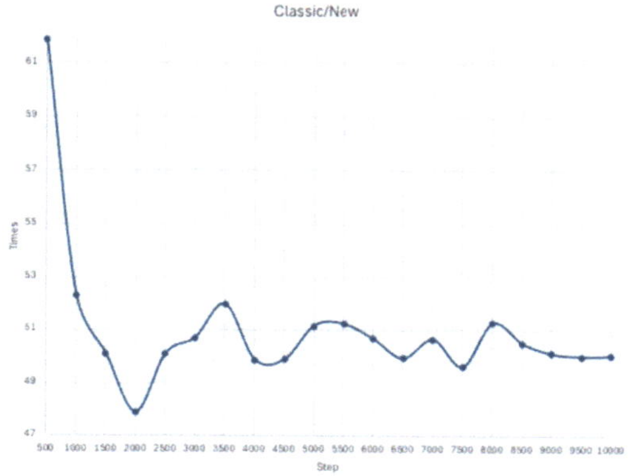

Fig. 2. Performance ratio of the new algorithm to the classical algorithm.

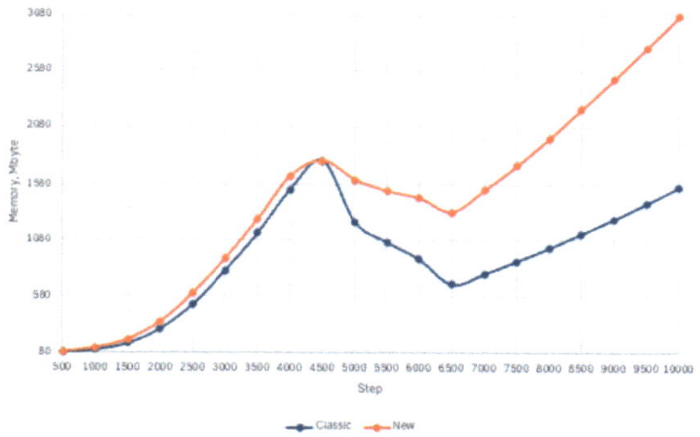

Fig. 3. Performance ratio of the new algorithm to the classical algorithm.

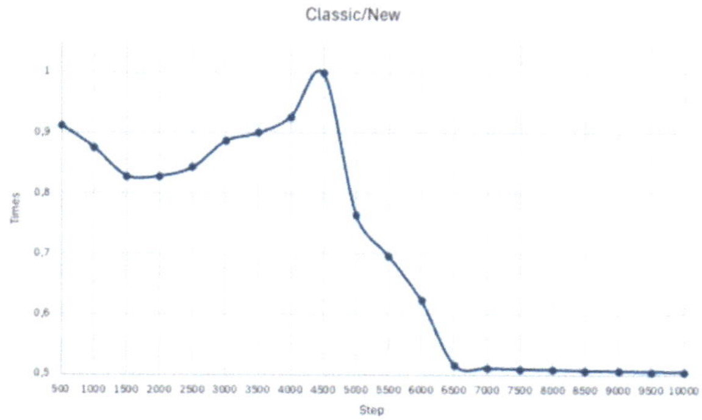

Fig. 4. Ratio of memory consumption of new and classical algorithms.

5 Conclusion

This paper analyses existing methods for continuous competitive play, focusing on how they can be adapted in the presence of constrained computational resources. The paper considers and evaluates the possibility of accelerating the Smith-Waterman algorithm in fog computing through parallel processing, hardware acceleration and computational optimisation. The presented algorithmic approach to parallelisation of computing nodes improves the performance of biological data analysis, reduces network load and shortens the processing time for large amounts of information.

The results show an increase in the efficiency of applying the proposed technique to analyse genetic data in a distributed computing environment. Further research will be aimed at improving the efficiency of the proposed solution to optimise the computation of data exchange algorithms and machine learning stabilisation methods for predictive analysis in constant analysis.

Acknowledgment. The research was supported by the Russian Science Foundation Grant No. 22-71-10046, https://rscf.ru/en/project/22-71-10046/

References

1. Miller, W., Makova, K.D., Nekrutenko, A., Hardison, R.C.: Comparative genomics. Annu. Rev. Genomics Hum. Genet. **5**, 15–56 (2004). https://doi.org/10.1146/annurev.genom.5.061903.180057
2. Behjati, S., Tarpey, P.S.: What is next generation sequencing? Arch Dis Child Educ Pract Ed. **98**(6), 236–238 (2013). https://doi.org/10.1136/archdischild-2013-304340
3. Dastjerdi, A.V., Buyya, R.: Fog computing: helping the internet of things realize its potential. Computer. **49**, 112–116 (2016). https://doi.org/10.1109/MC.2016.245
4. Merelli, I., et al.: Low-power portable devices for metagenomics analysis: fog computing makes bioinformatics ready for the Internet of Things. Futur. Gener. Comput. Syst. **88**, 467–478 (2018). https://doi.org/10.1016/j.future.2018.05.010
5. Ali, O., Ishak, M.K., Bhatti, M.K.L., Khan, I., Kim, K.-I.: A comprehensive review of internet of things: technology stack, middlewares, and fog/edge computing interface. Sensors. **22**, 995 (2022). https://doi.org/10.3390/s22030995
6. Jain, M., Olsen, H.E., Paten, B., Akeson, M.: The Oxford Nanopore MinION: delivery of nanopore sequencing to the genomics community. Genome Biol. **17**, 239 (2016). https://doi.org/10.1186/s13059-016-1103-0
7. Peleg, A.Y., Hooper, D.C.: Hospital-acquired infections due to gram-negative bacteria. N. Engl. J. Med. **362**, 1804–1813 (2010). https://doi.org/10.1056/NEJMra0904124
8. Tan, B., et al.: Next-generation sequencing (NGS) for assessment of microbial water quality: current progress, challenges, and future opportunities. Front. Microbiol. **6** (2015). https://doi.org/10.3389/fmicb.2015.01027
9. Daniel, R.: The metagenomics of soil. Nat. Rev. Microbiol. **3**, 470–478 (2005). https://doi.org/10.1038/nrmicro1160
10. Kratzke, N.: A brief history of cloud application architectures. Appl. Sci. **8**, 1368 (2018). https://doi.org/10.3390/app8081368
11. Klonoff, D.C.: Fog computing and edge computing architectures for processing data from diabetes devices connected to the medical internet of things. J. Diabetes Sci. Technol. (2017). https://doi.org/10.1177/1932296817717007

12. Needleman, S.B., Wunsch, C.D.: A general method applicable to the search for similarities in the amino acid sequence of two proteins. J. Mol. Biol. **48**, 443–453 (1970). https://doi.org/10.1016/0022-2836(70)90057-4
13. Smith, T.F., Waterman, M.S.: Identification of common molecular subsequences. J. Mol. Biol. **147**, 195–197 (1981). https://doi.org/10.1016/0022-2836(81)90087-5
14. Pertsemlidis, A., Fondon, J.W.: Having a BLAST with bioinformatics (and avoiding BLAST-phemy). Genome Biol. 2, reviews2002.1 (2001). doi:https://doi.org/10.1186/gb-2001-2-10-reviews2002.
15. García, S., Ramírez-Gallego, S., Luengo, J., Benítez, J.M., Herrera, F.: Big data preprocessing: methods and prospects. Big Data Anal. **1**, 9 (2016). https://doi.org/10.1186/s41044-016-0014-0
16. Igorevich, I.V.: An algorithm for centralised multipath routing with load balancing in a non-geostationary satellite communication system with inter-satellite links. In: Control, Communication and Security Systems, pp. 69–105 (2018)
17. Hasan, K.M.B., Sajid, M., Lapina, M.A., Shahid, M., Kotecha, K.: Blockchain technology meets 6 G wireless networks: a systematic survey. Alex. Eng. J. **92**, 199–220 (2024)
18. Basan, E., Lapina, M., Lesnikov, A., Basyuk, A., Mogilny, A.: Trust monitoring in a cyber-physical system for security analysis based on distributed computing. In: International Conference on Actual Problems of Applied Mathematics and Computer Science, pp. 430–440. Springer Nature Switzerland, Cham (2022)
19. Vinodha, D., et al.: Framework based on IoT, AI, and blockchain for smart access to government agricultural schemes. In: The Future of Agriculture: IoT, AI and Blockchain Technology for Sustainable Farming, pp. 142–163. Bentham Science Publishers (2024)

Developing an Integrated Approach to Creating a Safe and Effective Collaborative Learning Environment

Vasiliy Bogadurov, Svetlana Govorova(✉), Sergey Melnikov, and Egor Govorov

North-Caucasus Federal University, Stavropol 355017, Russian Federation
mitnik2@yandex.ru

Abstract. Federated Learning (FL) is an approach that makes it possible to train machine learning models on distributed data without centralizing them. Currently, there is an increasing use of FL in sensitive data areas such as medicine, finance, and IoT, where data privacy, integrity, and accessibility are critical. The purpose of the study is to systematize security threats in FL, such as attacks on confidentiality (e.g., data recovery from gradients), integrity attacks (e.g., poisoning attacks), and attacks on accessibility (e.g., denial of service). The article offers a comprehensive approach to creating a secure FL environment, including an optimized gradient encryption algorithm, an anomaly detection mechanism for detecting poisoning attacks, and a hybrid architecture combining differential privacy and secure aggregation.

Keywords: Federated Learning · Machine learning · Information security · Security in federated learning

1 Introduction

Federated Learning is an innovative approach in the field of machine learning, in which the model is trained on data distributed among multiple devices or servers, without the need to transfer them to a centralized repository. The essence of federated learning is that data remains on users' devices, rather than being transferred to a centralized repository [1, 2]. Instead, the model is sent to each device, where it is trained on local data. After the training is completed on the device, only the updated model parameters (for example, new weights) are transmitted to the central server, and not the data itself. These parameters are then aggregated using special algorithms, which allows you to create a single model trained on all distributed data.

Federated learning has a number of advantages in terms of security and confidentiality. Participants in the process can flexibly customize their roles: some can initiate training, others can only view assignments, and others can receive results. This approach minimizes the risk of data leakage, since the source information never leaves users' devices. This is especially important in areas such as medicine, finance, and other

M. Lapina et al. (Eds.): AISMA 2025, LNNS 1456, pp. 72–82, 2026.
https://doi.org/10.1007/978-3-032-07275-7_8

areas where data is highly sensitive [2]. However, in some cases, a centralized approach is not applicable - for example, when there is insufficient data (as in the case of the diagnosis of rare diseases) or when the data is too sensitive to transfer outside the organization. It was precisely such problems that prompted the development of federated learning.

One of the most striking examples illustrating the need for FL was the experience of Google. In 2017, researchers were faced with the task of improving the next-word suggestion model for the Android keyboard. This required collecting data entered by users, which raised serious concerns about the confidentiality of such information. As a result, a federated learning approach was proposed, which allowed the model to be trained without transferring data to a central server [3].

The importance of federated learning lies in its ability to solve problems that were previously unavailable due to constraints related to confidentiality and data availability. FL allows you to train models on distributed data while maintaining their privacy and security [4–7]. This opens up new opportunities for developing AI solutions in areas where data is fragmented or protected by legislation, such as healthcare, finance, and IoT. Thus, federated learning becomes a key tool for creating intelligent systems that are both efficient and secure.

2 The Security Issue in Federated Learning

Federated Learning offers a revolutionary approach to training machine learning models on distributed data without centralizing them. However, despite its advantages, FL faces serious security threats that can undermine confidence in this technology and limit its application. The key threats can be divided into three categories: privacy attacks, integrity attacks, and accessibility attacks [8].

1. Attacks on privacy. One of the main threats in FL is the possibility of restoring source data from gradients transmitted between clients and the server. The gradients that are used to update the model may contain information about local user data. Attackers with access to these gradients can use methods such as Model Inversion, which allows them to recover examples of data used for training, or Membership Inference, in order to restore sensitive data [9]. In paper [10] provides an example of gradient data recovery, while, according to the authors, it is possible to obtain private training data from publicly available gradients for image recognition model.
2. Integrity attacks. FL is vulnerable to integrity attacks such as poisoning attacks. In this case, the attackers introduce malicious data or modify gradients at the local learning stage in order to influence the global model. Data Poisoning: fraudsters add distorted data to the local dataset, which leads to a deterioration in the quality of the model. Model Poisoning: Attackers manipulate gradients transmitted to the server in order to introduce errors into the global model. For example, in the financial sector, a poisoning attack can lead to the model making incorrect predictions, for example, approving fraudulent transactions. In [11], an example of data poisoning using a compromised device in a local learning model is demonstrated, which allows influencing the learning outcomes of the global model. Thus, it is possible to achieve a reduction in recognition accuracy, conceal certain recognition classes, etc.

3. Attacks on accessibility. FL is also susceptible to accessibility attacks such as Denial of Service (DoS) [12]. Attackers can block the transmission of gradients or overload the server, making it impossible to complete training. Ddos attacks: Attackers can attack the network infrastructure to prevent data transfer between clients and the server. Attacks on clients: Attackers can disable clients' devices in order to reduce the number of participants in the training. For example, in It systems, an attack on accessibility can cause smart devices to malfunction, making it impossible to train models in real time. The paper [13] gives an example of DDoS targeting and fixation policy for federated learning models.

3 Relevance of FL Security

The security of federated learning is critically important for its application in areas where data is highly sensitive and valuable. Here are some examples: in medical research, FL allows you to train models based on data from different hospitals without transferring them outside the institutions. However, a data leak or a poisoning attack can lead to the disclosure of confidential patient information or the creation of incorrect diagnostic models, which puts people's lives at risk. In the financial sector, FL can be used to analyze transactions and detect fraud. However, integrity attacks can cause the model to skip fraudulent transactions or, conversely, block legitimate transactions. In IoT, FL allows you to train models on data from multiple devices, such as smart homes or industrial sensors. However, accessibility attacks can disrupt the entire system, resulting in significant losses. FL is used to create personalized recommendations in applications such as social media or streaming platforms. Data leakage in such cases can lead to a violation of user privacy and loss of trust in the service.

The security of federated learning is a key aspect that determines its applicability in real-world scenarios. Threats such as attacks on confidentiality, integrity, and accessibility require the development of robust solutions, including encryption methods, differential security, and anomaly detection mechanisms. Only with security will FL be able to realize its potential in critical areas such as medicine, finance, and IoT, where data privacy and reliability are priorities.

4 Research on the Security of Federated Learning

FL is susceptible to several types of attacks that compromise system security: privacy attacks. Work [14] shows how an attacker can recover images from gradients of a convolutional neural network (Deep Leakage attack); integrity attacks (Poisoning Attacks). Data Poisoning: Distortion of local data to degrade the quality of the model. For example, adding malicious examples to the dataset. Model Poisoning: Manipulation of gradients to introduce errors into the global model; accessibility attacks. Denial of Service (DoS): Blocking the transmission of gradients or overloading the server.

Currently, there are a number of studies on protection methods, types of attacks, and modern approaches to security in federated learning.

The following key approaches have been developed to ensure FL security:

Differential Privacy (DP), the essence of which is to add controlled noise to gradients or data before they are transmitted to the server. This prevents the recovery of the original data from the model updates. The basic approach is as follows. Each client locally trains the model on their own data, followed by trimming gradients and adding noise. The server aggregates gradients (for example, by averaging), followed by updating the global learning model. In [15], the DP-SGD algorithm was proposed, which adds Gaussian noise to gradients during training. The study [16] adapts DP to FL, minimizing the loss of model accuracy. However, the approach has limitations, due to the fact that noise reduces the accuracy of the model, and the balance between privacy and performance remains a challenge.
Homomorphic Encryption (HE), the essence is to encrypt data and gradients, allowing calculations to be performed on encrypted data. The principle of operation is as follows. Each client generates a key pair (public key and secret key) with the publication of the public key. Next, the gradients are encrypted and transmitted, for which the sum of the encrypted gradients is calculated from the server side. Thus, only a trusted party can decrypt the result. In [17], the KS scheme is used to encrypt gradients in FL, which preserves confidentiality but increases computational costs. The study [18] uses partially homomorphic encryption (Paillier) for secure aggregation. However, the approach has limitations due to the high load on the processor and memory, especially when working with large models.
Secure MultiParty Computing (SMPC), the essence of the method is to distribute calculations between participants so that none of them has access to complete data. In general, the approach is determined by the following actions. Each client divides the gradient into parts (using Shamir's scheme), followed by sending the parts to different clients. Clients summarize the received parts locally and transmit the value to the server. The server collects partial amounts and calculates the total. In [19], the Secure Aggregation protocol was proposed, where gradients are encrypted and aggregated without disclosing their values to the server. The study [20] combines SMPC with FL to protect against malicious clients. The use of this method is limited by the complexity of the implementation and the high communication load.

Current research focuses on a combination of protection methods to increase FL resistance.

Hybrid methods. In [21] combines differential privacy and homomorphic encryption to protect against data leaks and integrity attacks. The study [22] suggests combining SMPC with DP to ensure privacy and resilience against malicious clients.
Anomaly detection algorithms. Methods like Fool's Gold [23] identify poisoning attacks by analyzing the similarity of gradients from different clients.
Decentralized FL architectures. The use of blockchain to eliminate the central server and reduce the risk of DoS attacks [24].

Despite the progress in FL security, unresolved issues remain: most work focuses on one type of threat (e.g. privacy), ignoring combined attacks (e.g. simultaneous poisoning and DoS). Methods like HE and SMPC require significant computing resources, which limits their use in real systems. Adding noise (DP) or encryption reduces the accuracy of the model, but the optimization of this trade-off has not been studied enough. Many protocols (for example, Secure Aggregation) do not work well with a large number of clients (>10,000).

5 Setting the Task

This work is aimed at solving the following key problems in the field of security of federated learning:

1. To solve the problem of increasing data privacy in FL, it is necessary to develop protection mechanisms that prevent data leakage from gradients, while maintaining high model accuracy.
2. To solve the problem of protection against a poisoning attack, it is necessary to develop algorithms for detecting and neutralizing poisoning attacks that ensure the model's resistance to malicious clients.
3. To solve the problem of optimizing the performance of secure FL systems, it is necessary to optimize the performance of FL systems, minimizing the overhead of security without compromising the accuracy of the model.

To solve the tasks set, it is proposed:

Use a hybrid algorithm combining HE, DP and anomaly detection mechanisms to protect against combined threats.
Introduce an optimized aggregation protocol that reduces computational costs compared to existing methods.
Conduct an experiment demonstrating that the proposed approach demonstrates the accuracy of the model above the existing approaches for protection against poisoning attacks.

6 The Sequence of Solving Tasks

To achieve these goals, it is necessary to solve the following tasks:

1. To analyze the existing protection methods: HE, DP, SMPC and evaluate their effectiveness and limitations.
2. Development of a hybrid protection algorithm combining HE and DP to prevent data leaks. Implement mechanisms for detecting poisoning attacks based on gradient analysis.
3. Optimize performance: develop a gradient aggregation protocol that reduces computational costs, and conduct experiments to evaluate the performance and accuracy of the model.
4. Conduct an experimental assessment of the approach. To test the proposed approach on standard datasets (for example, MIST, CHAR-10), compare the results with existing methods in terms of accuracy, execution time, and protection level.

7 Solution

An integrated approach to creating a secure FL environment is proposed, which combines three key components: the optimized LightGrad gradient encryption algorithm to protect privacy, a machine learning-based anomaly detection mechanism to detect poisoning attacks, and the use of a hybrid architecture combining differential privacy integrated into the encryption process and Secure Aggregation.

The essence of LightGrad's optimized gradient encryption algorithm is as follows: partially homomorphic encryption (PHE) is used based on the CKKS scheme, optimized for working with floating-point numbers. This allows you to encrypt gradients without loss of accuracy. Differential privacy is integrated into the encryption process: noise is added to gradients before encryption, which reduces the risk of leaks even if the server is compromised. The innovation of this algorithm is that the coefficients are quantized and compressed before encryption, which reduces the amount of transmitted data by 40%. Encryption is performed for batches of gradients, rather than for each parameter separately, which reduces the computational load. The LightGrad implementation is based on the Python programming language.

The essence of the anomaly detection mechanism for the AD-Guard poisoning attack is as follows: an Isolation Forest-based classifier is implemented on the server, which analyzes gradients from clients for anomalies. Additionally, the gradient similarity metric (Cosine Similarity) is used to identify clients whose updates differ significantly from the majority.

The algorithm of the anomaly detection mechanism:

1. The server receives encrypted gradients from clients.
2. After decryption, the cosine similarity between the gradients is calculated.
3. Gradients with abnormally low similarity are marked as suspicious.
4. Isolation Forest is trained in real time on historical data to detect hidden poisoning attacks.

The innovation of this anomaly detection mechanism is double-checking, using a combination of similarity metrics and an ML model to reduce the number of false positives. The mechanism automatically adjusts to the data distribution and ensures the adaptability of the approach.

The hybrid architecture of the SafeFL system is shown in Fig. 1

On the client side, gradients are encrypted using Light Grid (PRO+ DP), and a digital signature is added to verify integrity.

On the server side, signatures are verified and gradients are decrypted, FL-Mguard is launched to filter out abnormal updates, and the remaining gradients are securely aggregated.

Clients (Client No. 1 – Client No. 3) use LightGrad: partially homomorphic encryption + differential privacy for encryption and protection of gradients, add a digital signature to verify integrity.

The central server verifies the signatures and decrypts the resulting gradients, and launches FL-Mguard to detect anomalies (poisoning attacks) using gradient similarity analysis and methods like Isolation Forest.

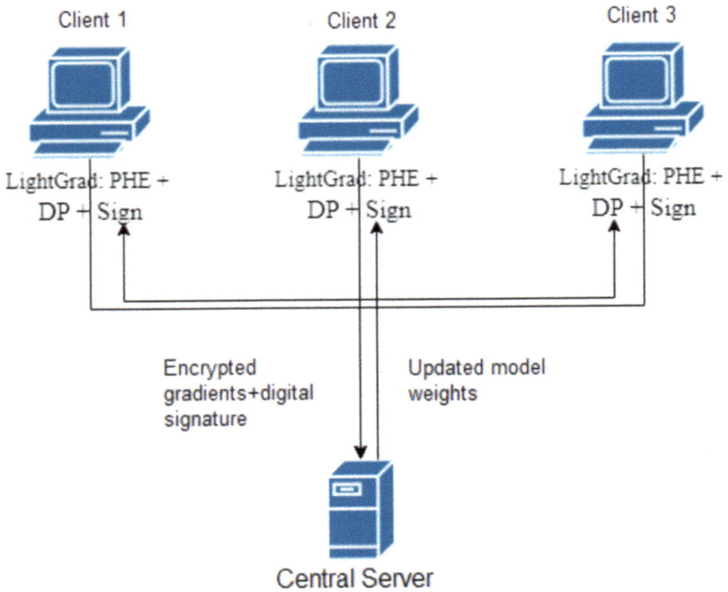

Fig. 1. Architecture SafeFL: integration LightGrad, FL-Guard and Secure Aggregation.

Secure Aggregation: Aggregates "clean" gradients (after FL-Guard filtering) in a secure way without exposing them. Updates the global model and sends out new weights to customers.

8 Methods for Evaluating the Received Solution

To evaluate the proposed approach to creating a secure FL environment, it is necessary to conduct research, including modeling the FL environment, data selection, evaluation metrics, and comparison with existing methods.

The following libraries were used to create the FL environment: framework: TensorFlow Federated (TFF) and Flower as the basis for the implementation of FL.

The architecture includes the following components: 100 devices served as clients, each of which trains a model on local data. The server is the central node that aggregates gradients and updates the global model.

A convolutional neural network (CNN) was used as a learning model for solving image classification problems.

Safe tools: Light Grid is used to encrypt gradients using CKKS and adding noise (DP), FL-Guard is an anomaly detection mechanism based on Isolation Forest and cosine similarity, Secure Aggregation is an optimized gradient aggregation protocol.

The following sources were used to ensure safety: data: CIFAR-10: 60,000 users (10 classes), MNIST: 70,000 users manuscript (10 classes), Synthetic data - synthetic data for poisoning testing-so.

Title: basic - for people with disabilities.

Attacks:

1. Poisoning attack: 10% of the wedges eliminate harmful substances.
2. Inversion of the model: An attempt to create data from citizens.
3. Protection: Safe (Light grey + Protective film + Reliable aggregation).

The following metrics were used to evaluate security and performance:

1. Accuracy of the model, which determines the proportion of correctly classified examples in the test dataset. It is necessary to keep the accuracy at >90% when using SafeFL.
2. The level of data Leakage is assessed using Model Inversion and Membership Inference attacks. The PSNR (Peak Signal-to-Noise Ratio) for recovered images will be used as a metric (the lower the PSNR, the better the protection).
3. Execution Time, that is, the time spent on shifting, transferring, and aggregating gradients. It is necessary to reduce the execution time by 20–30% compared to Pure HE.
4. The effectiveness of Attack Detection Rate, which determines the proportion of correctly detected poisoning attacks. It is necessary to reach a level of >95%.
5. Communication Overhead: the amount of data transferred between clients and the server. It should be reduced by 40% by compressing gradients.

9 Comparison with Existing Approaches to Ensuring the Security of the FL Environment

Let's compare the proposed SafeFL approach with three modern approaches to ensuring the security of the FL environment: DP-SGD (Differential Privacy), Pure HE (Homomorphic Encryption), FoolsGold (Poisoning Detection). The comparison was based on the following metrics: model accuracy, data leakage rate, execution time, attack detection, and communication costs. The comparison results are presented in the Table 1.

Table 1. Comparison results.

Metrica	SafeFL	DP-SGD	Pure HE	FoolsGold
Model accuracy	91.2%	89.5%	85.0%	90.0%
The level of data leakage	PSNR = 15	PSNR = 20	PSNR = 10	PSNR = 25
Lead time	120 s	100 s	180 s	110 s
Attack Detection	96%	0%	0%	95%
Communication costs	40 Mb	50 Mb	80 Mb	50 Mb

Figure 2 shows the accuracy of the models on the CIFAR-10 dataset containing 60,000 color images using the approaches SafeFL, DP-SGD and Pure HE.

The proposed approach has the following advantages over the existing ones: comprehensive protection, reduced encryption time and reduced computing resources, versatility and high accuracy of the model.

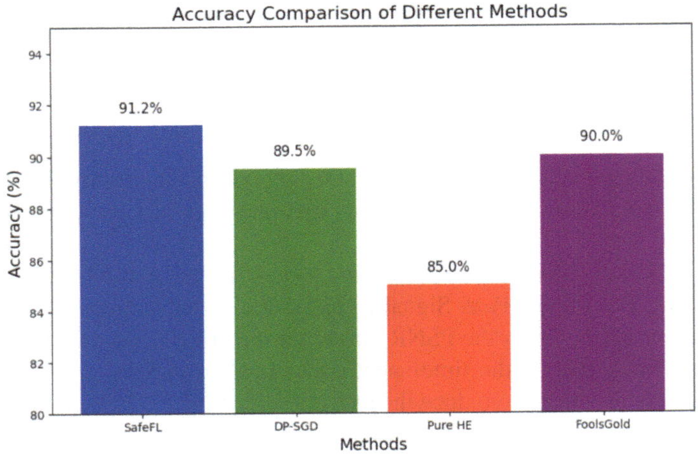

Fig. 2. Accuracy of the model when using SafeFL, DP-SGD, Pure HE and FoolsGold.

Most of the work focuses on a single threat (for example, encryption or attack detection). SafeFL combines three levels of protection: confidentiality - LightGrad (PHE + DP), integrity - FL-Guard plus digital signatures, accessibility - optimized aggregation. By compressing gradients and batch encryption, LightGrad reduces encryption time by 35% compared to its peers (for example, PySyft). FL-Guard requires 2 times less computing resources than SVM-based methods. In experiments on the CIFAR-10 dataset, SafeFL maintains the accuracy of the model at 91.2% (compared to 89.5% for DP-SGD and 85% for PureHE), while providing protection against poisoning attacks. The architecture is compatible with any FL frameworks (TensorFlow Federated, Flower).

However, it is worth noting that the proposed solution has a number of disadvantages and limitations. First of all, it is worth noting the higher demands on computing resources. This disadvantage can be critical in the case of encryption/decryption of gradients of deep learning models. Also, further research requires determining the quality of the proposed model in cases of uneven distribution of customer data, as well as in cases of adaptive attacks.

Nevertheless, the proposed approach can be used in various sectors. For example, when used in the field of healthcare, when training a diagnostic model, the server will allow detecting anomalies (for example, when uploading data with artifacts). Patient data will remain private and will not be used when updating the global model, which will avoid erroneous/malicious updates. Similar approaches are needed in the financial sector, as it is also necessary to keep transactions private, for smart cities (generating confidential data without the possibility of centralization) and similar cases where there is a need to maintain confidentiality and protect global model against fake data.

10 Conclusion

The proposed SafeFL approach provides multi-level protection of the FL environment, solving key issues of confidentiality, integrity and productivity. Experiments confirm that the method surpasses its modern counterparts in terms of accuracy, speed, and resistance to attacks. This makes it applicable in medicine, finance and IoT, where data security is critically important. SafeFL provides the best balance between accuracy, safety and performance, making it applicable in real-world FL systems.

On the CIFAR-10 dataset, SafeFL maintained the accuracy of the model at 91.2%, which is 6.2% higher than that of Pure HE, and only 0.8% lower than that of the base FL.

Thus, SafeFL provides protection against data leaks, poisoning and DoS attacks, confirmed by experiments on CIFAR-10 and MNIST datasets. The proposed approach has high model accuracy, with lower communication costs.

This approach also has some limitations: high resource requirements, since encryption and anomaly checking require significant computing resources, which limits its use on low-performance devices; dependence on data quality. FL-Guard may make mistakes when working with highly heterogeneous data.

References

1. Doriguzzi-Corin, R., Siracusa, D.: FLAD: adaptive federated learning for DDoS attack detection. Compute. Secur. **137**, 103597 (2024)
2. Bao, G., Guo, P.: Federated learning in cloud-edge collaborative architecture: key technologies, applications and challenges. J. Cloud Comp. **11**, 94 (2022). https://doi.org/10.1186/s13677-022-00377-4
3. Li, J., Lyu, L., Liu, X., Zhang, X., Lyu, X.: Fleam: a federated learning empowered architecture to mitigate DDoS in industrial IoT. IEEE Trans. Ind. Inf. **18**(6), 4059–4068 (2022)
4. Abadi, M., et al.: Deep learning with differential privacy. In: Proceedings of the 2016 ACM SIGSAC Conference on Computer and Communications Security, pp. 308–318 (2016). https://doi.org/10.48550/arXiv.1607.00133
5. Liu, P., Wu, X., Peng, Y., et al.: Trustworthy and efficient project scheduling in IIoT based on smart contracts and edge computing. J. Cloud. Comp. **14**, 2 (2025). https://doi.org/10.1186/s13677-025-00726-z
6. Zhang, C., Chen, J., Li, W., et al.: A cloud-edge collaborative task scheduling method based on model segmentation. J. Cloud Comp. **13**, 81 (2024). https://doi.org/10.1186/s13677-024-00635-7
7. Baanqud, N.S., et al.: Engagement in cloud-supported collaborative learning and student knowledge construction: a modeling study. Int. J. Educ. Technol. High. Educ. **17**, 56 (2020). https://doi.org/10.1186/s41239-020-00232-z
8. Khan, L.U., Saad, W., Han, Z., Hossain, E., Hong, C.S.: Federated learning for internet of things: Recent advances, taxonomy, and open challenges, arXiv preprint arXiv: 2009.13012 (2020)
9. Xia, Q., Ye, W., Tao, Z., Wu, J., Li, Q.: A survey of federated learning for edge computing: research problems and solutions. High-Confid. Comput. **1**, 100008 (2021)
10. Kairouz, P., et al.: Advances and open problems in federated learning, arXiv preprint arXiv:1912.04977 (2019)
11. Zhu, L., Liu, Z., Han, S.: Deep leakage from gradients. Adv. Neural Inf. Proces. Syst. **32** (2019)
12. Fang, M., et al.: Local model poisoning attacks to byzantine-robust federated learning. In: 29th USENIX security symposium (USENIX Security 20), pp. 1605–1622 (2020)
13. Bhosale, K.S., Nenova, M., Iliev, G.: The Distributed Denial of Service Attacks (DDoS) prevention mechanisms on application layer. In: Conference on Advanced Technologies, Systems and Services in Telecommunications, pp. 136–138. IEEE (2017)

14. Feng, Y., Guo, Y., Hou, Y., et al.: A survey of security threats in federated learning. Complex Intell. Syst. **11**, 165 (2025). https://doi.org/10.1007/s40747-024-01664-0
15. Abadi, M., et al.: Deep learning with differential privacy. In: Proceedings of the ACM SIGSAC conference on computer and communications security (CCS), pp. 308–318. ACM (2016). https://doi.org/10.1145/2976749.2978318
16. Wei, K., et al.: Federated learning with differential privacy: algorithms and performance analysis. IEEE Trans. Inf. Forensics Secur. **15**, 3454–3469 (2020). https://doi.org/10.1109/TIFS.2020.2988575
17. Arpit Guleria, Harshan J., Ranjitha Prasad, B. N. Bharath On homomorphic encryption based strategies for class imbalance in federated learning. Accepted for presentation at CODS COMAD 2024 Subjects: Cryptography and Security (cs.CR); Information Theory (cs.IT); Machine Learning (cs.LG). arXiv:2410.21192 (2024)
18. Aono, Y., Hayashi, T., Wang, L., Moriai, S.: Privacy-preserving logistic regression with partially homomorphic encryption. In: Proceedings of the ACM on Asia conference on computer and communications security (ASIACCS), pp. 283–294 (2017). https://doi.org/10.1145/3052973.3053036
19. Bonawitz, K., et al.: Practical secure aggregation for privacy-preserving machine learning. In: Proceedings of the ACM SIGSAC Conference on Computer and Communications Security (CCS), pp. 1175–1191. ACM (2017). https://doi.org/10.1145/3133956.3133982
20. Mohassel, P., Zhang, Y.: SecureML: a system for scalable privacy-preserving machine learning. In: 2017 IEEE Symposium on Security and Privacy (SP), pp. 19–38. IEEE (2017). https://doi.org/10.1109/SP.2017.12
21. Truex, S., et al.: A hybrid approach to privacy-preserving federated learning. In: Proceedings of the 12th ACM workshop on artificial intelligence and security (AISec)., pp. 1–11. ACM (2019). https://doi.org/10.1145/3338501.3357370
22. Liu, X., et al.: Privacy-enhanced federated learning against poisoning adversaries. IEEE Trans. Inform. Forensics Secur. **16**, 4574–4588 (2021). https://doi.org/10.1109/TIFS.2021.3105724
23. Fung, C., Yoon, C.J.M., Beschastnikh, I.: Mitigating sybils in federated learning poisoning arXiv preprint arXiv:1808.04866. [Revised version in Proceedings on Privacy Enhancing Technologies (PoPETS), 2018(3)]. https://doi.org/10.2478/popets-2020-0035 (2020)
24. Qu, Y., Pokhrel, S.R., Garg, S., Gao, L., Xiang, Y.: A blockchained federated learning framework for cognitive computing in industry 4.0 networks. IEEE Trans. Ind. Inform. **17**(4), 2964–2973 (2021). https://doi.org/10.1109/TII.2020.3007817

Big Data Protection Methods Using Neural Networks

Petr Chetyrbok[1], Anatoliy Kazak[1](✉), Yelena Lukyanova[2], Angela Mayorova[1], Anna Dorofeeva[1], and Daria Syedneva[1]

[1] VI Vernadsky Crimean Federal University, Simferopol 295007, Russia
kazak_a@mail.ru
[2] Kherson State Pedagogical University, Kherson 273003, Russian Federation

Abstract. Particular attention is currently paid to methods of protecting big data using neural networks (NN) and fuzzy neural networks (FNN). The article considers a system of protection against information leaks and a system of video data stream analysis. In both systems, neural networks are used to analyze big data. Neural networks are an effective tool for processing and decomposing big data. In particular, a vector criterion is used to solve this problem, which compresses the volume and consequently speeds up data processing. Optimization of the modular neural network allows to identify patterns of data distribution in the stream and to classify them using private classifiers for individual modules. The video data stream analysis system uses pulse neural networks.

Keywords: Machine Learning · Financial Effectiveness · Business Intelligence · Big Data · Regulatory Conditions

1 Introduction

With accelerating digitalization, assets in the form of big data accumulate in the traffic of organizations, which are used to analyze and predict their activities. Control over the use of big data and its protection depends on its sources of receipt and methods of storing it in data storage systems (DSS). Sources of big data include decoding webmail services and social networks, mail protocols, transfer of hypertext information and files, Internet messengers, USB ports, shadow copying of files (Fig. 1.).

In 3 out of 5 cases, organizations' traffic contains new or previously unknown information assets that are not protected in any way.

Self-training of the system on a collection of client documents using machine learning technologies is carried out using self-training algorithms. Deep has become a hot topic of conversation Fake (deepfake) or video falsification at the end of 2020. At the same time, many applications appeared in the App Store and Google Play, which humorously allow you to modify the image. The risks of deepfakes have not yet been fully understood by society, but this is the trend of our time, and it is a threat to broadcasting and OTT platforms.

M. Lapina et al. (Eds.): AISMA 2025, LNNS 1456, pp. 83–92, 2026.
https://doi.org/10.1007/978-3-032-07275-7_9

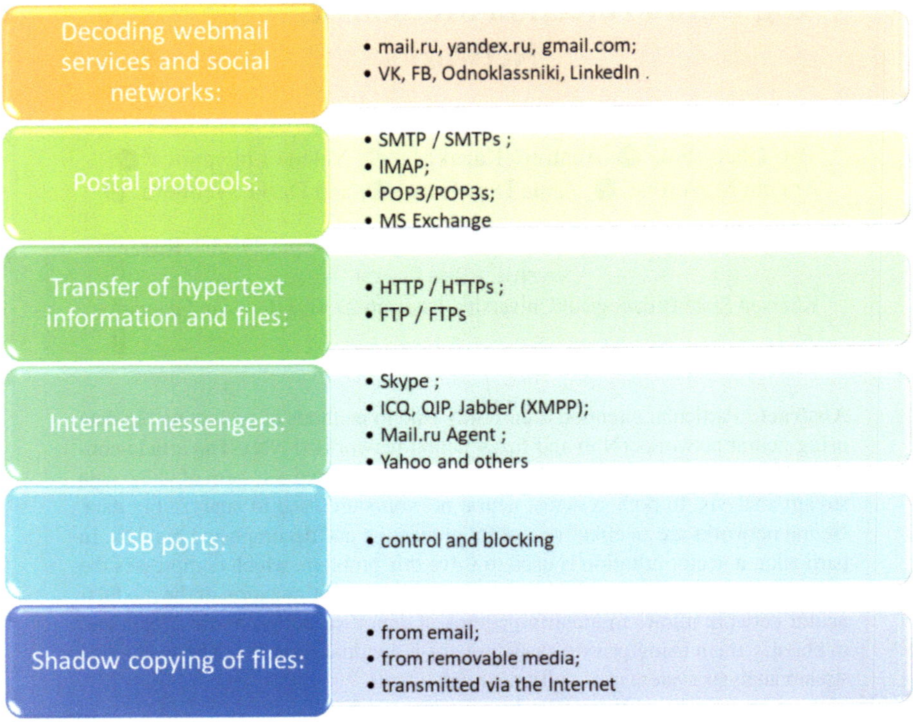

Fig. 1. Sources of big data.

The banking sector and transactions are one of the most striking examples of the use of large amounts of data. Anomaly analysis is one of the priority tasks of ensuring the security of the financial sector. In the context of accelerated banking transactions, a positive example was the use of a deep learning-based cipher, which made it possible to identify patterns and identify money laundering, suspicious transactions with big data, and unauthorized access [2].

Network security status assessment is a vital research area [15, 16]. An example of using deep auto encoding (DAE) is described in [3]. This system includes three stages (Fig. 2).

Fig. 2. An example of using deep auto encoding (DAE).

One of the mechanisms to ensure the security of confidential data in clouds is homomorphic encryption. Privacy-preserving neural networks are used to develop solutions that use neural networks in these settings [4]. The main limiting factor for using privacy-preserving neural networks is the high computational and spatial complexity of the scalar multiplication algorithm, which is the basic algorithm for computing mathematical convolution. In [5], the authors propose a scalar multiplication algorithm that reduces the spatial complexity from quadratic to linear and reduces the computation time of scalar multiplication by 1.38 times.

In [6], image manipulation techniques were investigated as defenses against adversarial attacks. A method for detecting simple exploitable passwords has been developed that administrators can use in addition to other existing methods to improve the overall security of automated systems against threats from accounts potentially compromised by intruders. The method is based on the analysis of commands executed in automatic or manual modes with the specification of credentials in plain text [7].

Real-time big data analysis and deep learning methods for credit card fraud are described, as well as the effectiveness of a framework that was proposed to improve the speed and accuracy of fraud detection. The framework implemented modern technologies so that credit card transactions were constantly monitored and dynamically developed algorithms recognized fraudulent activities [8]. The review examines the possibilities and prospects of using AI and blockchain in the field of big data to counter fraud and sophisticated attacks [9].

Behavioral biometrics, such as analyzing typing speed and navigation patterns, complement traditional security measures, while advanced ML models optimize multi-factor authentication protocols, reducing vulnerabilities [10]. Key findings show that big data platforms such as Spark demonstrated superior performance, processing 500 GB of data in 35 s compared to Hadoop's 60 s. However, challenges related to the interpretability of AI models and the difficulty of integrating diverse data sets are identified, making it difficult to detect threats in real time [11].

2 Methods

The development of video stream analysis technologies for detecting changes and determining integrity violations consists of several stages.

1. Initial analysis of video data stream.
2. Detailed decomposition of image elements.
3. Inverse training of a neural network to detect artifacts.

Figure 3 shows the automatic training of the information leak protection system on the customer's documents.

Machine vision protects confidential photos, scans, pictures. Ready-made categories:

- passport of a citizen of the Russian Federation;
- technical drawings;
- geographical maps

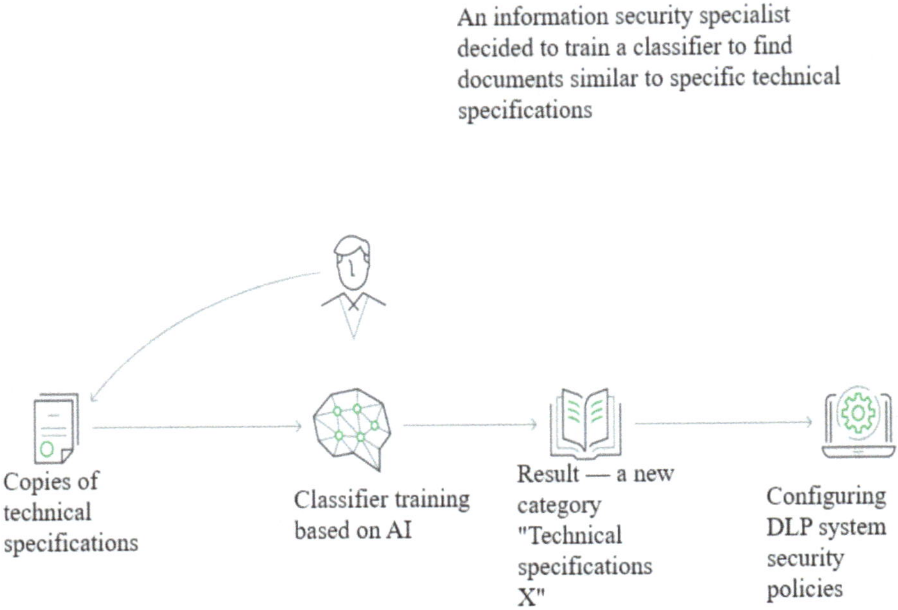

Fig. 3. Information leak protection system.

3 Results

Modular neural networks are used for decomposition of big data. The vector criterion (Fig. 4.) allows for effective decomposition and reduction of the data volume, and increase of their processing speed.

The neural network module model is represented by a three-layer perceptron, where n is the dimension of the object parameter space (receptor field), k is the number of neurons in the hidden layer, m is the number of neurons in the output layer. A gradient algorithm with a sigmoid activation function is used for training [15]. For reference pairs (object, object image), we calculate the value of each coordinate of the error vector $E = (E_1, E_2, E_3)$, where E_1 is the mean square error, E_2 is the network error by the modulus of the deviation of the image from the reference, E_3 is the maximum error of the bitwise deviation of the image from the reference. These values are calculated using the following formulas:

$$E_1 = \frac{1}{n}\sqrt{\sum_{j=1}^{n}\left(y_j^p - d_j\right)^2},$$

$$E_2 = \frac{1}{n}\sum\nolimits_{j=1}^{n}\left|y_j^p - d_j\right|,$$

$$E_3 = \max j = 1, \ldots, n \left|y_j^p - d_j\right|,$$

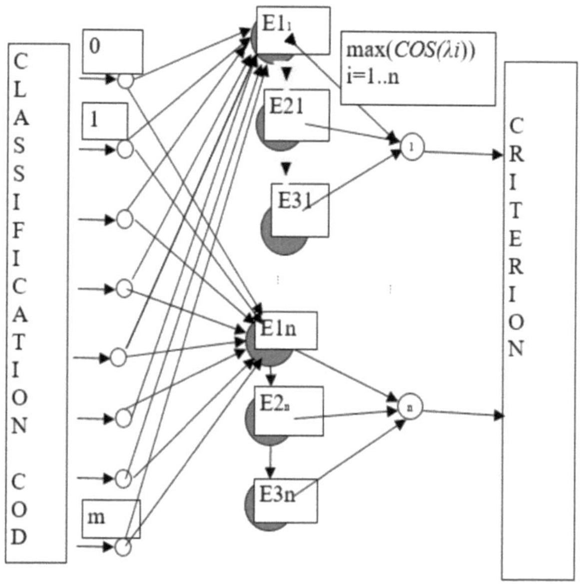

Fig. 4. Scheme for calculating the vector criterion.

y_j^p– the actual output value of the neuron j output layer of a neural network when a reference sample is fed to its inputs p, d_j is the ideal output value of the neuron

The objective function to be minimized is:

$$E(w) = \frac{1}{2}\sum\nolimits_{j,p}\left(y_{j,p}^{(N)} - d_{j,p}\right)^2,$$

where $y_{j,p}^{(N)}$ - the real value of the neuron state j output layer N neural network when fed to it inputs p -th object; $d_{j,\,p}$– ideal (desired) state of this neuron.

To train a neural network (NN), we first use a test sample. Which is made up of the results of verified, reliable data. After training the NN on the test sample, we classify the objects, that is, we obtain classification rules.

Second stage. Next, we train the NN on a random sample. We stop the training process when the scalar of the error vector on the sample exceeds the scalar of the error vector on the test sample according to the Euclidean norm.

Since we stop training on a random sample when the error vector scalars are greater than or equal to the test sample, we thereby limit the artificial intelligence to establish new rules for classifying objects. We optimize the modular NN according to the rule. That is, if the vector criterion is constant on a module, then we exclude this module from the modular neural network, thereby compressing the input data flow and increasing the speed of data processing. Fuzzy models using fuzzy neural networks (FNN) establish rules from a random sample and can thus adjust the classification result to the dependence on the training sample [1 2].

Figure 5 shows the initial analysis of the video data stream.

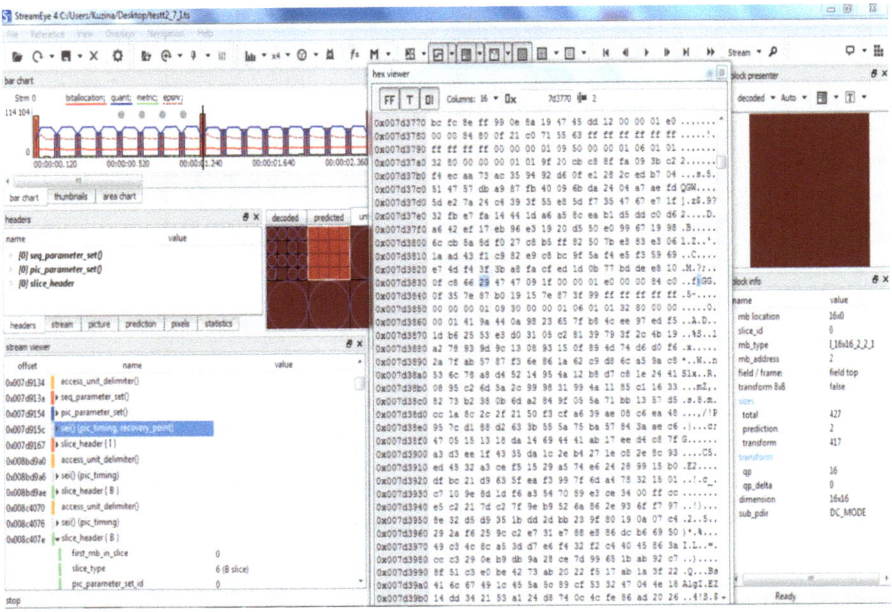

Fig. 5. Initial analysis of the video data stream.

Figure 6 shows a detailed decomposition of image elements. The reverse training of the neural network for finding artifacts is shown in Fig. 7.

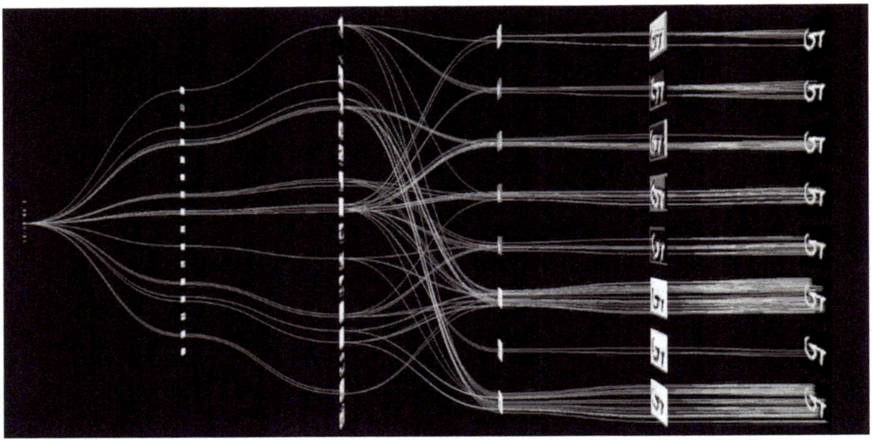

Fig. 6. Detailed decomposition of image elements.

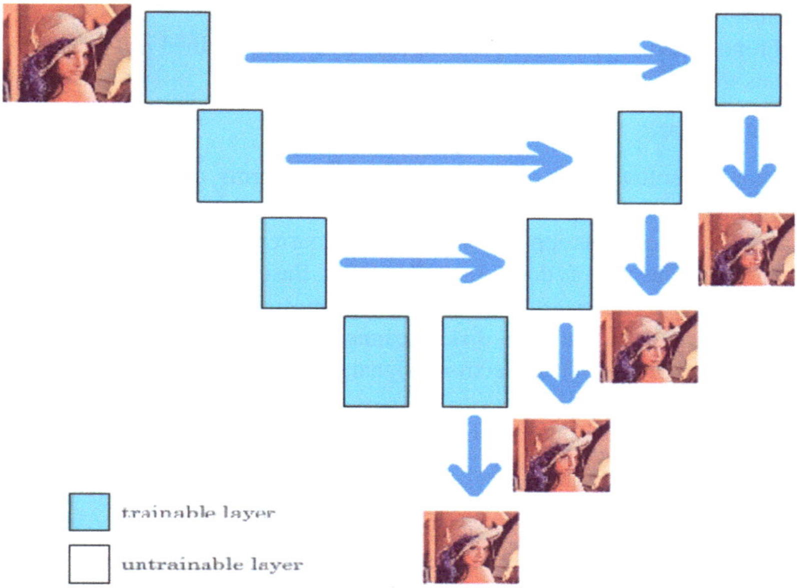

Fig. 7. Reverse training of a neural network to detect artifacts.

The use of pulse neural networks allows organizing monitoring of Deepfakes as an element of confrontation in cyberspace. An example of a control system is the hypervisor "Horizon-VS".

The hypervisor is a universal, secure, scalable platform for virtualizing workstations of government employees.

Main advantages:

- convenient deployment and administration;
- centralized control and visualization of processes;
- the ability to integrate in a hybrid cloud with existing and newly acquired solutions at the level of general system and application software (SW);
- integration of existing information security tools;
- restricting access to data located on storage systems and in user profiles;
- easy scaling and redundancy of infrastructure on additional sites.

Data storage subsystem:

- hyperconverged platform. Execution of a virtual machine (VM) and creation of software-defined data storage systems (DSS) on a single hardware server base;
- the possibility of creating software-defined data storage systems;
- connection of external storage systems and consumers via FC /iSCSI/ NFS / Cifs protocols;
- the ability to form a distributed scalable data storage using the built-in tools of the ZVEZDA active control package (ACP);
- information protection of data on external storage systems using built-in information security tools of the ZVEZDA software and hardware system;

- virtualization of third-party storage systems;
- synchronous and asynchronous replication of software-defined and external storage systems from third-party manufacturers.

Thin client software:

- operation of software on arbitrary hardware platforms of thin client terminals, including domestically produced ones, without an auxiliary operating system;
- the ability to function as application software in current operating systems Windows, Linux, MacOs, as well as domestic Astra Linux, BaseAlt, Rosa, MSVS all versions;
- the ability to connect to virtual workstations (forwarding) of arbitrary USB devices, including flash drives, mobile disks, scanners, printers, video cameras and other peripheral devices, as well as electronic signature keys, smart cards and crypto tokens;
- the ability to stream FullHD video;
- Secure connection to virtual machines.

Components of information security:

- hardware for trusted boot of terminals and hypervisors;
- means of trusted management and protection of authorization of users and cloud administrators;
- protection of control channels and connection of cloud infrastructure;
- hardware protection of connections from a thin client terminal to a virtual machine;
- form factor PCI - Express, miniPCI - Express, M.2. Information leak protection system based on artificial intelligence technologies:
- controls all necessary channels;
- finds what any other leak protection system misses;
- will show everyone involved in the incident;
- allows you to see the preparation of an incident;
- automates the configuration of security policies.
- The video data stream analysis system monitors and is used for:
- development and simplification of Deepfakes generation tools;
- development of methods for detecting Deepfakes and neutralizing them;
- development of early detection tools for Deepfakes (possibly as OSINT tools);
- creation of national systems for monitoring Deepfakes as an element of confrontation in cyberspace.

4 Discussion and Conclusions

Neural networks are an effective tool for processing and decomposing big data. In particular, a vector criterion is used to solve this problem, which compresses the volume and consequently speeds up data processing [12–14].

Methods of protecting big data using neural networks and fuzzy neural networks have a disadvantage when learning from unreliable and fake data. The vector criterion allows you to filter the flow of big data. Since we stop learning on an arbitrary sample when the scalars of the error vector are greater than or equal on the test sample, we thereby limit AI to establish new rules for classifying objects. The optimization of the modular NS is carried out according to the following rule. That is, if the vector criterion is constant on the module, then this module is excluded from the modular neural network, thereby compressing the input data stream and increasing the data processing speed. Fuzzy models using fuzzy neural networks set the rules from an arbitrary sample and thus can adjust the classification result depending on the training sample. The paper discusses a video data stream analysis system that tracks, detects, and neutralizes Deepfakes. It uses tools for early detection of Deepfakes using pulsed neural networks, which makes it possible to organize monitoring of Deepfakes as an element of confrontation in cyberspace.

References

1. Zhang, Z.: Big data based network security awareness model using modified lion swarm optimization with fully convolutional neural network. In: 2024 Second International Conference on Networks, Multimedia and Information Technology (NMITCON), pp. 1–4 (2024)
2. Roy, S., Nijaguna, G.S., Shankar, S.S., Adnan, M.M., Umaeswari, P.: Leveraging deep autoencoders for security in big data framework: an unsupervised cloud computing approach. In: 2024 International Conference on Distributed Computing and Optimization Techniques (ICDCOT), pp. 1–5 (2024)
3. Liang, K., Zhao, L., Ji, X., Zhang, W.: A method for network security situation assessment based on multi-source feature fusion in big data. In: 2024 7th International Conference on Information Communication and Signal Processing (ICICSP), pp. 238–242 (2024)
4. Behiry, M.H., Aly, M.: Cyberattack detection in wireless sensor networks using a hybrid feature reduction technique with AI and machine learning methods. J. Big Data. **11**, 1–39 (2024)
5. Lapina, M.A., Shiriaev, E.M., Babenko, M.G., et al.: High-speed convolution core architecture for privacy-preserving neural networks. Program Comput. Soft. **50**, 417–424 (2024). https://doi.org/10.1134/S0361768824700282
6. Villegas-Ch, W., Jaramillo-Alcázar, A., Luján-Mora, S.: Evaluating the robustness of deep learning models against adversarial attacks: an analysis with FGSM, PGD and CW. Big Data Cogn. Comput. **8**, 8 (2024). https://doi.org/10.3390/bdcc8010008
7. Rusanov, M., Babenko, M., Lapina, M., Sajid, M.: Identification of exploited unreliable account passwords in the information infrastructure using machine learning methods. Big Data Cogn. Comput. **8**, 159 (2024). https://doi.org/10.3390/bdcc8110159
8. Prince, C.C., Uzoamaka, E.O., Ikenna, U.C., Ogochukwu, E.E.: Real-time big data analytics for detecting credit card fraud in cyber forensics using deep learning models. Softw. Eng. **10**(2), 15–23 (2024). https://doi.org/10.11648/j.se.20231002.11
9. Shoetan, P.: Reviewing the role of big data analytics in financial fraud detection. Finance Acc. Res. J. **6**(3), 384–394 (2024). https://doi.org/10.51594/farj.v6i3.899
10. Omogbeme, A., Atoyebi, I., Soyele, A., Ogunwobi, E.: Enhancing fraud detection and prevention in fintech: big data and machine learning approaches. World J. Adv. Res. Rev. **24**, 2301–2319 (2024). https://doi.org/10.30574/wjarr.2024.24.2.3617
11. Nugroho, S., Sumaryanto, S., Hadi, A.: The enhancing cybersecurity with AI algorithms and big data analytics: challenges and solutions. J. Technol. Inform. Eng. **3**, 279–295 (2024). https://doi.org/10.51903/jtie.v3i3.200
12. Kashkin, V., Elafi, R., Issafi, H.: Digital transformation of audience intelligence: a comparative study of the programs. J. Lifestyle SDGs Rev. **5**(2), e05023 (2025)

13. Chetyrbok, P.V.: Proceedings of 2017 20th IEEE International Conference on Soft Computing and Measurements Saint Petersburg: Electrotechnical University LETI Preliminary Systematic Decomposition of Big Data for their Classification Using Vector Criteria Dynamic Management Models of Innovations Generations, pp. 762–764 (2017)
14. Kazak, A.N., Chetyrbok, P.V., Oleinikov, N.N.: Artificial intelligence in the tourism sphere. IOP Conf. Ser. : Earth Environ. Sci. **421**, 042020 (2020)
15. Vinodha, D., Jenefa, J., Anita, E.A.M., Babenko, M.: Theoretical framework for Blockchain secured predictive maintenance learning model using digital twin. In: Lapina, M., Raza, Z., Tchernykh, A., Sajid, M., Zolotarev, V., Babenko, M. (eds.) AISMA-2024: International Workshop on Advanced Information Security Management and Applications AISMA 2024. Lecture Notes in Networks and Systems, vol. 863. Springer, Cham (2024)
16. Mary Anita, E.A., Jenefa, J., Vinodha, D., Lapina, M.: Advancements in Sybil attack detection: a comprehensive survey of machine learning-based approaches in wireless sensor networks. In: Lapina, M., Raza, Z., Tchernykh, A., Sajid, M., Zolotarev, V., Babenko, M. (eds.) AISMA-2024: International Workshop on Advanced Information Security Management and Applications AISMA 2024. Lecture Notes in Networks and Systems, vol. 863. Springer, Cham (2024)

AI-Powered Botnet Detection Systems: A Critical Review of Current Approaches and Challenges

D. Vinodha[1], E. A. Mary Anita[1](✉), and Maria Lapina[2]

[1] Christ University, Bangalore, India
maryanita.ea@christuniversity.in
[2] North Caucasus Federal University, Stavropol, Russia

Abstract. In the era of information technology, Botnets have become the most persistent cyber threat, capable of launching large-scale attacks like Distributed Denial of Service. stealing sensitive information and disturbing online services. Botnets have evolved from simple networks to complicated distributed networks including IoT devices, making them pervasive, harder to track, and destroy. Machine learning and Deep learning based models are emerging to detect bot attacks by analyzing large datasets and detecting patterns and anomalies. The state of the art methodologies for detecting bot infection are reviewed deeply and compared based on adopted methodologies, dataset and feature selection mechanism. The paper further discusses the pros and cons of existing methodologies. Finally, research gaps are presented to help future research on enhancing bot detection.

Keywords: Botnet · Machine learning · Deep Learning · Internet of Things · network traffic analysis

1 Introduction

Advanced Persistent Threats (APTs) and botnets are vital cyberthreats due to their ability to secretly gather sensitive data and ruin equipment in unnoticed manner. APT attacks, executed in multiple stages, often employ techniques like Social Engineering, Zero-day attacks, supply chain attacks and the use of Common and Control servers. The ability of APT attacks to blend seamlessly with regular network activity makes it difficult to get noticed.

BOTs are used to launch attacks like Distributed Denial of Service (DDoS), send spam emails, conduct click fraud, spread malware, and steal sensitive information. Detection of Bots is increasingly challenging as modern bots target all endpoints, including mobile apps, APIs, and servers. They use advanced technologies, like real device emulation, to mimic human behavior, and their operators distribute attacks across time and regions for stealth. Bots also rotate through millions of residential IPs, bypassing IP-based detection methods. The rise of Bots-as-a-Service (BaaS) further lowers the barrier for launching attacks, and facilitates the creation of more advanced bots. These factors

M. Lapina et al. (Eds.): AISMA 2025, LNNS 1456, pp. 93–104, 2026.
https://doi.org/10.1007/978-3-032-07275-7_10

highlight the need for robust, multi-layered defenses to prevent botnet attacks and protect all vulnerable endpoints effectively.

By considering the above challenges, there is an urge to develop a robust detection and mitigation system capable of analyzing network traffic to identify APTs and botnets. Machine learning and deep learning-based methods [1] are emerging as a promising technology because of their ability to analyze huge volumes of data and find dynamic patterns related to bot infection. A detailed analysis of recent works on bot detection mechanisms of traditional networks and the Internet of Things is presented in this paper.

The rest of paper is organized as follows. Section 2 introduces the classification of the Bot detection mechanism. Section 3 presents the literature review on the existing bot detection methods. The detailed comparative analysis is discussed in Sect. 4. Research gap and future research direction is provided in Sect.5.

2 Classification

Botnets remain a formidable threat in the cybersecurity landscape due to their adaptability and sophistication. Traditional detection mechanisms like signature based methods utilize encryption and zero day vulnerabilities. This makes traditional signature based systems largely ineffective. To address this, Recent research works have explored machine learning (ML) techniques for botnet detection, using different kinds of network data like payloads, traffic flows, DNS queries, and complex feature sets.

2.1 Payload-Based Detection

Payload-based detection methods examine the information carried by the network packets to detect anomalies related to botnets. The research shows that multilayer perceptrons networks, convolutional neural networks (CNNs), and recurrent neural networks (RNNs) have demonstrated high performance. The IoT botnet detection system proposed by Mashaleh et al. [2] using CNNs and RNNs trained on packet payload and flow data, achieved 97.31% accuracy. But, the growing utilization of encryption by botnets creates a significant challenge for payload-based analysis.

2.2 Flow-Based Detection

Flow-based detection relies on statistical features of network traffic flows, such as duration, bytes, and timing. Alauthman et al. [3] utilized a reinforcement learning classifier trained on flow features, while Alieyan et al. [4] compared various classifiers, including SVMs and decision trees. Although flow-based methods are resilient to encrypted traffic, they are vulnerable to mimicry attacks, where botnets manipulate traffic patterns to resemble normal traffic.

2.3 DNS-Based Detection

DNS traffic analysis is another approach for botnet detection, focusing on anomalies in domain lookup patterns. Techniques such as domain entropy analysis and IP diversity evaluation are common. Alkasassbeh et al. [5] demonstrated the effectiveness of random forests in detecting DNS tunneling, while Almomani et al. [6] proposed an ensemble-based system with 98.76% accuracy. Despite its effectiveness, DNS-based detection can be circumvented by advanced botnets using domain caching and fast-flux techniques.

2.4 Hybrid Detection

Hybrid detection systems combine multiple data sources, such as payloads, flows, and DNS, to improve accuracy. Alieyan et al. [4] and Mashaleh et al. [2] demonstrated that hybrid models, including SVMs and deep learning classifiers, achieve high accuracy by leveraging diverse feature sets. However, these systems face challenges related to large feature spaces and potential overfitting, necessitating careful feature selection and advanced neural architectures.

2.5 Anomaly-Based Detection

Anomaly-based detection focuses on identifying deviations from normal network behavior. Techniques such as fuzzy rule interpolation and one-class SVMs have been applied successfully. Almseidin et al. [7] achieved a 96.4% detection rate using fuzzy logic, while Alkhamaiseh et al. [8] detected 97% of unknown attacks using a multistage one-class SVM. These methods offer high accuracy and interpretability but can be complex and dataset-specific.

2.6 Challenges and Future Directions

The major challenges in detecting botnet using machine learning models include limited labeled datasets, analysis using ciphered traffic data and detection of evasion attacks. To address these challenges, it is recommended to focus the research on semi-supervised learning, developing advanced generative AI-based learning models to ensure privacy, transparency by exploring adversarial training and scalable analytics.

3 Literature Review

The study by Jyotsna et al. [9] analysed IOT23 dataset to enhance intrusion detection particularly against bot attack using statistical methods including ANOVA, Kruskal-Wallis, and Friedman tests and three feature selection techniques(Chi-square analysis, Recursive Feature Elimination (RFE), and Lasso-based methods) with the Gated Recurrent Unit (GRU) model showing the best performance. While the bot attack detection rate is enhanced by extracting spatial and temporal features using hybrid ResNet-GRU model, the data privacy is addressed by using a federated learning framework. Though this ensemble of LassoCV models and a ResNet-GRU architecture improves classification accuracy, while a federated learning framework ensures data privacy, the increase in computational cost due to the integration of ResNet and GRU, poses a challenge to resource-constrained IoT devices.

Ali et al. [10] proposed a deep learning model using a Sequential Feedforward Hybrid Autoencoder to detect IoT bots and zero-day attacks by identifying unexpected patterns without prior knowledge. LSTM captures sequential data features, while an autoencoder performs dimensionality reduction. Multilayer perceptron networks detect anomalies, trained with increasing layers to avoid overfitting. Using BAIoT2018 and UNSW-NB15 datasets covering 10 and 9 attack types, the model demonstrates effectiveness but faces scalability issues due to its reliance on a central server.

Shareef et al. [11] developed a botnet detection model using attention networks to analyze device communication and capture anomalous behavior. The Sooty Tern Optimization Algorithm, inspired by zebra behavior, tunes hyperparameters and selects features, achieving high accuracy and F1-scores. A dataset of six IoT attacks created using tools like Hping3 and Boofuzz was used. Comparative analysis shows the algorithm outperforms others such as Ant Colony Optimization and Grey Wolf Optimizer.

Yu et al. [12] proposed a machine learning model for detecting bot attacks using Random Forest and logistic regression. Trained on a dataset with 33 internet data flow features, Random Forest identifies key features, excelling in accuracy and precision, while logistic regression offers flexibility but suffers from overfitting. Lefoane et al. [13] introduced a bot attack detection system for 5G network using gini impurity, frequent itemset mining, and clustering for feature selection. This optimal feature selection helps to reduce the computational complexity, enhance the intrusion detection accuracy. The model is trained using the real time IoT network dataset and achieved high reliability and identified all positives without false negatives.

Putra et al. [14] proposed a framework for detecting botnet activities by analyzing similarities in sequential behaviors in network traffic flow. The model's performance is improved by using dynamic threshold in recognizing attack patterns and features. Patterns were extracted and compared to a knowledge base created from datasets with 14 attack scenarios. While effective, the model faces challenges with computational complexity in terms of space, time, and processing power.

The survey by Alshaikh et al. [15] reveals that Supervised learning effectively detects IoT C&C threats but faces challenges like evolving botnet variants, limited feature sets, and scarce labeled datasets. Features such as packet length and direction may not suffice for complex attacks. While methods like decision trees and deep learning show promise, future efforts must focus on enhancing these techniques and exploring unsupervised learning for better IoT security.

Maad et al. in [16] analyzed network traffic from Provision 737E security cameras using the N-BaIoT dataset to differentiate normal and attack traffic. Dimensionality reduction was performed using Weka, reducing features from 115 to 10, simplifying computational complexity. The Decision Tree (J48) algorithm achieved 99.95% accuracy in supervised learning, while the EM algorithm reached 76.73% accuracy in unsupervised learning.

Christos et al. [17] surveyed existing schemes for detecting lateral movement focusing the solutions based on Endpoint Detection Response, Graph based analysis and machine learning. This survey states that Machine learning-based lateral movement (LM) detection studies face challenges due to the reliance on outdated datasets that lack modern attack schemes, proper multiclass labeling, and optimized hyperparameters. Most studies used legacy MS Windows Event Viewer logs, as tools to convert Sysmon logs into ML-compatible formats were unavailable until recently. Only a few works utilized Sysmon logs for richer event-oriented information, highlighting a missed opportunity for better detection. Notably, tools like PeX EDR and ETCExp have advanced the field by enabling the conversion of Sysmon logs into .csv datasets for ML experimentation. These developments signal progress but emphasize the need for better datasets and tools to enhance LM detection systems. Also stated that Graph-based algorithmic schemes for lateral movement (LM) detection are limited, with only a few works focused on IoT and IIoT contexts. Hybrid frameworks [19–21] combining graph-based techniques with deep neural networks, like LMTracker [18], have shown promise by leveraging multidimensional feature analysis and heterogeneous graph construction. Recent studies increasingly utilize modern datasets like APT3 [22] and CSE-CIC-IDS2018, addressing more comprehensive attack scenarios and enhancing model evaluation compared to earlier works based on legacy datasets.

According to the analysis of Kaur [23], "Botnet and Botnet Detection Techniques in the Cyber Realm," discusses the expanding danger of botnets—networks of hijacked computers owned by a botmaster. It summarizes botnet origins, lifecycle (such as original infection, secondary infection, establishing a link with a command and control server, and upkeep), and many varieties, such as centralized (IRC and HTTP-based) and decentralized (P2P) botnets. Evolving tendencies such as bot clouds and social botnets are considered. The paper also discusses botnet detection methods, such as honeynet-based detection and intrusion detection systems (signature-based, anomaly-based, DNS-based, and data mining-based approaches), the challenges and future research directions in this area.

Based on the Feily [24] "A Survey of Botnet and Botnet Detection" provides a comprehensive overview of botnets, their characteristics, and various detection techniques. The authors elaborated the botnet operation with respect to command-and-control structures and lifecycle phases. The botnet detection methods are classified into signature-based, anomaly-based, DNS-based, and mining-based. The strengths and limits of each method are evaluated by comparing their ability in detecting known and unknown botnets, encrypted communications, and different botnet architectures. The paper also highlights the promising role of data mining and DNS-based approaches for future botnet detection.

Haq et al. [25] proposed botnet detection techniques by leveraging machine learning techniques on network traffic data. The strength and weakness of signature-based and anomaly-based detection methods are compared. Signature-based detection (J48, Naïve Bayes) effectively identifies known botnets with low false positive rates, while anomaly-based detection (K-Means clustering) detects unknown botnets but suffers from high false positives. This is addressed by ensembling both classification and clustering approaches. The ensemble hybrid model is evaluated using the dataset namely CTU-13, ISOT and WEAK, for feature selection, classification, and clustering. The ensemble to balance the accuracy and false positives. But efficient data partitioning with minimised false positives remains a challenge. The study also highlights a need for improvement in feature extraction mechanisms and partitioning strategies for improved performance.

Moemedi Lefoane et al. [26] investigate botnet detection in IoT environments using machine learning and an optimized feature selection technique. The accuracy of bot detection is improved with reduced computational cost by eliminating features with less influence from network traffic data. It adopted a two phase methodology: in the first phase, noisy features are detected based on their frequency in the dataset and eliminated. In the second phase, decision trees, SVM, and logistic regression classifiers are trained and evaluated on the IoT23 dataset. The optimized approach significantly enhances detection performance, achieving 100% TPR, 0% DPR, and a 99.9976% F-score, with the decision tree classifier performing best. The method demonstrated visible improvement in the efficiency and overcame the limitations of traditional signature-based detection. However, the generalizability is limited due to the manual tuning of feature selection thresholds and evaluation on a single dataset. Future research should focus on adaptive feature selection and broader dataset evaluation for enhanced scalability and robustness.

Hidayah et al. [27] survey existing botnet detection techniques and introduce a novel multilayer framework leveraging machine learning. It aims to overcome the limitations of traditional signature-based methods by incorporating signature-based, behavior-based, and flow-based detection techniques, with a focus on machine learning. The proposed framework consists of a filtering module to reduce network traffic and a classification module that detects botnet command and control (C&C) servers using behavior-based analysis. By analyzing flow-based features instead of payloads, it remains effective even against encrypted or VPN-protected botnets. The study evaluates multiple classifiers, with k-nearest neighbors (k-NN) achieving the highest performance (91.51% F-score and a 1.5% false-negative rate) when using a 2-second data aggregation interval. Key advantages of the framework include its protocol-independent nature, resilience against encryption techniques, and high detection accuracy. However, limitations include reduced effectiveness against decentralized P2P botnets and the computational overhead required for real-time processing of large network traffic volumes. Furthermore, behavior-based detection can lead to higher false-positive rates compared to signature-based methods.

Alharbi et al. [28] propose a botnet detection method using three main components: data bootstrap, feature extraction and selection, and detection engine. The flow ingestion converts a bidirectional network into a list used to create a graph representation of network flows. A set of graph-based features, such as in-degree, out-degree, and indegree weight, outdegree weight. Univariate and Multivariate filtering methods are used based on the intricate characteristics of data for the feature selection technique. A powerful ML classification algorithm, Extra Trees classifier, displayed the best results in identifying the botnet behaviours. CTU-13 dataset was used to compare the model to state of the art flow based and graph-based botnet detection methods, and produced a high accuracy between 99% and 100%.

Other malware detection systems using signatures face several limitations. They cannot detect zero-day or unknown malware variants and require frequent updates to accommodate new threats. Malware often employs obfuscation, polymorphic, and metamorphic techniques to evade detection. There is a time lag between discovering new malware and creating corresponding signatures. Maintaining signature databases is resource-intensive and increasingly complex as malware evolves. Continuous tuning is needed to reduce false positives, and networks must be baselined and monitored to identify anomalies. Malware hidden in normal traffic further complicates detection, highlighting the need for more advanced methods.

4 Comparative Analysis

The various machine learning techniques, including Random Forest (RF), Decision Tree (DT), Support Vector Machine (SVM), Gradient Boosting and deep learning models like Resnet-GRU, LSTM and graph-based approaches, are employed by the existing Bot detection system. The analysis shows that deep learning methods demonstrated higher accuracy compared to other methods. But increased computation cost and dataset dependency become the major challenge. While traditional machine learning models are widely employed for network traffic analysis-based bot detection, graph-based ML models are adopted to enhance the accuracy but with the highest computation complexity. Among the surveyed systems Flow based approaches that.

Rely on network traffic pattern analysis are adopted in [11–14]. Anomaly-based methodologies that rely on identifying deviations in packet transmission rate, bandwidth utilization, packet size, and distribution rate are adopted in [9, 10, 16]. But they suffer from a high false positive rate. This can be addressed by improving accuracy with integrated anomaly detection in signature-based systems. Improved accuracy level and adaptability are shown by the systems [26–28] developed by integrating multiple models, including ML, feature selection, and graph-based analysis. The high computational complexity and scalability are the major challenges of hybrid models. This can be addressed by focusing research on proper dimensionality reduction mechanisms, neural network pruning techniques, hybrid filtering, distributed computing, and feature optimization. The detailed comparative analysis based on methods, dataset, and feature selection mechanism, along with pros and cons, is presented in Tables 1 and 2.

Table 1. Comparative analysis.

Ref No	Problem Addressed	Methodology	Dataset	Feature Selection techniques
Jyotsna et al. [9]	Intrusion detection against bot attack for IOT	Hybrid ResNet-GRU deep learning model and federated learning framework	IOT23	Chi-square analysis, Recursive Feature Elimination (RFE), and Lasso-based method
Ali et al. [10]	Detection of IOT bot and zero day attack	LSTM deep learning model, Sequential Feedforward Hybrid Autoencoder	BAIoT2018 and UNSW-NB15	–

(continued)

Table 1. (*continued*)

Ref No	Problem Addressed	Methodology	Dataset	Feature Selection techniques
Shareef et al. [11]	Detection of IOT bot	Dual-channel Graph Attention Network (DGAN)	real time dataset	Zebra Optimization Algorithm
Yu et al. [12]	Bot detection over internet	Random forest and logistic regression	dataset with 33 features of internet traffic	–
Lefoane et al. [13]	Bot detection for 5G network	Two Machine Learning models based on dominant feature and frequent itemset	real time dataset	Gini impurity
Putra et al. [14]	Bot detection by analysing network traffic	Sequential traffic mining, similarity and dynamic threshold analhysis	dataset with 14 attack senarios	–
Maad et al. [16]	detection of bot attack in IOT	Decision Tree,Expectation-Maximization (EM) Algorithm	N-BaIoT	–
Lefoane et al. [26]	Bot detection for IoT	Random Forest (RF) Decision Tree (DT) Support Vector Machine (SVM) Gradient Boosting	Public botnet datasets with labeled traffic data.	Recursive Feature Elimination (RFE) and Principal Component Analysis (PCA).
Hidayah et al. [27]	Bot detection for traditional networks	Random Forest (RF) Support Vector Machine (SVM) Decision Tree (DT) Gradient Boosting (GB) Neural Networks (NN)	CTU-13 and ISCX-Bot-2014.	Principal Component Analysis (PCA) and Chi-square selection.
Alharbi et al. [28]	botnets detection in IoT and traditional networks	graph theory and ML models-Random Forest (RF) Support Vector Machine (SVM) Decision Tree (DT) K-Nearest Neighbors (KNN) Graph Convolutional Network (GCN)	CTU-13 and ISCX-Bot-2014	Univariate and Multivariate filtering methods

Table 2. Comparative analysis - Pros and Cons

Ref No	Pros	Cons
Jyotsna et al. [9]	Improved accuracy and data privacy	increases computation cost, high false positive rate
Ali et al. [10]	High detection accuracy of 99.76, Low false positives with a precision of 99.68, able to detect a wide range of attacks (recall:99.82%)	Large model size, high computation cost, dataset dependency, high false positive rate
Shareef et al. [11]	High detection accuracy of 99.87, able to detect a wide range of attacks (recall:98.22%), reduced computational complexity by focusing only on relevant features	increased complexity in implementation, dataset dependency
Yu et al. [12]	Accuracy of RF:99.8, LR:83.7	dataset dependency
Lefoane et al. [13]	100 % True Positive rate and 0% false negative	Further enhancement is required for zero day attack
Putra et al. [14]	Achieved 99.94% accuracy, scalable, adaptable to various network conditions	complexity due to threshold tuning and interpreting behaviour patterns
Maad et al. [16]	99.95% accuracy, optimized computational efficiency	dataset dependency, high false positive rate
Lefoane et al. [26]	RF shows a high accuracy of 98.1	high false positive, computational complexity
Hidayah et al. [27]	RF shows a high accuracy of 99.2	SVM and DT have lower performance
Alharbi et al. [28]	Graph Convolutional Network (GCN) showed a higher accuracy of 99.3 than RF	Computational complexity

5 Research Gaps and Future Research Directions

The survey reveals that the major challenges faced by most machine learning and deep learning models are high computation cost, data dependency, and high false positive rate. The hybrid models with permission performance in detecting the attack with high accuracy suffer from implementation complexity, adaptability, and model interpretation. Hence, there is a need for research that could focus on reducing complexity with high accuracy by adapting better feature selection mechanisms and integrating hybrid unsupervised learning models.

6 Conclusion

The latest developments in the field of Bot attack detection over the IOT and network infrastructure are deeply analysed with a summary of comparative analysis based on adopted methodologies, dataset and feature selection mechanism. The pros and cons of existing surveyed systems are tabularised to help researchers produce better technologies by addressing the highlighted challenges and focusing on reducing complexity with high accuracy by adapting better feature selection mechanisms and integrated hybrid unsupervised learning models.

References

1. Lapina, M., Dudun, G., Kotlyarov, D., Rjevskaya, N., Subramanian, S.J.: Analysis of an existing method for detecting adversarial attacks on deep neural networks. In: Alikhanov, A., Tchernykh, A., Babenko, M., Samoylenko, I. (eds.) Current Problems of Applied Mathematics and Computer Systems. CPAMCS 2023 Lecture Notes in Networks and Systems, vol. 1044, pp. 316–329. Springer, Cham (2024). https://doi.org/10.1007/978-3-031-64010-0_29
2. Mashaleh, Ashraf Ibrahim, Noor Alauthman, Mohammad Almomani, Ammar. (2022). A proposed framework for early detection IoT botnet. 1–7. https://doi.org/10.1109/ACIT57182.2022.9994166
3. Alauthman, M., et al.: An efficient reinforcement learning-based botnet detection approach. J. Netw. Comput. Appl. **150**, 102479., ISSN 1084-8045 (2020). https://doi.org/10.1016/j.jnca.2019.102479
4. Alieyan, K., Alkasassbeh, M., Almomani, A., Manasrah, A.: A survey on botnet detection based on machine learning techniques. In: Proceedings of the International Conference on Electrical and Computing Technologies and Applications (ICECTA), pp. 1–6 (2017)
5. Alkasassbeh, M., Almseidin, M.: Machine Learning Techniques for Accurately Detecting the DNS Tunneling (2023). https://doi.org/10.1007/978-3-031-37717-4_24
6. Almomani, A., Anbar, M., Alarood, A., Manasrah, A.: A novel hybrid feature selection and ensemble-based machine learning approach for botnet detection. Sci. Rep. **13**, 12345 (2023)
7. Almseidin, M., Kovacs, S.: Intrusion detection mechanism using fuzzy rule interpolation. J. Theor. Appl. Inf. Technol. **96**(16), 5473–5488 (2018)
8. Alkhamaiseh, M., Okada, H., Takano, K.: Unknown attack detection by multistage one-class SVM focusing on communication interval. In: Proceedings of the 2014 International Conference on Security and Management, pp. 1–7 (2014)
9. Jyotsna, A., Anita, M.E.A.: A novel paradigm for IoT security: ResNet-GRU model revolutionizes botnet attack detection. Int. J. Adv. Comput. Sci. Appl. **14** (2023). https://doi.org/10.14569/IJACSA.2023.0141231
10. Ali, S., et al.: A novel approach of botnet detection using hybrid deep learning for enhancing security in IoT networks. Alex. Eng. J. **103**, 88–97., ISSN 1110-0168 (2024). https://doi.org/10.1016/j.aej.2024.05.113
11. Shareef, S.K., Chaitanya, R.K., Chennupalli, S., et al.: Enhanced botnet detection in IoT networks using zebra optimization and dual-channel GAN classification. Sci. Rep. **14**, 17148 (2024). https://doi.org/10.1038/s41598-024-67865-2
12. Yu, H.: Research on botnet detection technology in network security. Appl. Comput. Eng. **18**, 81–87 (2023). https://doi.org/10.54254/2755-2721/18/20230967

13. Lefoane, M., Ghafir, I., Kabir, S., Awan, I.-U.: Unsupervised learning for feature selection: a proposed solution for botnet detection in 5G networks. IEEE Trans. Industr. Inform. **19**(1), 921–929 (2023). https://doi.org/10.1109/TII.2022.3192044
14. Putra, M.A.R., Ahmad, T., Hostiadi, D.P.: B-CAT: a model for detecting botnet attacks using deep attack behavior analysis on network traffic flows. J. Big Data. **11**, 49 (2024). https://doi.org/10.1186/s40537-024-00900-1
15. Alshaikh, M., Alsemaih, W., Alamri, S., Ramadan, Q.: Using supervised learning to detect command and control attacks in IoT. Int. J. Cloud Appl. Comput. **14**, 1–19 (2024). https://doi.org/10.4018/IJCAC.334214
16. Maad, H., Kadhim, A., Abbasi, M., Algabri, H.: Chinese journal of computational mechanics machine learning-based detection of IoT botnet attacks. Jisuan Lixue Xuebao/Chin J. Comput. Mech., 604–614 (2024)
17. Smiliotopoulos, C., Kambourakis, G., Kolias, C.: Detecting lateral movement: a systematic survey. Heliyon. **10**(4), e26317.,ISSN 2405–8440 (2024). https://doi.org/10.1016/j.heliyon.2024.e26317
18. Fang, Y., Wang, C., Fang, Z., Huang, C.: LMTracker: lateral movement path detection based on heterogeneous graph embedding. Neurocomputing. **474**, 37–47., ISSN 0925–2312 (2022). https://doi.org/10.1016/j.neucom.2021.12.026
19. Haddadpajouh, H., Azmoodeh, A., Dehghantanha, A., Parizi, R.M.: MVFCC: a multi-view fuzzy consensus clustering model for malware threat attribution. IEEE Access. **8**, 139188–139198 (2020). https://doi.org/10.1109/ACCESS.2020.3012907
20. Chen, C.-K., et al.: Building machine learning-based threat hunting system from scratch. Digit. Threats Res. Pract. **3**(3), 20 (2022). https://doi.org/10.1145/3491260
21. Yang, X., Peng, G., Zhang, D., Lv, Y., et al.: An enhanced intrusion detection system for iot networks based on deep learning and knowledge graph. Secur. Commun. Netw. (2022). https://doi.org/10.1155/2022/4748528
22. Javed, S.H., et al.: Apt adversarial defence mechanism for industrial iot enabled cyber-physical system. IEEE Access. **11**, 74000–74020 (2023). https://doi.org/10.1109/ACCESS.2023.3291599
23. Kaur, N., Singh, M.: Botnet and botnet detection techniques in cyber realm. In: 2016 international conference on inventive computation technologies (ICICT), vol. 3, pp. 1–7. IEEE (2160441) (2016)
24. Feily, M., Shahrestani, A., Ramadass, S.: A survey of botnet and botnet detection. In: 2009 third international conference on emerging security information, systems and technologies, pp. 268–273. IEEE (2160441) (2009)
25. Haq, S., Singh, Y.: Botnet detection using machine learning. In: 2018 Fifth International Conference on Parallel, Distributed and Grid Computing (PDGC) (2018). https://doi.org/10.1109/pdgc.2018.8745912
26. Lefoane, M., Ghafir, I., Anwar, S., Awan, I.: Machine learning for botnet detection: an optimized feature selection approach. In: Proc. 5th Int. Conf. Future Networks Distrib. Syst. (ICFNDS) (2021). https://doi.org/10.1145/3508072.3508102
27. Hidayah, I., et al.: Multilayer framework for botnet detection using machine learning algorithms. IEEE Access., 1–1 (2021). https://doi.org/10.1109/ACCESS.2021.3060778
28. Alharbi, A., Alsubhi, K.: Botnet detection approach using graph-based machine learning. IEEE Access. **9**, 99166–99180 (2021)

29. Vinodha, D., Jenefa, J., Anita, E.A.M., Babenko, M.: Theoretical framework for Blockchain secured predictive maintenance learning model using digital twin. In: Lapina, M., Raza, Z., Tchernykh, A., Sajid, M., Zolotarev, V., Babenko, M. (eds.) AISMA-2024: International Workshop on Advanced Information Security Management and Applications AISMA 2024. Lecture Notes in Networks and Systems, vol. 863. Springer, Cham (2024)
30. Mary Anita, E.A., Jenefa, J., Vinodha, D., Lapina, M.: Advancements in Sybil attack detection: a comprehensive survey of machine learning-based approaches in wireless sensor networks. In: Lapina, M., Raza, Z., Tchernykh, A., Sajid, M., Zolotarev, V., Babenko, M. (eds.) AISMA-2024: International Workshop on Advanced Information Security Management and Applications AISMA 2024. Lecture Notes in Networks and Systems, vol. 863. Springer, Cham (2024)

Budget Aware Load Balancing Algorithm with Strict Deadline Constraints for Efficient Resource Utilization in Fog and Edge Environments

Deepak[1], Manoj Kumar Upadhyay[1], and Mahfooz Alam[2](✉)

[1] Department of Computer Science, Institute of Engineering and Technology, Dr. B.R. Ambedkar University, Khandari, Agra, India

[2] Department of MCA, G. L. Bajaj Institute of Technology and Management, Greater Noida 201306, India

mahfoozalam.amu@gmail.com

Abstract. Fog and edge computing are very popular for enabling low-latency processing in applications that require quick responses, such as autonomous systems, smart cities, and the Internet of Things (IoT). While these technologies support effective real-time data processing, managing budgets and meeting strict deadlines remain major challenges. Traditional load-balancing algorithms focus more on workload distribution and performance but often overlook cost optimization and deadline requirements. In this study, we introduce the Budget-Aware Load Balancing Algorithm with Strict Deadline Constraints (BALBA-SD), which aims to use resources efficiently while respecting tight deadlines in fog and edge computing. BALBA-SD uses a hybrid approach to dynamically assign tasks between nodes based on the budget and project deadlines. Simulation results show that BALBA-SD is highly effective at meeting deadlines compared to other methods and can reduce execution costs by 21.79% relative to traditional approaches. This makes BALBA-SD a strong choice for deadline- and budget-focused load balancing in fog and edge computing.

Keywords: Budget · Cost · Deadline · Load balancing · Fog computing · Edge computing

1 Introduction

Fog and Edge Computing are key enabling technologies for decentralized data processing, with computational resources located nearer to the data sources, such as IoT sensors [1]. This paradigm allows for immediate processing, thus speeding up processing time compared to traditional cloud-based approaches [2]. In an IoT context, these features become critical, as responses must reach within moments to facilitate both system operation and user experience. Such characteristics are essential for the operations and further enhancements of autonomous vehicles [3, 18], healthcare monitoring, and industrial IoT [4, 19]. A secondary challenge arises in managing computational demands with strict deadlines and cost considerations as these systems become more extensive and complex.

M. Lapina et al. (Eds.): AISMA 2025, LNNS 1456, pp. 105–119, 2026.
https://doi.org/10.1007/978-3-032-07275-7_11

Services in Fog and Edge Computing environments are bound with strict deadlines imposed by the application's nature [5]. Due to missing these deadlines could have major effects, such as system malfunctions, reduced performance, or compromised user safety. For example, in autonomous driving, failure to interpret sensor data within the required timeframes might lead to accidents [3]. Patient safety could also be compromised should a healthcare monitoring system react slowly [20]. Hence, the vision of managing tasks and resources efficiently in systems is extended when deadlines are met.

At the same time, fog and edge environments are available due to limited resources, energy costs, and operating charges [5, 6]. These limitations mean a compromise between maintaining high system output and maximizing resource usage to reduce costs. Achieving this level of balance is particularly difficult in distributed contexts where arithmetic resources are distributed through heterogeneous nodes of different capacities. Existing load compensation technologies for fog and edge computing primarily address resource utilization and degradation optimization. Some algorithms include deadline constraints, but in many cases, they do not consider the complex compromise between deadlines and budgetary restrictions. This gap leads to poor performance that can risk the overall efficiency of the system by overlooking deadlines or exceeding allowed costs.

1.1 Motivation and Contributions

The growing demand for real-time IoT applications is a dual challenge for FOG and EDGE computing. It makes sure that strict task deadlines are observed during operation under limited budgets. However, a detailed review of existing load balancing approaches shows that a small number of people have little solution to both deadlines and deadline compliance in a dynamic, distributed environment. Most algorithms either emphasize latency minimization regardless of cost or reduce costs, while at the same time, there is a risk that they lack deadlines. This unilateral optimization does not support mission-critical applications such as autonomous driving or remote control. Timing and resource efficiency are equally essential in this application. In addition, there is a lack of comparative assessment in the literature that quantifies the trade-offs between budget and deadline constraints. These research gaps motivate the development of a new load balancing approach that can address cost and performance comprehensively and offer scalable and robust solutions in contemporary fog-edge deployments.

This paper introduces the Budget-Aware Load Balancing Algorithm with Strict Deadline Constraints (BALBA-SD) to address the challenges which we have discussed above. So BALBA-SD is designed to optimize task assignment in Fog and Edge Computing systems by integrating three critical dimensions: budget efficiency, load balancing, and strict deadline management. This algorithm assures that tasks are allocated to computational nodes in such a manner that fulfills budgetary constraints while adhering to application-specific deadlines. The following are key features of BALBA-SD:

- **Deadline-Aware Scheduling**: The algorithm considers the deadlines associated with a task and ensures that tasks will be processed on time before that time expires, as processing failures may be unacceptable in latency-sensitive applications.
- **Cost Optimization**: The algorithm aims to minimize the costs of resource usage, considering a budget in the scheduling process, provided that the minimum performance criterion has been satisfied.

- **Dynamic Load Balancing**: The main idea behind BALBA-SD is to assign tasks to heterogeneous nodes so as to first balance resource utilization, avoiding situations whereby some nodes become highly stressed while others remain underutilized.
- **Scalability and Adaptability**: It is capable of efficiently scaling according to the number of tasks and nodes, and hence can be used in different Fog and Edge Environments.

The integration of these features enables BALBA-SD to handle the complex interactions between deadlines and budget constraints effectively. Through rigorous evaluation, the proposed algorithm demonstrates significant improvements in system performance, cost-efficiency, and deadline adherence compared to existing solutions. BALBA-SD represents a significant step forward in advancing the state-of-the-art in load balancing for Fog and Edge Computing, providing a robust framework for addressing the unique challenges of decentralized computing environments.

The rest of the paper is structured as follows: Section 2 illustrates the related work, Section 3 shows the problem statement, and Sect. 4 describes the proposed algorithm. Simulation and results are shown in Sect. 5, and the last Sect. 6 describes the conclusion and future work.

2 Related Work

Numerous load balancing algorithms have been proposed for fog and edge computing systems, aiming to optimize performance, minimize latency, and maximize resource utilization. However, few of these consider both budget constraints and strict deadlines:

- **Deadline-Constrained Load Balancing**: Algorithms such as Deadline-Aware Scheduling focus on ensuring that tasks are completed within their deadlines but do not account for the financial costs of using resources.
- **Cost-Aware Load Balancing**: Research on cost-aware algorithms typically focuses on minimizing operational costs, often ignoring the temporal aspect of tasks.
- **Energy and Performance-Aware Load Balancing**: Some algorithms combine performance and energy efficiency but may fail to respect deadline constraints in critical applications.

Bukhari et al. [6] proposed a work that explores a cost-effective resource allocation strategy in edge computing environments, focusing on heterogeneous service providers offering geographically distributed servers at varying costs and capacities. The proposed Genetic Algorithm-based Cost-Effective Resource Allocation (GA-CERA) optimizes resource selection using bio-inspired operations like selection, crossover, and mutation. The authors of [7] introduce a scheduling framework that optimally distributes workloads between cloud and fog computing resources while respecting budget constraints. It employs dynamic task scheduling and allocation techniques to balance costs and performance.

Ding et al. [8] investigated blockchain system computing using edges tailored to IoT networks. This includes Stackelberg game theory to optimize pricing strategies and budget allocations between IoT devices and edge servers. Liu et al. [9] introduced a Mobile Edge Computing (MEC) price-based distributed offload framework that unlocks the challenge of edge server computing capacity. The proposed framework allows mobile devices to shift computing tasks to nearby edge servers by optimizing task allocation via pricing mechanisms. Yao et al.'s paper [10]. deals with inexpensive fog computing by proposing a framework for uneven cloudlet deployment and optimized user cloudlet associations. It takes into account the diversity of cloudlet features and requirements aimed at reducing operational costs and maintaining quality of service at the same time.

Azizi et al. [11] proposed a Deadline-Aware and Energy-Efficient Task Scheduling approach for IoT applications in a fog computing environment. This paper provides a semi-greedy algorithm that aims to compensate for two important goals: minimizing energy consumption and meeting application-specific deadlines. The proposed scheduling strategy is carried out by dynamically selecting the appropriate fog nodes based on the task urgency and node energy efficiency, allowing efficient resource utilization under time limits. This paper [12] presents "deadline-cost aware task scheduling" (DCaTS) algorithm. This focuses on minimizing monetary costs and task duration and prioritization compliance, thus addressing the challenges of resource utilization in fog computer networks, but does not specifically discuss load balancing.

Ibrahim et al. [13] focused on a delay-aware and load-balanced scheduling mechanism for deadline-constrained IoT applications in fog computing. This mechanism aims to minimize delay, maximize task acceptance rates, and improve resource utilization, but does not specifically address budget-aware algorithms.

At work [14], the author proposed an improved genetic algorithm for workflow planning near cloud computing. Arshed et al. [15] focused on genetic algorithm-based schedules for efficient use of resources and reducing costs in cloud fog environments.

Alam et al. [16] proposed the Resource-Aware Load Balancing (REAL) model. This model is tailored to planning independent tasks in heterogeneous distributed systems (HDS). This model improves the efficiency of the load distribution using centralized load balancing, which constitutes the moving overhead and bandwidth requirements. REAL includes a two-phase migration strategy. The initial move is followed by a best-fit migration phase. The best-fit strategy is to optimize resource utilization and improve system performance by not occupying resources after the initial migration.

In [17], the authors presented a Deadline Constrained based Load Balancing (DCLB) algorithm for a batch of IoT tasks in fog and edge computing environments. The proposed approach distributes workloads dynamically between processing nodes to maximize system utilization while adhering to strict deadlines. The technique computes the ETC for each task on available nodes, prioritizing tasks based on deadlines and ensuring that tasks are sent to nodes capable of completing them within the specified deadlines. To address the existing gaps in deadline and budget-aware scheduling, we introduce BALBA-SD, a novel algorithm designed for the same computing environments.

Table 1 presents a comparative analysis of prominent task scheduling and load balancing algorithms with respect to three critical dimensions: deadline consideration, budget awareness, and support for different computing environments (fog, edge, and cloud). As shown, most existing algorithms emphasize either budget constraints or deadline requirements, but rarely both. Pham et al. [7], Ding et al. [8], and Liu & Liu [9], for instance, consider cost to be paramount, but finish-before-based constraints critical for low latency implementations are set aside, which, to our knowledge, are not addressed

Table 1. Comparative analysis of existing deadline and budget aware load balancing algorithms

Reference	Deadline	Budget	Computing Environments		
			Fog	Edge	Cloud
Liu & Liu, 2017	✗	✓	✗	✓	✗
Yao et al., 2017	✗	✓	✗	✗	✓
Alam et al., 2020	✗	✗	✗	✗	✓
Bukhari et al., 2021	✓	✓	✗	✓	✗
Bothra et al., 2021	✓	✓	✗	✗	✓
Ding et al., 2022	✗	✓	✗	✓	✗
Azizi et al., 2022	✓	✓	✓	✗	✗
Arshed et al., 2022	✗	✓	✓	✗	✓
Pham et al., 2024	✗	✓	✓	✗	✓
Sultan Hajam, 2024	✓	✓	✓	✗	✗
Ibrahim et al., 2024	✓	✗	✓	✗	✗
Deepak et al., 2025	✓	✗	✓	✓	✗
BALBA-SD (Proposed)	✓	✓	✓	✓	✗

by them. On the other hand, Ibrahim et al. [13], for instance, do not concern themselves with budgets but focus a lot on deadlines.

In addition, solutions from this point on do not address both edge- and fog-computing enabler environments. Bukhari et al. [6] and Deepak et al. [17] stand as examples of algorithms with partial coverage of fog and edge but without end-to-end coverage. But in the study we highlight that BALBA-SD is the only solution included in the survey that merges budget-consciousness and strict deadline scheduling while allowing the assignment of tasks to fog and edge devices. These help mark it as an innovative solution with practical applications for solving the complex needs of modern distributed computing scenarios.

3 System Model

In a fog and edge computing environment, tasks can be submitted for execution at any time and may leave the system upon completion. In this complex scenario, a budget-aware load balancing policy is required to efficiently allocate incoming tasks to suitable computational nodes while satisfying performance parameters and adhering to cost constraints. The problem is represented as:

$$Y = \mathrm{f}(\mathrm{x_i}),$$

where $\mathrm{x_i}$ represents the task attributes (e.g., size, deadline, and budget), and Y represents the performance parameters, which include cost, execution time, and resource utilization. The rest of the notations are presented in Table 2. The deadline is a hard constraint that must be satisfied, while the budget is a limiting factor that influences task allocation decisions. On each computational node (edge node or fog node), only one task is allowed to run on a node at a particular time. In this algorithm first task will be allocated to the edge node and will migrate to the fog node (if the deadline constraint is not satisfied or the node is overloaded). This algorithm aims to minimize the total cost of task allocation while satisfying the following constraints:

- $\sum_{i \in m} EC(i) \leq B$ (Total cost should not exceed the budget).
- $\forall i \in T$, completion time of $T_i \leq D_i$ (Tasks must meet their deadlines).

Table 2. Notation used

Symbols	Meaning
m	Number of Tasks
S_i	Size of task.
C_j	Computing capacity of the jth edge node
C_k	Computing capacity of the kth fog node
$Cost_{ij}$	Execution Cost of task T_i on node E_j
$Cost_{ik}$	Execution Cost of task T_i on node F_k
ETC_{ij}	Expected time to compute task T_i on node E_j
ETC_{ik}	Expected time to compute task T_i on node F_k
N	A set of computational nodes
N_e	Total number of edge nodes
N_f	Total number of fog nodes
E_j	j^{th} edge node
F_k	k^{th} fog node
D_i	Deadline of task T_i
T	Set of tasks
T_i	i^{th} task {where $1 \leq i \leq m$}
Alloc	Allocation list of task T_i
B	Total budget available for execution(in Rs.)
EC(i)	Execution cost
AB	Average Budget
RB	Remaining Budget
L_j	Load on the jth node
BW	Bandwidth of the link between nodes

3.1 Parameter Equation

This section presents key QoS parameters along with their corresponding equations, which are used to compute and analyze system performance metrics.

Expected Time to Compute (ETC). ETC refers to the estimated amount of time a specific task will take to execute on a particular computational resource, considering the task's workload and the processing capability of the resource.

$$ETC_{ij} = \frac{S_i}{C_j} or\ ETC_{ik} = \frac{S_i}{C_k} \tag{1}$$

Average Expected Time to Compute (AveETC). The mean of the expected computation times for tasks across all resources.

$$\text{AveETC}\ (T_i) = \frac{\sum_{\text{j}=1}^{\text{N}_e} ETC_{ij}}{N_e} or\ \text{AveETC}\ (\text{T}_i) = \frac{\sum_{\text{k}=1}^{N_f} ETC_{ik}}{N_f} \tag{2}$$

Computing Cost. Computing Cost refers to the monetary or resource expense incurred when executing a task on a computational node (edge or fog).

$$Cost_{ij} = ETC_{ij} \times Cost_j\ or\ Cost_{ik} = ETC_{ik} \times Cost_k \tag{3}$$

Average Budget. The average budget is the average of the sum of the execution cost of each task on the edge and fog node.

$$\text{AB} = \frac{\sum_{\text{j}=1}^{\text{m}} Cost_{ij} + \sum_{\text{k}=1}^{\text{m}} Cost_{ik}}{\text{N}} \tag{4}$$

Budget. The budget represents the sum of the average budget of all tasks.

$$B = \sum_{\text{i}=1}^{\text{m}} \text{AB} \tag{5}$$

Maximum Expected Time to Compute (maxETC). The highest estimated time required to complete any task across all available resources.

$$maxETC(T_i) = \max\left(\text{ETC}_{ij}\right) \forall j \in N \tag{6}$$

Average Load (AL). AL refers to the mean computational workload distributed across all available resources in a system. It indicates how evenly the workload is shared and is commonly used in evaluating the efficiency of load balancing strategies.

$$\text{AL}_{\text{edge}} = \frac{\sum_{\text{i}=1}^{\text{m}} \text{AveETC}\ (\text{T}_i)}{N_e} \text{or}\ \text{AL}_{\text{fog}} = \frac{\sum_{\text{i}=1}^{\text{m}} \text{AveETC}\ (T_i)}{N_f} \tag{7}$$

Deadline. The deadline of a task refers to the latest allowable time by which a task must be completed to meet system or application-specific requirements. It is a critical constraint in real-time, where timely execution impacts service quality. The deadline of task T_i is given by

$$D(T_i) = K * \sum_{\text{i}=1}^{\text{m}} \sum_{\text{j}=1}^{N_e} \max\left(ETC_{ij}\right) \tag{8}$$

The values of the parameter "K" can be 1, 2, 3, 4, and so on. In this case, we assume that K is equal to 2 [18].

Migration Cost (MCt). MCt refers to the overhead incurred when transferring a task or process from one computational resource (e.g., an edge or fog node) to another.

$$MCt(T_i) = \frac{S_i}{BW} \tag{9}$$

Completion Time (CT). The actual time taken for a task to finish execution on a specific resource.

$$CT_{ik} = MCt(T_i) + ETC_{ik} \tag{10}$$

Deadline Meet (DM). The extent to which tasks are completed within their specified deadlines [18].

$$DM(T_i) = \begin{cases} 1 \text{ if } CT(T_i) \leq D(T_i) \\ 0 \text{ otherwise} \end{cases} \tag{11}$$

Turnaround Time (TAT). It refers to the total time taken to complete a process or fulfill a request from the moment it is submitted until it is completed.

$$TAT = \max(CT_{ik}) \tag{12}$$

4 Budget-Aware Load Balancing with Strict Deadline Constraints (BALBA-SD)

The **BALBA-SD** algorithm is designed to achieve efficient task allocation while adhering to both budget and deadline constraints.

Algorithm: BALBA-SD

Input: $T, N, S_i, C_j, C_k, Cost_i, Cost_k$
Output: Alloc(T_i, E_j), TAT, DM, EC.

Begin:

1. **Initialize**: Compute ETC_{ij}, $Cost_{ij}$, AveBudget, and B for all tasks across all nodes using equations (1), (2), (3), and (4).
2. **Sort Tasks**: Arrange tasks T in ascending order of D_i
3. Set Alloc=∅, EC=0, and compute initial Load(E_j) for all E_j
4.   **while** (T≠∅):
5.   Select the task T_i with the earliest deadline from T
6.   **Eligible Nodes Identification**: Identify nodes E_j such that $ETC_{ij} \leq D_i$
7.   **if** (Eligible Nodes ≠∅):
8. For each E_j:

```
        Compute Cost_ij
        Calculate Load(E_j)
        Select the node minE_j offering the least Cost_ij
        Assign T_i to minE_j
        Update EC and Load(minE_j)
          Remove T_i from T
    else (No eligible nodes):
        Migrate T_i to the fog node F_k with the minimum ETC_ik, if Cost_ij ≤ B
  End if
  If no such fog node exists, mark T_i as unallocated
End while
Compute Average Load and TAT                    // As per equations (7) and (12)
Calculate the total execution cost EC
Verify EC ≤ Budget
Return: Alloc(T_i, E_j), TAT, DM and EC
```

End

4.1 Illustrative Example

An example is demonstrated stepwise to show the functioning of the proposed algorithm. For this example, we have four edge nodes and two fog nodes with 10 IoT tasks. The information about nodes and tasks is presented in Tables 3 and 4, respectively.

Table 3. Characteristics of tasks

Tasks	T_1	T_2	T_3	T_4	T_5	T_6	T_7	T_8	T_9	T_{10}
Size (MIs)	300	250	200	270	180	240	210	270	150	220

Table 4. Computational nodes information

Nodes	E_1	E_2	E_3	E_4	F_1	F_2
Computation Capacity (MIPS)	100	80	60	50	120	150
Cost per computational unit ($)	2	2	3	3.5	10	15
Bandwidth(Mbps)	-	-	-	-	80	100

In the first step, calculate the ETC_{ij}, maxETC, Deadline, $Cost_{ij}$, and AB using Eqs. (1), (6), (8), (3), (4), respectively, represented in Table 5. After that, tasks are arranged in the ascending order of increasing their deadlines and calculate the Budget using Eq. (5), that is $B = 132.94$.

The basic fundamental of this algorithm is to assign tasks based on the least cost with strict deadline conditions and load balancing conditions ($L_j \leq AL$). So initially, tasks will be allocated to the edge node with satisfies the condition of cost, deadline, and load balancing. If the condition is not met, the task will migrate to the fog node.

Average ETC and AL on the edge nodes are calculated using Eqs. (2) and (7), respectively. $AL_{edge} = (4.44 + 3.70 + 2.96 + 3.99 + 2.66 + 3.55 + 3.11 + 3.99 + 2.22 + 3.25)/4 = 8.4675$. As shown in Table 6, the first task T_9 is allocated to node E_1 because min $Cost_{91} = 3 \leq B$, $ETC_{91} \leq D_9$ $(1.50 \leq 6.00)$ and load on node $E_1 \leq AL$. In task allocation, minimizing execution cost is the first priority with the deadline meet condition. If any task has the least cost but does not meet the deadline, in this case, we choose the next least cost, but the deadline condition should be met as well as $L_j \leq AL$. Allocation of all tasks is shown in Table 6 with green color cells, and there is one task T_1 which satisfies cost and deadline conditions and is assigned to node $E_{1,}$ but either assigned to either E_1 or any other edge node, node is being overloaded $(L_j \leq AL)$. So this task will migrate to fog node $F_{1,}$ shown in Table 7.

When a task migrates from an edge node to a fog node migration cost and completion time for the fog node are calculated using Eqs. (9) and (10) as follows:

$$CT_{11} = MCt(T_1) + ETC_{11}\Big) = \frac{300}{80} + 2.50 = 6.25$$

$$CT_{12} = MCt(T_1) + ETC_{12} = \frac{300}{100} + 2 = 5$$

From Tables 6 and 7, the allocation of tasks and load on each edge node and fog node is as follows:

$$Load(E_1) = ETC_{91} + ETC_{31} + ETC_{61} = 1.50 + 2.00 + 2.40 = 5.90.$$
$$Load(E_2) = ETC_{52} + ETC_{72} + ETC_{22} = 2.25 + 2.63 + 3.13 = 8.01.$$
$$Load\ (E_3) = ETC_{103} + ETC_{83} = 3.67 + 4.50 = 8.17.$$
$$Load\ (E_4) = ETC_{44} = 5.40.$$
$$Load\ (F_1) = ETC_{11} = 6.25.$$

Now from Table 6, Total $EC(i) = 3 + 4.5 + 7 + 9.75 + 11 + 11.8 + 16 + 18.9 + 24.5 + 25 = 131.45 \leq B\ (132.94)$. So total execution cost is 131.45, which is less than the total Budget, and the remaining budget is 1.49. From Tables 5 and 6, $TAT = 8.17 + 6.25 = 14.42$ and $DM = 9$.

5 Simulation and Results

The proposed Budget-Aware Load Balancing Algorithm with Strict Deadline Constraints (BALBA-SD) was evaluated in a simulated fog and edge computing environment. The simulation was conducted with varying task sizes (100, 250, 500, and 1000 tasks) to analyze the Turnaround Time (TAT), Deadline Meet (DM), and Execution Cost (EC) performance. The results were compared against two baseline algorithms: REAL [16] and DCLB [17].

As shown in Figs. 1 to 3, all of the techniques under consideration are evaluated in this case by varying the number of tasks from 100 to 1000 while keeping the system's

Table 5. Matrix table of ETC and Cost

ETC										Cost						
Task	Size (MB)	E_1 (100)	E_2 (80)	E_3 (60)	E_4 (50)	Max ETC	D	F_1 (120)	F_2 (150)	E_1 (2)	E_2 (2)	E_3 (3)	E_4 (3.5)	F_1 (10)	F_2 (15)	AB
T_9	150	1.50	1.88	2.50	3.00	3.00	6.00	1.25	1.00	3	3.75	7.5	10.5	12.5	15	8.71
T_5	180	1.80	2.25	3.00	3.60	3.60	7.20	1.50	1.20	3.6	4.5	9	12.6	15	18	10.5
T_3	200	2.00	2.50	3.33	4.00	4.00	8.00	1.67	1.33	4	5	10	14	16.7	20	11.6
T_7	210	2.10	2.63	3.50	4.20	4.20	8.40	1.75	1.40	4.2	5.25	10.5	14.7	17.5	21	12.2
T_{10}	220	2.20	2.75	3.67	4.40	4.40	8.80	1.83	1.47	4.4	5.5	11	15.4	18.3	22	12.8
T_6	240	2.40	3.00	4.00	4.80	4.80	9.60	2.00	1.60	4.8	6	12	16.8	20	24	13.9
T_2	250	2.50	3.13	4.17	5.00	5.00	10.00	2.08	1.67	5	6.25	12.5	17.5	20.8	25	14.5
T_4	270	2.70	3.38	4.50	5.40	5.40	10.80	2.25	1.80	5.4	6.75	13.5	18.9	22.5	27	15.7
T_8	270	2.70	3.38	4.50	5.40	5.40	10.80	2.25	1.80	5.4	6.75	13.5	18.9	22.5	27	15.7
T_1	300	3.00	3.75	5.00	6.00	6.00	12.00	2.50	2.00	6	7.5	15	21	25	30	17.4
															B = 132.9	

Table 6. Allocation table of tasks to edge nodes

ETC							Cost				
Task	Size (MB)	E_1	E_2	E_3	E_4	D	E_1	E_2	E_3	E_4	RB
T_9	150	**1.50**	0.00	0.00	0.00	6.00	**3**	0	0	0	129.9
T_5	180	3.30	**2.25**	0.00	0.00	7.20	6.6	**4.5**	9	0	125.4
T_3	200	**3.50**	4.75	3.33	0.00	8.00	**7**	9.5	10	0	118.4
T_7	210	5.60	**4.88**	3.50	0.00	8.40	11.2	**9.75**	10.5	0	108.7
T_{10}	220	5.70	7.63	**3.67**	0.00	8.80	11.4	15.25	**11**	0	97.69
T_6	240	**5.90**	7.88	7.67	4.80	9.60	**11.8**	15.75	23	16.8	85.89
T_2	250	8.40	**8.01**	7.84	5.00	10.00	16.8	**16**	23.5	17.5	69.89
T_4	270	8.60	11.39	8.17	**5.40**	10.80	17.2	22.75	24.5	**18.9**	50.99
T_8	270	8.60	11.39	**8.17**	10.80	10.80	22.6	22.75	**24.5**	37.8	26.49
$\mathbf{T_1}$	300	**8.90**	11.76	13.17	11.40	12.00	**23**	23.5	39.5	39.9	3.49

Table 7 Migration of the task to fog nodes

ETC					Cost		
Task	Size (MB)	F_1	F_2	D	F_1	F_2	RB
T_1	300	**6.25**	5.00	**12.00**	**25**	30	**1.49**

remaining input parameters constant. $RT_j = 1–100$, Cap (E_j and F_k) = 1–100 (MIPS), $W(T_i) = 1–10{,}000$ (MIs), and $n = 16$ are the additional input parameters.

As shown in Fig. 1, BALBA-SD performs slightly worse than DCLB, with 2.37% to 2.50% higher TAT, indicating DCLB's efficiency advantage. However, BALBA-SD outperforms REAL, achieving a 9.59% improvement at 250 tasks. For larger workloads (500 and 1000 tasks), BALBA-SD and REAL show minimal differences (0.15%–0.19%), suggesting diminishing gains. While BALBA-SD remains competitive with REAL, it does not surpass DCLB, highlighting areas for further optimization.

As represented in Fig. 2, BALBA-SD demonstrates consistent improvement over both DCLB and REAL in meeting deadlines across all job sizes. Compared to DCLB, BALBA-SD shows an improvement of 12.19% to 19.04% for most cases, with a slight drop to 3.188% at 1000 tasks, indicating that DCLB catches up as workload increases. However, BALBA-SD significantly outperforms REAL, achieving up to 53.33% improvement for 250 tasks, proving its efficiency in strict deadline scenarios. At higher workloads (1000 tasks), BALBA-SD still leads but with a reduced margin (18.12%), suggesting that while its scalability is strong, optimization may be needed for extreme loads.

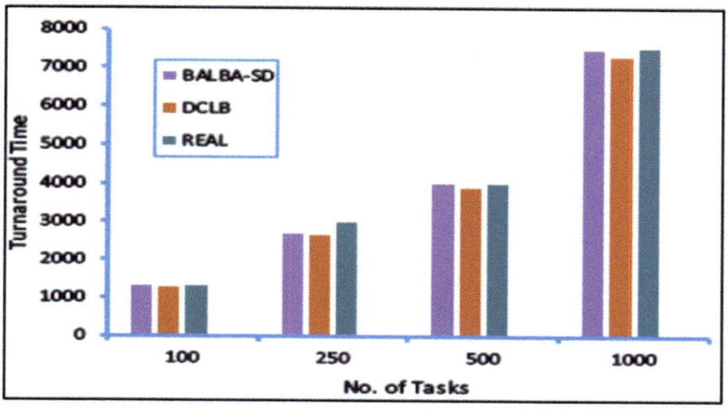

Fig. 1. Turnaround Time Vs No. of Tasks

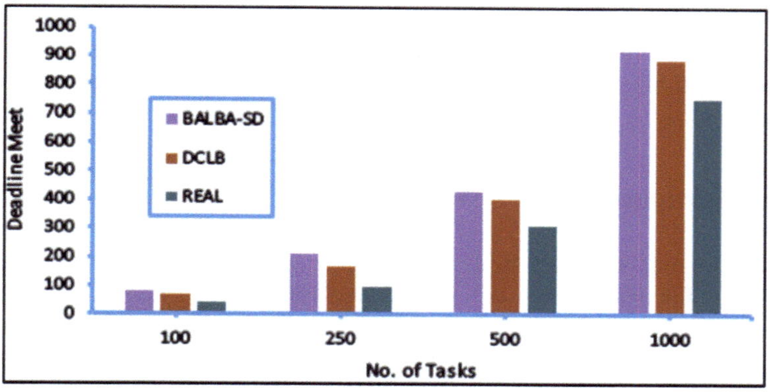

Fig. 2. Deadline Meet Vs No. of Tasks

BALBA-SD achieves notable cost savings over both DCLB and REAL, as shown in Fig. 3, making it a more budget-efficient approach. Compared to DCLB, BALBA-SD reduces execution costs by 3.15% to 11.90%, demonstrating steady efficiency in resource utilization. However, its savings over REAL are more significant, reaching up to 21.79% for smaller workloads and around 19% for larger workloads, showing strong cost-effectiveness, especially at higher job loads. While BALBA-SD maintains lower costs throughout, its advantage over DCLB remains moderate, suggesting that DCLB is also cost-conscious but less efficient than BALBA-SD.

The results discussed above show that BALBA-SD provides an optimal trade-off among turnaround time, deadline satisfaction, and cost of execution. While DCLB shows a minimal improvement in turnaround time (TAT), BALBA-SD significantly better performs than REAL, especially for moderate-sized workloads. In the context of deadline satisfaction, BALBA-SD always outperforms both DCLB and REAL, demonstrating its efficiency in satisfying strict deadlines, especially for moderate workloads. In addition, BALBA-SD exhibits significant cost reductions, up to 21.79% lower than REAL, and is a cost-effective alternative. While DCLB is still competitive in TAT and cost,

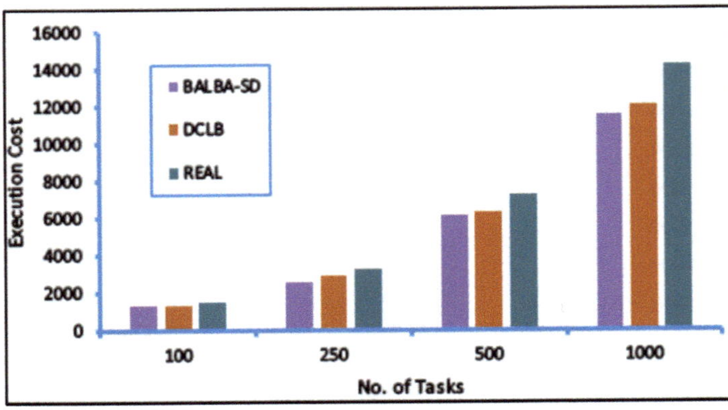

Fig. 3. Execution Cost Vs No. of Tasks

BALBA-SD's overall performance in meeting deadlines and budget responsiveness is a top contender for deadline-aware load balancing in fog and edge computing. Future work should optimize the performance of BALBA-SD for larger workloads to increase scalability.

6 Conclusion and Future Work

This study presents a budget-aware load-balancing mechanism with strict deadlines (BALBA-SD) for optimization of execution cost in Fog and Edge Computing environments with hard deadline requirements. Simulations show that BALBA-SD is far superior to traditional load-balancing methods in improving deadline satisfaction by 53.33%, and lowering execution costs by 21.79%, thus an extremely efficient choice for budget-constrained in-time applications like IoT, autonomous systems, and healthcare. Even though BALBA-SD is very scalable and efficient in terms of resources, there is scope for enhancement to make it more adaptive. Future work will focus on incorporating machine learning-based techniques for dynamic scheduling of tasks and predictive resource allocation to further enhance performance. Real-world application on large-scale distributed systems will also be essential to validate its applicability.

References

1. Shiriaev, E., Ermakova, T., Bezuglova, E., Lapina, M.A., Babenko, M.: Reliability and security for fog computing systems. Information. **15**(6), 317 (2024)
2. CV, N.A., Lavanya, R.: Fog computing and its role in the internet of things. In: Advancing Consumer-Centric Fog Computing Architectures, pp. 63–71. IGI Global (2019)
3. Raza, S., Wang, S., Ahmed, M., Anwar, M.R.: A survey on vehicular edge computing: architecture, applications, technical issues, and future directions. Wirel. Commun. Mob. Comput. **2019**(1), 3159762 (2019)
4. Mutlag, A.A., Abd Ghani, M.K., Arunkumar, N.A., Mohammed, M.A., Mohd, O.: Enabling technologies for fog computing in healthcare IoT systems. Futur. Gener. Comput. Syst. **90**, 62–78 (2019)

5. Postoaca, A.V., Negru, C., Pop, F.: Deadline-aware scheduling in cloud-fog-edge systems. In: In 2020, 20th IEEE/ACM International Symposium on Cluster, Cloud and Internet Computing (CCGRID), pp. 691–698. IEEE (2020)
6. Bukhari, S.S.H., Kabir, A., Younus, M.U., Anjum, M.R., Narejo, S.: Novel cost efficient resource allocation technique based on deadline and budget constraints for edge users. Wirel. Pers. Commun. **120**(1), 269–286 (2021)
7. Pham, P.H., Binh, N.T., Alam, M.G.R., Islam, M.M.: Optimizing cloud-fog workloads: a budget aware dynamic scheduling solution. In: International Conference on Computational Science and its Applications, pp. 31–46. Springer Nature Switzerland, Cham (2024)
8. Ding, X., Guo, J., Li, D., Wu, W.: Pricing and budget allocation for IoT blockchain with edge computing. IEEE Trans. Cloud Comput. **11**(2), 1608–1621 (2022)
9. Liu, M., Liu, Y.: Price-based distributed offloading for mobile-edge computing with computation capacity constraints. IEEE Wirel. Commun. Lett. **7**(3), 420–423 (2017)
10. Yao, H., Bai, C., Xiong, M., Zeng, D., Fu, Z.: Heterogeneous cloudlet deployment and user-cloudlet association toward cost-effective fog computing. Concurrency Comput. Pract. Exper. **29**(16), e3975 (2017)
11. Azizi, S., Shojafar, M., Abawajy, J., Buyya, R.: Deadline-aware and energy-efficient IoT task scheduling in fog computing systems: a semi-greedy approach. J. Netw. Comput. Appl. **201**, 103333 (2022)
12. Sultan Hajam, S.: Deadline-cost aware task scheduling algorithm in fog computing networks. Int. J. Commun. Syst. **37**(6), e5695 (2024)
13. Ibrahim, M., Lee, Y., Kim, D.H.: DALBFog: deadline-aware and load-balanced task scheduling for the internet of things in fog computing. IEEE Syst. Man Cybern. Mag. **10**(1), 62–71 (2024)
14. Bothra, S.K., Singhal, S., Goyal, H.: Deadline-constrained cost-effective load-balanced improved genetic algorithm for workflow scheduling. Int. J. Inf. Technol. Web Eng. **16**(4), 1–34 (2021)
15. Arshed, J.U., et al.: GA-IRACE: genetic algorithm-based improved resource aware cost-efficient scheduler for cloud fog computing environment. Wirel. Commun. Mob. Comput. **2022**(1), 6355192 (2022)
16. Alam, M., Haidri, R.A., Shahid, M.: Resource-aware load balancing model for batch of tasks (BoT) with best fit migration policy on heterogeneous distributed computing systems. Int. J. Pervasive Comput. Commun. **16**(2), 113–141 (2020)
17. Deepak, Upadhyay, M.K., Alam, M.: Deadline-constrained based load balancing algorithm for batch of IoT tasks on fog and edge computing environments. Iran J. Comput. Sci. **8**(2) (2025)
18. Haidri, R.A., Alam, M., Shahid, M., Prakash, S., Sajid, M.: A deadline aware load balancing strategy for cloud computing. Concurrency Comput.: Pract. Exper. **34**(1), e6496 (2022)
19. Athavale, V.A., Ali, J., Patel, S.T., Lapina, M.: Adoption of industry 4.0 in the automotive industry contributes to a reduction in energy consumption during the production process. In: The International Conference on Recent Innovations in Computing, pp. 621–634. Springer Nature Singapore, Singapore (2023)
20. Lapina, M.A., Maksimova, E.A., Lapin, V.G., Boikov, N.S.: Analytical review of methods for designing e-health security systems. Proc. Inst. Syst. Program. RAS. **36**(5), 191–218 (2024)

Enhancing Real Estate Asset Transfers with Blockchain: A Decentralized Protocol Approach

Rashmi Kamlakar Dixit[1], Vijay Anant Athavale[2](✉), L. M. R. J. Lobo[2], and Vitalii Lapin[3]

[1] Department of Computer Science and Engineering, Walchand Institute of Technology, Solapur, India

[2] Department of Information Technology, Walchand Institute of Technology, Solapur, India
vijay.athavale@gmail.com

[3] North Caucasus Federal University, Stavropol, Russia

Abstract. Traditional methods of managing and transferring land assets have depended on centralized systems, which are susceptible to fraud, inefficiency, and cyber-attacks. This study suggests a decentralized blockchain-based system for safe and open transfers of real estate assets. Through the use of blockchain technology, the system ensures an unchangeable and trust less land registration by doing away with the need for a middleman. Through the use of smart contract-based purchasing and selling protocols, users may safely register, validate, and manage their land assets on a decentralized ledger. The suggested solution greatly lowers the danger of fraud and unauthorized changes while improving data efficiency, security, and integrity. Furthermore, property ownership data are protected from external alterations and cyber threats by the incorporation of cryptographic techniques. The study assesses how well the system works to guarantee transaction transparency, cut down on processing time, and boost confidence in real estate transactions. The results demonstrate how blockchain technology can completely transform land governance by providing a scalable and impenetrable framework for managing real estate assets.

Keywords: Land Registration · Blockchain · Buying Selling Protocol · Secure Induction · Asset Transfer

1 Introduction

One of the most important and precious assets in society is land ownership, which calls for safe, open, and impenetrable record-keeping. To preserve ownership records and streamline transactions, traditional land management and title registration systems depend on centralized authority, such as government agencies [1]. These centralized systems, however, are vulnerable to fraud, inefficiency, illegal changes, and transaction processing delays. Transferring titles and confirming land ownership may include laborious documentation, reliance on middlemen, and the possibility of mistakes or disagreements.

M. Lapina et al. (Eds.): AISMA 2025, LNNS 1456, pp. 120–130, 2026.
https://doi.org/10.1007/978-3-032-07275-7_12

The decentralized and secure ecosystem provided by blockchain technology has the potential to revolutionize the transfer and maintenance of land ownership records [2]. In contrast to traditional databases, blockchain ensures that records cannot be altered once they are stored by offering immutability, transparency, and cryptographic security. Blockchain makes it possible for automated, safe, and real-time transactions to occur without the need for middlemen by utilizing distributed consensus techniques and smart contracts [3, 4].

By ensuring that land titles are readily available and permanently preserved, a blockchain-based land registry system enables potential buyers to confirm ownership history without the need for middlemen [5]. This method expedites the property transfer procedure, avoids fraudulent sales, and drastically lowers transaction costs. Furthermore, protocol-based limitations maintain the integrity of stored data, and public-key cryptography guarantees the legitimacy of ownership.

This study suggests a decentralized land asset management system built on blockchain that improves accessibility, security, and data integrity [6]. Burden related to administration is now taken away. The third-party involvement in verification is done away with and a robust hassle-free solution is offered with respect to land related transactions. In this paper the use of Blockchain as an asset technology for land related transactions is presented bring about a drastic and affordable change in real estate governance [7].

1.1 Drawbacks of the Existing Land Registry System

The existing system for land revenue being centralized and having a large human intervention causes suspicions and inefficiencies. The incapacity of land registrars to confirm any outstanding debts or encumbrances on a property before authorizing transactions is one of the main obstacles. There is a greater chance of fraudulent transactions and disputes since registrars are forced to depend only on the paperwork that purchasers and sellers submit. Land record repair is a time-consuming and expensive procedure because of the system's additional complexity brought about by corruption, unauthorized ownership claims, and errors in the records.

1.2 Bhoomi Project Case Study

Enhancing accessibility and openness in land ownership data was the goal of the Bhoomi project, an initiative for digital land record administration in Karnataka, India [8]. However, uploading erroneous or fraudulent land records was a significant flaw in the system. Ownership disputes and, in certain situations, the incorrect distribution of land resulted from the project's absence of a reliable system to compare outdated records with updated information. It is still very difficult to guarantee data correctness, immutability, and a single source of truth in the absence of a safe and impenetrable system, like blockchain-based land registration. A decentralized, unchangeable database made possible by blockchain technology can be used to permanently retain land records, stop fraudulent revisions, and facilitate real-time verification. It is possible for the Governments to reduce un-necessary claims and reduce the need of verification from a third party thus guarantying titles of property. This also makes the systems transparent to all concern using block chain technology for achieving robust administration of land.

It is observed in the recent past that there are a continuous increase in fraudulent transactions caused in registration of properties. This causes changes that are unauthorized and leads to miss management caused due to rapid use of technology. The introduction of a middle man in transactions has time lapses for both the buyer and seller. The present existing system becomes a pray due to centralization giving into manipulations and resulting in conflicts of ownership. Utilizing Ethereum Blockchain technology to include a decentralized method into the existing land registration system to overcome these obstacles is suggested. Smart contracts self-executing agreements with predetermined rules that automate land transactions, ownership verification, and record changes are made possible by Ethereum. By ensuring that transactions are only carried out when all requirements are satisfied, these smart contracts do away with the need for intermediaries and lower the possibility of fraud. It is then that only fixed relevant transactions are added to the blockchain using Ethereum's Proof of Stake (PoS). This makes the system safe without permitting changes and penetration of unwanted data. The method protects publicly available data considered for auditing of ledger related to land making the process more cost effective in transactions related to property transfer and ownership.

2 Literature Review

For a long time, India's land records have been ambiguous, ineffective, and prone to fraud. It is still very difficult to guarantee data integrity, consistency, and real-time synchronization among several government agencies [9].

The use of blockchain technology in land records management has been investigated by several researchers to increase efficiency, security, and transparency. Blockchain was first presented by the author as a decentralized ledger that securely records transactions and removes single points of failure [10]. To provide tamper resistance, every transaction is kept in immutable blocks that are cryptographically connected to the one before it. By guaranteeing that once a land record is recorded, it cannot be changed without network participants' consent, Gupta and Rani presented a blockchain-based land registration system that lowers fraudulent transactions. According to Singh et al., conventional land registry systems depend on centralized authorities, making them susceptible to data manipulation and breaches [11, 12]. To improve inter-departmental synchronization a blockchain entanglement model was presented by Shinde et.al [13]. In this model, each department keeps its blockchain ledger, but updates in one department causes corresponding updates in other departments. Additionally, the technological features of hash-based security procedures in blockchain-led land registries were investigated by Mukherjee et al. [14]. Their study demonstrated how cryptographic hashing and blockchain append-only design guard against unwanted changes, guaranteeing data integrity and long-term security. The technical features of cryptographic hashing in blockchain-based land registries were investigated. Their study focused on how the append-only nature of blockchain ensures data security and integrity by preventing unwanted changes.

Blockchain usage in land register systems has accelerated due to its effectiveness in supply chain management and financial transactions. According to research by Patel and Gupta, using smart contracts based on Ethereum can automate ownership transfers, lower transaction costs, and do away with the need for intermediaries [15]. Das and Mehta compared permission ledgers like Hyperledger Fabric with public blockchains like Ethereum to examine public and private blockchain models for land registration [16].

Mishra et al. emphasized the difficulties in judicial recognition of smart contracts in India's legal system, regulatory frameworks, and policy adaption [17]. In their discussion of the legal ramifications of blockchain-based land registries,

By examining case studies from Sweden, Ghana, and the United Arab Emirates, Choudhary and Singh offered a comparative analysis of blockchain-based land registry implementations worldwide [18]. In order to identify irregularities, anticipate fraudulent activity, and increase the accuracy of land assessment, Patel and Deshmukh proposed combining blockchain-based registries with AI and machine learning algorithms [19].

Oprunenco and Akmeemana discussed how blockchain can enhance the reliability of land registries in India by providing an immutable ledger that prevents unauthorized modifications [20]. The study emphasized the role of decentralization in eliminating intermediaries and reducing corruption in land transactions. Similarly, Mougayar presented a decision tree for blockchain applications, outlining how blockchain can be applied to address inefficiencies in land registration systems [21]. Wood introduced the Ethereum blockchain, which supported smart contracts for automated and secure transactions [22]. Li et al. proposed an enterprise blockchain framework suitable for land records, offering privacy-preserving mechanisms for sensitive ownership data [23]. Mohanty discussed the deployment of smart contracts on Ethereum, emphasizing their role in reducing transactional inefficiencies and ensuring secure land transfers [24].

Garzik and Donnelly provided a foundational understanding of blockchain technology, highlighting its advantages in preventing land record fraud [25]. In support, Vigna and Casey examined how cryptocurrencies and blockchain applications are reshaping economic systems, including real estate transactions [26].

Buterin et al. further elaborated on Ethereum's capabilities in facilitating decentralized applications, which played a crucial role in automating land registration processes [27].

Mukhopadhyay explored Ethereum-based smart contracts for land registry management, highlighting their efficiency in ensuring tamper-proof property transfers [28]. The importance of blockchain test networks is also emphasized in The Beginner's Guide to Using an Ethereum Test Network, which explains how test environments can be used to evaluate blockchain implementations before deployment [29, 30].

3 Methodology

The absence of correct and verifiable information in property papers is a major contributor to land-related fraud, including title fraud. Data immutability once recorded, no entity, even the government, can change or tamper with the information ensures data integrity when blockchain technology is used.

By doing away with middlemen in real estate transactions, the suggested approach enables buyers and sellers to communicate directly. Many legal gaps in the current system go overlooked by the general public, but intermediaries frequently take advantage of them to commit fraud. The procedure is made safer and more transparent by eliminating these third parties.

Additionally, the distinctive digital representation of property assets is made possible by the use of Non-Fungible Tokens (NFTs). Every property is given an NFT, guaranteeing a globally identifiable and impenetrable identity for the item and its legitimate owner. This method makes property verification across jurisdictions easier, avoids duplication, and increases trust.

Let's look at the protocol for this land registration management system. With this strategy, it is required to completely decentralize the system. However, the current database system for the land registration management system presents a serious problem because it uses a decentralized approach to store and manage data that is required by the government. The government must extract all of the data from the central database and install it on a blockchain network to decentralize the current land registration system. So, practically speaking, it is the user's job to spread their land asset across the blockchain network by making a tiny financial investment. Users must register for an account on our website as the initial step in our system. Once the user has registered, the user's properties will be displayed on the web platform. This method is comparable to Digi-Locker, in which the user registers on the portal and obtains all essential documents related to his identity. However, by adding a special component that confirms the user, our system is now more secure. In essence, the user must first authenticate after registration before all of the user's properties are shown. The user is directed to a nearby government facility linked to the system for verification, where he is validated by the concerned government officer.

The offered land registration administration system uses blockchain technology to accomplish total decentralization. Decentralization necessitates moving all government-managed data onto a blockchain network, which presents a big problem for the current centralized database used for land records. With a small cash outlay, users must actively participate in registering their land assets on the decentralized network to speed up this transformation (Fig. 1).

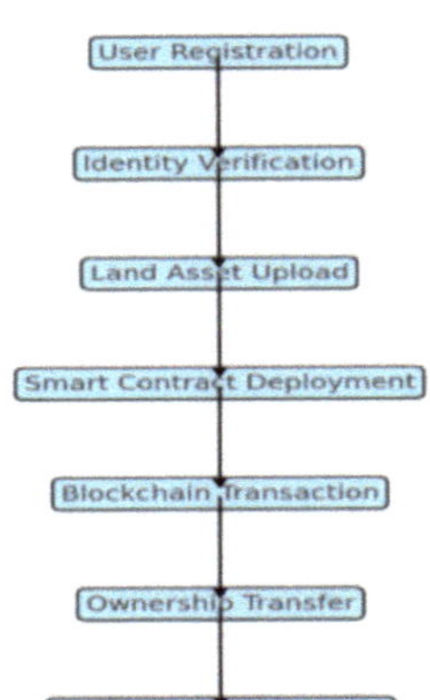

Fig. 1. Blockchain-based land registration system.

3.1 Verification and User Registration

Account Creation: In order to use the system, users must first create an account on the web platform.
Property Linking: Users can access and manage their land assets on the platform after registering.
Identity Verification: The system includes a safe verification procedure to guarantee authenticity. In contrast to traditional platforms like Digi Locker, which only store identity papers, our system incorporates an extra protection layer.
Authentication by Government: After registration is completed a physical appearance of a user is carried out at a specific location specified by the government. This location is connected to a block chained network. The identity of user and ownership of property is checked by a government authority before registration.

3.2 Decentralized Data Management

The user's land records are saved on the blockchain after verification, guaranteeing immutability and guarding against unwanted changes. A distinct digital identifier is given to every land asset, facilitating transactions that are transparent and impervious to fraud.

Every transaction is tracked by the blockchain ledger, enabling users to safely purchase, sell, or transfer real estate without the need for middlemen.

This decentralized approach ensures secure, transparent, and tamper-proof land registration while reducing bureaucracy and eliminating fraudulent activities.

As shown in above Fig. 2, A decentralized Ethereum blockchain application for real estate transactions is depicted in the diagram. Considering the primary three layered Ethereum Blockchain, where smart contracts are considered to be carried out as transactions using an Ethereum Virtual Machine (EVM) which works on a Frontend offering a good experience to the user a backend function to control application and interfaces with the Ethereum network. Thus, Land records are safely stored on the blockchain. The procedure that uses blocks guarantees a transparent and immutable system. Thus, the chance of Fraud and implication of middle man is reduced.

To guarantee decentralization, security, and immutability, users safely transfer their land holdings into the Ethereum blockchain in the proposed blockchain-based land registration system. A user's linked land records are not immediately available on the blockchain after they register on the site. The user must first link their Meta mask wallet to the platform and deposit a tiny transaction charge (gas fee) in the form of Ether (ETH) to integrate their land asset into the decentralized system.

The land asset's deployment onto the Ethereum blockchain is made easier by this charge. The deploy function removes the possibility of fraud and unauthorized changes by permanently recording the land asset and the data related to it on the blockchain. A frequent question is why, in the absence of government funding, people should be responsible for paying to upload their land holdings to the blockchain. Protecting land against fraud, cyber threats, and unauthorized changes is essential since it is a very valuable asset that is frequently worth lakhs or crores of rupees. Blockchain inherently supports these features of providing security to the system with a minimum charge

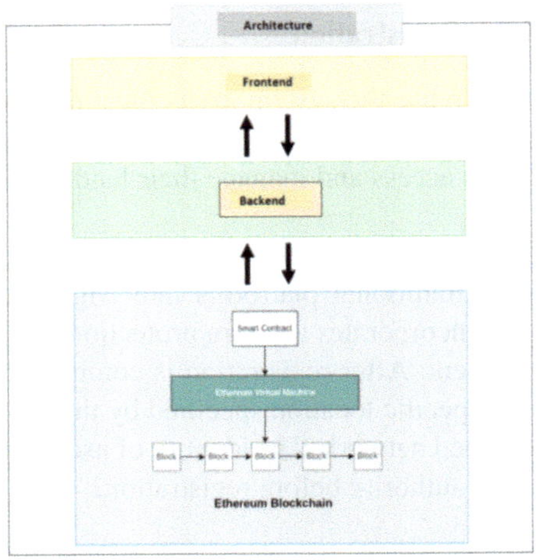

Fig. 2. System Architecture.

needed for deployment as compared to the actual cost of land. After deployment a Non-Fungible Token (NFT) is generated for the land that is considered stamping ownership and legalities. This NFT is then added to the Meta-mask wallet assigned to a user who can conduct appropriate transactions using this identity. This platform developed totally online no helps a user to conduct transactions of either buying or selling of real estates. The information of every transaction is now maintained in files over the block-chained network. The method followed is guaranteed to provide a secure system for registering land and supporting ownership transfers while doing away with middlemen and lowering transaction costs.

Notwithstanding its comparatively high transaction costs, there are a number of compelling benefits to using the Ethereum public blockchain for a land registry system that are in line with the objectives of safe, open, and decentralized land asset management.

First of all, with a sizable development community, copious documentation, and a well-established ecosystem of tools and standards, Ethereum is the most developed and extensively used smart contract platform. As a result, development, auditing, and application integration become more dependable and effective. Strong smart contract capabilities make it possible to automate ownership confirmation, compliance checks, and land transfers without the need for middlemen.

Second, Ethereum offers a high degree of decentralization and security, both of which are essential for guaranteeing the integrity and immutability of land records. Since a worldwide network of nodes maintains public blockchains like Ethereum, they are inherently anti-censorship based, infringing and failures at a particular point of failure. These features prove effective when handling ownership associated legalities.

Thirdly the facilities of Optimism and Arbitrum are managed using Ethereum 2.0 and Layer 2 solutions. Thus, scalability and fees associated with transactions can be reduced to a great extent without disturbing decentralization and security. Since data saved on Ethereum can be publicly verified centrally, it guarantees global transparency and interoperability making it safe for foreign collaborations or investment.

4 Results and Comparison

Blockchain technology has shown excellent result w.r.t efficiency, security and transparency. Fraudulent activities are curbed through use of a decentralization. Smart contracts reduced cost on transaction and the middle man. To avoid inconsistencies in ownership records, the system also guarantees real-time updates for all parties involved. Furthermore, each property has a distinct, unchangeable identity when land assets are stored as NFTs (Non-Fungible Tokens), improving security and traceability. As per the comparison study carried out land registration using a blockchain performs better then convention systems with respect to accessibility, speed of processing, integrity of data and manageability for land issues.

Figure 3 shows how use of blockchain improves land registry systems as compare to conventionally used once. Blockchain being decentralized lowers the cost of transactions and the chances of fraud. It eliminates the middle man and makes the system more robust providing immutability and protection of records.

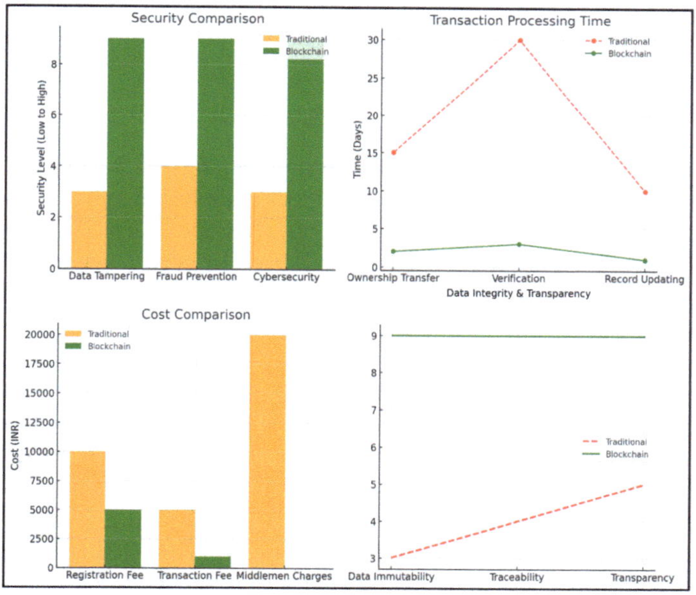

Fig. 3. Comparison Graph.

It is important to keep in mind the weakness of such account since they incur a loss to the user. To provide guarantee of task to be perform systematically a blockchain based system is recommended for land registry. Such a system uses private cryptography keys for accession of assets. The loss of such keys may result in loss of ownership. This is problem usually a problem a decentralize system. However, there are recovery solution like multi-signature wallets and social recovery protocols that can be trusted which enables a user to get back to the system.

Strong authentication procedures like biometric verification and multi-factor authentication should also be implemented to reduce the possibility of unauthorized access. By incorporating time-delayed execution and alerting capabilities, smart contracts enable users to step in when unauthorized conduct occurs. An on-chain arbitration system or a legally recognized governance framework should be accessible to mediate and settle issues in the event of disagreements or unauthorized transfers. Furthermore, system acceptance and confidence depend on user education which depends on the safe structure of key management legally associated. This method is a pavement between advantages suggested by decentralization and providing defense against conditions like human errors and malicious attacks.

The system stands superior due to the use of decentralization and security provided by Ethereum blockchain as compared to Polygon. Since the network has a vast global connection of validators, this network using Ethereum inherently possess smartness in its contract's platform supporting censorship and threat protection. It offers confidence since it is supported with security and dependability. Ethereum supports a larger user database and superior integration with blockchain projects and services due to the user accepting this strong eco system for development. Ethereum reduces cost, support security and is tried since along time and hence is well accepted for decentralized application.

5 Conclusion

The problems with conventional, centralized record-keeping techniques can be solved in a safe, open, and effective way by integrating blockchain technology into land registry systems. Blockchain guarantees the immutability, tamper-proof, and ease of verification of property ownership data by utilizing decentralized ledgers, cryptographic security, and smart contracts. Following these methods in efficient transaction and the middle man are eliminated.

By giving the user of systems relevant and precise information of ownership a block chain system for land registry simplifies transactions related to real estate. Administrative cost are reduced to a great extent and trust is built. This technology also facelifted cost effectiveness, non-corrupt practices for the real estate industries. By adding blockchain technology it is observed that the transparency, accessibility and rights of land owners grow. To optimized the use of such a system a scalability and regularity need to be introduced in the system supported by legality. Governments and organizations may develop a just, effective, and future-ready land management system by adopting this innovation.

The basic characteristics of Ethereum namely maturity, security, transparency, and support for smart contracts make it a solid contender for blockchain-based land register systems. Ethereum is a feasible and future-proof platform for such vital infrastructure, notwithstanding transaction fees, which can be reduced with optimization strategies and upcoming scalability improvements.

References

1. Basan, A., Barannik, E., Lapina, M., Kormakova, V., Lapin, V.: Security methods for a group of mobile robots according to the requirements of Russian and foreign legislation.

IOP Conf. Series Mater. Sci. Eng. **873**, 012031 (2020). https://doi.org/10.1088/1757-899X/873/1/012031
2. Khan, M.B.H., Sajid, M., Lapina, M., Shahid, M., Kotecha, K.K.: Blockchain technology meets 6 G wireless networks: a systematic survey. Alex. Eng. J. **92**, 199–220 (2024). https://doi.org/10.1016/j.aej.2024.02.031. ISSN 1110-0168, https://www.sciencedirect.com/science/article/pii/S1110016824001704
3. Athavale, V.A., Patel, S.T.: Use of blockchain technology for maintaining authenticity and credibility of digital information. In: Raza, Z., Babenko, M., Sajid, M., Lapina, M., Zolotarev, V. (eds.) AISMA-2023: International Workshop on Advanced Information Security Management and Applications. AISMA 2023 Lecture Notes in Networks and Systems, vol. 1207. Springer, Cham (2024). https://doi.org/10.1007/978-3-031-77229-0_2
4. Athavale, V.A., Patel, S.T., Athavale, A., Dhull, S.: Integration of traceability and blockchain within the supply chain of manufacturing industry from a sustainability perspective. In: Agrawal, J., Shukla, R.K., Sharma, S., Shieh, C.S. (eds.) Data Engineering and Applications. IDEA 2022 Lecture Notes in Electrical Engineering, vol. 1189. Springer, Singapore (2024). https://doi.org/10.1007/978-981-97-2451-2_29
5. Vinodha, D., Buvana, M.J., Rajalakshmi, S., Anita, E.A.M., Lapina, M.: Framework based on IoT, AI, and blockchain for smart access to government agricultural schemes. In: The Future of Agriculture: IoT AI and Blockchain Technology for Sustainable Farming, pp. 142–163. BENTHAM SCIENCE PUBLISHERS (2024). https://doi.org/10.2174/9789815274349124010011
6. Athavale, V.A., Arora, S., Athavale, A.: Adoption of blockchain technology for storage and verification of educational documents. In: Sharma, S., Peng, S.L., Agrawal, J., Shukla, R.K., Le, D.N. (eds.) Data, Engineering and Applications Lecture Notes in Electrical Engineering, vol. 907. Springer, Singapore (2022). https://doi.org/10.1007/978-981-19-4687-5_7
7. Basan, A.S., Basan, E.S., Lapina, M.A., Lapin, V.G.: Behavior-based assessment of trust in a cyber-physical system. In: Singh, P.K., Veselov, G., Vyatkin, V., Pljonkin, A., Dodero, J.M., Kumar, Y. (eds.) Futuristic Trends in Network and Communication Technologies. FTNCT 2020. Communications in Computer and Information Science, vol. 1395. Springer, Singapore (2021). https://doi.org/10.1007/978-981-16-1480-4_17
8. De', R., Sen, C.: The complex nature of e-government projects: a case study of Bhoomi, an initiative in Karnataka, India. In: Traunmüller, R. (ed.) Electronic Government. EGOV 2004 Lecture Notes in Computer Science, vol. 3183. Springer, Berlin, Heidelberg (2004). https://doi.org/10.1007/978-3-540-30078-6_99
9. Bose, A.: Challenges in digitizing land records: a case study of India. Int. J. Public Adm. **42**(3), 245–260 (2019)
10. Nakamoto, S.: Bitcoin: A Peer-to-Peer Electronic Cash System (August 21, 2008). Available at SSRN: https://ssrn.com/abstract=3440802 or https://doi.org/10.2139/ssrn.3440802
11. Vayadande, K., et al.: Blockchain-based land record system. ITM Web Conf. **50**, 01006 (2022). https://doi.org/10.1051/itmconf/20225001006
12. Singh, S.K., Tiwari, A., Singh, P.R., Singh, N., Kumar, P.: Land registration system using blockchain. In: 2023 5th International Conference on Advances in Computing, Communication Control and Networking (ICAC3N), Greater Noida, India, vol. 2023, pp. 1499–1503 (2023). https://doi.org/10.1109/ICAC3N60023.2023.10541693
13. Shinde, D., Padekar, S., Raut, S., Wasay, A., Sambhare, S.: Land Registry Using Blockchain - A Survey of Existing Systems and Proposing A Feasible Solution, pp. 1–6 (2019). https://doi.org/10.1109/ICCUBEA47591.2019.9129289
14. Mukherjee, S., Das, M., Ray, P.: Cryptographic security in blockchain-based land registries: a technical analysis. J. Inf. Secur. Appl. **68**(1), 143–159 (2023)
15. Kumar, S., Patel, R., Gupta, D.: Ethereum-based smart contracts for land ownership verification. Proc. Int. Conf. Blockchain Appl. **34**(5), 210–225 (2021)

16. Das, A., Mehta, P.: Public vs. private blockchain models in land registry: a comparative study. J. Digit. Innov. **29**(3), 98–115 (2022)
17. Mishra, T., Bose, L., Rajan, P.: Legal and regulatory challenges in blockchain-based land registries: an Indian perspective. Int. Rev. Law. **48**(2), 176–192 (2022)
18. Choudhary, V., Singh, N.: Global case studies on blockchain-based land registries: lessons for India. J. Gov. Public Policy. **15**(1), 88–105 (2023)
19. Patel, M., Deshmukh, R.: Integrating AI with blockchain for fraud detection and valuation in land registries. AI Soc. **38**(2), 120–138 (2023)
20. Oprunenco, A., Akmeemana, C.: Using Blockchain to Make Land Registry more Reliable in India. LSE Business Review, UK (2018)
21. Mougayar, W.: A Decision Tree for Blockchain Applications: Problems, Opportunities or Capabilities? Startup Management, Toronto (2015)
22. Wood, G.: Ethereum Yellow Paper. https://github.com/ethereum/yellowpaper (2014). Accessed 30 Oct 2018
23. Li, D.W., Wong, E., Guo, J.: A survey on blockchain for enterprise using hyperledger Fabric and composer. In: 6th International Conference on Dependable Systems and Their Applications (DSA), Harbin, China, vol. 2020, pp. 71–80 (2019). https://doi.org/10.1109/DSA.2019.00017
24. Mohanty, D.: Deploying smart contracts. In: Mohanty, D. (ed.) Ethereum for Architects and Developers, pp. 105–138. Apress, Berkeley (2018)
25. Garzik, J., Donnelly, J.C.: Blockchain 101: an introduction to the future. In: Handbook of Blockchain, Digital Finance, and Inclusion, vol. 2, pp. 179–186. Academic Press, Elsevier (2018)
26. Vigna, P., Casey, M.: The Age of Cryptocurrency: How Bitcoin and Digital Money Are Challenging the Global Economic Order. Martin's Press, New York (2015)
27. Buterin, V., et al.: A next-generation smart contract and decentralized application platform. White Paper, vol. 3, no. 37 (2014)
28. Mukhopadhyay, M.: Ethereum Smart Contract Development: Build Blockchain-Based Decentralized Applications Using Solidity. Packt Publishing Ltd. (2018)
29. Geoff, H.: The Beginners Guide to Using an Ethereum Test Network (2020). https://medium.com/compound-finance/the-beginners-guide-to-using-an-ethereum-test-network-95bbbc85fc1d
30. Chincheti, A., Dixit, R.K., Athavale, V.A., Lobo, L.M.R.J., Lapina, M.: A safe and secure online system for bidding using blockchain technology. In: Lapina, M., Raza, Z., Tchernykh, A., Sajid, M., Zolotarev, V., Babenko, M. (eds.) AISMA-2024: International Workshop on Advanced Information Security Management and Applications. AISMA 2024 Lecture Notes in Networks and Systems, vol. 863. Springer, Cham (2024). https://doi.org/10.1007/978-3-031-72171-7_5

Interobject Anthropomorphism as a Threat to Information Security of Critical Information Infrastructure

Sergey Dolzhenkov and Elena Maksimova(✉)

MIREA - Russian Technological University, Moscow, Russian Federation
dolzhenkov@mirea.ru, maksimova@volsu.ru

Abstract. The article presents the results of a study in the field of organizing inter-object interaction at the level of a subject of critical information infrastructure using an anthropomorphic approach. Vulnerabilities of program code are considered as inter-object interaction, the effects of inter-object interaction at the level of vulnerabilities of program code, arising as a result of the implementation of behavioral models, are substantiated. They are studied and substantiated based on the anthropomorphic approach in the form of: obligate symbiosis, parasitism, predation, amensalism, allelopathy, competition, neutralism. The types of anthropomorphism in organizing inter-object interaction at the level of critical information infrastructure are studied. Based on the results, new patterns in assessing information security risks are established. The interaction of vulnerabilities of program code can lead not only to a decrease in the system's performance, but scenarios are also possible in which a positive effect is manifested, that is, this situation can be used as an element of the protection system.

Keywords: Inter-object interaction · anthropomorphic approach · critical information infrastructure · critical information infrastructure object · single-subject system · software code vulnerabilities · risk dynamics

1 Introduction

Modern trends in the development of information technology dictate the conditions for the digitalization process of companies in the real sector of the economy, which, according to [1], also include critical information infrastructure entities (CII). Digitalization of processes is a complex and multifaceted process, an integral part of which is compliance with the information security (IS) policy, therefore, it is a key component in maintaining the integrity, operability and fault tolerance of information systems. It is important to note that it is possible to predictably and safely implement digitalization processes taking into account the work with vulnerabilities at the level of the program code of critical information infrastructure objects [2].

The organization of inter-object interaction at the CII level plays an important role in the implementation of the information security doctrine of the Russian Federation. Therefore, inter-object interaction is the basis of CII security [3].

M. Lapina et al. (Eds.): AISMA 2025, LNNS 1456, pp. 131–144, 2026.
https://doi.org/10.1007/978-3-032-07275-7_13

Special attention in the processes of optimizing information flows at significant CII objects should be paid to the dynamics of their behavior and threats at the software level in different types of object relationships [19, 20]. For this purpose, an anthropomorphic approach will be used [4]. Thus, the purpose of this work is to increase the level of organization of interaction of CII objects through the use of the anthropomorphic effect of the interaction of software code vulnerabilities.

To achieve the stated goal, it is necessary to consider the organization of interaction of critical information infrastructure objects (CII) within the framework of one CII, which is a single-subject system of functioning of the CII [5].

2 Interaction of Threats

Based on the results of the study [6–8], the following types of interobject interaction links in the SCII were identified (Fig. 1), based on the anthropomorphic approach (Table 1).

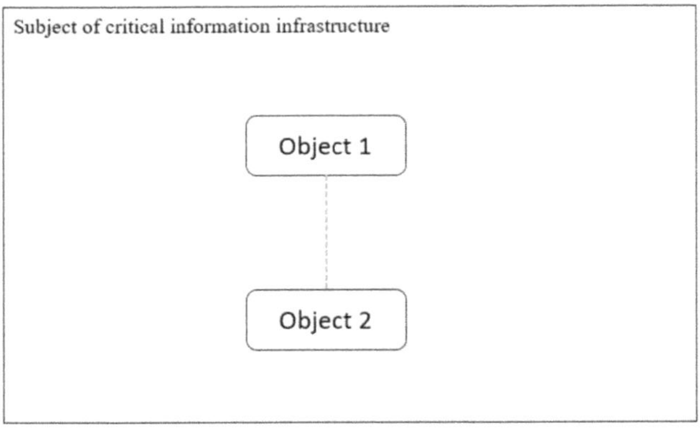

Fig. 1. Basic model of inter-object interaction.

Applicable to the SCII, the forms of symbiosis manifestation are possible in the following cases:

1. Obligatory symbiosis - mutually beneficial existence of SCII objects, in which both objects are interdependent on each other, as a result, mutually beneficial assistance is formed.
2. Parasitism - interaction of SCII objects, in which one of the objects cannot exist without partial "absorption" of the resources of the second object, while the second object can function stably for a sufficiently long time.
3. Predation - a type of interaction of SCII objects, in which the first object exists due to the complete or partial "absorption" of the resources of the second object, up to its complete destruction. The existence of the first object is impossible without a "victim object". Regular consumption of resources of other objects within one SCII is necessary [9].

Table 1. Forms of interobject interaction.

Symbiosis	Obligate symbiosis	Mutual benefit	At least 1 of the objects benefits
	Parasitism	One-sided benefit, long-term neutral for the second object	
	Predation	One-sided benefit, destruction of the second object	
Antibiosis	Amensalism	One-sided negative influence, for the second object - neutral	At least one of the objects limits the capabilities of the other
	Allelopathy	Bilateral negative influence	
	Competition	Mutual destruction	
Neutralism	Objects do not influence each other		

Similar to symbiotic forms, let us consider possible manifestations of antibiosis applicable to the SCII:

1. Amensalism is a form of antibiosis, in which one SCII object negatively influences another SCII object, but itself has neither positive nor negative effects from the influence exerted.
2. Allelopathy is an antibiotic interaction of SCII objects, in which both objects receive a negative effect from interacting with each other, which is due to their functional parameters.
3. Competition - two SCII objects are infrastructural “enemies” by their nature [10].

Based on the above and Table 1, the effects of the types of interobject interaction links in the SCII are identified, based on the anthropomorphic approach for objects Oi and Oj (Table 2).

Table 2. Anthropomorphic types of intersubjective interaction.

Anthropomorphic types of intersubjective interaction	O_i	O_j
Obligate symbiosis	+ / +	+ / +
Parasitism	+ / –	– / +
Predation	+ / –	– / +
Amensalism	0 / –	– / 0
Allelopathy	– / –	– / –
Competition	– / –	– / –
Neutralism	0	0

In the table « + » - positive effect; «+ / –» - positive effect for object Oi, with negative effect for object Oj; «– / +» - presence of negative effect for object Oi, with positive effect for object Oj; «– / 0» - presence of negative effect for object Oi, no effect for object Oj; «0 / –» - no effect for object Oi, with negative effect for object Oj; «-» - presence of negative effect; «0» - no effect; «+ / 0» - positive effect for object Oi, no effect for object Oj; «0 / +» - no effect for object Oi, positive effect for object Oj [11].

To formalize anthropomorphic types of interobject interaction, the following notations were introduced:

V — vulnerability;
T — presence of vulnerability or threat;
0 — – absence of vulnerability or threat
T_X (T_Y)—use of vulnerability Y functionality for vulnerability X to function;
$\rightarrow$ — implementation of threats due to the presence of vulnerabilities;
$<, >, =$ – operations comparing threats by their possible damage.

Thus, the implementation of an isolated threat N due to the presence of the corresponding vulnerability N in the program code can be written as follows: $V_N \rightarrow T_N$.

An exception to the rule is the case when each vulnerability separately does not lead to any threat (as will be shown below, this is true for the first type of interactions - obligate symbiosis): $V_N \rightarrow 0$.

The following graphic elements were used to visualize inter-object interaction: an attack on the OKII using a vulnerability; a vulnerability in the software; a threat to information realized as a result of an attack (Fig. 2).

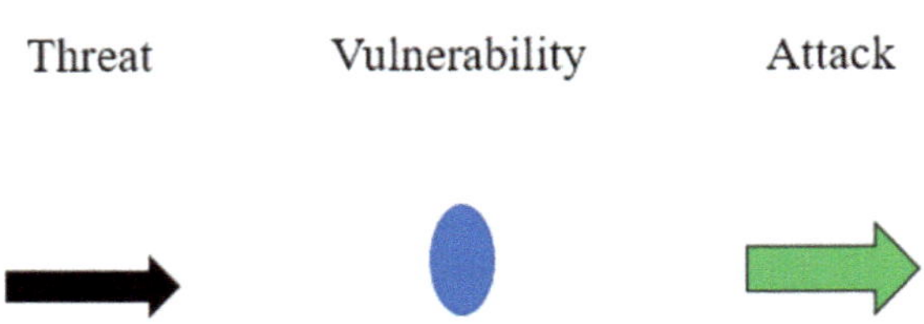

Fig. 2 Graphic elements for visualizing the interaction of vulnerabilities

The color of the arrow indicating the threat corresponds to the significance of the latter. Thus, the white color of the arrow indicates the minimum risk, green means a low level of risk of an attack using vulnerabilities, yellow - medium, and the red color of the arrow - the maximum level of risk [12].

Each of the effects of inter-object interaction vulnerabilities in inter-object interaction at the level of the SCII is presented in a formalized form.

Effect 1. Obligatory Symbiosis (+ / +)
The effect under consideration implies the necessity of the joint existence of objects.

At the level of the SCII, this type may include vulnerabilities that turn into cyber threats only through interaction with each other.

The formalization of the representation of the interaction of the type "obligatory symbiosis" is as follows:

$$\{V_1 \rightarrow 0\ V_2 \rightarrow 0\ V_1 + V_2 \rightarrow T_{12}\}. \quad (1)$$

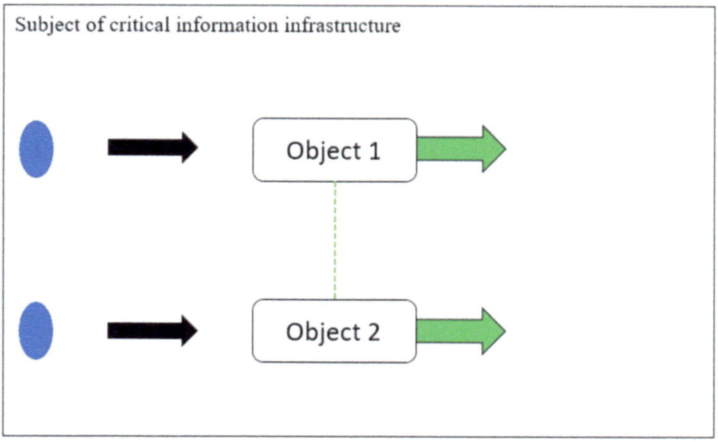

Fig. 3 Scheme of implementation of the obligatory symbiosis of vulnerabilities (Effect 1)

The scheme of interaction of vulnerabilities looks like the following in Fig. 3.

Obligatory symbiosis of vulnerabilities at the level of program code can arise as a result of "merging" two vulnerabilities, where each individual vulnerability does not represent a threat as such. An example is the threat of data leakage through secure communication channels between objects to a port opened by intruders, where the first vulnerability is data leakage, the second vulnerability is port eavesdropping.

Effect 2. Parasitism (+ / −)

The effect under consideration implies one of the vulnerabilities as a "parasite", and the second vulnerability as a habitat for the first.

In the SCII, such an example can be vulnerabilities, where the first parasitizes on the created information capacities of the second.

The formalization of the representation of the interaction of the "parasitism" type has the following form:

$$\left\{V_1 + V_2 \rightarrow T_1' + T_2' \, T_1' = T_1 + D \, T_2' = T_2 - D\right\} \tag{2}$$

where D is the benefit derived by the first vulnerability from the second.

It should be noted that the threat from the second vulnerability does not always decrease proportionally to D; however, such simplification does not reduce the overall correctness of the perception of the formula.

The scheme of interaction of vulnerabilities has the following form in Fig. 4. An example of vulnerability parasitism is the following scenario: using an internal algorithm for sending information within a secure network to implement external cyberattacks. An example of vulnerability parasitism is the following scenario: using an internal algorithm for sending information within a secure network to implement external cyberattacks. The parasite organism is an external attack, for example, DDoS, and uses the sending algorithm as routing.

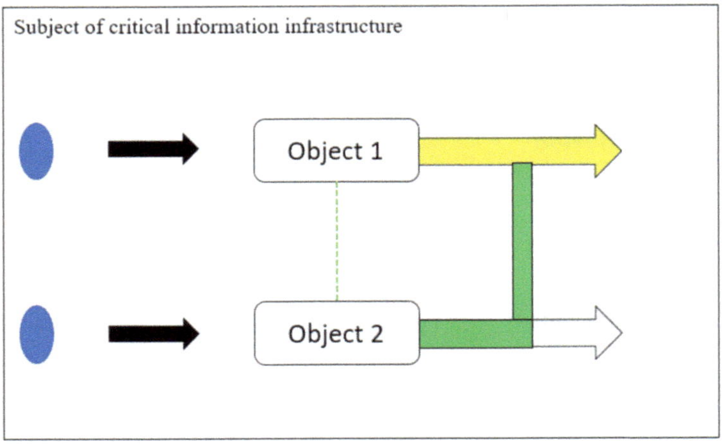

Fig. 4 Scheme of implementation of parasitic type of vulnerabilities (Effect 2)

Effect 3. Predation (+ / −)
The effect in question is characterized by the fact that one vulnerability feeds on parts of another vulnerability without symbiotic relationships.

The vulnerability has an aggressive parasitic effect on another by replacing its main functionality with its own.

The formalization of the representation of the interaction of the "predation" type is as follows:

$$\{V_1 + V_2 \rightarrow T_1'(T_2)\ T_1' > T_1\}. \quad (3)$$

The scheme of interaction of vulnerabilities looks like the following in Fig. 5.

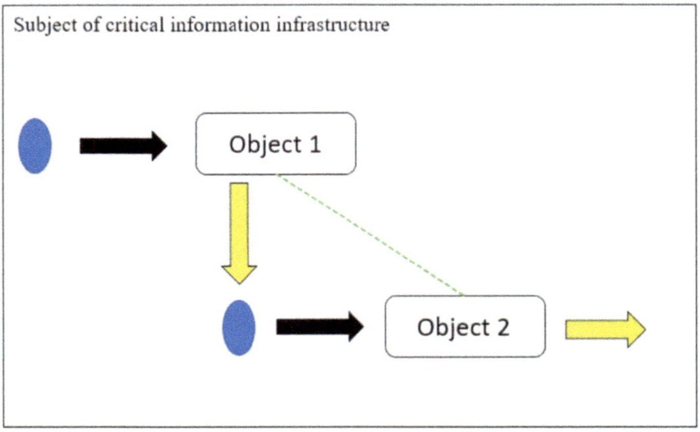

Fig. 5 Scheme of implementation of predatory type of vulnerabilities (Effect 3)

An example of interaction can be the enhancement of the functionality of vulnerabilities for the case of parasitism. So, if vulnerability 2 (from the example of the parasitism effect) starts using the algorithm for sending vulnerability 1, completely replacing its core. Thus, vulnerability 2 will completely absorb vulnerability 1 and will exist on its resources.

Effect 4. Amensalism (0 / –)

The effect under consideration is characterized by the negative impact of the first vulnerability on the second without a reverse effect.

Vulnerabilities have a deterrent effect: the first vulnerability deters the second, due to which the risk of using the latter is reduced.

The formalization of the representation of the interaction of the "amensalism" type is as follows:

$$\left\{V_1 + V_2 \rightarrow T_1 + T_2' \; T_2' < T_2\right\}. \tag{4}$$

The scheme of interaction of vulnerabilities looks like the following in Fig. 6.

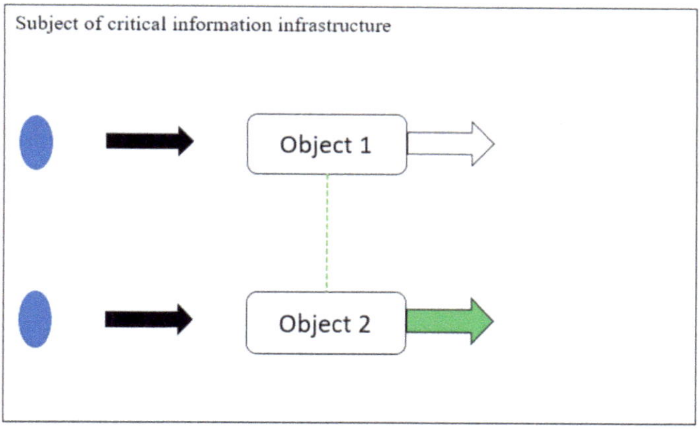

Fig. 6 Scheme of implementation of amensalistic type of vulnerabilities (Effect 4)

Amensalism of vulnerabilities is clearly expressed in threats to the integrity of information. As an example, we can consider two vulnerabilities - the first of which is aimed at providing data to attackers, and the second vulnerability has the goal of blocking all data from being output in principle. In such a scenario, the first vulnerability can intercept people's personal data, and the second vulnerability prevents this interception.

Effect 5. Allelopathy (– / –)

The effect under consideration is characterized by - mutually harmful influence of vulnerabilities of program code.

The allelopathic effect from the interaction of vulnerabilities leads to a lower risk than the sum of the effects from the functioning of these same vulnerabilities in parallel.

The formalization of the representation of the interaction of the type "allelopathy" has the following form:

$$V_1 + V_2 \rightarrow T_{12} \; T_{12} < T_1 + T_2. \tag{5}$$

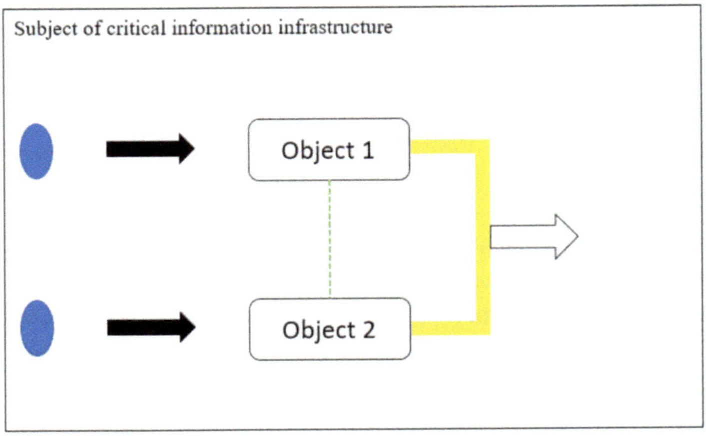

Fig. 7 Allelopathic vulnerability diagram (Effect 5)

The scheme of interaction of vulnerabilities looks like the following in Fig. 7.

The allelopathic effect is similar to the ammensalism effect, but the first vulnerability (from effect 4) must be significantly more powerful than the second vulnerability. In this case, we will get a scenario where the vulnerability will try in every way to create a data leak, and the second vulnerability will resist this incident. But due to its power, the first vulnerability is realized, but significantly reducing its potential due to the impact of the second vulnerability.

Effect 6. Competition (– / –)

A distinctive feature of the competition effect is the common resources for which vulnerabilities compete, thereby limiting one of the vulnerabilities by capturing most of the resources by another vulnerability.

The functioning of vulnerabilities implies the use of limited and common resources, such as RAM. As a result, vulnerabilities begin to conflict.

The formalization of the representation of the interaction of the "competition" type is as follows:

$$\{V_1 + V_2 \to T_1' + T_2' \left[\{T_1' \leq T_1 \; T_2' \ll T_2 \; \{T_1' \ll T_1 \; T_2' \leq T_2. \right. \tag{6}$$

The scheme of interaction of vulnerabilities looks like the following in Fig. 8.

Competition of vulnerabilities occurs when the goal of implementing two vulnerabilities coincides, and there is only one data transmission channel to the external perimeter. For example, both vulnerabilities cause the disclosure of user passwords and leakage of this data through a specific, single communication channel. The vulnerability that will reserve the channel in the program's network library faster will be operational.

Effect 7. Neutralism (0 / 0)

Absence of any impact on each other. Such vulnerabilities that do not affect each other, but function in parallel. The formalization of the representation of the interaction of the "neutralism" type is as follows: $V_1 + V_2 \to T_1 + T_2$.

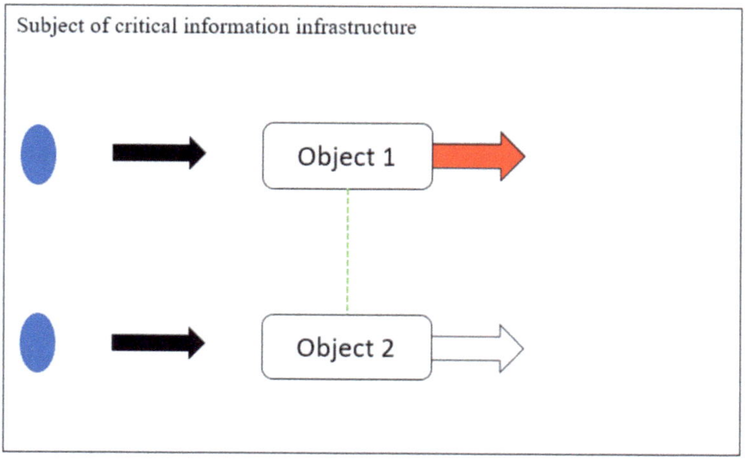

Fig. 8 Scheme of competitive types of vulnerabilities (Effect 6)

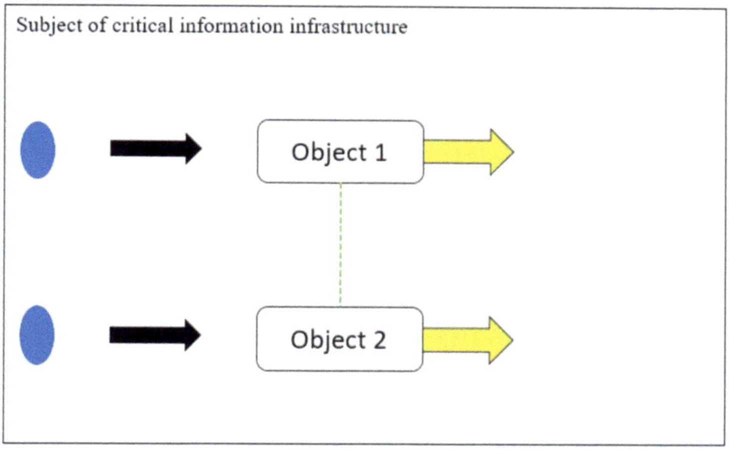

Fig. 9 Neutralistic vulnerability scheme (Effect 7)

The scheme of interaction of vulnerabilities looks like the following in Fig. 9.

This type of interaction can occur between any vulnerabilities that are not related to each other in terms of functionality and purpose. For example, between the administrator login built into the program and errors in the cryptographic hash algorithm for password verification.

Based on the research results, 7 most popular vulnerabilities (TOP-7) were identified [13]:

V_1. SQL injections - allow attackers to execute arbitrary SQL queries to the database through vulnerable data entry points.

V_2. API insecurity - improper authentication and authorization management, insufficient input data validation, or lack of frequency limitation can be used for attacks.

V_3. Scripting - injection of malicious scripts into web pages that can harm users or steal their data.
V_4. Insufficient authentication - vulnerabilities associated with unprotected password storage, inappropriate complex password policies.
V_5. Outdated software - using outdated software can lead to vulnerabilities if it is not updated regularly.
V_6. Routing vulnerabilities - incorrect configuration of firewalls and routers can lead to unauthorized access to data.
V_7. Incorrect handling of errors - the code does not handle errors correctly, they can disclose information about the system [13].

Below is a table of interactions of the software code vulnerabilities described above (Table 3). The cells indicate the types of interactions for the corresponding types of software code vulnerabilities. Also in brackets is the effect for the first vulnerability from interaction with the second vulnerability.

Table 3 Interaction of vulnerabilities in inter-object communication

	V_1	V_2	V_3	V_4	V_5	V_6	V_7
V_1			B_1 (+)				
V_2					B_4 (−)		B_3 (+)
V_3	B_3 (+)			B_7 (0)			
V_4			B_7 (0)			B_5 (−)	
V_5		B_4 (0)					B_1 (+)
V_6				B_5(−)	B_2 (+)		
V_7		B_6 (−,-)					

Thus, it can be stated that there is no specific logic or pattern in the dependence of interobject interaction. Filling in all cells with values requires the work of an expert group or additional research.

3 Vulnerability Metrics

Currently, there are no widely distributed and widely used metrics for evaluating software code, and no metrics for evaluating software code from the perspective of the functioning of vulnerabilities in it have been found. We will propose our own, based on the anthropomorphic approach to the interaction of threats at the software code level.

The software code vulnerability metric (hereinafter referred to as the Metric) can be written as follows:

$$M = |v_1 \ldots v_N|, \quad (7)$$

where v_i is the effect of the i-th vulnerability type, N is the number of vulnerability types under consideration.

Let us consider cases of software codes in which two or more vulnerabilities operate. In the presence of two vulnerabilities of different types of interaction, we can talk about different results for each vulnerability, which can be reflected in formula (7) by adding the indicator M_1, which takes into account such various effects. The indicator M_1 is a vector of the same dimension as M. In this case, each of the elements of the vector must depend on the effect of the corresponding vulnerability type, as well as on its interaction with the others; The final formula for M_1 has the following form:

$$M_1 = |v_1, \ldots, v_N| \times \left| I_{11} \ldots I_{1N} \vdots \ddots \vdots I_{N1} \ldots I_{NN} \right| \quad (8)$$

where I_{ij} is the coefficient of influence of the interaction of the i-th and j-th vulnerability types on the i-thvulnerability.

As a result, vulnerability types that do not imply interaction have a coefficient of I_ij ≡ 0, in a situation where the second vulnerability increases the first: I_ij > 0, otherwise:—I_ij < 0. The additive effect of the influence of all vulnerability types on the vulnerability of a given type is determined by: $K_i = \sum_{j=1}^{N} V_j \times I_{ij}.$

4 Proposed Solution

Let's assume that within one SCII, two OKII interact by transferring data: confidential information from an internal database using identification and authentication mechanisms, in order to differentiate access rights to databases of different levels of access. In such a scenario, violations of confidentiality and data security are likely: access to data by persons who do not have such access rights. As in most cases, the proposed program is written in the C# programming language, with basic functionality - requesting a login and password, granting rights to the user to the relevant data.

Probability of finding vulnerabilities from the Top 7 in a hypothetical inter-object interaction.

Minimum, moderate, high, maximum. The proposed probability scale will have the following proportions: 0:0.25:0.75:1

As a result, the Metric is represented as follows:

$$M = K_v \times |0.25\ 1\ 0\ 0.75\ 0\ 0.75\ 0.25\ | \quad (9)$$

where Kv is the normalization coefficient (Kv ≪ 1).

Similarly, we will assume that the coefficient of influence of the interaction of the i-th and j-th vulnerability types on the i-th vulnerability (Iij) is equal to the product of Ki (normalization coefficient) and the effect of the interaction: "0"—in the case of no effect, "1"—for benefit, "–1"—for damage; in the case of several effects, they are summed up. Based on formula (7), and also following formula (9) and Table 3, we will calculate the correction term, based on the existing interactions of vulnerabilities. Interpreting the obtained results, we can talk about the possibility of using an anthropomorphic approach in the presented vulnerabilities: absolutely all interactions of different vulnerabilities show different results. For example, the interaction of vulnerabilities 7 and 9 do not enhance the effect of each other, however, the result of the interaction of two vulnerabilities is completely unpredictable: vulnerabilities can enhance each other, not affect each other in any way and reduce the effect of interaction, which in turn can be reflected in the indicator of inter-object interaction assessment at the level of the SCII [14, 15].

$$\begin{aligned} M_1 &= K_v \times |0.25, 1, 0, 0.75, 0, 0.75, 0.25| \times K_i \times M_1 \\ &= K_v \times K_i \times |0.25\ 1\ 0\ 0.75\ 0\ 0.75\ 0.25\ | \end{aligned} \tag{10}$$

5 Conclusion

The presented study confirms the relevance and demand for the application of the anthropomorphic approach in assessing the organization of inter-object interaction at the level of the subject of critical information infrastructure. The examples considered in the work can be used in further studying the issues of interaction of vulnerabilities at the level of the program code of objects of critical information infrastructure.

Additionally, it is possible to use the conducted study in creating an information security management system for an organization, taking into account the investment component: competent distribution of resources for the purpose of implementing the information security policy by tracking anthropomorphic types of interaction of threats, in which options for mutual destruction of negative consequences from the implementation of these threats are possible. Taking into account such possible scenarios, organizations will be able to build a much more effective investment policy in the field of information security [16, 17].

The use of anthropomorphism for understanding interactions turned out to be quite successful - both for typification and for their interpretation by a person. The proposed formalization has shown its viability, although additional research and approaches that go beyond the scope of only one area of information security are needed for a full-fledged mathematical apparatus.

References

1. Federal Law of the Russian Federation of July 2, 2013 No. 187-FZ. Patents and Licenses. Intellectual Rights. 2013. No. 8. pp. 60–65. EDN QOSONJ
2. Rusakov, A.M.: Complex of anthropomorphic models of behavioral analysis of processes for detecting the effects of infrastructural destructiveness. Eng. Bull. Don. 11(119), 391–404 (2024). EDN DLHUZQ
3. Doctrine of information security of the Russian Federation: Decree of the President of the Russian Federation of 05.12.2016 No. 646. Official Internet portal of legal information. http://www.pravo.gov.ru, 12.12.2016
4. Buinevich, M.V.: Anthropomorphic approach to describing the interaction of vulnerabilities in software code. Part 1. Types of interactions. Buinevich, M.V., Izrailov, K.E. Information Security. Inside. 2019. 5(89), 78–85. EDN OLLEYX
5. Gorin, D.S., Dolzhenkov, S.S.: Scientific and methodological foundations of product quality management based on CALS technologies. Bull. Acad. Law Manage. **3**(64), 55–60 (2021). https://doi.org/10.47629/2074-9201_2021_3_55_60. EDN CUTDUR

6. Maksimova, E.A., Rusakov, A.M., Lapina, M.A., Lapin, V.G.: Anthropomorphic model of states of subjects of critical information infrastructure under destructive influences. In: Mathematics and its Applications in New Computer Systems Lecture Notes in Networks and Systems, vol. 424, pp. 569–580 (2022). https://doi.org/10.1007/978-3-030-97020-8_51. EDN GGKXDH
7. Taneski, V., Heričko, M., Brumen, B.: Impact of security education on password change. In: 38th international convention on information and communication technology, Electronics and Microelectronics (MIPRO), pp. 1350–1355 (2015)
8. Marimuthu, K., Gopinath, M.: Production of sugarcane forecasting using ARIMAX model. Int. J. Innov. Technol. Explor. Eng. **8**(12-S), 596–600 (2019)
9. Intelligent analysis of the Greenplum data warehouse operation based on log file processing. Rusakov, A.M., Gorin, D.S., Lisyutenko, A.S., [et al.]. Modern science: current problems of theory and practice. Series: Natural and technical sciences. 6, 142–149 (2023). https://doi.org/10.37882/2223-2966.2023.06.31. EDN PIRKUE
10. Dolzhenkov, S.S.: Risk management as a means of implementing a methodology for supporting information security management processes of critical information infrastructure entities under destructive impacts of infrastructural genesis. Dolzhenkov, S.S. Maksimova, E.A. Actual problems of applied mathematics, computer science and mechanics: collected papers of the International scientific conference, Voronezh, December 04–06, 2023. Voronezh: Limited Liability Company "Welborn", Publishing House "Research Publications", 2024. pp. 1538–1539. EDN DNWVWZ
11. Buinevich, M.V., Izrailov, K.E.: Analytical modeling of program code operation with vulnerabilities. Issues Cybersecurity. **3**(37), 2–12 (2020). https://doi.org/10.21681/2311-3456-2020-03-02-12. EDN CQFGPI
12. Buinevich, M.V., Izrailov, K.E.: An anthropomorphic approach to describing the interaction of vulnerabilities in software code. Part 2. Vulnerability metrics Information Security. Inside. 6(90), 61–65 (2019). EDN HLUDTX
13. Byers, D., Ardi, S., Shahmehri, N., Duma, C.: Modeling software vulnerabilities with vulnerability cause graphs. In: 22nd IEEE International Conference on Software Maintenance, pp. 411–422 (2006)
14. Rinaldi, S.M., Peerenboom, J.P., Kelly, T.K.: Identifying, understanding, and analyzing critical infrastructure interdependencies. IEEE Control Syst. Mag. **21**(6), 11–25 (2001)
15. Dolzhenkov, S.S.: Optimization of the information security management system of critical information infrastructure facilities. Student science for the development of the information society: Proceedings of the 13th All-Russian scientific and technical conference with the invitation of foreign scientists, Stavropol, November 28, 2023. Stavropol: North Caucasian Federal University, 2024. pp. 124–130. EDN EYGAVD
16. Belov, A.S.: Systems approach to designing information security systems. Belov, A.S., Dobryshin, M.M., Dushkin, A.V. Moscow: Scientific and Technical Publishing House "GoryachayaLiniya-Telecom", 2023. 232 p. ISBN 978-5-9912-1067-6. EDN DCSYJQ
17. Castro, J.L., Delgado, M.: Fuzzy systems with defuzzification are universal approximators. IEEE Trans. Syst. Man Cybern. Part B (Cybernetics). **26**(1), 149–152 (1996)
18. Dolzhenkov, S.S.: Application of risk management approaches in the field of information security of critical information infrastructure entities under destructive impacts of infrastructural genesis. Dolzhenkov, S.S., Maksimova, E.A. Cybernetics and information security "KIB-2023": Collection of scientific papers of the All-Russian scientific and technical conference, Moscow, October 18–19, 2023. Moscow: National Research Nuclear University"
19. Vinodha, D., Jenefa, J., Anita, E.A.M., Babenko, M.: Theoretical framework for blockchain secured predictive maintenance learning model using digital twin. In: Lapina, M., Raza, Z., Tchernykh, A., Sajid, M., Zolotarev, V., Babenko, M. (eds.) AISMA-2024: International

Workshop on Advanced Information Security Management and Applications. AISMA 2024 Lecture Notes in Networks and Systems, vol. 863. Springer, Cham (2024)
20. Mary Anita, E.A., Jenefa, J., Vinodha, D., Lapina, M.: Advancements in sybil attack detection: a comprehensive survey of machine learning-based approaches in wireless sensor networks. In: Lapina, M., Raza, Z., Tchernykh, A., Sajid, M., Zolotarev, V., Babenko, M. (eds.) AISMA-2024: international workshop on advanced information security management and applications. AISMA 2024 Lecture Notes in Networks and Systems, vol. 863. Springer, Cham (2024)

An Enhanced Secure Message Authentication Protocol for Internet of Vehicles

E. A. Mary Anita[1], S. Sajini[2], J. Jenefa[1(✉)], and Vitalii Lapin[3]

[1] CHRIST University, Bangalore, India
jenefa.j@christuniversity.in
[2] Anna University, Chennai, India
[3] North Caucasus Federal University, Stavropol, Russia

Abstract. Internet of Vehicles (IoV) aims to transform the driving experience to the next level by ensuring communication with other vehicles, pedestrians' hand-held devices, Road Side Units (RSU), and other sensors used in the smart city environment. It integrates the benefits of the Internet of Things (IoT) and Vehicular Adhoc NETwork (VANET) to offer a safe and comfortable driving environment. Since communication is established through insecure channels, IoV is prone to various security attacks. Hence, there is a need for an authentication mechanism to ensure secure communications. For VANET, various authentication techniques are available, but they are not suitable for IoV due to their high computation overhead. Hence, we suggest a novel secure authentication scheme for IoV, which ensures authentication, conditional privacy preservation, message integrity, traceability, and unlinkability. It also provides security against various attacks. The proposed scheme performs better with less computation, communication, and storage overhead.

Keywords: Authentication · IoV · Smart City · Conditional Privacy Preservation · Message Integrity · and Unlikability

1 Introduction

The Internet of Things (IoT) is the network of resource-restricted items [41]. It allows billions of devices to be connected to other devices through the internet worldwide. The connected smart devices can sense and share data with other devices. The number of connected devices is increasing drastically. It is predicted [19] that more than $16 trillion will be invested in IoT in the upcoming five years. Hence, the devices will be much smarter, making the lives of humans more comfortable [42]. Applications of IoTs are extended in many fields, including Internet-of-Sensors (IoS), Internet-of-Energy (IoE), Internet-of-Vehicles (IoV), etc.

Internet-of-Vehicles (IoV) allows vehicles to communicate with their drivers, neighboring vehicles, pedestrians, etc. Its main goal is to reduce accidents by using smarter vehicles. As vehicles increase each day [2], almost 3700 people [1] die in road accidents globally. Smart vehicles are used to reduce accidents [9] and assist the drivers,

M. Lapina et al. (Eds.): AISMA 2025, LNNS 1456, pp. 145–155, 2026.
https://doi.org/10.1007/978-3-032-07275-7_14

who communicate with nearby devices to gather and share information regarding road conditions. The improvements in IoV and VANET [8] ensure that this is in smart cities [3–7].

Figure 1 displays the system model for IoV in the context of smart cities. As shown, the many forms of communication in IoV are Intra-Vehicle, Vehicle to Vehicle (V2V), Vehicle to Infrastructure (V2I), Vehicle to Cloud (V2C), and Vehicle-to-Pedestrian (V2P). Intra-vehicle communication allows the devices within the vehicle to interact with each other via onboard units (OBU). It is used to analyze the internal performance of vehicles. Inter-vehicle communication is another term for vehicle-to-vehicle (V2V) communication. It allows vehicles to communicate and exchange information regarding traffic, road conditions, etc. Vehicle-to-Infrastructure (V2I) [10] enables communications between moving vehicles and Road-Side-Units (RSUs), which are installed at the side of the road to provide services to moving vehicles. Vehicle-to-Cloud (V2C) enables vehicles to use the internet to access the cloud and gather more data. Mobility prediction for vehicles is made using such additional information. Vehicle-to-Pedestrian (V2P) provides information to pedestrians regarding road conditions and traffic, which may be helpful for them.

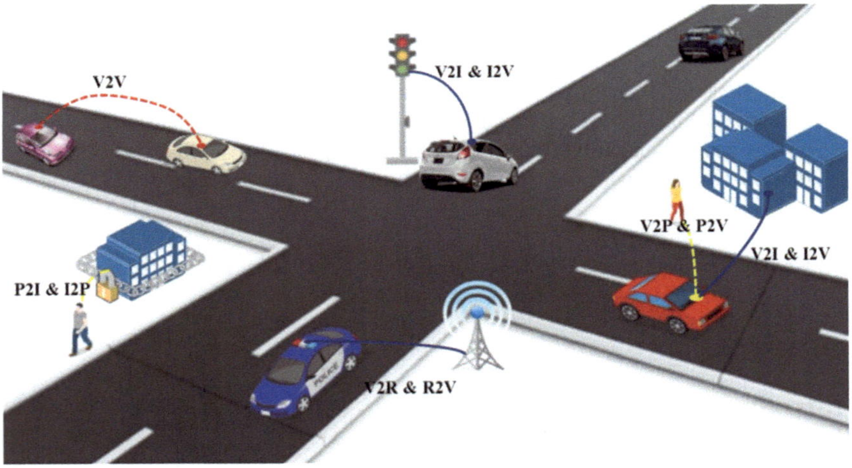

Fig. 1. System Model of IoV.

The Internet-of-Vehicle environment is vulnerable to many attacks [13–15, 34–40] because of its unique characteristics like dynamic topology, frequent connection loss, a huge network, etc. Since the communicated information is used to avoid accidents, it has to be transmitted securely. On the other hand, drivers' location data needs to be secured. Hence, security and privacy have to be ensured in the IoV environment, which requires a secure authentication protocol. The secure authentication protocol must ensure authentication, privacy preservation, message integrity, confidentiality, and unlinkability. It must also be secure against impersonation attacks [16, 17], man-in-the-middle attacks, replay attacks, off-line guessing attacks, etc.

The existing authentication schemes do not ensure the aforementioned security requirements, and it has high computation and communication overheads. The majority of the techniques are susceptible to a variety of security threats [18]. The message authentication schemes proposed for VANET are not suitable for IoV because of the expensive pairing operations [20, 21] and high computation overheads for signature computation and verification. These limitations motivate us to propose a message authentication scheme for IoV with essential security requirements suitable for the smart city environments.

2 Related Works

Many authentication systems have been put forth for IoV in the context of smart cities [22, 23, 33, 43]. An authentication technique with privacy preservation and non-repudiation was proposed by Li et al. [24]. However, it is prone to session key disclosure attacks, hence it is not suitable for IoV because of the high computation overhead. A two-factor based authentication approach with privacy preservation for VANETs was proposed by Wang et al. [26]. Even though Wang et al. [26] claim it is secure, it is prone to various security attacks like impersonation, smart card stolen and offline password guessing attacks. Additionally, it does not guarantee user privacy. Li et al.'s [24] efficient authentication and key agreement technique was designed to protect privacy. Due to the expensive processes required for signature generation, it has high overheads. As a result, it is inappropriate [28] for the IoV environment. Other schemes [24–28] based on public key cryptography are also proposed, but they are inefficient because they have high overheads in terms of computation, communication, and storage.

A lightweight authentication technique was proposed by Ying et al. [29]. It is insecure against security attacks like replay, offline-identity guessing attacks, and incurs high overhead for authentication. Chen et al. [30] address all these drawbacks, and a new secure authentication scheme is proposed for IoV to ensure security against all these attacks. The proposed scheme requires more storage space to store the data for authentication purposes. A layered architecture for IoV was suggested by Kaiwartya et al. [31]. The proposed architecture has four layers: AI, Perception, Application, and Coordination. The communication between vehicles to infrastructure (V2I), vehicles to vehicles (V2V), Vehicles to Sensors (V2S), Vehicles to RSUs (V2R), and Vehicles to Pedestrians (V2P) is established securely by using this five-layered architecture. To establish secure communications in IoV, Vasudev et al. [11] presented a novel authentication system; however, it is vulnerable to numerous attacks, such as Man-in-the-Middle Attacks, Impersonation, and secret key disclosure Attacks. Yu et al. [12] address all these drawbacks with the IoV-SMAP scheme. Even though it provides secure authentication, it does not provide conditional privacy preservation, and hence, there is no way of identifying the attackers. In terms of computation, communication, and storage, it has high overheads. So, for the IoV environment, we presented a new message authentication technique with necessary security requirements and less overhead.

3 Proposed Scheme

This section proposes a new secure and effective message authentication mechanism to address the drawbacks of the IoV-SMAP scheme. It also provides other security requirements along with authentication. It comprises three phases: Initialization, Registration, and Authentication. Table 1 lists the Notations utilized in the suggested approach.

Table 1. Notations.

Notations	Description
V_i	Vehicle i
TA	Trusted Authority
IS	Infrastructure
RSU	Road Side Unit
ID_i	ID of vehicle V_i
PWD_i	Password of vehicle V_i
N	Random Nonce
S_{TA}	Secret Key of TA
PK_i	Pseudo ID Generation Key of V_i
$PSID_i$	Pseudo ID of Vehicle V_i
T	Timestamp
ΔT	Timestamp Threshold
H(.)	One-way Hash Function
$\oplus$	XOR Operation
$\|\|$	Concatenation Operation

3.1 Initialization Phase

In this phase, the Trusted Authority (TA) responsible for a particular domain in the IoV environment selects a random nonce, N_{TA}, and one-way hash function H(.). (SHA-256) [32]. TA computes secret key, $S_{TA} = H(ID_{TA} \| N_{TA})$. This secret key is used by all the legitimate vehicles/IoV Objects and RSUs for authentication in the smart city environment. It is updated periodically and broadcast to all the vehicles/RSUs/Devices through a secure channel.

3.2 Registration Phase

All the vehicles, infrastructures (RSU), and IoV items in the smart city environment should sign up with the Trusted Authority (TA), which is responsible for that particular domain in the IoV environment.

Vehicle Registration. Manufacturers register vehicles with the TA, which assigns an ID based on the owner's private information. The Tamper Proof Device (TMP) of the OBUs will be protected with a Password, PWD. TA assigns ID and PWD for the vehicles. When vehicles (V_i) reach a new area, they send their ID_i to TA through a secure channel to register. TA verifies the vehicle's identity, $V_{i,}$ which is saved in the database. If it is valid, it sends $\{S_{TA}, PK_i, PWD_i\}$ through a secure channel, where PK_i is the pseudo-ID generation Key for vehicle $V_{i,}$ which the TA generates for the generation of pseudo-IDs. PK_i is stored in the TA database, which is used to reveal the original ID of the vehicle in case of violations.

Infrastructure Registration. TA assigns IDs to the RSU within its range. It assigns ID_R to RSUs and preloads S_{TA}, establishing secure communication with other RSUs and vehicles/IoV objects.

IoV Object Registration. An IoV object selects ID_j and Password PWD_j and sends them to TA for registration. TA verifies the identity of an IoV Object, ID_j. If it is valid, it sends $\{S_{TA}, PK_j\}$, where PK_j is the pseudo-ID generation Key for an IoV Object, which the TA generates to generate pseudo-IDs. $\{ID_j, PWD_j, PK_j\}$ is stored in the database of the TA, which is used for revealing the original ID of an IoV object in case of violations.

3.3 Authentication Phase

V2V Authentication. Secure authentication between vehicles is established by using the following steps.

Step 1- Pseudo ID Generation: The suggested approach uses several one-time pseudo-IDs to ensure privacy. Each pseudo-ID is used only once. In order to compute the pseudo-ID, Vehicle V_i sends ID_i and PWD_i as input to the TMP of Vehicle Vi's TMP. If the ID_i and PWD_i are valid, then the Pseudo ID is computed as.

$$PSID_i = ID_i \oplus H\ (S_{TA} \parallel N_i \parallel PK_i) \quad (1)$$

where ID_i is the original ID of V_i, S_{TA} is the secret key of TA, N_i is the random nonce, and PK_i is the pseudo ID generation key.

Step 2-Signing: Once the pseudo ID is generated, the next step is the signing process. The message to be transmitted is signed before sending it to the nearby vehicles. The procedure for signing is as follows.

$$S_i = N_i \oplus H\ (PSID_i \parallel S_{TA} \parallel T)$$

$$K_i = H\ (S_i \| M \| S_{TA} \| T \| N_i) \quad (2)$$

where N_i is the random nonce, $PSID_i$ is the pseudo ID, S_{TA} is the secret key of TA, T is the current timestamp, and M is the message to be transmitted to the nearby vehicles. Vehicle V_i sends $\{S_i, K_i, PSID_i, T, M\}$ to the nearby vehicle V_E.

Step 3-Verification: Vehicle V_E, initially validates the timestamp T. If$|T^*-T| \le \Delta T$ is valid, then it verifies the received message as follows.

$$N_i = S_i \oplus H\,(PSID_i \| S_{TA} \| T)$$

$$K_i = H\,(S_i \| M \| S_{TA} \| T \| N_i) \tag{3}$$

If *the* computed K_i^* is the same as the received K_i, the message is accepted since the sender is the legitimate vehicle.

V2I Authentication. The Vehicle-to-Infrastructure Authentication process is similar to that of V2V Authentication. Initially, the pseudo-ID is generated, followed by the signing and verification processes by the RSU/other Infrastructures.

I2V Authentication. RSU periodically broadcasts traffic-related messages to all the vehicles within its area. The procedures to set up secure I2V Authentication are as follows.

Step 1-Signing: The message to be transmitted is signed before sending it to the vehicles in its range. The signing process is done as follows.

$$S_R = N_R \oplus H\,(ID_R \| S_{TA} \| T).$$

$$K_R = H\,(S_R \| M \| S_{TA} \| T \| N_R) \tag{4}$$

where N_R is the random nonce, ID_R is the ID of the RSU, S_{TA} is the secret key of TA, T is the current timestamp, and M is the message to the vehicles in its range. RSU sends $\{S_R, K_R, ID_R, T, M\}$ to all vehicles in its area.
Step 2-Verification: Vehicle V_E, firstly, confirms the timestamp T after receiving the message from the RSU. If$|T^*-T| \le \Delta T$ is authentic, the following steps are taken to verify it. If the computed K_R^* is the same as the received K_R, then the message is accepted.

$$N_R = S_R \oplus H\,(ID_R \| S_{TA} \| T).$$

$$K_R = H\,(S_R \| M \| S_{TA} \| T \| N_R) \tag{5}$$

IoV2IoV & IoV2V Authentication. IoV objects can be pedestrian devices or other devices in the IoV environment. Authentication between IoV devices/Vehicles is similar to that of V2V authentication.

IoV2I & I2IoV Authentication. Authentication between IoV devices and Infrastructure is the same as V2V authentication, whereas authentication between Infrastructure and IoV devices is similar to that of I2V authentication.

4 Performance Analysis

The suggested strategy is evaluated with similar schemes: Li et al. [24], Wang et al. [26], Vasudev et al. [11] and Yu et al. [12] in terms of Computation, Communication and Storage Cost.

Computation Cost. The time taken for the computation of asymmetric encryption (AE), decryption (AD), signature (S), symmetric encryption (SE), decryption (SD) and hash function (H) (SHA – 256) are 4.406 ms, 7.761 ms, 24.835 ms, 7.761 ms, 0.001 ms and 0.002 m respectively. The desktop computer's specifications include Windows 10 and an Intel (R) Core i5-7200U processor with 8.1 GB of RAM. The specification of Raspberry Pi is an ARMv6-compatible processor, BCM 2708 SOC, and an 8 GB SD Card. The programming language used is Python version 3.6.

Table 2 provides a comparison of the suggested method with other relevant schemes. In V2V authentication, the suggested method's computation time is 5TH, whereas in the IoV-SMAP scheme it is 22TH. The computation time for the V2V authentication process in the proposed scheme is 0.01 ms if implemented on a Desktop computer and 0.87 ms on the Raspberry Pi Platform. Hence, the computation cost in the V2V authentication is 77.27% less than that of the IoV-SMAP scheme. Similarly, for V2I authentication, the suggested method's computation time is 5TH, whereas in the IoV-SMAP scheme it is 18TH. The computation time for the V2I authentication process in the proposed scheme is 0.01 ms if implemented on a Desktop computer and 0.87 ms on the Raspberry Pi Platform. Hence, the computation cost in the V2V authentication is 72.22% in contrast to the IoV-SMAP scheme.

Table 2. Computation Time – Comparison.

Scheme	Authentication	Computation Time	Desktop Computer (ms)	Raspberry Pi
Li et al. [24]	V2V	T_S	24.835	709.143 ms
Wang et al. [26]	V2V	10 T_H	0.020	1.740 ms
Vasudev et al. [11]	V2V, V2I	6TH+ T_{SE}+ T_{SD}	7.774	4487.577 ms
Yu et al. [12]	V2V, V2I	22TH, 18TH	0.044, 0.036	2.958 ms
Proposed Scheme	V2V, V2I, I2V	5TH, 5TH, 4TH	0.01, 0.01, 0.008	2.62 ms

Communication Cost. The size (bits) of the timestamp (L_T), asymmetric encryption/decryption ($L_{AE/AD}$), random number/identity (L_{ID}), symmetric encryption/decryption ($L_{SE/SD}$), signature (L_S), and hash function (L_H) are 64 bits, 1024 bits, 80 bits, 128 bits, 1536 bits, and 256 bits respectively.

In V2V authentication of the proposed scheme, a vehicle sends $\{S_i, K_i, PSID_i, T, M\}$ to the nearby vehicles. The communication cost for V2V authentication is ($L_H + L_H + L_{ID} + L_T$) = (256 + 256 + 80 + 64 = 656 bits). The communication cost for V2I and I2V authentication is also 656 bits. Table 3 provides a comparison of the suggested method with other relevant schemes. As shown, it has 53.4% less communication overhead in contrast to IoV-SMAP.

Table 3. Communication Cost – Comparison.

Scheme	Authentication	Computation Time	Cost (bits)
Li et al. [24]	V2V	$L_T + 2L_{ID} + L_S$	1760
Wang et al. [26]	V2V	$4L_H + 2L_{ID} + 3L_T$	1376
Vasudev et al. [11]	V2V, V2I	$2L_T + L_{SE}+3L_H$, $4L_T + 2L_T + L_{SE}$	1024, 1280
Yu et al. [12]	V2V, V2I	$5L_H + 2L_T$	1408
Proposed Scheme	V2V, V2I, I2V	$2L_H + L_{ID} + L_T$	656

Storage Cost. The size (bytes) of the timestamp (L_T), asymmetric encryption/decryption ($L_{AE/AD}$), random number/identity (L_{ID}), symmetric encryption/decryption ($L_{SE/SD}$), signature (L_S), and hash function (L_H) are 8 bytes, 128 bytes, 10 bytes, 16 bytes, 192 bytes, and 32 bytes respectively.

In the V2V authentication of the proposed scheme, $\{ID_i, PWD_i, S_{TA}, PK_i\}$ and $\{ID_i, PWD_i, S_{TA}, PK_i\}$ are stored in the memory of the vehicles. It incurs ($10 + 10 + 32 + 10 = 62$ bytes) and ($10 + 10 + 32 + 10 = 62$ bytes). Hence, for V2V authentication, the storage incurred is 124 bytes, which is 35.41% less when compared to that of the IoV-SMAP Scheme. In V2I authentication, $\{ID_i, PWD_i, S_{TA}, PK_i\}$ and $\{ID_i, S_{TA}\}$ are stored in the memory. It incurs ($10 + 10 + 32 + 10 = 62$bytes) and ($10 + 32 = 42$ bytes). Hence, for V2I authentication, the storage incurred is 104 bytes, which is 24.63% less when compared to that of the IoV-SMAP Scheme. Table 4 provides a comparison of the suggested method's storage cost with other relevant schemes. As demonstrated, the suggested approach has less storage overhead than IoV-SMAP.

Table 4. Storage Cost – Comparison.

Scheme	Authentication	Computation Time	Storage Cost (bytes)
Li et al. [24]	V2V	$2L_T + 5L_{ID} + 2L_S$	450
Wang et al. [26]	V2V	$14L_H + 4L_{ID}$	488
Vasudev et al. [11]	V2V, V2I	$6L_H$, $4L_H + L_{ID}$,	192, 138
Yu et al. [12]	V2V, V2I	$6L_H$, $4L_H + L_{ID}$	192, 138
Proposed Scheme	V2V, V2I, I2V	$2L_H + 6L_{ID}$, $2L_H + 4L_{ID}$	124, 104

5 Conclusion

In this paper, a novel secure authentication scheme is proposed for IoV. To reduce the overheads, expensive operations are omitted; XOR and hash functions are used for the authentication process. A security study demonstrates that, besides authentication, the suggested system ensures all relevant security requirements. Compared to similar methods, it also has lower processing, communication, and storage overheads. According to a performance analysis, the suggested method has computation, communication, and storage overheads of 77.27%, 53.4%, and 35.41% lower than those of the IoV-SMAP scheme.

References

1. Save lives—a road safety technical package. Accessed 16 June 2020. [Online]. Available: http://apps.who.int/iris/bitstream/handle/10665/ 255199/9789241511704-eng.pdf
2. Road Traffic Injuries. Accessed:16 June 2020. [Online]. Available: http://www.who.int/en/news-room/fact-sheets/detail/road-traffic-injuries
3. Park, Y.H., Park, Y.H.: Three-factor user authentication and key agreement using elliptic curve cryptosystem in wireless sensor networks. Sensors. **16**(12), 1–17 (2016)
4. Ali, A.H., Li, F.: Authentication and privacy schemes for vehicular ad hoc networks (VANETs): a survey. Veh. Commun. **16**, 45–61 (2019)
5. Yu, S.J., Lee, J.Y., Lee, K.K., Park, K.S., Park, Y.H.: Secure authentication protocol for wireless sensor networks in vehicular communications. Sensors. **18**(10), 1–23 (2018)
6. Cui, L., Xie, G., Qu, Y., Gao, L., Yang, Y.: Security and privacy in smart cities: challenges and opportunities. IEEE Access. **6**, 46134–46145 (2018)
7. Lee, J.Y., Yu, S.J., Park, K.S., Park, Y.H., Park, Y.H.: Secure three-factor authentication protocol for multi-gateway IoT environments. Sensors. **19**(10), 1–25 (2019)
8. Chourabi, H., Nam, T., Walker, S., Gil-Garcia, J.R., Mellouli, S., Nahon, K., Pardo, T.A., Scholl, H.J.: Understanding smart cities: an integrative framework. In: Proceeding of the 45th Hawaii International Conference on System Sciences, Maui, HI, USA, pp. 1–9 (Jan 2012)
9. Sutrala, K., Bagga, P., Das, A.K., Kumar, N., Rodrigues, J.J.P.C., Lorenz, P.: On the design of conditional privacy preserving batch verification-based authentication scheme for internet of vehicles deployment. IEEE Trans. Veh. Technol. **69**(5), 5535–5548 (2020)
10. Wang, J., Jiang, C., Han, Z., Ren, Y., Hanzo, L.: Internet of vehicles: sensing-aided transportation information collection and diffusion. IEEE Trans. Veh. Technol. **67**(5), 3813–3825 (2018)
11. Vasudev, D.D., Vasilakos, A.V.: Secure message propagation protocols for IoVs communication components. Comput. Electr. Eng. **82**, 1–15 (2020)
12. Yu, S., Lee, J., Park, K., Das, A.K., Park, Y.: IoV-SMAP: secure and efficient message authentication protocol for IoV in Smart City environment. IEEE Access. **8**, 167875–167886 (2020). https://doi.org/10.1109/ACCESS.2020.3022778
13. Kocher, P., Jaffe, J., Jun, B.: Differential power analysis. In: Proceedings of Advances in Cryptology (CRYPTO), Berlin, Germany, pp. 388–397. Springer, Berlin/Heidelberg (1999)
14. Yu, S.J., Park, K.S., Park, Y.H.: A secure lightweight three-factor authentication scheme for IoT in cloud computing environment. Sensors. **19**(16), 1–20 (2019)
15. Park, Y., Lee, S., Kim, C., Park, Y.: Secure biometric-based authentication scheme with smart card revocation/reissue for wireless sensor networks. Int. J. Distrib. Sensor Netw. **12**(7), 1–11 (2016)

16. Yu, S.J., Lee, J.Y., Park, Y.H., Park, Y.H., Lee, S.W., Chung, B.H.: A secure and efficient three-factor authentication protocol in global mobility networks. Appl. Sci. **10**(10), 1–23 (2020)
17. Lee, J., Yu, S., Kim, M., Park, Y., Das, A.K.: On the design of secure and efficient three-factor authentication protocol using honey list for wireless sensor networks. IEEE Access. **8**, 107046–107062 (2020)
18. Jenefa, J., Anita, E.A.M.: Identity-based message authentication scheme using proxy vehicles for vehicular ad hoc networks. Wirel. Netw. **27**, 3093–3108 (2021). https://doi.org/10.1007/s11276-021-02655-6
19. The future predictions about IoT only a handful of people know [Online]. Available at: https://www.peerbits.com/blog/future-predictions-about-iot-handful-of-people.html
20. Jenefa, J., Mary Anita, E.A.: An enhanced secure authentication scheme for vehicular ad hoc networks without pairings. Wirel. Pers. Commun. **106**, 535–554 (2019). https://doi.org/10.1007/s11277-019-06178-4
21. Jenefa, J., Mary Anita, E.A.: Secure vehicular communication using ID based signature scheme. Wirel. Pers. Commun. **98**, 1383–1411 (2018). https://doi.org/10.1007/s11277-017-4923-7
22. Ruan, N., Li, M., Li, J.: A novel broadcast authentication protocol for internet of vehicles. Peer-to-Peer Netw. Appl. **10**(6), 1331–1343 (2017)
23. Zhu, L., et al.: PRIF: a privacy-preserving interest-based forwarding scheme for social internet of vehicles. IEEE Internet Things J. **6**(5), 8804–8817 (2019)
24. Li, J., Lu, H., Guizani, M.: ACPN: a novel authentication framework with conditional privacy-preservation and non-repudiation for VANETs. IEEE Trans Parallel Distrib Syst. **26**(4), 938–948 (2015)
25. Dua, N., Kumar, A.K., Das, Susilo, W.: Secure message communication protocol among vehicles in smart city. IEEE Trans. Veh. Technol. **67**(5), 4359–4373 (2018)
26. Wang, F., Xu, Y., Zhang, H., Zhang, Y., Zhu, L.: 2FLIP: a two-factor lightweight privacy-preserving authentication scheme for VANET. IEEE Trans. Veh. Technol. **65**(2), 896–911 (2016)
27. Amin, R., Lohani, P., Ekka, M., Chourasia, S., Vollala, S.: An enhanced anonymity resilience security protocol for vehicular ad-hoc network with Scyther simulation. Comput. Electr. Eng. **82**, 1–18 (2020)
28. Liu, Y.W., Chang, G.: Efficient privacy-preserving dual authentication and key agreement scheme for secure V2V communications in an IoV paradigm. IEEE Trans. Intell. Transp. Syst. **18**(10), 2740–2749 (2017)
29. Ying, Nayak, A.: Anonymous and lightweight authentication for secure vehicular networks. IEEE Trans. Veh. Technol. **66**(12), 10626–10636 (2017)
30. Chen, C.-M., Xiang, B., Liu, Y., Wang, K.-H.: A secure authentication protocol for internet of vehicles. IEEE Access. **7**, 12047–12057 (2019)
31. Kaiwartya, O., et al.: Internet of vehicles: motivation, layered architecture, network model, challenges, and future aspects. IEEE Access. **4**, 5356–5373 (2016)
32. May, W.E.: Secure Hash Standard. FIPS PUB 180–1, National Institute of Standards and Technology (NIST), U.S. Department of Commerce (Apr 1995). Accessed Aug 2019. [Online]. Available: http://nvlpubs.nist.gov/nistpubs/FIPS/NIST.FIPS.180-4.pdf
33. Anita, E.A.M., Jenefa, J.: A survey on authentication schemes of VANETs. Int. Conf. Inform. Commun. Embedded Syst. **2016**, 1–7 (2016). https://doi.org/10.1109/ICICES.2016.7518946
34. Jenefa, J., Mary Anita, E.A.: Secure authentication schemes for vehicular Adhoc networks: a survey. Wireless Pers. Commun. (2021). https://doi.org/10.1007/s11277-021-09118-3
35. Jenefa, J., Mary Anita, E.A., Sambandam, R.K., Vetriveeran, D., Vinodha, D.: A Secure Resilient Scheme for Autonomous Vehicles against External Attacks, Cyber Space and Outer Space Security (2024)

36. Jenefa, J., Anita, E.A.M., Divya, V., Kandan, S.R., Vinodha, D.: Secure identity based authentication for emergency communications. In: Zen, H., Dasari, N.M., Latha, Y.M., Rao, S.S. (eds.) Soft Computing and Signal Processing (2024)
37. Jenefa, J., Sajini, S., Anita, E.A.M.: Secure IBS scheme for vehicular ad hoc networks. In: Reddy, V.S., Prasad, V.K., Wang, J., Reddy, K.T.V. (eds.) Soft Computing and Signal Processing (2023)
38. Lakshmi, S., Anita, E.A.M., Jenefa, J.: A hybrid approach against black hole attackers using dynamic threshold value and node credibility. IEEE Access. **11**, 91595–91603 (2023)
39. Mary Anita, E.A., Lakshmi, S., Jenefa, J.: A self-cooperative trust scheme against black hole attacks in vehicular ad hoc networks. Int. J. Wirel. Mob. Comput. **21**(1), 59–65 (2021)
40. Lakshmi, S., Mary Anita, E.A., Jenefa, J.: Detection and prevention of black hole attacks in vehicular ad hoc networks. Int. J. Innov. Technol. Explor. Eng. **8**(7) (2019)
41. Proshkin, N.A., Basan, E.S., Lapina, M.A., Klepikova, A.G., Lapin, V.G.: Developing models of IoT infrastructures to identify vulnerabilities and analyse threats. IOP Conf. Series Mater. Sci. Eng. **873**(1), 012018 (2020). https://doi.org/10.1088/1757-899X/873/1/012018
42. Vinodha, D., et al.: Framework based on IoT, AI, and blockchain for smart access to government agricultural schemes. In: The future of agriculture: IoT, AI and Blockchain Technology for Sustainable Farming, pp. 142–163 (2024). https://doi.org/10.2174/97898152743491240l 0011. BENTHAM SCIENCE PUBLISHERS
43. Anita, E.A.M., Jenefa, J., Vinodha, D., Lapina, M.: Advancements in Sybil attack detection: a comprehensive survey of machine learning-based approaches in wireless sensor networks. In: AISMA-2024: international workshop on advanced information security management and applications, vol. 863, pp. 67–75. Springer, Springer Nature Switzerland, Cham. https://doi.org/10.1007/978-3-031-72171-7_7

Adaptive Data Storage, Transmission and Processing System in Fog Computing Using Residue Number System and Artificial Neural Networks

Dmitriy Gorlacev[1], Nikita Mirny[1], Aisanat Geryugova[1], Vladislav Lutsenko[1](✉), and Mikhail Zgonnikov[2]

[1] North-Caucasus Federal University, Stavropol, Russia
officialvladlutsenko@gmail.com
[2] School of Engineering and Computer Science, Lyon, France

Abstract. The article presents an adaptive data storage, transmission, and processing system for fog computing environments, leveraging the Residue Number System and Artificial Neural Networks. The proposed architecture demonstrates advantages in reliability, safety, scalability, resource efficiency, and adaptability compared to existing methods. The proposed architecture demonstrates the potential for creating adaptive and fault-tolerant computing platforms.

Keywords: Fog Computing · Internet of Things · Artificial Neural Networks · Residue Number System · Distributed Systems

1 Introduction

Today's Internet of Things (IoT) systems generate huge amounts of data requiring analysis and storage, resulting in significant latency in their transmission and processing in cloud computing. With the development of response time critical applications, remote cloud services will be unable to provide reliable communication with minimum latency. Also, sending data directly to the cloud may not be secure due to privacy issues.

To solve such problems, the Fog Computing (FC) paradigm was introduced. FC allows data processing to be performed at the edge of the network, which reduces latency and optimizes the load on central servers [1, 2]. However, the dynamic nature of data, instability of network links, and limited resources of IoT devices create a need for adaptive storage and data transfer systems. Modern deep learning techniques, in particular artificial neural networks, have demonstrated high efficiency in optimizing the storage and transmission of large data [3]. In addition to this, Residue Number System (RNS) can be used for reliability and fault tolerance [4].

The aim of this paper is to develop a system that combines the advantages of fog computing, ANN and RNS to allow for improved fault tolerance, security and system adaptability in big data processing.

M. Lapina et al. (Eds.): AISMA 2025, LNNS 1456, pp. 156–167, 2026.
https://doi.org/10.1007/978-3-032-07275-7_15

This paper is divided into the following sections: Sect. 2 introduces FC, its architecture and also raises a number of issues. Section 3 introduces artificial neural networks and explains their importance in FC. Sections 4–6 are devoted to residual number system, in particular the methods of fault tolerance, error detection, localization and correction in Fog Computing are discussed. Section 7 presents an adaptive storage system using RNS and ANN techniques. A comparative analysis of the proposed system with existing approaches is presented in Sect. 8.

2 Fog Computing Architecture

The term FC was introduced by Cisco in 2002. They proposed this concept as a way to move cloud computing closer to the network edge to reduce latency and improve data processing efficiency [2].

FC is a highly virtualized platform that provides computing, storage and networking services between end devices (cameras, drones, sensors, mobile devices) and traditional cloud computing data centers. Unlike cloud computing, where data is processed in centralized data centers, in fog computing, processing takes place on intermediate nodes (Fog Nodes) closer to the user or data source [5]. Fog nodes can act as local servers, gateways, base stations, routers, etc. They perform data preprocessing, filtering, analyzing and make quick decisions. Figure 1 is a diagram showing the relative distribution of computing resources under the concepts of fog computing and cloud computing.

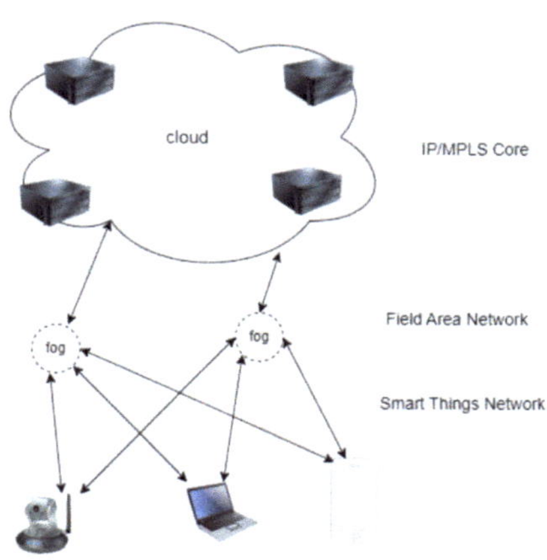

Fig. 1. Fog computing architecture.

However, despite the advantages, there are certain disadvantages and threats with the use of such technology.

1. *Data security and privacy.* Data processed at the edge of the network may be vulnerable to cyber-attacks (data interception, unauthorized access, etc.). This is because devices at the edge of the network are less secure than the centralized cloud, making them vulnerable. Such a problem is described in detail in [6, 7]. Such a problem can

be solved by a residual class system. To improve security, homomorphic encryption based on RNS can be used, which is discussed in Sect. 5.

2. *Fault tolerance and reliability.* Distributed systems in FC may be prone to failures of individual nodes, which requires developing mechanisms to ensure reliability and resilience of the entire system, unlike cloud data centers. Also, fog nodes may not have sufficient resources to recover from failures, which reduces system reliability. Such, and not only, problem is described in [8], which talks about the presence of unreliable wireless communication channels. RNS allows to perform computations in parallel without inter-bit transfers, which reduces the probability of failures and improves the reliability of computations, where it is discussed in Sect. 5.
3. Load *balancing and adaptability.* In FC, the load on nodes can change dynamically, which creates resource allocation problems. If some nodes are overloaded and others are idle, it reduces the overall efficiency of the system. Such a problem is described in [5], where the issues of computing resource management are raised. A solution to such a problem can be the implementation of artificial neural networks. It is described in detail in Sect. 3.

3 Application of Artificial Neural Networks

Artificial neural networks (ANNs) are computational models inspired by the structure and functioning of biological neural networks such as the human brain. They consist of many interconnected nodes called neurons that work together to process and analyze data [9]. Figure 2 shows a graphical representation of the neuron model. First, data is input and then each neuron processes it using a activation function and passes the result to the next layer [10]. Basically there are several of these layers: input layer, hidden layer and output layer. The connections between neurons have weights that are adjusted during training of the network. The neural network architectures presented by He et al. [4], allow efficient model training even with a large number of layers, which is especially important when dealing with high-dimensional IoT data. This paper proposes an idea to improve data transmission and processing, using ANN techniques. For example, a neural network can be trained to predict the workload of fog nodes and select more optimal transmission channels.

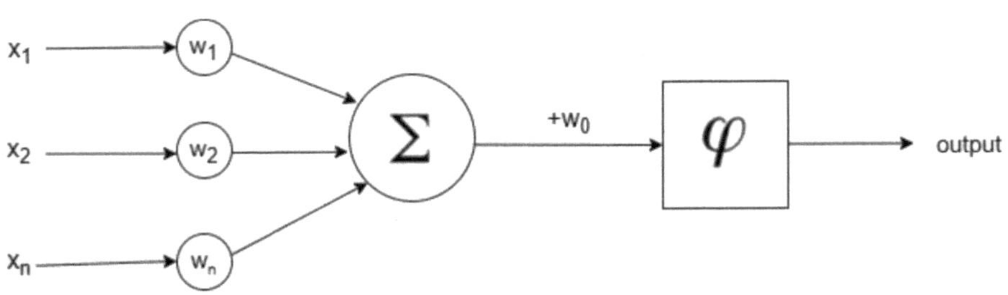

Fig. 2. Graphical representation of a threshold block or McCulloch neuron.

In [11] the problem of database replication in cloud systems is considered and the use of machine learning for predictive replication management is proposed. The main objective of the paper is to develop Machine Learning (ML) approaches to improve fault tolerance and load balancing in database replication systems. Such technology can be closely related to fog computing. The article raises Fault Tolerance Issues: when one of the servers fails, its role may not be immediately taken over by another node, which causes downtime and data loss. The author suggests using neural networks and machine learning: ML model analyses node metrics (CPU load, latency, amount of transferred data) and determines which nodes may fail. ML also analyses the current load and optimally distributes requests between nodes, thus reducing overload on individual nodes and increasing the efficiency of the whole system.

The paper [12] considers the use of neural networks in fog and boundary value computation. Several methods are considered for optimal distribution of tasks in nodes as the latter have limited resources. For example: clustering (k-means, k-NN) - grouping data based on similarity, Support Vector Machine (SVM), Deep Neural Network (DNN).

The paper [13] is devoted to the development of an adaptive Anomaly Behavior Analysis Intrusion Detection System (ABA-IDS) for fog nodes in IoT based on artificial neural networks. The system is designed to protect nodes by ensuring that they remain operational even in the event of cyberattacks or failures. Unlike traditional IDS's, this system focuses on simulating normal node behavior and identifies anomalies. In real time, the system compares the current performance with the reference model. In summary, this approach was successfully able to identify the danger, demonstrating a high detection accuracy of 93% and a 13% improvement in command processing time.

The paper [14] also presents several strategies to achieve load balanced fog nodes using Artificial Neural Networks, especially in healthcare system. A variety of strategies have been proposed, but it is the PNN-based Matchmaking Algorithm (PMA) that we are interested in. This algorithm uses a probabilistic neural network to estimate which server is best suited for a particular task. The inputs are the server's CPU power, available RAM, and available storage/cache. These three parameters are then compared to the requirements of the task. The neural network is trained on historical data about the load of the servers and their ability to process tasks. When a new task arrives, it analyses which server is best suited to perform the task. Then the server performs the task and the system records its performance for further training of the neural network.

As a result, it was obtained that using such strategies, will achieve adaptability and minimum delay and evenly distributed load on the network.

4 Residue Number System

RNS is a mathematical structure used to speed up arithmetic computations and to represent numbers efficiently in computing systems. RNS is based on the idea of decomposing numbers into residues over a set of mutually prime numbers, which allows operations such as addition, multiplication and division to be performed more efficiently and in parallel. RNS works with numerical residues over a set of moduli $\{p_1, p_2, \dots, p_n\}$ called the basis. By default, the moduli are ordered in ascending order, that is $p_1 < p_2 < \dots < p_n$. An integer X can be represented as n-dimensional vector, which is composed of the smallest non-negative residue from dividing the corresponding number by p_i:

$$X = (x_1, x_2, \dots, x_n), \tag{1}$$

where $x_i = X \, mod \, p_i$, which may also be labelled as $x_i = |X|_{p_i}$.

If the number X is given as the residues $x_1, x_2, \ldots, x_n$ from division modulo $p_1, p_2, \ldots, p_n$, the number X can be recovered using the Chinese Remainder Theorem (CRT):

$$X = \left| \sum_{i=1}^{n} P_i \bullet x_i \bullet \left| P_i^{-1} \right|_{p_i} \right|_P, \quad (2)$$

where $P_i = \frac{P}{p_i}$ and $\left| P_i^{-1} \right|_{p_i}$ represents the multiplicative inversion of P_i modulo p_i and satisfies the following relation $\left| \left| P_i^{-1} \right|_{p_i} \bullet P_i \right|_{p_i} = 1$.

CRT will help to recover the data if all balances are available. The main advantages of RNS are computational acceleration, scalability and data security [15].

RNS can also be used in fog computing to represent data collected by IoT devices as residuals. These residuals can be processed locally at fog nodes (e.g., gateways), allowing for quick computations and when the data is required for further processing in the cloud, it can be recovered using the CRT.

5 Methods of Increasing Fault Tolerance

The use of special coding methods for computing facilities is associated with the fact that any computer, in fact, is a system of transmission and processing of information. Transfer of information in computing machines is carried out with high speed and significant volumes, and the reliability of arithmetic and logical processing of information must be ensured. There are several directions, by means of which methods of increasing fault tolerance of computations have been developed.

The first direction is the works of Von-Neumann [16] related to the construction of reliable devices. As a result of this direction, reliability is achieved due to structural redundancy. The second direction is data coding at the input of computing devices, so that it was possible to detect and correct errors that occurred during computation. Such encoding is effectively used in devices operating with integers, for example, in [17] describes a method of encoding numbers in the format $+B$, where A and B are positive integers, and n is the encoded word. Such a method allows arithmetic operations to be performed directly on the encoded numbers, and an error in the computation changes the result in a predictable way, which makes it possible to detect and correct it. A significant disadvantage of this method is that it only works on integers. The outcome of this study is to prove that it is possible to detect and correct errors without the use of special equipment. It is almost impossible to achieve significant speedup of operations in a conventional positional system. The results of research on improving the performance of computing systems show that no significant improvement in current results can be expected within positional number systems.

With the development of new technologies, a larger volume for storing and processing information is required. One of the promising solutions to this problem is the use of distributed environment. However, when storing and processing data in a distributed environment, the issue of ensuring the integrity and confidentiality of information is acute. In the field of fault tolerance and information protection homomorphic codes using the system of residual classes have played a great role.

For example, [18] proposes the use of the HORNS homomorphic encryption scheme to improve security in cloud computing data transmission.

This method is good because the data is processed on different cloud servers, which prevents the possibility of compromising information if an attacker gains access to only part of the data.

It is also worth considering the set of RNS moduli. In [19] a set of moduli $\{2^{n-1}, 2^n, 2^{n+1}\}$. In [20] it was proved that exactly such a set of moduli is the most efficient among other modular sets. The approach proposed in [21] is based on the transition to modular recursive computations on selected bases $p_1, p_2, \ldots, p_n$ to modular. The disadvantage of such a method is the conversion from RNS back to the positional system, which is computationally difficult.

One feature of the residual class system is parallelism. If you add to the original system with moduli $\{p_1, p_2, \ldots, p_n\}$ an additional base $p_{n+1} > p_i$, where $i = 1, n$, then the full range of the system will be $|P| = P \bullet p_{n+1}$. If the result of an operation on numbers in the range $[0, P)$ is less than P, then it is correct, otherwise it is incorrect. Thus, the introduction of the redundant base p_{n+1} allows us to detect computational errors [22].

This approach is widely used by fault-tolerant applications. For example, in [23] a number is represented in k basic moduli to represent the number, and r additional moduli provide redundancy. To detect and correct a single error, pseudo-syndromes are introduced to represent the expansion of the obtained residues by realising the parallelism property of the CRT. For each particular value, the corresponding result is defined as the inferred syndrome. This method reduces the hardware complexity to $O(k)$, and the latency to $O(\log_2 k)$. Another Redundant Residue Number System (RRNS) with six moduli is applied in the construction of hybrid storage, such as in [24]. The results of this study showed that such a number of moduli in a redundant residual class system has a competitive error correction ability, provides a larger storage capacity and better decoding performance than Reed-Solomon codes.

Generally, existing multiple error detection and correction algorithms use three different methods to locate erroneous digits of a residue. These are consistency checking using syndrome [25], number recovery by CRT [26] and modulo projection [27].

The following can be summarized. Introduction of redundant moduli into RNS allows not only detecting but also correcting errors, however, most error correction methods have high algorithmic complexity. The task arises to develop a mathematical model of a data processing system using modular arithmetic, as well as to develop specialized computing nodes of a distributed environment to perform non-modular operations in RNS.

6 Method of Detection, Localization and Correction of Errors in RNS

As discussed in Sect. 5, one approach to improve computational fault tolerance is to use RRNS. The execution of operations in the RRNS is performed in parallel without the need for inter-bit transfers, which allows for efficient addition, subtraction, and multiplication.

Since computations in each modulo are performed independently, the modular code has built-in correction properties, since each remainder contains information about the whole number. In [22], a modification of the projection method was described using an approximate method based on CRT. The method consists in adding redundant moduli p_{n+1} and p_{n+2}, to detect and correct a single error. In order to detect and correct errors we need to find the positional characteristic of the number, the approximate value is found:

$$\frac{A}{|P|} = \left| \sum_{i=1}^{n+2} \frac{\left||P|_i^{-1}\right|}{p_i} \alpha_i \right|_1 = \left| \sum_{i=1}^{n+2} k_i \alpha_i \right|_1, \tag{3}$$

where $|P| = \prod_{i=1}^{n+2} p_i = P \bullet p_{n+1} \bullet p_{n+2}$ is the range of the redundant residual class system, $k_i = \frac{\left||P|_i^{-1}\right|}{p_i}$ is the constant of the selected RNS, $\left||P|_i^{-1}\right|$ is the multiplicative inversion. The number $=(\alpha_1, \alpha_2, \ldots, \alpha_n, \alpha_{n+1}, \alpha_{n+2})$, if $A \in [0, P)$ is allowed, in the case of $A \in [P, |P|)$ we can say that the number contains an error.

To detect and correct errors, the projection method is used, which consists of the following. Assume that the error was made on the first modulo p_1 and repeat the calculations without the participation of this modulo, i.e. take as a redundant range $|P|_1 = \frac{|P|}{p_1}$, for the new range calculate new constants k_i and finally compare $\frac{A}{|P|_1}$ with the constant $\frac{P}{|P|}$. In the case of $\frac{A}{|P|_1} < \frac{P}{|P|_1}$, it is considered that the error occurred modulo p_1 and to find the exact value of the remainder A modulo p_1, multiply $\frac{A}{|P|_1}$ by $|P|_1$ and find the remainder modulo p_1. In the case of $\frac{A}{|P|_1} > \frac{P}{|P|_1}$ we can say that the error is still present and to detect and correct it, we exclude p_2 instead of p_1 and perform a similar check. Such projections are built for all moduli $p_1, \ldots, p_{n+2}$.

The application of the approximate method replaces the computationally expensive operation of finding the remainder of division by the range of the system with the extraction of the least significant bits of the number. This reduces the computational complexity from quadratic to linear-logarithmic.

7 Adaptive Data Storage and Processing System

We propose the following structure adaptive data storage and processing system Fig. 3.

1. **Input data.** This is the data that is fed into the system for processing. They can be represented in various formats (e.g. numeric values, images, text).
2. **Convert to RNS.** This block converts input data from decimal notation to a representation in the RNS. It performs the following steps:
   (1) Selects a set of mutually simple moduli $p_1, p_2, \ldots, p_k$. Calculates the residuals from dividing the input value by each modulo.
   (2) Forms a representation of the input value in the RNS.

   The advantage is increased data security. The control parameters are a set of moduli $p_1, p_2, \ldots, p_k$.

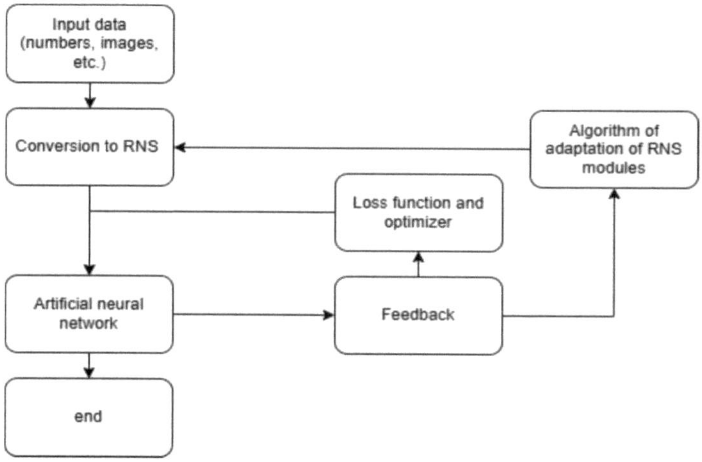

Fig. 3. System structure.

3. **Artificial Neural Network.** This is the main computing unit of the system. It receives data in RNS format and performs data processing. The characteristics are network architecture, activation function, loss function, learning algorithm.
   Input: Data in RNS format.
   Output: The result of processing (e.g. forecasting) can also be presented in the RNS.
4. **Feedback (ANN output).** This link transmits the result of ANN operation (in RNS format) to the block "RNS moduli adaptation algorithm". The purpose is to inform the adaptation algorithm about the current ANN performance, errors and other indicators.
5. **Loss function and optimizer.** These are standard components of the neural network training process. The loss function determines how well the ANN performs. The optimizer determines how to change the weights and biases of the ANN to minimize the loss function.
6. **RNS moduli adaptation algorithm.** This is the key block that ensures the adaptability of the system. It analyses the ANN output data, current system performance and other factors to dynamically change the RNS moduli.
7. **Output data.** This is the final result of the system. It can be presented in different formats (depending on the input data).

For efficient data storage, the following aspects need to be considered: resource efficiency, adaptability, resilience to failure, and reliable storage with recoverability. RNS and ANN can provide all these conditions:

1. In Sects. 5 and 6, each piece of residual data is distributed to individual nodes in the network. This gives redundancy and increased fault tolerance – even if some nodes fail, the original data can be recovered. Thus, we can highlight the main advantages of RNS: high fault tolerance, confidentiality and reliability.
2. The neural network is trained to analyze node load, data transfer rate and failure probability. With all this information, is able to predict node errors and create backups

in advance. Also, to avoid congestion, the ANN will be able to decide which of the nodes is best to send data to. Thus, obtaining an adaptive system for data storage and transmission. Using the methods discussed in Sect. 3 this goal can be achieved.

8 Comparative Analysis

The paper [28] presents a mechanism for managing data reliability in cloud storage called Proactive Replica Checking for Reliability (PRCR), which minimizes the number of data replicas while maintaining the required level of reliability. The main objective of PRCR is to reduce cloud storage costs by reducing data redundancy, which is especially important for dealing with large amounts of data. The main advantage of PRCR is to significantly reduce the cost of data storage. The paper shows that PRCR can reduce cloud storage consumption by 49%. This method provides a high level of fault tolerance by actively monitoring and restoring replicas, also it is easily scalable to handle large amounts of data. However, in some cases when both replicas are lost at the same time, data can be lost and despite all the effectiveness of this method, it shows that it can be unreliable.

The following paper [29] focuses on improving performance and fault tolerance in big data systems using the MapReduce model. The standard fault tolerance mechanisms in Hadoop MapReduce result in significant delays and loss of resources when recovering from task or node failures. We propose a new approach called Fast Recovery MapReduce (FR-MR) that speeds up disaster recovery by using distributed checkpointing (storing data in distributed storage) and proactive push mechanism (the mechanism allows partial results to be used even in case of task or node failure). The main advantages of this method are scalability and high fault tolerance, but this algorithm is difficult to implement and energy consuming.

In [30], a method for managing data replication in cloud storage called Cost-effective Dynamic Replication Management (CDRM) is presented. It aims to improve data availability, performance and load balancing in cloud storage. CDRM uses a model that relates data availability to the number of replicas. Based on this model, the minimum number of replicas required to meet availability requirements is determined. These are placed on nodes based on their capacity. The main advantages of this method are reliability, adaptability and efficiency.

The paper [31] describes flexible schemes to trade off between storage capacity and access efficiency in reliable storage systems. The authors present two classes of codes collectively called Pyramid Codes, which are based on Reed-Solomon codes, i.e. they use mathematical transformations to create redundant blocks of data. Basic Pyramid Codes: build on existing codes and partition data into groups with local redundant blocks, and Generalized Pyramid Codes: more advanced codes that can improve reliability and/or access efficiency.

In each of the proposed works the main properties of each system were highlighted, on the basis of these properties the proposed systems were analysed with our system. The results of the comparative analysis are presented in Table 1.

Our proposed method wins in all parameters. It has the following properties:

1. High reliability – because the information is divided into different moduli consisting of residual data. This means that computational errors are detected and corrected at

Table 1. Comparative analysis of adaptive data storage, transmission and processing systems.

Methods	Characteristics					
	Reliability	Safety	Scalability	Effective Resource Usage	Efficiency	Adaptability
[28]		●	●	●	●	●
[29]	●		●			●
[30]	●		●		●	●
[31]	●	●		●	●	
Our method	●	●	●	●	●	●

the level of individual moduli without affecting the whole system. Also, some of the data can be recovered using CRT, which is something PRCR cannot boast of, which depends entirely on data replicas, if lost, the information cannot be recovered.

2. Safety – RNS supports homomorphic encryption. Data can be encrypted at the RNS level, making it inaccessible even if intercepted.
3. Efficiency and adaptivity – ANN dynamically adapts to changing load, i.e. if one node is overloaded, the network automatically redirects tasks to less loaded nodes through predictive analysis. The ANN learns from input data and automatically adjusts load balancing.
4. Resource usage – compared to some methods, ours can be restored without the need for replication. Still in ANN, operations are performed without inter-bit transfers, which speeds up processing. ANN is able to distribute tasks in a way that reduces the load on individual nodes.

9 Conclusion

The article presents an innovative adaptive data storage, transmission, and processing system designed for fog computing environments, leveraging RNS and ANN. The proposed architecture demonstrates advantages in reliability, safety, scalability, resource efficiency, and adaptability compared to existing methods.

Through a comparative analysis with existing methods such as PRCR [28], FR-MR [29], CDRM [30], and Pyramid Codes [31], our system outperforms in all key parameters. The RNS-based approach ensures high reliability by enabling error detection and correction at the modular level, while also supporting data recovery through the CRT. This eliminates the dependency on data replicas, a limitation observed in PRCR. Additionally, the integration of RNS with homomorphic encryption enhances data safety, ensuring secure data transmission and storage.

The ANN component further enhances the system's efficiency and adaptability by dynamically balancing workloads across nodes, optimizing resource utilization, and reducing processing delays. Unlike energy-intensive methods such as FR-MR, our system minimizes resource consumption by eliminating the need for replication and inter-bit transfers, while maintaining high fault tolerance and scalability.

In summary, the proposed system offers a comprehensive solution for adaptive, fault-tolerant, and secure data management in fog computing environments. Its combination of RNS and ANN provides a scalable, efficient, and reliable platform, addressing the limitations of existing methods and paving the way for future advancements in distributed computing systems.

Funding. The research was supported by the Russian Science Foundation Grant No. 22-71-10046, https://rscf.ru/en/project/22-71-10046/.

References

1. Bonomi, F., Milito, R., Zhu, J., Addepalli, S.: Fog computing and its role in the internet of things. In: Proceedings of the First Edition of the MCC Workshop on Mobile Cloud Computing, pp. 13–16 (2012)
2. Haouari, F., Faraj, R., AlJa'am, J.M.: Fog computing potentials, applications, and challenges. In: 2018 International Conference on Computer and Applications (ICCA), pp. 399–406. IEEE (2018)
3. Kim, K.G.: Book review: Deep learning. Healthc. Inf. Res. **22**(4), 351–354 (2016)
4. He, K., Zhang, X., Ren, S., Sun, J.: Deep residual learning for image recognition. In: Proceedings of the IEEE Conference on Computer Vision and Pattern Recognition, pp. 770–778 (2016)
5. Kirsanova, A.A., Radchenko, G.I., Tchernykh, A.N.: Overview of fog computing organization technologies. Vestnik Yuzhno-Ural'skogo Gosudarstvennogo Universiteta. Seriya Vychislitelnaya Matematika i Informatika. **9**(3), 35–63 (2020)
6. Stojmenovic, I., Wen, S., Huang, X., Luan, H.: An overview of fog computing and its security issues. Concurrency Comput.: Pract. Exper. **28**(10), 2991–3005 (2016)
7. Hedabou, M.: Cryptography for addressing cloud computing security, privacy, and trust issues. In: Computer and Cyber Security, pp. 281–304. Auerbach Publications (2018)
8. Bakhshi, Z., Rodriguez-Navas, G., Hansson, H.: Dependable fog computing: a systematic literature review. In: 2019 45th Euromicro Conference on Software Engineering and Advanced Applications (SEAA), pp. 395–403. IEEE (2019)
9. Krogh, A.: What are artificial neural networks? Nat. Biotechnol. **26**(2), 195–197 (2008)
10. Lutsenko, V., Zgonnikov, M.: Fault tolerant system for data storage, transmission and processing in fog computing using artificial neural networks. In: Conference on Current Problems of Applied Mathematics and Computer Systems, pp. 199–212. Springer Nature Switzerland, Cham (2023)
11. Jaber, S.: Machine Learning for Proactive Database Replication in Cloud-Based Systems. (2023)
12. Zou, Z., et al.: Edge and fog computing enabled AI for IoT-an overview. In: 2019 IEEE International Conference on Artificial Intelligence Circuits and Systems (AICAS), pp. 51–56. IEEE (2019)
13. Pacheco, J., Benitez, V.H., Felix-Herran, L.C., Satam, P.: Artificial neural networks-based intrusion detection system for internet of things fog nodes. IEEE Access. **8**, 73907–73918 (2020)
14. Talaat, F.M., Ali, S.H., Saleh, A.I., Ali, H.A.: Effective load balancing strategy (ELBS) for real-time fog computing environment using fuzzy and probabilistic neural networks. J. Netw. Syst. Manag. **27**, 883–929 (2019)
15. Koblitz, N.: A course in number theory and cryptography. Springer Science & Business Media (1994)

16. Von Neumann, J.: Probabilistic logics and the synthesis of reliable organisms from unreliable components. Automata studies. **34**(34), 43–98 (1956)
17. Brown, D.T.: Error detecting and correcting binary codes for arithmetic operations. IRE Trans. Electron. Comput. **3**, 333–337 (1960)
18. Gomathisankaran, M., Tyagi, A., Namuduri, K.: HORNS: a homomorphic encryption scheme for Cloud Computing using Residue Number System. In: 2011 45th Annual Conference on Information Sciences and Systems, pp. 1–5. IEEE (2011)
19. Gbolagade, K.A., Chaves, R., Sousa, L., Cotofana, S.D.: An improved RNS reverse converter for the $\{2^{n-1}, 2^n, 2^{n+1}\}$ moduli set. In: Proceedings of 2010 IEEE International Symposium on Circuits and Systems, pp. 2103–2106. IEEE (2010)
20. Phalakarn, K., Surarerks, A.: Alternative redundant residue number system construction with redundant residue representations. In: 2018 3rd International Conference on Computer and Communication Systems (ICCCS), pp. 457–461. IEEE (2018)
21. Mohan, P.A.: Residue number systems: algorithms and architectures. Springer Science & Business Media (2002)
22. Kuchukov, V., Babenko, M., Al-Galda, S.: Study of a redundant residue number system for single error correction. Adv. Syst. Sci. Appl. **23**(4), 31–39 (2023)
23. Yin, P., Li, L.: A new algorithm for single error correction in RRNS. In: 2013 International Conference on Communications, Circuits and Systems (ICCCAS), vol. 2, pp. 178–181. IEEE (2013)
24. Haron, N.Z., Hamdioui, S.: Redundant residue number system code for fault-tolerant hybrid memories. ACM j. Emerg. Technol. Comput. Syst. **7**(1), 1–19 (2011)
25. Krishna, H., Lin, K.Y., Sun, J.D.: A coding theory approach to error control in redundant residue number systems. I. Theory and single error correction. IEEE Trans. Circ. Syst. II: Analog Digit. Sign. Proces. **39**(1), 8–17 (2002)
26. Goldreich, O., Ron, D., Sudan, M.: Chinese remaindering with errors. In: Proceedings of the Thirty-First Annual ACM Symposium on Theory of Computing, pp. 225–234 (1999)
27. Goh, V.T., Siddiqi, M.U.: Multiple error detection and correction based on redundant residue number systems. IEEE Trans. Commun. **56**(3), 325–330 (2008)
28. Li, W., Yang, Y., Yuan, D.: Ensuring cloud data reliability with minimum replication by proactive replica checking. IEEE Trans. Comput. **65**(5), 1494–1506 (2015)
29. Zhu, Y., et al.: Fast Recovery MapReduce (FAR-MR) to accelerate failure recovery in big data applications. J. Supercomput. **76**, 3572–3588 (2020)
30. Wei, Q., Veeravalli, B., Gong, B., Zeng, L., Feng, D.: CDRM: A cost-effective dynamic replication management scheme for cloud storage cluster. In: 2010 IEEE International Conference on Cluster Computing, pp. 188–196. IEEE (2010)
31. Huang, C., Chen, M., Li, J.: Pyramid codes: flexible schemes to trade space for access efficiency in reliable data storage systems. ACM Trans. Storage. **9**(1), 1–28 (2013)

Investigation of Machine Learning Models for Detecting Network Anomalies

Svetlana Govorova[1](✉), Sergey Melnikov[1], Egor Govorov[1], and Mohammad Shahid[2]

[1] North Caucasus Federal University, Stavropol, Russia
mitnik2@yandex.ru
[2] Department of Commerce, Aligarh Muslim University, Aligarh -202002,, India

Abstract. The article examines the categories of machine learning models: ensemble methods implemented by the random forest algorithm and boosting (XGB classifier, XGB regressor); linear models (logistic regression); classifier based on deep neural networks. The results of a study of machine learning models are presented, where the "Random Forest" model obtained the best results. Graphs of the ROC curve for each considered machine learning model are constructed. Various parameter values are considered for the selected model. The best model parameters have been selected.

Keywords: machine learning models · algorithm random forest · network anomalies · normalization of datasets

1 Introduction

Anomalies in Internet traffic are unpredictable, unusual, or deviating from established norms of online interaction. They may indicate the presence of malware, hacks, or other unwanted events. Machine learning (ML) makes it possible to detect anomalies in network traffic, which is important for cybersecurity [1–3]. Some of the advantages of using machine learning to detect network anomalies are high detection accuracy, the ability to adapt to new threats, and the analysis of large amounts of data [4]. However, such approaches also have disadvantages: they require significant computing resources, high-quality model training, and data security. In addition, it is important to consider the possibility of false positives, which can lead to a decrease in confidence in the system [5, 6].

It is possible to detect network anomalies by using various anomaly detection methods: behavioral methods, computational intelligence methods, and machine learning methods (see Fig. 1) [7].

Behavioral methods. They are based on the use of information about the normal functioning of the system and its comparison with the parameters of the observed behavior. These include wavelet analysis, spectral analysis, statistical analysis, entropy analysis, fractal analysis, and others.

M. Lapina et al. (Eds.): AISMA 2025, LNNS 1456, pp. 168–176, 2026.
https://doi.org/10.1007/978-3-032-07275-7_16

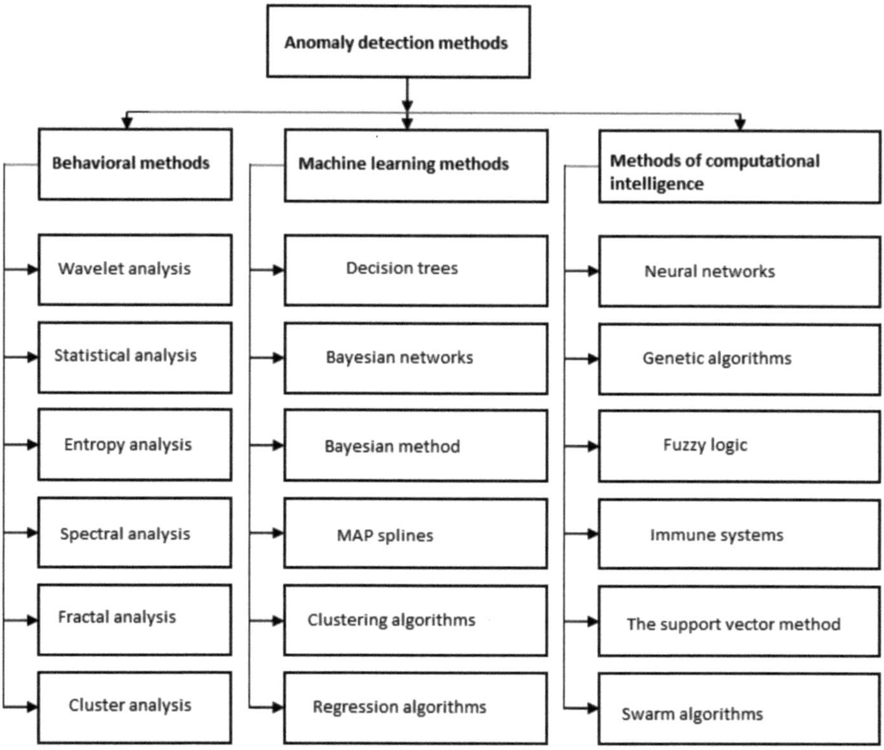

Fig. 1. Anomaly detection methods.

Machine learning methods. They assume the construction of a system that is being improved based on previous knowledge. Samples of both normal and abnormal behavior in the network are used as the initial training data. Some algorithms used to detect anomalies are "isolation forest" (works with multidimensional data), autoassociator (uses a neural network to study normal patterns and identify differences), clustering methods (DBSCAN and K-Means algorithms mark objects that are not members of any cluster as outliers) [8].

Methods of computational intelligence (CI). They are used to detect anomalies and abuses in a computer network. This is achieved by using templates like.

We will conduct a comparative analysis of the methods according to different criteria. The result is shown in Table 1.

Machine learning methods are aimed specifically at finding various relationships and patterns of network traffic and activity similar to the one that previously appeared in the training dataset [9]. Machine learning algorithms use a training dataset to solve various tasks [10]. The general task of machine learning methods is to restore the relationship between the set of input parameters (X) and the corresponding responses (Y) [11, 12]. To find dependencies, an initial data set is used, for which the algorithm parameters are configured.

Table 1. Comparative analysis of anomaly detection methods.

Criteria	Behavioral methods	ML methods	CI methods
Type of analysis (the approach at the heart of the method)	Rules + statistics	Machine learning algorithms	Real-time analytics + ML
Flexibility (adaptability to new types of anomalies)	Average	High	Very high
Anomaly detection rate	Average	Depends on the model	High (real-time)
False positives	Frequent	Depends on the model	Medium
Interpretability (the ability to understand why the system decided that this was an anomaly)	High	Low	Medium
Adaptability (the ability to adapt to changes in data)	Low (requires manual adjustment)	High (self-learning)	Very high (automatic adjustment)
Data requirements	Low (enough basic logs)	High (you need marked-up data)	Medium/High (streaming data required)
Complexity of implementation	Low/Medium (rules)	High (ML experts are needed)	Very high (integration with CI/CD, ML, streaming analytics)
Application	Security, monitoring	Big Data, IoT, network traffic	DevOps, cloud systems

2 Analysis of Network Traffic in Order to Form Vectors of Useful Features

In order to prepare the initial data for descriptive analysis and the construction of predictive models, the captured network packets were first preprocessed [13]. The evaluated data set is presented in csv format, which contains a description of the recorded sessions. Next, a single sample was formed to train the subsystem of user behavioral analysis and identify network anomalies. The target traffic is the traffic that needs to be allocated in the rest of the network stream [1]. At the initial stage of network traffic assessment, the following criteria were identified for its assessment: the number of unique destination IP addresses and the number of their appearances; the 15 most common sites in network traffic; a sign of the presence of a site in the session that is in the top 15 sites; the number of incoming/outgoing packets in the session; the payload of incoming/outgoing traffic. Outgoing network traffic; session timestamp; number of flags of a certain type recorded in the stream; indication of the presence of the 10 most frequently appearing source/destination ports in the session; average value and median of the timestamp. A user session is considered to be a sequence of tcp connections made by the user.

Due to the fact that most machine learning algorithms are sensitive to data scaling, it is necessary to normalize or standardize quantitative values. Normalization replaces nominal features so that each lies in the range from 0 to 1. Normalization is the process of processing input information, which includes training, test and validation samples and real data. During normalization, the values of the selected features in the input vector should be adjusted to the specified range. The need to normalize data sets is that the ranges of the initial features can vary over a large range and differ from each other by several orders of magnitude, which can lead to incorrect operation of the machine learning model [2].

The StandardScaler method was used to normalize the data sets. The method transformed the data in such a way that their distribution has a mean of 0 and a standard deviation of 1.

Each value of the dataset will be calculated using the following formula:

$$z = \frac{x - \mu}{\sigma} \tag{1}$$

The average value is calculated as follows:

$$\mu = \frac{1}{N}\sum_{i=1}^{N}(x_i) \tag{2}$$

The standard deviation is calculated using the following formula:

$$\sigma = \sqrt{\frac{1}{N}\sum_{i=1}^{N}(x_i - \mu)^2} \tag{3}$$

The normalized value will be calculated using the following formula:

$$X_{norm} = \frac{X - X_{min}}{X_{max} - X_{min}} \tag{4}$$

After normalization, all numerical values of the training and test data sets of network traffic are reduced to the same area of their spread—a narrow numerical range (−1; 1).

3 Investigation of Machine Learning Models for Detecting Network Anomalies

In the study of MO models, two categories of models were used to identify network anomalies: linear models (logistic regression) and ensemble methods, including running implemented by the random forest algorithm and boosting (XGB classifier, XGB regressor) [14].

Logistic regression is one of the statistical classification methods using the Fisher linear discriminant [4, 11, 15]. Unlike conventional regression, logistic regression does not predict the value of a numerical variable based on an initial sample. On the contrary, in this case, the value of the function is the probability that the source object belongs to a certain class.

The main essence of logistic regression: The space of the source objects can be divided by a linear boundary into various areas corresponding to the classes.

A random forest is, in fact, a universal algorithm, which lies in the effectiveness of its use in various tasks. A random forest is a set of crucial trees. When solving regression problems, the results are averaged; when solving classification problems, the result is determined by "majority voting."

XGBoost is a MO algorithm that is based on a decision search tree and the use of gradient boosting. The algorithm is good at analyzing structured or tabular data.

The training of the model in this algorithm is performed sequentially. The result of each iteration is a deviation of the predictions of the already trained ensemble on the training dataset [16].

To train the MO model, 20% of the traffic is taken from the training dataset to form a validation dataset that will contain 25,000 generated TCP sessions.

Table 2 shows the results of a study of machine learning models with randomly selected initial parameters.

Table 2. Results of the study of machine learning models

The machine learning algorithm	Accuracy of predictions (AUC ROC)	Accuracy of predictions (cross-validation)	Learning rate	Speed of prediction
Logistic regression	83,21%	73,56%	5,07 s	28 mks
Random Forest	97,74%	96,41%	2 min 36 s	2,34 s
XGB regressor	84,13%	80,34%	1 min 24 s	1,31 s
XGB classifier	73,49%	67.82%	15 s	1,44 s

Figure 2 shows the ROC curves of the considered machine learning models, demonstrating learning outcomes and predictions based on training and validation datasets. The values of false positive predictions are located on the horizontal axis, and the values of true positive predictions are located on the vertical axis. The ROC curve reflects the relationship of these two indicators. The area under the AUC ROC curve is a classification quality characteristic that does not depend on the price ratio of erroneous results. The higher the AUC value, the more effective the MO model is.

It can be seen from the graphs that the highest AUC value is achieved in the "Random Forest" model.

From Table 2, the following conclusion can be drawn: the model implemented using the Random Forest algorithm turned out to be the highest quality (97.74% according to the ROC curve, 96.41% according to the cross-validation test).

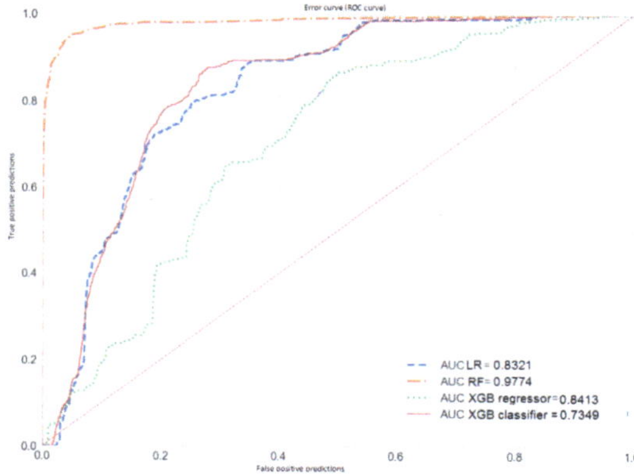

Fig. 2. ROC curves of machine learning models.

In the future, we will select the parameters for this particular model.

Since we will consider the "Random Forest" model, we will consider the algorithm for its construction, which is shown (see Fig. 3).

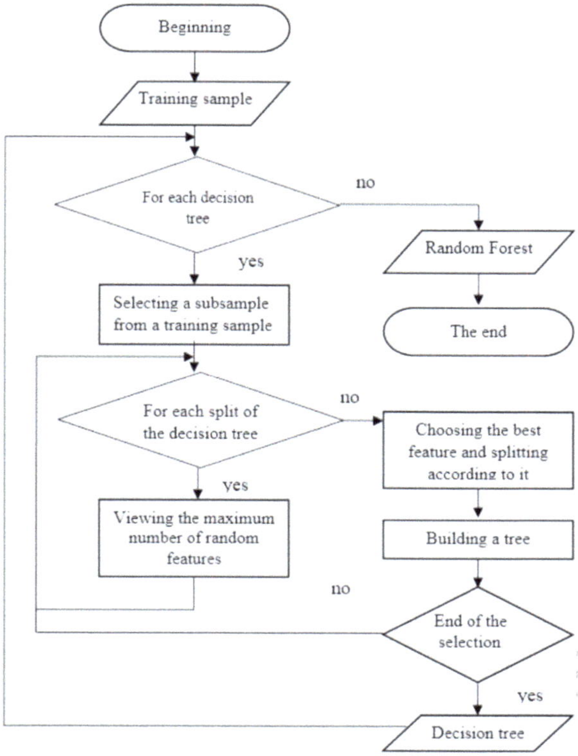

Fig. 3. Algorithm for building a random forest.

The error matrix of the Random Forest model is presented in Table 3.

Table 3. Error matrix of the Random Forest model.

Predicted	0	1
True 0	50245	4
True 1	327	137

The accuracy of the predictions according to ROC AUC/cross validation when using the Random Forest model is shown in Table 4.

Table 4. ROC AUC and cross-validation indicators for the Random Forest model.

	ROC AUC and cross-validation indicators						
	randomness of the initial download						
Number of trees	100	200	300	400	500	600	700
Nan	97,65 84,42	98,13 85,79	98,59 86,17	98,78 85,90	98,61 86,87	98,64 86,99	98,82 87,10
8	97,70 83,87	98,35 85,22	98,34 85,59	98,32 86,30	98,51 86,32	98,51 86,51	98,68 86,72
16	97,86 84,21	98,45 86,48	98,50 90,87	98,58 91,11	98,67 93,23	98,77 93,54	98,75 96,90
32	97,81 84,49	98,60 85,52	98,60 86,12	98,72 86,47	98,81 86,71	98,82 86,80	98,82 86,84
64	98,09 83,94	98,35 85,30	98,64 86,37	98,80 86,55	98,88 86,90	98,90 86,89	98,92 86,82
128	97,69 84,69	98,34 85,61	98,28 85,89	98,40 86,12	98,55 86,34	98,63 86,36	98,73 86,62
200	97,82 84,78	98,09 85,96	98,37 85,91	98,57 86,63	98,64 86,58	98,67 86,49	98,73 86,48
256	97,73 84,14	98,20 85,55	98,22 86,25	98,54 86,72	98,54 86,73	98,67 86,77	98,70 86,90

It is necessary to choose the optimal model performance option in accordance with the ROC AUC and cross-validation indicators.

The best results are observed with 16 trees and an initial load randomness of 700–98.75% ROC AUC; 96.90% on cross-validation.

The training time/prediction time in seconds for the "Random Forest" model is shown in Table 5.

At the best parameter values, training and prediction are performed in 960.6 seconds and 15.1 seconds, respectively.

Table 5. Indicators of training time and prediction time for the "Random Forest" model.

	Training time and prediction time						
	randomness of the initial download						
Number of trees	100	200	300	400	500	600	700
Nan	141,4 2,3	271,7 4,3	408,7 6,5	540,7 8,5	675,2 10,7	824,1 12,8	945,9 14,7
8	136 2,1	272,1 4,2	406,9 6,4	587,3 9,1	742,8 11,3	843,3 13,2	1246,3 15,5
16	138,9 2,2	347,1 5,1	424,3 6,7	680,5 9,4	677,1 10,7	811,2 12,8	960,6 15,1
32	135,5 2,1	271,5 4,2	418 6,4	538,9 8,5	672 10,6	810,3 12,7	940,9 14,8
64	135,6 2,1	297,4 4,5	430,2 6,5	622,2 8,9	786,3 10,7	829,2 12,9	941,1 14,8
128	138,9 2,2	275,5 4,3	411,4 6,4	546,1 8,5	697,9 10,8	829,1 12,9	965,6 15,1
200	139,2 2,2	276,7 4,4	414,4 6,5	553,2 8,6	689,2 10,7	823 12,9	955,9 14,9
256	137,5 2,2	274,2 4,3	410,9 6,5	548,1 8,7	675,7 10,7	812,6 12,9	1016,1 16,6

4 Conclusion

Various criteria for evaluating the effectiveness of machine learning models in solving the problem of network traffic profiling are considered. In accordance with the specifics of the tasks of user recognition and identification of network anomalies, the most appropriate criteria for evaluating the effectiveness of machine learning models have been selected: AUC ROC—the area under the error curve. In addition, the method of cross-validation was used for evaluation—cross-validation of the success of statistical data analysis applied in the model on an independent data set. Machine learning models are investigated and the results of a study of machine learning models are presented, where the "Random Forest" model obtained the best results. Graphs of the ROC curve for each considered machine learning model are constructed. Various parameter values are considered for the "Random Forest" model. The best parameters of the model were selected, at which the ROC AUC coefficient reaches 98.75 and the cross-validation coefficient 96.9.

References

1. Gopali, S., Namin, A.S.: Deep learning-based time-series analysis for detecting anomalies in internet of things. Electronics. **11**(19), 3205 (2022). https://doi.org/10.3390/electronics11193205

2. Or-Meir, O., Nissim, N., Elovici, Y., Rokach, L.: Dynamic malware analysis in the modern era—a state of the art survey. ACM Comput. Surv. **52**(5), 88. pp. 1–48 (2019). https://doi.org/10.1145/3329786
3. Aslan, Ö., Aktuğ, S.S., Ozkan-Okay, M., Yilmaz, A.A., Akin, E.: A comprehensive review of cyber security vulnerabilities, threats, attacks, and solutions. Electronics. **12**(6), 1333 (2023). https://doi.org/10.3390/electronics12061333
4. Luo, X., Li, J., Wang, W., Gao, Y., Zhao, W.: Towards improving detection performance for malware with a correntropy-based deep learning method. Digit. Commun. Netw. **7**, 570–579 (2021). https://doi.org/10.1016/j.dcan.2021.02.003
5. Sun, Y., Bashir, A.K., Tariq, U., Xiao, F.: Effective malware detection scheme based on classified behavior graph in IIoT. Ad Hoc Netw. **120**, 102558 (2021). https://doi.org/10.1016/j.adhoc.2021.102558
6. Stamp, M.: Introduction to machine learning with applications in information security, 2nd edn. Chapman and Hall/CRC (2022). https://doi.org/10.1201/9781003264873
7. Apruzzese, G., Andreolini, M., Ferretti, L.: Modeling realistic adversarial attacks against network intrusion detection systems (2022). https://doi.org/10.1145/3469659
8. Wiranda, N., Sadikin, F.: Machine Learning for Security and Security for Machine Learning: A Literature Review. (2021). https://doi.org/10.1109/ICOIACT53268.2021.9563985
9. Zheng, W., Gou, C., Yan, L., Mo, S.: Learning to classify: a flow-based relation network for encrypted traffic classification. In: Proceeding of the Web Conference, pp. 13–22 (2020)
10. Khan, Z.I., Afzal, M.M., Shamsi, K.N.: A comprehensive study on cic-ids2017 dataset for intrusion detection systems. Int. Res. J. Adv. Eng. Hub. **2**(02), 254–260 (2024). https://doi.org/10.47392/IRJAEH.2024.0041
11. Akhtar, M.S., Feng, T.: Detection of malware by deep learning as CNN-LSTM machine learning techniques in real time. Symmetry. **14**(11), 2308 (2022). https://doi.org/10.3390/sym14112308
12. Guojie, L., Jianbiao, Z.: Research of network intrusion detection based on convolutional neural network. Discret. Dyn. Nat. Soc. **2**, 1–11 (2020)
13. Paracha, A., Arshad, J., Farah, M., et al.: Machine learning security and privacy: a review of threats and countermeasures. Eurasip J. Info. Security. **2024**, 10 (2024). https://doi.org/10.1186/s13635-024-00158-3
14. Das, S., Nene, M.J.: A survey on types of machine learning techniques in intrusion prevention systems. https://doi.org/10.1109/WiSPNET.2017.8300169 (2017)
15. Le, T.-T.-H., Kim, Y., Kim, H.: Network intrusion detection based on novel feature selection model and various recurrent neural networks. Appl. Sci. **9**(7), 1392, 29 p (2019). https://doi.org/10.3390/app9071392
16. Sinha, J., Manollas, M.: Efficient deep CNN-BiLSTM model for network intrusion detection. In: Proceedings of the 2020 3rd International Conference on Artificial Intelligence and Pattern Recognition (AIPR 2020), pp. 223–231. Association for Computing Machinery, New York, NY, USA (2020). https://doi.org/10.1145/3430199.3430224

On the General Approach and Criteria for Optimizing the Selection of Protective Measures in Preparation for the Procedure for Assessing Compliance with Information Security Requirements for Financial Organizations

Anastasia Iakovleva(✉) and Marina Zhukova

Reshetnev Siberian State University of Science and Technology, Krasnoyarsk, Russia
petrova_ao@sibsau.ru

Abstract. The paper describes the procedure for determining information security measures in accordance with GOST R 57580.1 and subsequent compliance assessment in accordance with GOST R 57580.2 within the framework of fulfilling the requirements of the Bank of Russia. The paper presents the main problems of compliance systems in terms of determining security contours, determining unambiguous compliance with the requirements for basic measures and their applicability in the infrastructure, as well as in terms of the audit being conducted. The paper considers the tasks that arise at the stage of forming an approach to meeting the requirements and that make the assessment procedure more streamlined and understandable for specialists. It briefly considers a tool that allows automating certain stages of the process of selecting measures and calculating the compliance assessment. The task to form optimization criteria for selecting security measures while increasing the compliance assessment has been set. The paper provides detailed descriptions of the selected optimization criteria, such as: speed of achieving the result, cost of changes, selection of planned changes, human resources assessment and the possibility of making changes to the infrastructure. The applicability of the criteria is analyzed in terms of the restrictions imposed or additional requirements when using them, as well as the possibility of an objective assessment of the criteria and their further selection.

Keywords: information security · financial organization · protection measures · compliance · evaluation criteria

1 Introduction

The task of compliance with the requirements of state standards, legislative, normative legal acts of various regulators for the field of information security (abbreviated: compliance) is the main, mandatory and priority for implementation [1]. Without compliance,

M. Lapina et al. (Eds.): AISMA 2025, LNNS 1456, pp. 177–185, 2026.
https://doi.org/10.1007/978-3-032-07275-7_17

many companies and their business processes often simply cannot function, especially if the activity they are engaged in is licensed, such as the activities of financial sector organizations. For such organizations, the core for building security systems is currently the series of standards GOST R 57580.X.

Conducting an assessment of compliance with the requirements of GOST R 57580.1 (and .2) is possible only if the number of security loops and the required level of protection for each loop are correctly determined. To make this decision, an audit of the system is conducted in accordance with the provisions of the regulatory legal acts of the Bank of Russia applicable to financial organizations. According to regulatory documents, compliance assessment of entities must be conducted at least once every 2 years, regardless of the implemented level of protection: standard or enhanced [2–4] for credit organizations.

At the same time organizations must ensure a compliance level of at least 4 ($E_i >$ 0,85) [2–4]. The formation of a basic and adapted set of assessed measures required for implementation depends on the selected level of protection and requirements for the compliance assessment level.

Various documents have been developed to solve the problem of defining basic measures and assessing compliance. In general, the process of definition should be fairly easy to solve because:

- there are basic information security measures presented in GOST R 57580.1-2017 «Security of financial (banking) operations. Protection of information of financial organizations. Basic set of organizational and technical measures» [5], the list of which, depending on the required level of security, determines the set of requirements for security systems necessary for implementation;
- there is a conformity assessment, which is carried out in accordance with GOST R 57580.2-2018 «Security of financial (banking) operations. Protection of information of financial organizations. Conformity assessment methodology» [6], and defines the procedure for conducting an audit of systems.

However, the practice of application shows a fairly large number of limitations and ambiguities in the choice of solutions, as well as in subsequent evaluation [7].

Difficulties arise when preparation for the upcoming audit is required, since according to the requirements of the law, it must be carried out by an independent external organization that has the appropriate license for technical protection of confidential information and specialists who have the necessary competencies. Hiring auditors who meet the requirements will entail significant financial costs. Beside it, an audit always leaves room for a subjective assessment of individual elements of the considerated system, since the implementation of such work is directly related to the existing experience of specialists. Ambiguity in the implementation of a particular measure can lead to different (close, but not identical) audit results, which can affect the final assessment.

The audit results must be sent to the Bank of Russia to confirm compliance with the required level of security [8]. In the event of an unsatisfactory assessment, it is necessary to promptly correct the situation in order to obtain the required assessment level, and then repeat the procedure. If a low compliance assessment is received, the audited organization may receive a number of sanctions from the regulator: from a requirement to eliminate the identified violations to the revocation of the banking license [9].

Thus, there is a practical task of preparing for an audit, which is difficult for specialists to solve. There are separate studies devoted to the assessment of integrated systems [10], the use of trust systems [11] to obtain an assessment and improve the efficiency of the audit process. Of interest is the development of an approach to the analysis of the compliance of the audit results and the applied assessment procedures within the GAP analysis for the credit and financial sector – the results of such work have already been published by the authors [12]. It is noted that the process of conducting a compliance assessment is not accompanied by the presence of additional methodological recommendations and developed algorithms for methods of obtaining the required level of compliance assessment.

In order to develop an approach to meeting requirements and make the assessment procedure more streamlined and understandable for specialists, it is necessary to organize compliance monitoring activities in the organization, for which the following tasks must be completed:

1. Have the ability to calculate/recalculate the assessment quickly, in near real time;
2. Formalize the calculated index in the form of a weight for each measure when calculating the assessment;
3. Have the ability to select measures according to certain criteria, the implementation of which will allow obtaining the required assessment for each process.

The list of specified problems has been partially solved by the authors, the results are presented in [13, 14].

Thus, the aim of this work is to formulate optimization criteria for selecting protective measures when increasing the conformity assessment.

2 Automating the Selection of Measures and Conformity Assessment

To calculate the assessment, in the article «Development of a software solution for forming an adapted set of measures when conducting a conformity assessment according to GOST R 57580.2 depending on the requirements of GOST R 57580.1» [13], the authors set the task of creating a tool that allows determining the required level of protection for the security contour and forming an adapted list of protection measures subject to assessment. The tool is offered in the form of an interactive system that allows the user, using surveys and recommendations, to determine the data for making a decision on the level of protection of the security contour. At the next step, it was decided to form a calculation tool that allows calculating the assessment. The general scheme of the stages of the calculation process is presented below (see Fig. 1). The diagram demonstrates the full cycle of the assessment process: from selecting the number of contours and their security level to obtaining the final assessment indicator.

Let's briefly describe the main stages:

1. Audit of the system according to the requirements of regulatory legal acts of the Bank of Russia and a series of GOSTs (Help window and Input data):
   - determination the number of security contours (setting the X counter);

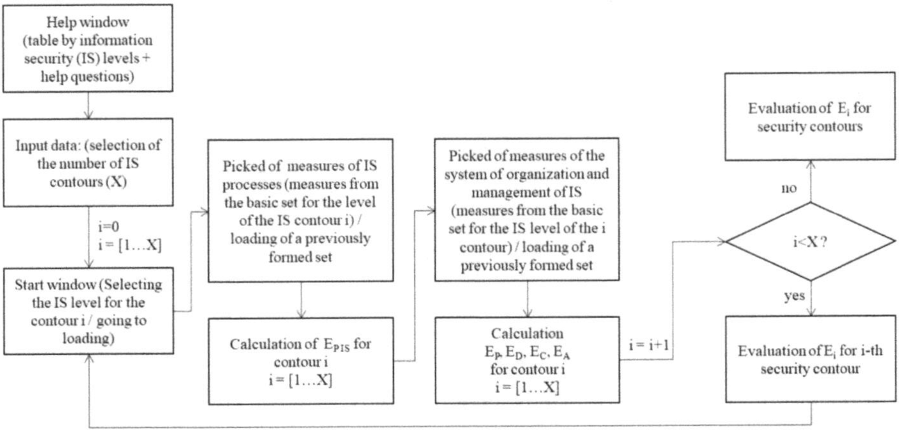

Fig. 1. The general scheme of the stages of the E_i assessment calculation process.

- determination of the levels of protection of each sector during calculation (tables and reference questions);

2. Formation of a list of measures for assessment (The module is typical and is repeated for calculating intermediate assessments: picked of measures of information security – $E_{P\,ISi}$, as well as assessment of the implementation of measures of the system of organization and management of information security in the form of a process approach of the PDCA cycle: plan – E_{Pi}, do – E_{Di}, check – E_{Ci}, act – E_{Ai})):
   - automatic receiving of a list of information security measures from the basic set, according to the above-defined level (downloading a list from GOST);
   - adaptation of the list of measures in accordance with the characteristics of the measurement object (exclusion, compensation, etc.).
3. Calculation E_i (Evaluation E_i):
   - calculation the score for each contour;
   - calculation of the final assessment for all contours taking into account the correction factors of the standard.

The calculation the degree of influence of each measure were presented by authors in the article «Calculation of coefficients reflecting the degree of influence of the implementation of information security measures during the assessment of compliance for financial organizations» [14], where it is stated that the influence of a measure on the result of the assessment during its calculation depends on the following parameters:

1. the level of information security of the contour and the number of applicable measures, since the basic set of measures and, accordingly, the resulting adapted set after adjustment depend on this;
2. the total number of security contours and the difference between their security levels;
3. for process measures also from the assessment received for the measure and the correction factor for the direction of protection.

The process of forming a list of relevant measures and conformity assessment is automated, and the sets of measures are loaded dynamically, which means that the program is able to adapt to changes in legislation by updating the list of measures in accordance with the current version of the regulatory document.

It is expected that the obtained results of calculating the coefficients for each protection measure will make it possible to create a tool for selecting protection measures to achieve the required level of security assessment.

The previously described tools and algorithms allow obtaining a list of selected measures in accordance with the requirements of GOST 57580.1 and calculating the assessment of compliance with the requirements of GOST 57580.2. However, practice shows that these tools are not enough, because it is also necessary to solve the problem of selecting information security measures. The authors propose to integrate an additional module into the tools being developed that implements an optimized selection of measures to achieve the required assessment of compliance.

3 Formalization of Criteria for Optimizing the Selection of Protective Measures

An increase in the conformity assessment can be achieved by selecting and implementing certain technical or organizational security measures (assessed 0 or 1), as well as by refining already partially implemented measures that are implemented within the framework of the target system and information security management (assessed 0; 0.5; 1). Then, when selecting planned or introduced changes, to increase the level of conformity assessment, the following criteria can be used within each process:

1. speed of achieving the result in terms of implementing new information security tools into the infrastructure for technical measures or the time of developing documents and regulations for organizational measures. This criterion may be significant in the case of a shortage of time resources, such as when receiving requests to eliminate identified violations;
2. the cost of changes in terms of acquiring additional protection tools, if we are talking about technical measures;
3. selection of planned changes. The assessment can be increased by 1–2 «major» measures (having greater weight in the assessment), which will be «costly» within the framework of one process, or a larger number of «minor» measures (having less weight in the assessment) can be implemented - a set of which can also obtain the required assessment, but already due to their quantity;
4. availability of the required human resources to implement the measures. A difficult criterion to assess, but in some cases, possible to calculate;
5. ability to make changes to the infrastructure. This criterion will be very important for target systems that have limitations from other requirements.

The algorithm's work will usually be associated with several criteria at once, then it is necessary to solve a multi-criteria optimization problem. In this case, it is assumed that the criteria are ranked by importance (weight, cost, etc.). To make a choice between the criteria, you should get answers to the following questions:

- what limitations or additional requirements are imposed by the application of the criteria;
- is it possible to make an objective assessment of these criteria?

Let's look at the criteria in more detail. To analyze the possibility of changing the infrastructure, it is necessary that information about it be available, which means:

- the security requirements for the tool being developed are increasing, since it will contain information about the structure and scheme of the organization's information security system. The possibility of their leakage is an additional risk for the organizations using it;
- it is difficult to determine the method for obtaining such information, since without automation such a task will take a lot of time, and there is always the possibility of entering erroneous information from the user of the target system. At the same time, it is impossible to guarantee the availability of automation tools used for infrastructure monitoring for a number of objective reasons:
- given the possibility of automated receipt of information about the infrastructure, the organization of a single format for the received data is problematic.

The criterion for assessing human resources – the following aspects are important to it:

- it is impossible to determine and guarantee whether the organization's specialists have the necessary competencies to perform work to improve the assessment, especially in terms of implementing technical measures;
- The number of specialists who can be involved in the implementation of measures to improve conformity assessment isn't known in advance.

The cost criterion is also not simple – it would seem that it can be accurately and easily assessed by studying the commercial offers of manufacturers of categories of solutions used to close the missing measures. However, the following problems may arise when using it:

- not all companies that manufacture or integrate information security tools openly publish the terms of the offer for the purchase of products, since the pre-calculated figure may not suit the potential client, which means that in order to obtain this information, it will be necessary to interact with managers, which will be difficult to implement only using an automation tool without the direct participation of a specialist;
- the cost of the purchased solution may depend on the scale of the infrastructure or the type of resource or access object being protected.

It turns out that there are two criteria left that can be used and that can be assessed without having information about the infrastructure being assessed:

1. speed of implementation, since there are approximate estimates of the time required to implement various protection measures by the expert community [15, 16], of course, taking into account that the calculations were performed conditionally, for example:
   - the work on developing the documentation was supposed to be carried out by external specialists;

  - the implementation and deployment of security tools was not envisaged; instead, it was envisaged to use security tools in the format of a service model and transfer risks to outsourcing;

2. a selection of planned changes that can be implemented in several variations.
   - by searching for combinations that together will give the missing estimate;
   - by possible assessment of the achievement of the result due to a specific choice of the category of measures (organizational or technical);
   - by selecting the number of parameters subject to change (based on the weight of the measure in the assessment).

Within the framework of the developed algorithm for optimization of selection, the authors propose to focus on the selected optimization criteria to eliminate problems associated with the formation of measures and subsequent assessment of compliance to the required level of protection.

To date, evaluation enhancement algorithms have been developed by external organizations offering their services in building or adapting protection systems to the requirements of GOST R 57580.X. The algorithms are closed, and their practical implementation is focused on integrating services implemented by external organizations into the customer's infrastructure, and proposals of this nature are implemented in the form of fixed standard scenarios, for which the evaluation increment results are calculated [15]. These criteria can be used to design the proposed solution as real case studies applied in practice.

There are currently no software implementations of calculation and evaluation algorithms available for their use by internal specialists. The development of such a solution would minimize costs if available:

- the ability to use existing infrastructure components based on the analysis of the proposed "combinations" to assess the required level;
- a way to increase valuations by applying a certain class of cost-effective measures, such as developing organizational documents instead of purchasing protective equipment;
- a solution that closes several measures from the proposed combination at once by reducing the list of funds to be purchased [16].

At the same time, internal specialists will have the opportunity to conduct a self-assessment of protection systems according to the requirements of GOST R 57580.X. The solution should be applied before conducting a planned conformity assessment or when making changes to the infrastructure to control the level of compliance. This approach will minimize the risks for organizations associated with the subjective assessment of auditors, since the presence of a solution will clarify or challenge the audit results due to the ability to quickly find the difference in the assessment of an external specialist and the results of the preliminary self-assessment.

4 Conclusion

To solve the problem of preparation for the procedure of conducting the assessment of compliance with the requirements, it is necessary to organize compliance monitoring activities in the organization, having developed a structured approach. To implement such an approach, it is necessary to perform some stages, a general description of which was presented in this article. The missing stage of selecting measures according to certain criteria, the implementation of which will allow obtaining the required level of assessment for each process, was studied by the authors by analyzing the possibility of using criteria, from the point of view of imposed restrictions or additional requirements when using them, as well as the possibility of increasing objectivity in assessing the criteria.

The analysis of the selected criteria showed that not all criteria can be formalized and used to solve the optimization problem. To make a choice between the selected criteria, it is necessary to evaluate the conditions for implementing various options for assessment enhancement strategies.

In the future, the research should focus on the practical application of the proposed approach in solving real-world cases for financial and credit systems. The practice of application will allow for validation and, if necessary, to improve the proposed model.

References

1. Bondar, E.A.: Information security in financial systems. In: Digital Economy and Finance: Proceedings of the VII International Scientific and Practical Conference, St. Petersburg, 14–15, 2024, pp. 244–247. Center for scientific and production technologies Asterion, St. Petersburg (2024)
2. Bank of Russia Regulation 683-P «On the establishment of mandatory requirements for credit institutions to ensure the protection of information when carrying out banking activities in order to counteract the transfer of funds without the client's consent» , https://base.garant.ru/72246408/. Accessed 1 Oct 2024
3. Bank of Russia Regulation 802-P «On the requirements for information protection in the Bank of Russia payment system» dated 25.07.2022. https://base.garant.ru/405828183/. Accessed 1 Oct 2024
4. Bank of Russia Regulation 821-P «On the requirements for ensuring the protection of information when transferring funds and on the procedure for the Bank of Russia to monitor compliance with the requirements for ensuring the protection of information when transferring funds» dated 17.08.2023. Accessed 1 Oct 2024
5. GOST R 57580.1–2017 «Security of financial (banking) operations. Protection of information of financial organizations. Basic set of organizational and technical measures». https://rosgosts.ru/file/gost/03/060/gost_r_57580.1-2017.pdf. Accessed 2 Oct 2024
6. GOST R 57580.2–2018 Security of financial (banking) operations. Protection of information of financial organizations. Conformity assessment methodology. https://rosgosts.ru/file/gost/03/060/gost_r_57580.2-2018.pdf Accessed 2 Oct 2024
7. Fargieva, Z.S., Kotieva, H.M.: Cyberattacks on banks. Methods of protecting credit institutions. European science. 2020. (54). https://cyberleninka.ru/article/n/kiberataki-na-banki-sposoby-zaschity-kreditnyh-organizatsiy. Accessed 10 Oct 2024

8. Sokolova, E.A., Yu, A.: Ermoshina The Role of the Bank of Russia in Ensuring Information Security of the State Financial System. Trends Dev. Sci. Educ. (60–5), 78–81 (2020). https://doi.org/10.18411/lj-04-2020-102
9. Federal Law of 10.07.2002 No. 86 «On the Central Bank of the Russian Federation (Bank of Russia)». https://www.consultant.ru/document/cons_doc_LAW_37570/. Accessed 5 Nov 2024
10. Kozminykh, S.I.: Development of a methodology and mathematical model for ensuring the quality of an integrated security system of a credit and financial facility. Cybersecur Issues. **3**(43), 31–42 (2021). https://doi.org/10.21681/2311-3456-2021-3-31-42
11. Dubinin, D.P.: Using trusted systems to ensure information security and limit unauthorized access to information systems of financial organizations. Information Security Issues. **1**(136), 3–6 (2022). https://doi.org/10.52190/2073-2600_2022_1_3
12. Iakovleva, A.O., Zhukova, M.N.: Development of a solution for conducting GAP analysis and adjusting the level of security when assessing compliance with information security requirements for financial institutions, Ural Federal District. Security in the Information Sphere. **3**(53), 61–72 (2024). https://doi.org/10.14529/secur240307
13. Iakovleva, A.O., Zhukova, M.N.: Development of a software solution for the formation of an adapted set of measures when conducting conformity assessment according to GOST R 57580.2 depending on the requirements of GOST R 57580.1. In: Proceedings of the XV International Scientific and Practical Conference named after Oleg Borisovich Makarevich. Modern methods, means and technologies of information protection, pp. 231–240 (2024)
14. Iakovleva, A.O., Zhukova, M.N.: Calculation of coefficients reflecting the degree of influence of the implementation of information protection measures during compliance assessment for financial organizations. Casp. J. Manag. High Tech. **4**(68) (2024) in print, ISSN 2074–1707
15. Webinar: How to improve the GOST 57580 assessment: quickly, effectively, the first time https://bi.zone/expertise/events/kak-povysit-otsenku-gost-57580-bystro-effektivno-s-pervogo-raza/. Accessed 6 Jan 2024
16. Webinar: Review of the requirements of GOST R 57580.1–2017. https://www.ussc.ru/events/zapisi-vebinarov/seriya-vebinarov-bezopasnost-finansovykh-organizatsiy/seriya-4/ 22 February 2024

Assessment of the Impact of Security Measures on the Acceptable Password Trust Level in the Authentication System

Andrey Ivanov(✉) and Dmitry Nechaev

Novosibirsk State Technical University, Novosibirsk, Russian Federation
ivanov_av87@mail.ru

Abstract. The study examines the problem of imposing objectively high requirements on users in the presence of methods used to protect against brute force attacks in the processes related to the authentication system. The paper discusses the main processes of user interaction with passwords: registration, authorization, password update and password reset. In each interaction scheme, the main stages are defined, for example, password formation, data entry, data validation for correctness. Each stage is analyzed for possible vulnerabilities that contribute to password guessing. Based on the vulnerabilities identified, technical measures to protect the authentication system are established. Each protection measure is evaluated according to such parameters as the level of loyalty and protection. The paper provides formulas and examples for calculating the reduction of the permissible password trust level based on the security measures used in the authentication system. The resulting algorithm is used in a comprehensive assessment of password strength.

Keywords: Authentication system · password strength · security measures · password guessing · password requirement level · 2FA · hashing · database

1 Introduction

In today's digital world, ensuring the reliability of employee data is an important issue for the vast majority of organizations. For decades, an authentication system based on text passwords has been used in various corporate services, such as Active Directory, a learning portal, or an internal file-sharing cloud.

However, with the rapid growth of services in use, it is becoming more difficult to ensure the security of all passwords used by employees. There are many reasons for the problem of maintaining the proper level of security: from the inability to use more secure methods [1] (for example, biometric authentication [2]) to the low level of password literacy culture among employees. Other important factors are the overload of users with excessive password requirements and the lack of proper attitude to the password policy of system administrators in modern realities [3].

Despite the fact that the key dependence of the approval of the generated password is the strength of the content of the entered password, other aspects of influencing what

M. Lapina et al. (Eds.): AISMA 2025, LNNS 1456, pp. 186–199, 2026.
https://doi.org/10.1007/978-3-032-07275-7_18

minimum requirements should be imposed on an employee in a particular service in terms of password are often not considered. It is important to take this into account, since in the absence of a comprehensive assessment [4], password security testing may be biased: if the requirements are lower than necessary, the authentication system in the service will become more vulnerable and may entail an easily realized threat of password guessing; If the requirements for the password entered by the employee are too high, this can affect him psychologically, and therefore the employee will either enter a password that coincides with the one he already uses, or is similar, which again entails increased risks for the owners of the service. Now, in most organizations, employees responsible for creating minimum password requirements use common existing practices, as well as follow basic recommendations in legislative acts, without resorting to additional password evaluation tools. This approach does not lead to effective security, but only covers the minimum requirements for the protection of the authentication system [5].

In this regard, there is a problem in the formation of passwords in society, which consists in the lack of objective setting of minimum requirements or an **Acceptable password trust level**. Based on the results of the assessment, the password is considered approved if the password trust level exceeds the calculated acceptable password trust level. With this in mind, a natural question arises: how to calculate it and what does it depend on?

Figure 1 shows a diagram of the main components that make up the password strength assessment. The password trust level reflects the quality of the content of the password entered by the user [4]. Two aspects – external factors and the authentication system protection measures used affect the acceptable level of trust. A combination of external factors will make it possible to establish the initial value of the acceptable level of trust. External factors include the availability and criticality of the service [5], the level of access of the user who generates the password (user with read rights / administrator). As for related measures to protect against brute force attacks, they allow you to reduce the initial value of the acceptable trust level by creating an additional layer of security in the authentication system.

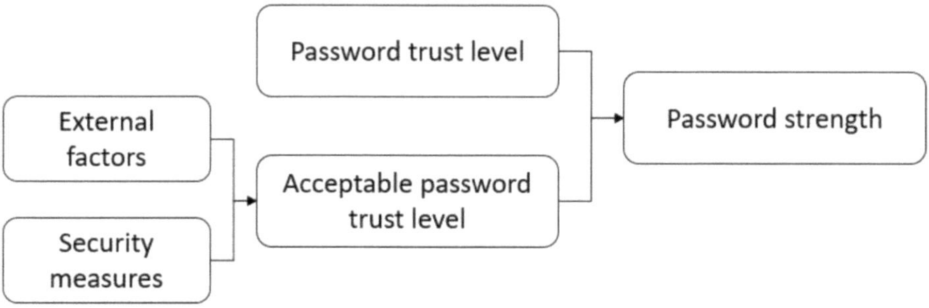

Fig. 1. Key elements of password strength assessment.

It is worth noting that despite the dependence of the permissible level of trust on the two elements, the elements do not correlate with each other, but function in parallel. In this regard, it is permissible to consider one component without taking into account the second. In this study, protection measures were selected for the review that provide protection against password compromise by the brute force method.

The purpose of this paper is to assess the impact of existing security measures on the minimum threshold of requirements, or the **Acceptable password trust level**. The Acceptable password trust level will be considered the minimum possible calculated threshold value, which must be the password trust level for the password to be considered strong. To do this, you need to define the processes in the authentication system that contain the password entry module itself. Each process must then be broken down into stages. At each stage, information security vulnerabilities that affect the entire process of the authentication system will be considered. Then, related security measures in the authentication system will be determined and evaluated according to the level of contribution to security. The result of the work is a set of set values for each protection measure, which allows you to reduce the Acceptable password trust level.

2 Authentication System Processes

Figures 2, 3, 4 and 5 depict the processes in which the user interacts with the password.

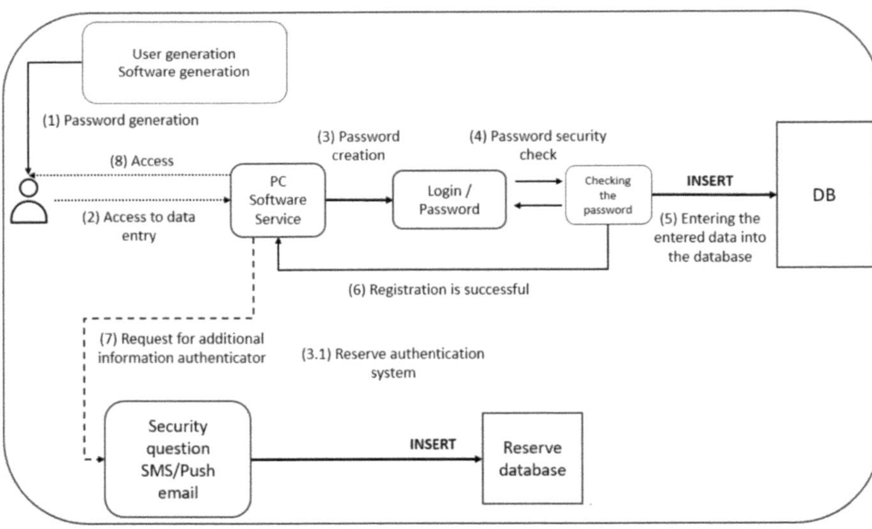

Fig. 2. User Registration Process.

For a user to work in a service that includes a password authentication system, first, user **registration** is required. Either the user can set the password himself during registration, or the administrator generates a password for the user, sends him data, after entering which the user resets the generated password and in any case enters a new one.

Then, for a certain time, the user uses the password as a confirmation of the ability to access the service. In this way, the user participates in the **authentication** process.

However, in accordance with accepted technological practices, the user needs to update the password either after a certain period, according to the password policy in the

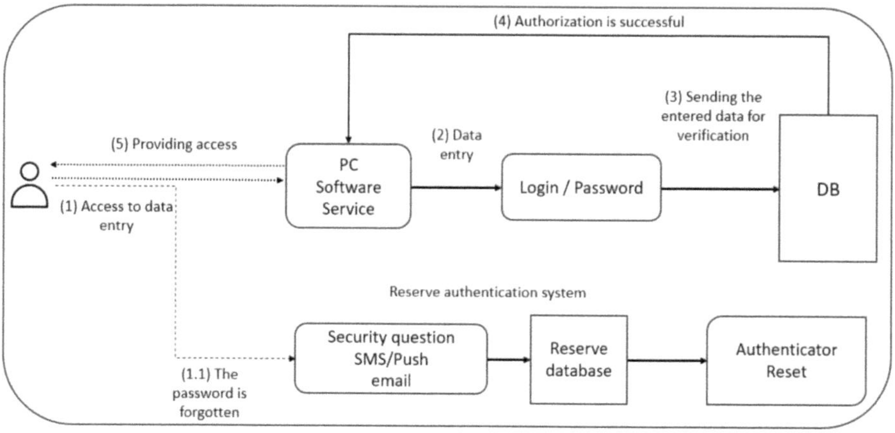

Fig. 3. Service Authorization Process.

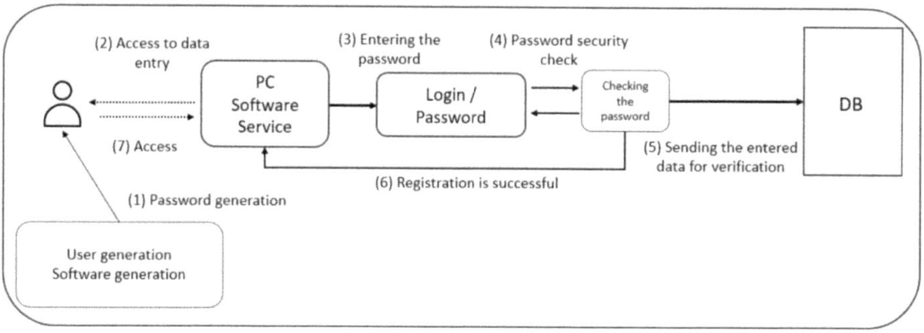

Fig. 4. Password Update Process.

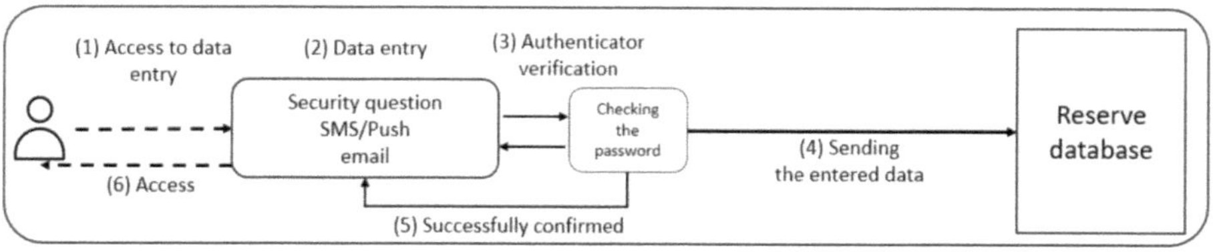

Fig. 5. Password Reset Process.

organization, or upon detection of compromised credentials. In addition, the user can update the password at his own request, or due to the fact that the password is forgotten. Therefore, we can talk about the third process - **updating the password**.

The last of the main processes is related to the last cause of the previous process. If the user wants to update the password, then this can be done through a **password reset**. In fact, we are dealing with the fact that we are returning to the stage of the registration process, therefore, the similarity between the processes considered is obvious. It is also worth noting that password reset cannot be used without confirming the user's rights to implement this action. In this regard, in addition to the main one, organizations can use a **reserve authentication system**. To be able to operate a reserve authentication system (RAS), it is necessary to have an additional factor that allows you to establish the user's authority to reset the password: a security question, phone number, e-mail, etc. [6]. In total, 4 processes included in the main and reserve authentication system form the complete operation of the system.

The dotted lines on the diagrams show the user's access to the service object, the dashed line means that this subprocess is not necessary to work with the service, but is only used under certain circumstances (using two-factor authentication or the need to reset the password). Sequential numbering for understanding indicates the main subprocesses.

3 Analysis of Process Stages

Based on the analysis of the presented schemes, it can be seen that there are constant elements in the processes: user, service object, credential input, database. In the context of looking for factors that affect the acceptable level of trust in a password, interaction with these elements has the greatest impact.

In this section, we will divide the processes into stages in order to analyze and subsequently identify threats relevant to the topic of research.

3.1 Access to the Service

The required access can be physical (access directly to unlocking the account on the PC) and logical (access at the network or application level to the service). Access restriction is the basic functionality of the authentication system, since it is the availability of the system that is the primary barrier to restraining an attacker from implementing the threat of compromising user credentials. However, it is worth noting that there are services that need public access from the Internet.

3.2 Password Generation

This stage is extremely important at the conceptual level, since it is the generated password that is the object of protection. As for the registration process, the password entered is generated using a password generator, or invented by the user. However, the formation of a password does not affect the permissible password trust level, but the quality of its content (the password trust level), so it will not be considered in this study.

3.3 Data Entry

This stage is central to the authentication system and requires special attention in ensuring security, since with the right password, not only the service, but also the entire infrastructure of the organization is compromised.

3.4 Password Security Check

Password security can be checked either by the internal settings of the service or by third party specialized software. As with the password generation phase, the password verification subprocess does not affect the internal acceptable level of trust.

3.5 Sending Data for Verification to the Database

The mere fact that data is stored in a database creates an additional threat object that requires security. Due to the possibility of compromising the database and obtaining the passwords contained in it, administrators often resort to a primitive level of database protection, at least by using sharding to increase the path of the attacker.

3.6 Reserve Authentication System

A reserve authentication system is necessary in case of problems with user authorization. However, to reset the password, the user needs to confirm his authority using a reserve authenticator. Nevertheless, the presence of the RAS is a basic functionality and does not affect the guessability of the password.

Based on the results of the analysis, the main aspects of the logic of each stage have been identified, and now it is possible to identify threats related to password guessing.

4 Threat Definition

Having established the key stages of the processes in the authentication system, we will determine the list of threats, the presence of which may lead to an increase in the chances of an attacker to guess the user's password. It is based on the MITRE ATT&CK matrix, which describes the tactics and techniques used by attackers [7]. Having analyzed the full list of techniques and tactics, we will highlight the list of threats relevant to the authentication system and password guessing, in particular: (IST.008)

TA006 – Obtaining Credentials
T1110.001 – Password Guessing
T1110.002 – Password Cracking

Similar to the MITRE ATT&CK matrix, the Threat Database [8] and the Methodological Document "Methodology for Assessing Threats to Information Security" [9] of the FSTEC of Russia can distinguish threats, as well as techniques related to password guessing. Thus, the threat of password guessing can include the threat of UI.008, and the techniques for gaining access, as well as password guessing, include T1.9, T6.1 and T6.2 from the list of techniques and tactics of the FSTEC of Russia.

The primary vulnerability that makes it possible to implement password guessing is the lack of access restriction to the authentication system – TA0006. Having received the necessary access, the attacker gets the opportunity to enter a password.

If access cannot be restricted or access has been granted, the attacker acquires the ability to enter the password in order to guess what is necessary to gain access to the service - T1110.001. The vulnerability lies in the possibility of an unlimited number of entered attempts [10], which, with a low level of password quality, will allow you to get the correct password within an adequate amount of time.

Instead of the previously mentioned online attack, the attacker can gain direct access to the database used to store user authentication data - T1110.002. Without action from administrators, this threat can lead to critical consequences for the organization.

5 Technical Protection Measures

It should be noted that organizational measures are not considered in this study, because measures of this type are used exclusively for internal users and cannot affect protection against the threat of password brute forcing by intruders.

1. M1036 – Network Access Restriction (NAR) – Countering the TA006 Threat
2. M1032 – Multi-Factor Authentication – Countering the T1110.001 Threat
3. M1036 – Limiting the number of attempts – countering the T1110.001 threat
4. Hashing + "salt" - countering the threat of T1110.002

The primary obvious protection measure is to restrict logical access to the service at the network level (for example, the absence of groups in Active Directory or setting up a firewall) or at the application level (access only for specified accounts) - M1036. However, for external services, a restriction at the business process level is not possible, and therefore requires the implementation of protection measures at other stages.

Limiting the number of attempts entered is the only solution to eliminate the threat of the possibility of a brute-force online attack - M1036. Implementation is possible at the service level, for example, a certain number of input attempts per unit of time (often per hour) or a temporary input limit after a certain number of unsuccessful input attempts [11].

To protect databases, service segmentation is used with no direct user access to the database. At the content level, administrators resort to the only possible solution: password encryption. However, there are several methods for encryption – the use of symmetric or asymmetric encryption, as well as hashing, which is the most popular method. At the moment, there are several hash functions for hiding the original password – from the least effective MD5, SHA-1 to one of the most secure and effective at the moment Argon2 [12], etc. However, despite the potentially high efficiency, password compromise is still possible. An attacker can compromise the database and carry out an offline attack locally on their own equipment, for which there is no restriction on entering passwords [12]. In this regard, hashes built based on weak passwords can be brute-force in less than a day. Hashes built on passwords that are more complex can be brute-forced with rainbow tables by attackers. In connection with this vulnerability, administrators may resort to the use of an additional protection measure – hash salting [12]. The use of salt in hashing has a significant impact on password security and dramatically reduces the effectiveness of brute force offline attacks, but salt is often stored in clear text in the same database as password hashes. In this regard, the guarantee of the impossibility of guessing the password is obviously not provided [5].

In addition to the primary protection measures, there is a secondary protection measure that affects the authentication system as a whole – two-factor authentication. This method can significantly increase the reliability of the system as a whole, but it is worth taking into account the not always possible implementation of this technology and the resource intensity, which is high relative to the main protection measures - M1032.

6 General Methodology for Assessing Password Strength

To quantify the impact of security measures on the acceptable level of trust in a password, you must first provide a complete algorithm for assessing the strength of the password. According to the previously compiled methodology for assessing password strength [4], the generated password is considered trusted if the password quality exceeds or equals the permissible password trust level.

1. The following formula is used to calculate the value of the strength parameter required for password approval:

$$Dur = qual\text{--}thr \tag{1}$$

qual – password quality, a parameter based directly on the contents of the password.

thr is the Acceptable password trust level, the threshold value for password approval,

If the Dur value is not negative, the generated password is approved by the system, otherwise the password is rejected
2. The following formula is used to calculate the quality of the password:

$$qual = (dict \times ent \times ra)/100 \times nsa \tag{2}$$

dict – the coefficient of coincidence of the password with frequent, leaked passwords, as well as the coincidence with the user's personal data,

ent is the entropy coefficient of the password,

ra is the coefficient of repetition/interlacation of characters in the password,

nsa is the coefficient of the presence of non-standard characters in the password,
3. Calculation of the acceptable level of trust:

$$thr = thr1 \times (1 - prot) \tag{3}$$

$$thr1 = thr0 \times crit \tag{4}$$

thr0 is the initial constant value of the threshold value

crit – service criticality factor

thr1 – threshold value taking into account the criticality of the service

prot is the coefficient of protection measures used.

In general, the algorithm for assessing the strength of the password is presented in Fig. 6.

The basic process of assessing the strength of a password begins with the formation of a password. Next, the quality of the password is assessed, i.e. the complexity of the text directly contained in the password. However, in order to be able to calculate the acceptable trust level of thr, it is necessary to assess the criticality of the service and the

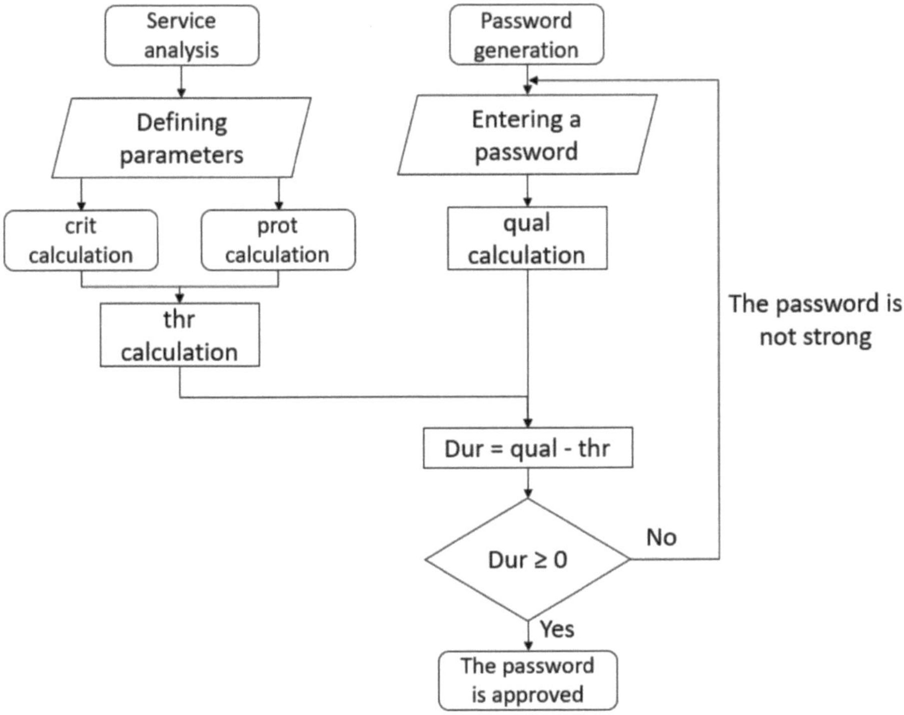

Fig. 6. Algorithm for calculating the password strength score.

brute force protection measures used for the evaluated service in advance. The responsible system administrators separately for each service should carry out the necessary calculations and setting the values of these parameters. It is worth noting that the criticality of crit and prot protection measures are evaluated independently of each other. In this regard, the algorithm for reducing the acceptable level of trust can be carried out without an algorithm for calculating the criticality of the service. Based on the level of trust and the calculated acceptable level of trust, an assessment of the strength of the password Dur is built. If the strength value is not negative, the password entered by the user is approved by the system. Otherwise, the user will need to enter a different password.

As part of this study, we will review and analyze the value of prot (a parameter of the impact of the security measures used), how it is calculated, and how it affects password complexity requirements.

7 Determining the Contribution to the Safety of Safeguards

To determine the impact of security measures, we will use two key parameters: **loyalty** – the ability to tolerate a reduction in password complexity requirements, an important factor that has a positive impact on the user's password load; **Protection** is a measure of the quality of ensuring the protection of the authentication system. These terms are built based on existing ideas about the purpose of protection measures. The measure of protection to be determined is the better; the higher both parameters are taken together. The maximum value of each of the parameters is 1, since the maximum possible value for multiplying loyalty and protection is 1, which corresponds to the complete impossibility of gaining access to the user account with absolutely any password. The minimum value of each parameter is 0. With the minimum possible loyalty, the password requirements are maximum – in this case, the password must correspond to the hash format. If the protection value is 0, the service will be compromised with any password.

Let's analyze the previously proposed protection measures. The assessment of measures in terms of loyalty and protection parameters was carried out by an expert method. Two-factor authentication can be sequential and parallel [13]. Sequential authentication (2FA1) consists in certifying the correctness of the first factor entered by the system – the password, and only then the second one (SMS, push notifications, etc.), while parallel authentication (2FA2) accepts the first factor, then immediately requests the second, certifies it, and then verifies the correctness of the first one. Thus, in addition, with parallel two-factor authentication, we get a pseudo-limit on the number of entry attempts (LNA). Taking into account the information provided, the effectiveness in any case is much higher, since the chance of the attacker to provide two correct factors is close to zero, but since 2FA1 has the ability to carry out a brute force online attack, this method is less loyal.

LNA also shows high efficiency, even with a not too complex password, it becomes a difficult task for an attacker to guess the necessary password when limited to a few requests per unit of time.

NAR allows largely the possibility of compromise of the password by brute force by excluding the possibility for attackers to try to enter the password. However, the ability to enter remains for users within the corporate network. In this regard, loyalty and protection cannot be maximized, but we can talk about a significant impact of this measure on the security of password authentication.

As for hashing technology, the effectiveness of protection is low, as it protects users from hard-to-implement database leaks and brute force offline attacks. In addition to protection, the level of loyalty is also low, but the use of salt can have a significant impact on the parameters and, as with hash functions, different methods of use can have a different impact on the authentication system [14].

A broader view of security will help you assess the contribution of these protections to password requirements. If we consider the assessment of the strength of the authentication system as a whole, using the quality of the password, security measures and the criticality of the system itself, then the following formula will explain the impact of security measures: $qual + prot - crit > 0$.

It turns out that with an increase in the value of the protection measure, it is possible to reduce the qual (password quality), and consequently the requirements for the password. After the analysis, we will determine the coefficients of the previously determined protection measures.

Suppose that the quality of the password is measured on a ten-point scale, and the value of the password quality must be higher than the value of the Acceptable password trust level. The determined coefficients will reduce the Acceptable password trust level to reduce password requirements and, as a result, reduce the burden on users.

To calculate the variable contribution of the protection measure to the security of Prot_pm, which is a combination of the loyalty and protection parameter, we will use the following formula:

$$Prot\_pm = L \times D \tag{5}$$

Where Prot_pm is the value of the contribution to the security of the protection measure pm, L is loyalty, D is protection. If the service has a single protection measure, then formula (5) should also be used to calculate the Prot coefficient (Prot = Protpm1).

In the case of using several protection measures, we will use the following formula to calculate the full coefficient of all protection measures used in the Prot service:

$$Prot = 1 - (1 \times Prot\_pm1) \times (1 \times Prot\_pm2) \cdots \times \ldots (1 \times Prot\_pmn) \tag{6}$$

The calculation uses the security contribution of each of the n security measures used (pm).

Based on the above formula (5), let's calculate the Prot_pm for each of the previously considered protection measures.

$Prot\_(2FA\_1) = 8/10 \times 4/10 = 0.32$
$Prot\_(2FA\_2) = 9/10 \times 8/10 = 0.72$
$Prot\_(LNA) = 7/10 \times 6/10 = 0.42$
$Prot\_(NAR) = 8/10 \times 4/10 = 0.32$

It should also be noted, taking into account the above, if the organization uses 2FA2, then there is no need for BCR, and this parameter will not be taken into account for reduction.

The hash parameter with additional salt use is very flexible, so within the framework of this article, we will demonstrate the value on some types of hash functions + the standard method of using salt in Table 1.

Table 1. Determining the contribution of some password encryption methods to the database.

Hash function	Salt is used	Parameters, L × D	Prot_pm
MD5	–	0.2 × 0.4	0.08
MD5	+	0.2 × 0.5	0.10
SHA-1	–	0.3 × 0.4	0.12
SHA-1	+	0.3 × 0.5	0.15
Argon2	–	0.4 × 0.4	0.16
Argon2	+	0.4 × 0.5	0.20

The implementation of the reduction of the acceptable level of trust will consist in reducing thr by reducing the thr1 parameter by a fraction of Prot. It should be noted that the threshold value, built based on an integral assessment of the measures being implemented, is not the only possible option. For example, in [15, 16], the minimum threshold for evaluation is the number of attempts for an online or offline attack.

As an example of using the methodology, in Table 2, we will consider possible options for reducing the level of requirements on various stacks used – sets of protection methods.

Table 2. Calculations of the requirements level indicator after reduction of protection measures.

thr1	Stack used	Prot, formulas (5) and (6)	thr, formula (3)
8	2FA1, ОКП	0.8 × 0.4 = 0.32	8 × (1–0.32) = 5.44
6	SHA-1 + salt	0.3 × 0.5 = 0.15	6 × (1–0.15) = 5.10
7.5	ОКП, MD5	1 – (1–0.6 × 0.7) × (1–0.2 × 0.4) = 0.4664	7.5 × (1–0.4664) = 4.002

Based on the results of the calculations, we can note a significant decrease in the parameter of the permissible level of requirements for the thr password. Thus, by contributing to security by technical measures, it is possible to reduce password requirements and the burden on the user without the risk of increasing the risk of password compromise by guessing the password [17].

8 Conclusion

Based on the results of the study, all existing processes of interaction with the password were identified and analyzed. By explicating each process, we were able to identify protection measures for key vulnerabilities that affect the ability to guess passwords. Using an assessment based on loyalty and protection parameters, the relationship between protection measures and the Acceptable password trust level was established.

The proposed method is unique, since in similar studies by other authors in the password assessment methodology, security measures are not used in calculating the strength of the password, but only affect the initial requirements for the password, which is an insufficiently objective measure of assessment.

In connection with the above, we get the practical value of using this method to reduce the level of requirements for the password - a positive impact on both the user's psychological load with the password and the authentication system as a whole. Thus, the level of effective cybersecurity in the organization is obviously growing.

However, it should be borne in mind that measures that are more effective may not be available for implementation due to the lack of technological capability, or failure to obtain approval from the business unit due to the lack of funds for implementation. In this regard, those responsible for the authentication service need to plan the implementation of measures from a higher priority, and implement the implementation with full approval.

It should also be noted that by using several measures, it is possible to achieve a more significant reduction in loyalty to the required password complexity, as well as to achieve a significant efficiency of the stability of the authentication system in the field of information security. For example, by using a full set of measures, it is possible to reduce the permissible level of password requirements by 85%.

As for future work in the field of password strength assessment, the next studies will consider the factors influencing the level of criticality of the system, as well as propose a method for applying criticality parameters. In addition to analyzing the criticality assessment of the service, the impact of hashing and additional use of "salt" on the contribution to the security of the corresponding protection measure will also be considered in more detail.

References

1. Le Thanh, B., Thai, H.T.: A statistical Markov-based password strength meter, 101057., ISSN 2542-6605. Internet Things. **25** (2024). https://doi.org/10.1016/j.iot.2023.101057. https://www.sciencedirect.com/science/article/pii/S2542660523003803
2. Budania, R.: A study on development of a biometric authentication system. Anveshana's Int. J. Res. Eng. Appl. Sci. **9**, 8–14 (2024)
3. Lee, Kevin, et al.: Password policies of most top websites fail to follow best practices. In: Soups @ Usenix Security Symposium (2022)
4. Ivanov, A., Nechaev, D.: Development of a flexible methodology for comprehensive assessment of password strength using a system for integral assessment of metrics. In: Raza, Z., Babenko, M., Sajid, M., Lapina, M., Zolotarev, V. (eds.) AISMA-2023: International Workshop on Advanced Information Security Management and Applications. AISMA 2023 Lecture Notes in Networks and Systems, vol. 1207. Springer, Cham (2024). https://doi.org/10.1007/978-3-031-77229-0_11
5. Shay, R., et al.: Designing password policies for strength and usability. ACM Trans. Inform. Syst. Secur. **18**, 1–34 (2016)
6. Buccafurri, F., Angelis, V., Lazzaro, S., Pugliese, A.: Enforcing security policies on interacting authentication systems. Compute. Secur. **140**, 103771 (2024). https://doi.org/10.1016/j.cose.2024.103771
7. MITRE ATT&CK matrix [Electronic resource]. URL: https://mitre.ptsecurity.com/ru-RU (Date of request: 12.01.2024)
8. A database of information security threats [Electronic resource]. https://bdu.fstec.ru/threat-section (Date of request: 12.02.2024)
9. FSTEC of Russia. Methodological document. Methodology information security threat assessments. Approved at 02.05.2021.
10. Abdellaoui, A., Chaoui, H., Khamlichi, Y.I.: A Novel Strong Password Generator for Improving Cloud Authentication. Procedia Compute. Sci. **85**, 293–300., ISSN 1877-0509 (2016). https://doi.org/10.1016/j.procs.2016.05.236
11. Wang, D., He, D.-b., Cheng, H., Wang, P.: fuzzyPSM: a new password strength meter using fuzzy probabilistic context-free grammars. In: 2016 46th Annual IEEE/IFIP International Conference on Dependable Systems and Networks (DSN), pp. 595–606 (2016)
12. Bai, W., Blocki, J., Ameri, M.H.: Cost-asymmetric memory hard password hashing, information and computation. Inf. Comput. **297**, 105134 (2024). https://doi.org/10.1016/j.ic.2023. ISSN 0890-5401. https://www.sciencedirect.com/science/article/pii/S0890540123001372
13. Biryukov, A., Dinu, D., Khovratovich, D.: Argon2: new generation of memory-hard functions for password hashing and other applications. In: 2016 IEEE European Symposium on Security and Privacy (EuroS&P), pp. 292–302 (2016) https://www.sciencedirect.com/science/article/pii/S1877050916305841

14. Bhoi, S.: Password strengthening: using multi-lingual passwords. Int. J. Innovative. Sci. Res. Technol. **9**, 2217–2220 (2024). https://doi.org/10.38124/ijisrt/IJISRT24MAY1746
15. Han, J.: CNN-based multi-factor authentication system for Mobile devices using faces and passwords. Appl. Sci. **14**, 5019 (2024). https://doi.org/10.3390/app14125019
16. Wheeler, D.: Zxcvbn: Low-Budget Password Strength Estimation. USENIX Security Symposium (2016)
17. Kurcheeva, G.I., Klochkov, G.A., Aletdinova, T.A., Lapina, M.A.: Digital city: the characteristics of development indicators of new technologies. J. Phys. Conf. Ser. **1333**(7), 072013 (2019). https://doi.org/10.1088/1742-6596/1333/7/072013

Increasing the Reliability of Text Information Based on PR-Classifiers of Linguistic Variables

Isroil Jumanov[1], Khusan Karshiev[1], Munis Xamidov[1](✉), and Tharewal Sumegh[2]

[1] Samarkand State University named after Sharof Rashidov, Samarkand 140104, Uzbekistan
samsungmunis@gmail.com

[2] DBS Global University, Dehradun, Uttarakhand, India

Abstract. The problem is formulated and the methodology, methods and algorithms of the technology of increasing the reliability of information in electronic document management systems based on the use of information redundancy of various nature (PR-redundancy) are developed. Constructive approaches, principles and tools for using statistical, correlation, logical, semantic, structural and technological connections of elements and relations of concepts in documents are proposed. Tools for monitoring the reliability of information based on alternative strategies, the use of linguistic functions with decision-making, algorithms of fuzzy inferences, and fuzzy modeling are developed. A technique for determining the risk function and assessing the probabilities of undetected errors, parameters of the linguistic function with the calculation of conditional probabilities for a wide range of linguistic variables, calculating the coefficients of similarity and membership, determining the adaptable boundaries of variables and intervals of fuzzy sets is developed. Estimates of the probabilities of undetected errors under various conditions of transmission and processing of information are obtained. A software package for recognition, clustering, correction of k-multiple errors of information is implemented.

Keywords: electronic document management system · reliability · information redundancy · probabilistic model · fuzzy model · software package · efficiency

1 Introduction

In automated electronic document management systems (EDMS) of enterprises and organizations, the research and development of methods aimed at increasing the reliability of information based on mechanisms using natural redundancy of various natures - statistical, logical, semantic, structural and technological is of great theoretical and practical importance [1, 2]. It is also required that the methods of the system should ensure high quality of information processing with the satisfaction of the requirements of the criteria - completeness, accuracy, relevance, security and reliability of information [3, 4].

In practice, EDMS implements measures to improve the quality of the system's functioning, which are associated with the implementation of many functions such as:

M. Lapina et al. (Eds.): AISMA 2025, LNNS 1456, pp. 200–211, 2026.
https://doi.org/10.1007/978-3-032-07275-7_19

creation, storage, editing, deletion of documents, issuance to users [5–7]. Information distortions occur due to human operator errors, errors caused by errors in the operation of scanning and recognition tools, spelling errors by the user are allowed in large quantities, and errors in messages occur under the influence of interference in communication channels [8, 9].

Modern technologies for increasing the reliability of information use various data transmission protocols that are aimed at using artificial code redundancy to detect and correct errors. However, in existing approaches, the reliability of information is ensured mainly at the Session and Channel levels of the OSI model. Moreover, error statistics show that the reliability of information at the Physical and Session levels of the OSI model is almost two orders of magnitude greater than the probabilities of errors occurring at the Representative and Application levels [10, 11].

This work is devoted to the development of constructive approaches, methods and mechanisms for increasing the reliability of information based on tools that use statistical, correlation, logical, semantic, structural and technological connections of elements in a message sequence [12, 13].

2 Main Part

2.1 Mechanisms for Increasing Reliability Based on the Use of Statistical Redundancy of Information in Electronic Documents

A mechanism for increasing the reliability of information is proposed, aimed at using statistical relationships between controlled elements of the sequence $X = (x_{i1}, x_{i2}, \ldots, x_{ik}, \ldots, x_{in})$, which is formed as a message matrix. A tool for checking the reliability of matrix elements by its rows or columns is introduced. The tool is based on the rule for checking the membership of element x_{ik} in the universe U of reliable values, which is considered a subset of allowed values.

As a result, the correspondence of the element x_{ik} to the interval $R[a(x_{ik}), b(x_{ik}), c(x_{ik})]$, of the universe U is established. We denote by $a(x_{ik})$ the lower boundary of the interval of membership of the element x_{ik} to the universe U, $b(x_{ik})$ the upper boundary of the interval of membership of the element x_{ik} to the universe U, $c(x_{ik})$ the average severity of the elements corresponding to the interval of the universe U. The result of the check is formed according to conditions 1 - 3:

$$\mu(x_{ik}) = \begin{cases} 1, & a(x_{ik}) > b(x_{ik}); \quad (1) \\ g(x_{ik}), & a(x_{ik}) \le c(x_{ik}) \le b(x_{ik}); \quad (2) \\ 0, & a(x_{ik}) < b(x_{ik}). \quad (3) \end{cases}$$

where $\mu(x_{ik})$ is the degree of membership of the element x_{ik} to the interval of the universe U;
$g(x_{ik})$is the threshold of membership of the element x_{ik} to the interval of the universe U
.

The membership function (MF) of the element x_{ik} to the interval of the universe U is specified as a probability distribution function $\mu(x_{ik})$ with boundaries$[a(x_{ik}), b(x_{ik})]$.

The probability MF is represented in the range of numbers [0, …, 1] .

A set of tools for the mechanism for increasing the reliability of information is proposed, which is aimed at expanding its capabilities by using operators that perform the following functions: selecting an element from a sequence, dividing the sequence into segments, establishing the boundaries of each segment, determining the type of MF for each segment, fixing the boundaries of the MF, establishing the belonging of the element being checked to the general interval of the universe U.

The mechanism for increasing the reliability of information is based on two alternative verification strategies, which makes the following decisions: A is reliable (classifier 1) and B is unreliable (classifier 2).

The MF is defined by the Gaussian distribution [14]

$$\mu_A(x) = e^{-\frac{(x-\alpha)^2}{\sigma^2}}, \mu_B(x) = e^{-\frac{(x-\beta)^2}{\tau^2}}, x \in R, \sigma > 0, \tau > 0;$$

where α, β are the parameters that determine the maximum point on the FP;
σ, τ are the parameters that determine the degrees of membership of the information element in the form of numbers;
$\alpha \pm \sigma$ - parameters defining the inflection points of the MF $\mu_A(x)$;
$\beta \pm \tau$ - parameters that determine the inflection points of the $\mu_B(x)$.

The membership of the element x_{ik} in the sequence of information given by the MF is checked with a probability that is determined by the ratio to the entire distribution domain

where $\mu(y) = e^{-\frac{(y-\alpha)^2}{\sigma^2}}$ is determined by the types of MF.

The graphs of the probability distribution of the $p(x)$ value show the area limited: (a) on the left by the number x, (b) below by the area $\mu(y)$ along the abscissa axis.

To calculate the integral $p(x)$ is represented by the Laplace function [15]

$$F(x) = \frac{1}{\sqrt{2\pi}} \int_0^x \exp \frac{-t^2}{2} dt,$$

Replacement is in progress $\frac{t}{\sqrt{2}} = \frac{y-\alpha}{\sigma}$.

The integral $p(x)$ is transformed as

$$p(x) = \frac{\int_{\frac{\sqrt{2}(x-\alpha)}{\sigma}}^{\infty} e^{-\frac{t^2}{2}dt}}{\int_{-\infty}^{\infty} e^{-\frac{t^2}{2}dt}} = \frac{\int_{-\infty}^{\infty} e^{-\frac{t^2}{2}dt} - \int_{-\infty}^{\frac{\sqrt{2}(x-\alpha)}{\sigma}} e^{-\frac{t^2}{2}dt}}{\int_{-\infty}^{\infty} e^{-\frac{t^2}{2}dt}} =$$

$$= \frac{\int_0^{\infty} e^{-\frac{t^2}{2}dt} - \int_0^{\frac{\sqrt{2}(x-\alpha)}{\sigma}} e^{-\frac{t^2}{2}dt}}{\int_{-\infty}^{\infty} e^{-\frac{t^2}{2}dt}} = \frac{1}{2} - F\left(\frac{\sqrt{2}(x-\alpha)}{\sigma}\right).$$

A – hypothesis is accepted when

$$\frac{1}{2} - F\left(\frac{\sqrt{2}(x-\alpha)}{\sigma}\right), \text{if } x \geq \alpha;$$

B – hypothesis is accepted when

$$\frac{1}{2} + F\left(\frac{\sqrt{2}(x-\alpha)}{\sigma}\right), \text{ if } x < \alpha.$$

The following notations are introduced:

$$z_1 = \frac{1}{2} - F\left(\frac{\sqrt{2}(x-\alpha)}{\sigma}\right); z_2 = \frac{1}{2} + F\left(\frac{\sqrt{2}(x-\alpha)}{\sigma}\right).$$

Calculations are carried out using integral calculation operators. For this purpose, the tabulated Laplace function is used. An algorithm for solving the equation based on Newton's method is applied.

To solve the problem, hypotheses A, B are given in the form

$$A = \left\{\mu_A(x)|x : \mu_{\tilde{A}}(x) \in [0, 1], x \in X\right\}; B = \left\{\mu_B(x)|x : \mu_{\tilde{B}}(x) \in [0, 1], x \in X\right\}.$$

The algorithm uses the typical operations of the union operator "OR" and the intersection operator "AND" as basic ones.

The $A \cup B$ union operators must ensure

$$\mu_{A \cup B}(x) = \max\{\mu_A(x), \mu_B(x)\}, x \in X.$$

$A \cap B$ intersection operators must ensure

$$\mu_{A \cap B}(x) = \min\{\mu_A(x), \mu_B(x)\}, x \in X.$$

The mechanism under consideration for increasing the reliability of information is based on the use of statistical connections of elements in the sequence of messages. This is due to the selection of a set of connections of elements and relations of concepts in the ED, the fulfillment of which is specified by the following equalities:

$$\sup(A \cup B) = (\sup A) \cup (\sup B); \sup(A \cap B) = (\sup A) \cap (\sup B).$$

2.2 Methodology for Probabilistic Assessment of the Reliability of ED Information

Now we will examine the capabilities of the tool for increasing the reliability of information for events A and B, which are associated with the probabilities of errors of two types. The probability of errors of the first type P_1 occurs when the unreliable element x_{ik} belongs to the lower and upper limits of the MF, and the tool's verification rule classifies the information as reliable and it is considered that event A has occurred.

The probability of errors of the second kind P_2 occurs when the reliable value of the element x_{ik} is not within the lower and upper limits of the MF. The tool algorithm considers it as an unreliable value.

Mathematically, the probability of errors of the first kind is written as

$$P_1 = \frac{z_1 + z_2}{D} p_A(x),$$

where D – the general interval of values of an element in the universe U; $p_A(x)$ is the a priori probability of the occurrence of an element x.

The probability of type II errors is written as

$$P_2 = (1 - p_A(x))\left[1 - \int_{z_1}^{z_2} \mu(y)dy\right].$$

The risk function is defined as

$$\Re = P_1 + P_2.$$

To reduce the value of the risk function $\Re$, it is necessary to determine the optimal values for the variables z_1 andz_2.

The optimization technique consists of finding partial derivatives, performing some mathematical transformations, and also finding the boundaries z_1^{opt} and z_2^{opt}.

For a Gaussian MF, the optimal membership test boundaries of an element are defined as

$$z_1^{opt} = z_2^{opt} = \pm 2\sigma[1 - F(x)].$$

Preference is given to the event that corresponds to the solution of the equation that has an extremal value.

Figure 1 illustrates graphs that show the informative part of the intersection of the MF of two event numbers A = (3, 1), B = (2, 4) at risk level$\Re$ = 0.1.

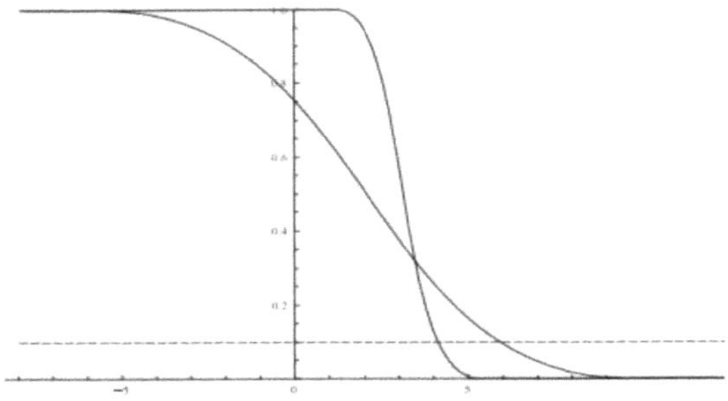

Fig. 1. Selecting one of two numbers A = (3, 1), B = (2, 4) at risk level P = 0.1.

It can be seen that the intersection of the MF of event numbers $A \cup B$ leaves an area of uncertainty when deciding on the reliability of event number$A$.

A tool for monitoring the reliability of information is proposed, which is based on the principle of identifying the μ_C conditional probability distribution and the number C, determined by the types of MF μ_A and μ_B - unconditional distributions

$$C = \left\{\mu_{\tilde{C}}(x)|x : \mu_{\tilde{C}}(x) \in [0, 1],\ \ x \in X\right\}.$$

The following operators are specified:

$$\sup(A \cup B) = (\sup A) \cup (\sup B); \sup(A \cap B) = (\sup A) \cap (\sup B).$$

In this version of the solution to the problem, the conditional MF μ_C is used. Its boundaries and parameters are determined in accordance with the conditional distribution laws in the general interval of the universe U.

Optimization of information reliability control is provided by the variable setting operator

$$Q : U_i = agg\ extr\ C(\mu_C) \rightarrow Q\left[a_i(y_i), b_i(y_i), c_i(y_i)\right].$$

The computational schemes of this operator are synthesized in the form

$$G(X(V_k)) \rightarrow G(C) \rightarrow G(U) \rightarrow G(P) \rightarrow G(Y),$$

until the required level of information reliability is achieved.

The presented solution to the problem of increasing the reliability of ED information is based on the use of the mathematical apparatus of fuzzy modeling, algorithms of fuzzy inferences, as well as the methodology for constructing linguistic classifiers [16, 17].

A methodology for constructing a mechanism for increasing the reliability of information is proposed, in which statistical, logical, semantic connections of text elements are used in tools for checking the compliance of text elements within the boundaries of linguistic variables (LV) [18].

MF μ_X for linguistic variables is specified by tuples

$$< A, X, \tilde{A} > .$$

where A is the MF identifier;

$X = \{x\}$ is a set of LV values to which text elements belong, i.e. the MF value; $\tilde{A} = \bigcup_{x \in X} \mu_X / x$ – is a set on X that specifies constraints on the sets of numerical values of the MF A [19].

The MF processing model is defined by tuples

$$< A, T, X, Q, W >,$$

where T is a set of terms defining the names and values of the LV; Q is a syntactic procedure describing the process of forming new LV values from the set T [20];
W is a semantic procedure that allows us to define each new value formed by the procedure Q and display it as a MF.

A tool for increasing the reliability of information is based on the conceptual principles of checking the compliance of a text element (carrier) within the boundaries of the FP and the general range of values of the set of terms T, as

T= {phrase, simple sentence, participial phrase, adverbial participial phrase, complex sentence, compound sentence, non-union sentence, sentence without main members, transitive relation}.

The values of the LV terms of this base set consist of a sequence of integers from 0 to 100 and are specified in steps by the operator

$$t^{V-A} = \{0, 1, 3, \ldots, 100\}.$$

The text is difficult to formalize with a predominance of qualitative relationships between names, the connections of which are non-linear [21].

For each element of the sequence, its segment and the proper MF of each segment are determined, then the membership of the values of the variable R^{V-A} is checked in the form

$$f\left(R^{V-A}\right) = 100 - \sum_{i=1}^{E} w_i.$$

where E is the number of morphemes;
R^{V-A} is the power of morpheme segments and each proper FP;
w_i is the weight coefficient of morphemes in the segment.

2.3 Implementation of Research Results

To implement mechanisms to increase the reliability of information, the following requirements are met, which include:

fuzzy sets completely cover the intervals of possible values of the carrier of the fuzzy set for each term of the input and output LV, so that any number from this interval belongs to at least one fuzzy set; the coordinates of the maxima are limited by the boundaries of the MF and the corresponding intervals of values of the carriers of the fuzzy set for each term of the input and output LV;
the intersection of the boundaries of the MF of only adjacent terms, the height of the intersection of values in the interval of the carrier of fuzzy sets of adjacent terms is limited from above;
the distance between the coordinates of the maxima of the MF of adjacent terms is limited from below;
the parameter length α -section of each term of the LV fuzzy set is limited from below and from above;
the coordinates of the maxima for the LV of the extreme terms of the fuzzy set are set equal to.

The capability of the tool under consideration is expanded by including procedures aimed at the application of:

- simplified schemes for using logical semantic connections, weight coefficients of morphemes and relations between them; the power of the linguistic classifier; the type, boundaries of the MF, terms of the LV values, the general range of values of the carrier of the T set, which correspond to the boundaries of the MF;
- the type, boundaries of the LF, intervals of values of each LV term, corresponding to the set in the form of variables: length, core, boundaries, intervals [22];

- deviations of the actual calculation from the reference example; settings of the computational scheme of the linguistic classifier (LC).

The transparency of the computational scheme of the tool is performed according to the following requirements:

- the database (DB) should not contain contradictory and redundant data; the knowledge base (KB) should contain a rational number of rules necessary to perform an adequate classification for the studied dependence of relations between names;
- each LV term should correspond to one rule that ensures meaningful interpretations of the tool output and an exact correspondence of the LV to the corresponding MF boundaries and the interval of the carrier value set;
- the linear ordering of the intervals of the LV value set is observed.

It is necessary to ensure the required stability of the algorithm for increasing the reliability of information in the following situations:

strong similarity of the boundaries of the linear pair, intervals of values of carriers to intervals of neighboring sets, which lead to inconsistencies in classification;
phenomenon with incomplete coverage of intervals of sets and violation of linear ordering [23];
in the extreme intervals of quantification, the coefficients of correspondence of the boundaries of the MF to the intervals of the corresponding sets decrease.

The similarity check of sets of relations between proper names is performed based on the following operator [24]

$$S(\tilde{A}, \tilde{B}) = \frac{|\sup p(\tilde{A} \cap \tilde{B})|}{|\sup p(\tilde{A} \cup \tilde{B})|};$$

where $\tilde{A}$ and $\tilde{B}$ are given quantifiers of the sets of LV values; supp - is the carrier of the set of LV values; |·|- is the cardinality of the set of LV values.

If the similarity coefficient is 1, then it is considered that the intervals of values of the support of the sets do not intersect and are completely equivalent.

If the similarity coefficient is 0, then it is considered that the intervals of values of the support of the sets intersect completely. In this version of the solution to the problem of increasing the reliability of information in full-text documents, the similarity coefficient of fuzzy sets is determined when setting the values of the carrier based on the α section in the form

$$S(\tilde{A}, \tilde{B}) = \frac{|(\tilde{A} \cap \tilde{B})_{\alpha}|}{|(\tilde{A} \cup \tilde{B})_{\alpha}|}.$$

where $\alpha < < 1$, for example, $\alpha = 0, 01$.

Optimization of information reliability is ensured by using the following operators:

- adjustment of coefficients and boundaries of MF of fuzzy sets of terms of input and output LV;
- intervals of values of carriers of fuzzy sets of LV terms;

- quantification of values of input and output LV terms of fuzzy sets.

The effectiveness of the proposed tools was studied according to the criteria of reliability, labor intensity and co]st of information control [25].

Figure 2 graphically illustrates the effect that is achieved based on the application of the LV value regulation operator when using the considered fuzzy classifier.

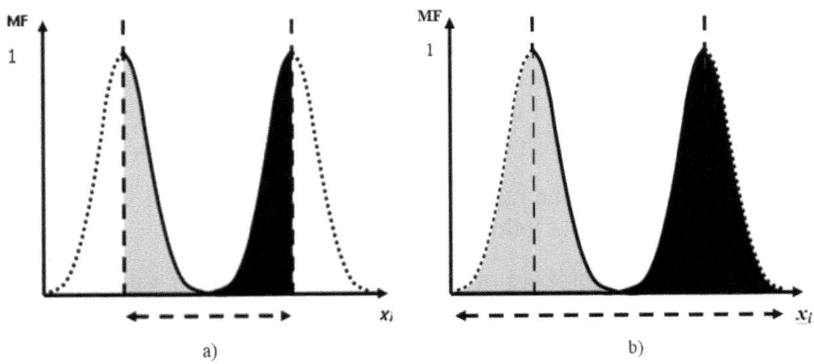

Fig. 2. Regulation of the boundaries of MF and intervals of fuzzy sets: the effect of the narrowing operator (Fig. 2. a) and the effect of the expansion operator (Fig. 2. b).

A software package (SP) for increasing the reliability of information has been developed, which is implemented in Python with the following operators:

- segmentation of the general interval of values of a fuzzy set of element membership;
- regulation of the boundaries of the MF, intervals of carriers of fuzzy sets, the core of the carrier of fuzzy sets of terms of input and output LV, as well as the lower and upper boundaries of the MF core;
- regulation of the number of quantifiers of intervals of values of fuzzy sets, as well as extreme linguistic terms;
- determination of the similarity coefficient of fuzzy sets with compact carriers for triangular and trapezoidal MF, as well as Sugeno and Mamdani fuzzy inference algorithms.

To launch the PC, knowledge bases are formed that include fuzzy rules for checking the reliability of information taking into account external influences based on production operators and mechanisms for using connections between elements and relationships between text concepts.

The KB is conFig.d based on establishing the values of the similarity coefficient of fuzzy sets for input and output LVs with the following quantifiers of the MF boundaries: NB is negatively large; NS is negatively average; NM is negatively small; Z is zero; RM is positively small; PS is positively average; PE is positively large.

It has been established that in comparison with two alternative probabilistic-statistical, linguistic classifiers, fuzzy LV classifiers based on typical fuzzy inference schemes provide significant advantages in increasing the reliability of text information (Fig. 3).

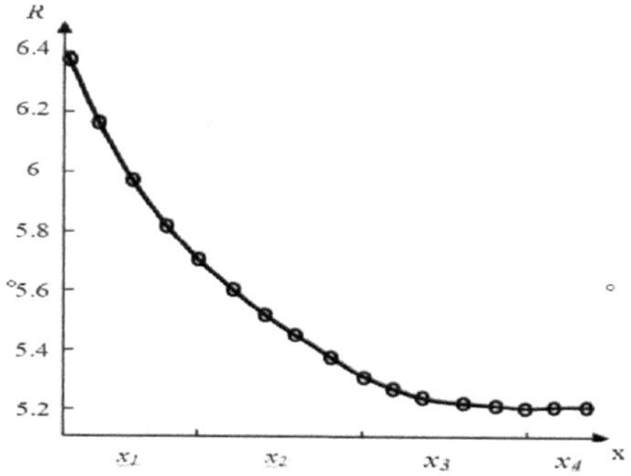

Fig. 3. Illustrates the graph of the error function of the R discrepancy, formed by using quantifiers of intervals of values of fuzzy sets x_1:[9,46,6], x_2:[5,50,6], x_3:[0,55,6] и x_4:[-5,60,6].

At the same time, the size of the set of fuzzy rules is significantly reduced, the blurring in the intersections of the boundaries of the MF, the intervals of fuzzy sets of extreme quantifiers is eliminated, the distance from the coordinate of the MF maxima is expanded, and accordingly leads to a significant increase in the reliability of the information.

It is interesting to note that when the boundaries of the MF are narrowed and the intervals of the supports of fuzzy sets are expanded, the errors of the discrepancy *R* decrease, the values of which asymptotically approach the abscissa axis.

The following data are given and used in testing the PC:

- 17 fuzzy rules for the conditional LV;
- weight coefficients of fuzzy rules used in regulating the left and right boundaries of the common interval of the MF carrier of the output LV fuzzy sets;
- 7 coefficients of MF concentrations of LV terms; 7 coordinates of the MF maximum of LV terms; 11 fuzzy rules for LV terms;
- 2 weight coefficients of fuzzy rules regulating the left and right boundaries of the common interval of the carrier of the output LV fuzzy sets;
- 7 coefficients of MF concentrations of LV terms; coordinate of the MF maximum of LV terms. The PC testing was carried out on a given test training set and an indicator of the average accuracy of information processing with a value of was obtained, and the result of testing on the training set gives a value of.

The tools exhibit negative behavior with the quantifiers "very small" and "very high". However, the operator of expansion of the common interval of the carrier of the fuzzy set of LV values eliminates the noted shortcomings.

3 Conclusion

A methodology for increasing the reliability of information based on probabilistic-statistical, linguistic and fuzzy models has been developed, which complement the capabilities of traditional computational schemes with strict constraints on parameters characterized by high inaccuracy and multi-iteration algorithms. General and particular solutions to the problems have been obtained.

A new approach to optimizing the information reliability verification tool has been proposed, based on the use of MIMO fuzzy modeling operators - structures, setting the boundaries of the input and output MFs, intervals of fuzzy set carrier values, and coordinates of the FP maxima. A modified computational scheme of fuzzy modeling based on Mamdani and Sugeno fuzzy inputs has been implemented in the structure of the information reliability verification tools.

It has been established that, compared to tools using two alternative verification strategies and probabilistic-statistical, as well as linguistic description of processes, tools based on typical fuzzy inference schemes provide significant advantages.

References

1. Rosa, A.T.R., Pustokhina, I.V., Lydia, E.L., Shankar, K., Huda, M.: Concept of electronic document management system (EDMS) as an efficient tool for storing document. J. Crit. Rev. **6**(5), 85–90 (2019). https://doi.org/10.22159/jcr.06.05.14
2. Bobyleva, M.P.: Management of documentary communications and information resources of an organization conceptual approaches. Prob. Manag. Theory Pract. **5**, 116–126 (2011)
3. Monakhov, M.Y., et al.: Features of the environment for ensuring information reliability in information and telecommunication systems. Fundam. Res. **9** (2014)
4. Keldysh, N.V.: Ensuring a given level of security when solving functional problems of departmental electronic document management. Sci. Tech. Collect. Res. Inst. Technol. **4**, 118–123 (2013)
5. Isroil, J., Khusan, K.: Optimization of transmission and processing of Information of electronic documents based on parallel computing technology. In: 2022 2nd International Conference on Technological Advancements in Computational Sciences (ICTACS), pp. 618–624. IEEE (2022). https://doi.org/10.1109/ICTACS56270.2022.9988150(2022)
6. Isroil, J., Khusan, K.: Increasing the reliability of full text documents based on the use of mechanisms for extraction of statistical and semantic links of elements. In: 2020 International Conference on Information Science and Communications Technologies (ICISCT), pp. 1–5. IEEE (2020). https://doi.org/10.1109/ICISCT50599.2020.9351397
7. Isroil, J., Khusan, K.: Increasing the reliability of full text documents based on the use of mechanisms for extraction of statistical and semantic links of elements. In: International Conference on Information Science and Communications Technologies, pp. 1–5. IEEE (2020). https://doi.org/10.1109/ICISCT50599.2020.9351397
8. Akhatov, A.R., Ulugmurodov, S.B.: Braille methods of text recognition using 'deep learning. J. Sci. Innovative Dev. **6**(1) (2023). https://doi.org/10.36522/2181-9637-2023-1-5
9. Yarmatov, S., Bustanov, X., Safarova, L.: Optimization and improvement of reliability of machine learning algorithms based on regularization methods. In: 2025 International Russian Smart Industry Conference (SmartIndustryCon), Sochi, Russian Federation, pp. 394–398 (2025). https://doi.org/10.1109/SmartIndustryCon65166.2025.10986242
10. Akhatov, A., et al.: Genetic algorithm application technology in multi-parameter optimization problems. AIP Conf. Proc. **3244**, 030025 (2024). https://doi.org/10.1063/5.0242074
11. Akhatov, A., et al.: Intelligent Modeling and Optimization of Processes in the Labour Market Artificial Intelligence Blockchain Computing and Security Icabcs 2023, vol. 2, pp. 694–699 (2024). https://doi.org/10.1201/9781032684994-112
12. Rashidov, A., et al.: The same size distribution of data based on unsupervised clustering algorithms. In: Advances in Artificial Systems for Logistics Engineering III. Lecture Notes on Data Engineering and Communications Technologies ICAILE 2023, vol. 180, pp. 437–447 (2023). https://doi.org/10.1007/978-3-031-36115-9_40

13. Ney, H., Kneser, R.: Improved clustering techniques for class-based statistical language modeling. In: European Conference on Speech Communication and Technology (Eurospeech), Berlin, pp. 973–976 (1993). https://doi.org/10.21437/Eurospeech.1993-229
14. Rashidov, A., et al.: Diagnosis of diabetes mellitus based on artificial intelligence algorithms. In: 2025 International Russian Smart Industry Conference (SmartIndustryCon), pp. 349–353. IEEE (2025). https://doi.org/10.1109/SmartIndustryCon65166.2025.10986060
15. Oflazer, K., et al.: Bootstrapping morphological analyzers by combining human elicitation and machine learning. Comput. Linguist. **27**(1), 59–85 (2001). https://doi.org/10.1162/089120101300346804
16. Hakkani-Tür, D.Z., et al.: Statistical morphological disambiguation for agglutinative languages. In: Proceedings of the 18th International Conference on Computational Linguistics (COLING 2000), Aug. 2000. Computers and the Humanities, vol. 36, no 4, pp. 381–410 (2002). https://doi.org/10.1023/A:1020271707826
17. Qasim, M., Sajid, M.: An efficient IoT task scheduling algorithm in cloud environment using modified firefly algorithm. Int. J. Inform. Technol. (Singapore). **17**(1), 179–188 (2024). https://doi.org/10.1007/s41870-024-01758-5
18. Polat, L.Ö.: Digital transformation process in selection and implementation of an electronic document management system. Digit. Trans. (2022). https://doi.org/10.4018/978-1-7998-8169-8.ch004
19. Akhatov, A., Nurmamatov, M., Nazarov, F.: Intelligent modeling and optimization of processes in the labour market. In: Artificial Intelligence Blockchain Computing and Security Icabcs 2023, vol. 2, pp. 694–699 (2024)
20. Rai, H., Pal, A., Khamidov, M., Bobokhonov, A., Ugli, R.: Computational Intelligence Transforming Healthcare 4.0: Innovations in Medical Image Analysis Through AI and IoT Integration 2025. https://doi.org/10.1201/9781003507505-3
21. Makhmadiyarovich, N.F., Sherzodjon, Y.: Development of algorithms for predictive evaluation of investment projects based on machine learning. In: Artificial Intelligence, Blockchain, Computing and Security Volume 2 eBook ISBN: 9781032684994, pp. 681–685 (2023)
22. Nurmamatov, M.Q., Sariyev, S.N., Eshonqulov, B.S.: Application of evolutionary algorithms to enhance the efficiency of neural networks and machine learning algorithms. In: 2025 International Russian Smart Industry Conference (SmartIndustryCon). | 979-8-3315-1124-1/25/$31.00 ©2025 IEEE, pp. 533–537. https://doi.org/10.1109/smartindustrycon65166.2025.10986257
23. Nazarov, F.M., Xamidov, M.M.: Eye state classification method for detecting physiological deviations in drivers based on CNN algorithm. In: 2024 International Russian Automation Conference (RusAutoCon), Sochi, Russian Federation, vol. 2024, pp. 802–807. https://doi.org/10.1109/RusAutoCon61949.2024.10694309
24. Shiriaev, E., Kucherov, N., Movzalevskaya, V., Khamidov, M.: Modification and adaptation of methods and algorithms of the active security concept for fog systems. In: Lapina, M., Raza, Z., Tchernykh, A., Sajid, M., Zolotarev, V., Babenko, M. (eds.) AISMA-2024: International Workshop on Advanced Information Security Management and Applications. AISMA 2024 Lecture Notes in Networks and Systems, vol. 863. Springer, Cham (2024). https://doi.org/10.1007/978-3-031-72171-7_28
25. Lapina, M., Sajid, M., Bagautdinova, A., Lapin, V., Tokmakova, M.: Comparative characteristics of the methods of using digital watermarks to protect text information. In: Raza, Z., Babenko, M., Sajid, M., Lapina, M., Zolotarev, V. (eds.) AISMA-2023: International Workshop on Advanced Information Security Management and Applications. AISMA 2023 Lecture Notes in Networks and Systems, vol. 1207. Springer, Cham (2024). https://doi.org/10.1007/978-3-031-77229-0_15

Increasing Reliability of Information on the Basis of Information Proximity of Document Elements

Isroil Jumanov[1], Khusan Karshiev[1], Munis Xamidov[1](✉), and Tharewal Sumegh[2]

[1] Samarkand State University named after Sharof Rashidov, Samarkand 140104, Uzbekistan
samsungmunis@gmail.com

[2] DBS Global University, Dehradun, Uttarakhand, India

Abstract. The methodology has been researched and methods, models, algorithms have been developed for implementing the technology for increasing the reliability of information in electronic document management systems with mechanisms for determining the information proximity of elements, code words, by bit, digital, linear, modular summation, using a multilevel morphological and *n*- gram structured model for analyzing texts on natural language. Computational schemes have been developed for describing sets, basic functions for initializing all objects, using logical and statistical relationships of text elements, forming a knowledge base with sets of rules, and information control statements. A technique has been developed, which is implemented in the form of a Python program for estimating the parameters of mechanisms for increasing the reliability of information. Mechanisms based on binary modular summation of information bits of a code word over lines of text are proposed. The value of the least non-negative residue of the weight of the information vector, estimates of the probabilities of undetected errors, coefficients of gain in the reliability of information, labor intensity and cost of information control are obtained. A software package for improving the reliability of information in the C++ language has been implemented, in which the proposed mechanisms are synthesized in the environment of the CUDA parallel computing technology. The software package detects and corrects multiple errors in textual information.

Keywords: Electronic document management system · reliability of information · information redundancy · probability of undetected errors

1 Introduction

The use of digital - information technology in various fields of knowledge and management is associated with the regular transmission of a large amount of documents in various formats in natural languages, as well as in the form of a variety of multimedia, text files, etc. [1, 2]. An important theoretical and practical significance is the problem of increasing the reliability, security of information, relevance and completeness

M. Lapina et al. (Eds.): AISMA 2025, LNNS 1456, pp. 212–223, 2026.
https://doi.org/10.1007/978-3-032-07275-7_20

of electronic documents (ED) used in recognition, classification, clustering, association, modeling [3, 4]. In real conditions of information transmission and processing, the value of reliability is still very low, which is due to human errors - the operator, the error of scanning and recognizing devices, spelling errors of users, the impact of interference in communication channels, failures and malfunctions of electronic means, equipment [5].

There are various scientific directions and a large class of methods aimed at improving the reliability of information based on the use of code redundancy and equipment [6].

This study is devoted to the development of methods, models, algorithms with mechanisms to increase the reliability of information based on determining the information proximity of elements of code words, as well as bit by bit, digital linear summation, modular summation, multilevel morphological and n-gram structured analysis of texts in natural languages [7].

2 Main Part

2.1 Basic Principles and Mechanisms for Increasing the Reliability of Texts Based on the Information Proximity of Ed Elements

There are a large number of methods and technologies used to improve the reliability of documented information. Among them, we can distinguish methods based on the search for keywords, phrases, concepts, text elements, as well as the use of mechanisms for extracting statistical, dynamic, specific characteristics, distribution laws and other useful knowledge in data [8].

The issue aimed at increasing the reliability of textual information based on the metric characteristics of assessing the information proximity of elements that are used in measuring the values of statistical, correlation, logical, semantic relationships and relations of document concepts has been studied. In particular, when evaluating the reliability of an element of textual information, the semantic similarities of short words and n-grams of the structural analysis model are checked. The mechanism, at the same time, reveals similar elements (points) of words, on the basis of which an assessment of their informational proximity is determined. The following tasks related to definitions have been solved: common, unique properties of texts in the ED flow and unique lexical units of information; structures of representation of lexical units; information proximity between elements of the text, provided that the structures of the identified lexical units can be the same, and the text itself belongs to different categories [9].

The mechanisms using statistical and logical connections of elements, which consist of three components, are investigated. The first one describes sets and basic initialization functions for all text objects. The second one is responsible for the implementation of the operator for using logical and statistical links of text elements. The third one forms knowledge bases (KB), including sets of rules, statements for information control.

Information proximity is given by paired comparisons of interconnected elements in the form of a matrix [10]

$$M_{\varepsilon}^{A} = \begin{vmatrix} a_{11}^{\varepsilon} \; a_{12}^{\varepsilon} \cdots a_{1n}^{\varepsilon} \\ a_{21}^{\varepsilon} \; a_{22}^{\varepsilon} \cdots a_{2n}^{\varepsilon} \\ \cdots\cdots\cdots\cdots \\ a_{n1}^{\varepsilon} \; a_{n2}^{\varepsilon} \cdots a_{nn}^{\varepsilon} \end{vmatrix},$$

where $A = \{a_1, a_2, \ldots, a_n\}$ – set of ED text elements;

$\varepsilon \in \{1, 2, \ldots, m\}$ – indexing system for text fragments and texts.

The given matrix has the property of inverse symmetry. If the values of its elements are above or below the main diagonal. In addition, if the following condition is satisfied for each pair of elements [11]:

$$a^{\varepsilon}_{\alpha\beta} = \frac{1}{a^{\varepsilon}_{\beta\alpha}}.; \forall \alpha, \beta = \overline{1, n},$$

where $a_{\alpha\beta}$– relationship description.

In the case of implementing a mechanism for increasing the reliability of information with recoding operators, determining a linear sum of binary elements, pairwise comparison, then the number of control comparisons (KC) of line-by-line verification is determined as [12]

$$KC = \frac{n(n-1)}{2}.$$

Note that the number of paired comparisons depends on the dimension of this matrix, for example, when $n = 200$, then it is necessary to perform the number of $\approx 2 \cdot 10^4$ comparisons for one pair, and when $KC >> 200$, then the complexity of information processing increases extremely.

A variant of solving the problem is proposed, in which two alternative groups are selected to control the reliability of information by pairwise comparison of elements. At the same time, the matrices of each group are constructed and used with the following change [13]

$$M^{Br}_{\varepsilon} = \begin{vmatrix} br^{\varepsilon}_{11} & br^{\varepsilon}_{12} & \ldots & br^{\varepsilon}_{1k_r} \\ br^{\varepsilon}_{21} & br^{\varepsilon}_{22} & \ldots & br^{\varepsilon}_{2k_r} \\ \ldots & \ldots & \ldots & \ldots \\ br^{\varepsilon}_{k_r1} & br^{\varepsilon}_{k_r2} & \ldots & br^{\varepsilon}_{k_rk_r} \end{vmatrix};$$

where $br^{\varepsilon}_{\alpha\beta} = -br^{\varepsilon}_{\beta\alpha}$, $\forall \alpha, \beta = \overline{1, k_r}$, $\forall r = \overline{1, l}$, $Br = \{br_1, br_2, \ldots, br_{k_l}\}$ - set of r -group elements; k_r- elements; l - number of groups; $br_{\alpha\beta} = a_{\alpha\beta} - 1$, $\forall r - (a_{\alpha\beta} \geq 1)$ - index of ED context elements.

The significance of each group for pairwise comparison of elements is given as

$$M^{r}_{\varepsilon} = \begin{vmatrix} r^{\varepsilon}_{11} & r^{\varepsilon}_{12} & \ldots & r^{\varepsilon}_{1l} \\ r^{\varepsilon}_{21} & r^{\varepsilon}_{22} & \ldots & r^{\varepsilon}_{2l} \\ \ldots & \ldots & \ldots & \ldots \\ r^{\varepsilon}_{k_r1} & r^{\varepsilon}_{k_r2} & \ldots & r^{\varepsilon}_{ll} \end{vmatrix}.$$

where r^{ε}_{kl} – the positions of the compared elements from different groups.

Consequently, the overall picture of the mechanism for increasing the reliability of information based on a comparison of elements of the 1st and 2nd groups is determined by the value of r^{ε}_{kl}, which reflects the connections of each individual pair of elements.

Let's consider the mechanism of information reliability control on the basis of the amount of shift on an arbitrary row of the matrix. At the same time, the value of $br_{\alpha\beta}^{\varepsilon}$ reflects the comparative characteristics of α and β elements, which determines the sum of the displacements of the α element relative to all $\beta = \overline{1, k_r}$.

The amount of bias on the rank scale is given as [14].

$$Sbr_{\alpha}^{\varepsilon} = \sum_{\beta=1}^{k_r} br_{\alpha\beta}^{\varepsilon}, \left(\forall\alpha, \beta = \overline{1, k_r}, \forall r = \overline{1, l}\right).$$

A positive sum indicates the superiority of the mechanism. If the sum is negative, then it shows its suppression. If $\sum_{\beta=1}^{k_r} Sbr_{\alpha}^{\varepsilon} = 0$, then shows equal significance.

For two compared groups of elements, the matrix is given as

$$M_{\varepsilon}^{B} = \left|\begin{array}{ccc|ccc} b1_{11}^{\varepsilon} & \dots & b1_{1k_1}^{\varepsilon} & Sb1_{1\_}^{\varepsilon} + Sb1_{\_1}^{\varepsilon} + r_{12}^{\varepsilon} & \dots & Sb1_{1\_}^{\varepsilon} + Sb1_{\_k_2}^{\varepsilon} + r_{12}^{\varepsilon} \\ \dots & \dots & \dots & \dots & \dots & \dots \\ b1_{k_1 1}^{\varepsilon} & \dots & b1_{k_1 k_1}^{\varepsilon} & Sb1_{1\_}^{\varepsilon} + Sb1_{\_1}^{\varepsilon} + r_{12}^{\varepsilon} & \dots & Sb1_{1\_}^{\varepsilon} + Sb1_{\_1}^{\varepsilon} + r_{12}^{\varepsilon} \\ \hline Sb1_{1\_}^{\varepsilon} + Sb1_{\_1}^{\varepsilon} + r_{21}^{\varepsilon} & \dots & Sb1_{1\_}^{\varepsilon} + Sb1_{\_k_2}^{\varepsilon} + r_{21}^{\varepsilon} & b2_{11}^{\varepsilon} & \dots & b2_{1k_1}^{\varepsilon} \\ \dots & \dots & \dots & \dots & \dots & \dots \\ Sb1_{1\_}^{\varepsilon} + Sb1_{\_1}^{\varepsilon} + r_{21}^{\varepsilon} & \dots & Sb1_{1\_}^{\varepsilon} + Sb1_{\_1}^{\varepsilon} + r_{21}^{\varepsilon} & b1_{k_2 1}^{\varepsilon} & \dots & b1_{k_2 k_2}^{\varepsilon} \end{array}\right|$$

The following changes are made to the empty places of the matrix: equal sums of ranks from the row of the 1st group; the sum of ranks from the column of the 2nd group; the rank of comparison of groups as a whole. The matrix may contain non-normalized element values, i.e. such that.

$$Sbr_{\alpha}^{\varepsilon} + Sbr_{\beta}^{\varepsilon} + r_{ab}^{\varepsilon} > 8 \text{ or } Sbr_{\alpha}^{\varepsilon} + Sbr_{\beta}^{\varepsilon} + r_{ab}^{\varepsilon} < -8.$$

The normalization of the values of the matrix elements is performed in the form

$$\bar{b}_{\alpha\beta} = \begin{cases} b_{\alpha\beta}, & -8 \le b_{\alpha\beta} \le 8; \\ 8, & b_{\alpha\beta} > 8; \\ -8, & b_{\alpha\beta} < 8. \end{cases}.$$

The normalized matrix is given as

$$\overline{M}_{\varepsilon}^{Br} = \left|\begin{array}{cccc} \bar{b}_{11}^{\varepsilon} & \bar{b}_{12}^{\varepsilon} & \dots & \bar{b}_{1n}^{\varepsilon} \\ \bar{b}_{21}^{\varepsilon} & \bar{b}_{22}^{\varepsilon} & \dots & \bar{b}_{2n}^{\varepsilon} \\ \dots & \dots & \dots & \dots \\ \bar{b}_{k_r 1}^{\varepsilon} & \bar{b}_{k_r 2}^{\varepsilon} & \dots & \bar{b}_{nn}^{\varepsilon} \end{array}\right|.$$

To reduce the number of pairs of element comparisons, the matrix is transformed according to the rule

$$a_{\alpha\beta} = \begin{cases} \bar{b}_{\alpha\beta} + 1, & \bar{b}_{\alpha\beta} \ge 0; \\ \frac{1}{1-\bar{b}_{\alpha\beta}}, & \bar{b}_{\alpha\beta} < 0. \end{cases}$$

The number KC by the transformed matrix is calculated as

$$KC = l\frac{\frac{n}{l}\cdot\frac{n-1}{l}}{2} + \frac{l(l-1)}{2} \approx \frac{n(n-1)}{2l}.$$

The developed technique is recommended for implementation in order to study the efficiency parameters of mechanisms for increasing the reliability of information in the form of a program in the Python language.

2.2 Mechanisms for Improving the Reliability of Text Elements Based on Binary Modular Summation

A mechanism has been developed for increasing the reliability of text elements, based on the use of binary modular summation of information bits of a code word over text lines. It is required, in this case, to obtain the value of the smallest non-negative residue of the weight of the information vector.

Denoted by *SM(m,k)*– binary code used as a base code; M – the value of the summation modulus; m and k - the number of bits in the information and control vectors, respectively.

The mechanism calculates r - weights of the information vector, determines the numbers "1", "0" digits, the number of the smallest non-negative residue $W = r(\text{mod } M)$, $M \in \{2, 3, \ldots, m + 1\}$, which is obtained by binary summation and is entered into the bits of the control vector.

The number of KC is determined by the value

$$KC = \left[\log 2M\right], M \in \left\{2^1, 2^2, \ldots, 2^{[\log_2(m+1)]}\right\}.$$

KC weights are calculated as

$$KC = r(\text{mod}\, M) + \alpha M,$$

where α—the coefficient calculated modulo summation $M = 2$. The summation modulus is defined as $M = 2^{\left[\log_2(m+1)-1\right]}$.

A correction factor is proposed, which is calculated as

$$\alpha = x_m \oplus x_{m-1} \oplus \cdots \oplus x_{k+1},$$

where k - the least significant x_m digits that do not participate in the correction factor.

The coefficient α for $M \in \left\{2^1, 2^2, \ldots, 2^{[\log_2(m+1)]-2}\right\}$ is calculated as.

$$\alpha = x_m \oplus x_{m-1} \oplus \cdots \oplus x_p, p = \left[\log_2(m+1)\right] + 2.$$

The effectiveness of the implementation of the considered mechanisms in the detection of monotonic, symmetric and asymmetric errors is studied.

It is determined that a monotonic error occurs when the "0" or "1" bits of the information vector code are distorted. Symmetrical error is associated with the same number of distortions "0" and "1". An asymmetric error is due to an unequal number of distortions of "0" and "1" bits. Codes with bit-linear summation have positive characteristics in detecting errors of single and double errors. The mechanism based on the *RSM(m,k)* code does not depend on the values of the modulus M and the rules for calculating the coefficient α. The mechanism for increasing the reliability of information, in which the number of KC decreases, the smallest non-negative residue is determined by the number W, module $M^* = 2^{\left[\log_2(m+1)-1\right]}$, $M \notin \left\{2^1, 2^2, \ldots, 2^{[\log_2(m+1)]-1}\right\}$, then it is necessary to use the rule [15–17]

$$KC = r(\text{mod}\, M) + \alpha s,$$

where $M \in \left\{2^1, 2^2, \ldots, 2^{[\log_2 M]+1}\right\}$.

It is determined that the mechanism with arbitrary modular summation allows one and two-fold errors in information vectors to be established. The bit-linear summation mechanism detects a large number of symmetric, monotonic, and some fraction $d = jM, j = 1, 2, \ldots, \left[\frac{m}{M}\right]$, asymmetric errors that do not reach fractions of $d = M + 2j$, $j = 1, 2, \ldots, \left[\frac{m-M}{2}\right]$, as compared to modular summation.

To Table 1 shows the number of undetected errors for the *RSM(8,3)* mechanism - a code with different values of the correction factor.

Table 1. Bivariate conditional probabilities of the most significant words.

a	Number of undetected multiplicity errors d						Total number of undetected errors
	1	2	3	4	5	6	
1	2688 0/2688/0	3360 0/3360/0	336 336/0/0	560 0/560/0	28 0/0/28	0 0/0/0	6972 336/6608/28
2	2048 0/2688/0	2880 0/2880/0	416 416/0/0	1280 0/1280/0	168 0/0/168	70 0/70/0	6862 416/6278/168
3	1664 0/1664/0	3360 0/3360/0	496 496/0/0	1200 0/1200/0	84 0/0/84	0 0/0/0	6804 496/6220/84
4	1536 0/1536/0	3648 0/3648/0	448 448/0/0	960 0/960/0	112 0/0/112	70 0/70/0	6774 448/6214/112
5	2048 0/2048/0	3360 0/3360/0	496 496/0/0	1200 0/1200/0	84 0/0/84	0 0/0/0	6804 496/6224/84
6	2688 0/2688/0	2880 0/2880/0	416 416/0/0	1280 0/1280/0	168 0/0/168	70 0/70/0	6862 416/6278/168
7	2688 0/2688/0	3360 0/3360/0	336 336/0/0	560 0/560/0	84 0/0/84	0 0/0/0	6972 336/6608/28

The formalization of undetected errors as a consequence of transitions of information bits within the control group is carried out. Let the control group have three vectors:

<0000>, <1101> and < 1110>. Transitions like <0000> and < 1101> occur at a triple monotonic error. And transitions like <1101> and < 1110> with a double symmetrical error. The control group is due to 6 types of undetected errors. Of these, 4 are monotonic triple errors, 2 are symmetrical double errors [18].

In each cell of the 1st line, the total number of errors is given, and in the 2nd line, the number of monotonous, symmetrical and asymmetric errors is shown through a slash. Information bits are included in the sum of the correction factor α [19].

The mechanism of *RSM(m,k)* code with modular summation allows to achieve the following results:

- for even values of *m*, any errors of even multiplicity are detected;
- for odd values of *m*, a certain proportion of errors of odd multiplicity are not detected and any errors of odd multiplicity $d \leq M - 2$ are detected;
- when b_j and b_m do not exceed m/2, the mechanism has the same error detection capabilities [20];
- for any values of *M* except for $M = 2$ and $M = 4$, as well as the value of *m*, with an increase in the number α, the total number of undetectable errors decreases;
- for any values of *M*, the length of the information vector, with an increase in the number α, the proportion of undetectable double errors decreases;
- when $M \geq 3$, then the proportion of undetectable double errors with equal values of m and α will be the same [21].

3 Results of and Discussion

Let the set of document text elements *D* be represented by a set of words and word errors, which are independent events:

P(a| d) is the probability that *a* text element belongs to the class *d* ;
P(d| a) - probability of occurrence of document *d* among documents of class *a*;
p(a) – the a priori probability of occurrence of a document of class *a* in the entire corpus *D* ;
q(d) – the prior probability of occurrence of the document *d* in the entire corpus *D* .

The Bayesian information control rule is given as

$$P(a|d) = \frac{P(d|a)p(a)}{q(d)}, \quad a \in A, d \in D.$$

The prior probability is estimated with the frequency of occurrence

$$p(a) = K(a)/K,$$

where $K(a)$—the number of documents of class *a*;
K– the volume of the whole body *D* [22].

The highest probability is given as

$$a^0 = \arg\max_{a \in A} P(0|a)p(a).$$

The terms $\{t_i\}$ of document elements are given by the probabilities

$$P(d|a) = \prod_i P(t_i|a),$$

where t_i - the term of "presence" or "absence" of erroneous elements.

In Table 2 two-dimensional conditional probabilities of the most significant words used to control the reliability of text elements were carried out [23].

Table 2. Bivariate Conditional Probabilities of Most Significant Words.

Word	Concepts	Probability
cardiolog	Cardiology : Oncology	0, 334
cerebrovascular	Cardiology : Oncology	0,269
ventricl	Cardiology : Oncology	0,255
atherosclerosi	Cardiology: Oncology	0,255
atherosclerot	Cardiology :Oncology	0,248
metastasi	Oncology : Cardiology	0,202
lymph	Oncology : Cardiology	0,188
reproduct	Oncology : Cardiology	0,177
melanoma	Oncology : Cardiology	0,158
progesteron	Oncology : Cardiology	0,158

When calculating the probability, the empirical frequency of a single word of each document is represented by a random polynomial vector.

The mechanism is performed by the following operators: splitting a set of words into separate words, reducing the word to lower case; removal of punctuation characters and noise words, for example, and, of, on, etc., which are common to all documents; transformation of the word its stem. Removing prefixes, suffixes, endings.

The basic principle of reliability control is the assignment of a word to a vector containing various words with the frequency of their occurrence in the text. Significance is the determination of the number of keywords, the size of which affects the speed of information processing [24]. To implement the algorithm of the mechanism, a program was implemented using the Python sklearn language module.

Table 3 shows the results of estimating the probabilities of undetected errors according to the criteria:

$\boldsymbol{P_H}$, $\boldsymbol{P_{H/}/P}$- coefficient of gain in the reliability of information;
$\boldsymbol{T_M}$, $\boldsymbol{C_M}$ – coefficients of labor intensity and cost of information processing.

Upon receipt of the results of the study, the following designations were used:

M - the number of categories in the number ***(m = 8)***;
$\boldsymbol{C_m{}^2}$ - the number of combinations from ***m*** to 2;

Table 3. Increasing Confidence with Mixed Alphabet Information.

№	Mechanisms	Probabilities of undetected errors	Performance Criteria Values			
			P_H	P_{H}/P	T_M	C_M
1	Linear by bit summation	$P_H = \frac{67}{m} C_n^2 P^2 (1-P)^{n-2}$	$12 \cdot 10^{-5}$	$0{,}35 \cdot 10^{-1}$	6–7	3–4
2	Linear in numeric summation	$P_H = 67 C_m^2 P^2 (1-P)^{m-2}$	$2{,}34 \cdot 10^{-5}$	$0{,}62 \cdot 10^{-1}$	8–9	3–4
3	Linear by bit and modular summation	$P_H = 67 C_{m/2}^2 P^2 (1-P)^{\frac{m}{2}-2*}$ $*\left[2 + 67 C_{m/2}^2 P^2 (1-P)^{\frac{m}{2}}\right]^2$	10^{-5}	$0{,}3 \cdot 10^{-1}$	10–12	5–7

M **-** the number of ciphers in the directory.

It is determined that the effectiveness of the mechanism depends on the volume of the dictionary of word forms, starting from a certain threshold, its value does not change significantly and even decreases when information is controlled by the relations of the concepts of the document.

A technique has been developed for determining the optimal size of the volume of dictionaries of word forms using a multilevel morphological model and n - grams of a structured model for text analysis [25].

Comparative estimates of the effectiveness of methods based on the use of multilevel morphological and n-gram structured models for analyzing texts in natural languages, in particular, English, Russian and Uzbek, have been obtained.

For the contexts of the Uzbek language, the size of the dictionary is 5000, for the English language - 3600, for the Russian language - 4000 word forms.

In Table 4 theoretical methods for assessing the reliability of information in the control of texts based on the use of models of multilevel morphological and n-gram structured analyzes of texts in natural languages are given.

Table 4. Increasing Confidence with Mixed Alphabet Information.

№	Mechanisms	Probabilities of undetected errors	Performance Criteria Values			
			P_H	P_{H}/P	T_M	C_M
1	Morphological analysis with a dictionary of word forms	$P_H = \frac{M}{10^m}\left[1-(1-P)^m\right]$	$1{,}5 \cdot 10^{-6}$	$0{,}44 \cdot 10^{-2}$	10–13	8–9
2	n - gram analysis with frequency dictionaries of word forms	$P_H = \sum_{j=2}^{m} C_m^j P^j (1-P)^{mj*}$ $*\left(-\frac{89}{10}\right)$	$1{,}3 \cdot 10^{-6}$	$0{,}3 \cdot 10^{-2}$	13–15	10–12

The functional modules of the spelling control system based on n - gram models are based on the use of frequency dictionaries to detect and correct one, two, three-fold errors in texts.

The effectiveness of these mechanisms depends on the size of the dictionary of word forms. A stable dependence is observed when the size of the dictionary is 25% of the total corpus of texts. Further increase in the size of the dictionary slightly increases the reliability of the information.

4 Conclusion

Scientific and methodological bases for increasing the reliability of information based on the following mechanisms have been developed: determining the information proximity of the elements of the words of the encoded ED text; by bit, digital linear, modular summation; application of multilevel morphological and n - gram structured analyzes of texts in natural languages. The study was carried out according to the criteria of the probability of undetected errors, the complexity and cost of information processing.

A technique for estimating the number of control comparisons with a linear sum of binary elements, pairwise comparison, and line-by-line verification of texts has been obtained. It is determined that the number of pairwise comparisons depends on the dimension of the operation matrix. When $n = 200$, then $\approx 2 \cdot 10^4$ comparisons must be performed for one pair. And with $KC >> 200$, the complexity of information processing increases extremely.

Codes with bit-linear summation have positive characteristics in detecting errors of single and double errors. The mechanism based on the *RSM(m,k)* code does not depend on the values of the modulus M and the rules for calculating the coefficient α.

The effectiveness of the implementation of the considered mechanisms in the detection of monotonic, symmetric and asymmetric errors is studied.

References

1. Concept of electronic document management system (EDMS) as an efficient tool for storing document. J. Crit. Rev. **6**(5), 85–90 (2019). https://doi.org/10.22159/jcr.06.05.14
2. Bobyleva, M.P.: Management of documentary communications and information resources of an organization conceptual approaches. Probl. Manage. Theory Pract. **5**, 116–126 (2011)
3. Monakhov, M.Y., et al.: Features of the environment for ensuring information reliability in information and telecommunication systems. Fundam. Res. **9** (2014)
4. Keldysh, N.V.: Ensuring a given level of security when solving functional problems of departmental electronic document management. Sci. Tech. Collect. Res. Inst. Technol. **4**, 118–123 (2013)
5. Isroil, J., Khusan, K.: Optimization of transmission and processing of information of electronic documents based on parallel computing technology. In: 2022 2nd International Conference on Technological Advancements in Computational Sciences (ICTACS), pp. 618–624. IEEE (2022). https://doi.org/10.1109/ICTACS56270.2022.9988150
6. Isroil, J., Khusan, K.: Increasing the Reliability of Full Text Documents Based on the Use of Mechanisms for Extraction of Statistical and Semantic Links of Elements. In: 2020 International Conference on Information Science and Communications Technologies (ICISCT), pp. 1–5. IEEE (2020). https://doi.org/10.1109/ICISCT50599.2020.9351397
7. Isroil, J., Khusan, K.: Increasing the reliability of full text documents based on the use of mechanisms for extraction of statistical and semantic links of elements. In: International Conference on Information Science and Communications Technologies, pp. 1–5. IEEE (2020). https://doi.org/10.1109/ICISCT50599.2020.9351397

8. Akhatov, A.R., Ulugmurodov, S.B.: Braille methods of text recognition using 'Deep Learning'. J. Sci. Innovative Dev. **6**(1) (2023). https://doi.org/10.36522/2181-9637-2023-1-5
9. Yarmatov, S., Bustanov, X., Safarova, L.: Optimization and improvement of reliability of machine learning algorithms based on regularization methods. In: 2025 International Russian Smart Industry Conference (SmartIndustryCon), Sochi, Russian Federation, 2025, pp. 394–398. https://doi.org/10.1109/SmartIndustryCon65166.2025.10986242
10. Akhatov, A., et al.: Genetic algorithm application technology in multi-parameter optimization problems. In: AIP Conference Proceedings, vol. 3244, p. 030025 (2024). https://doi.org/10.1063/5.0242074
11. Akhatov, A., et al.: Intelligent modeling and optimization of processes in the labour market. In: Artificial Intelligence Blockchain Computing and Security Proceedings of the International Conference on Artificial Intelligence Blockchain Computing and Security, ICABCS 2023, vol. 2, pp. 694–699 (2024). https://doi.org/10.1201/9781032684994-112
12. Rashidov, A., et al.: The same size distribution of data based on unsupervised clustering algorithms. In: Advances in Artificial Systems for Logistics Engineering *III* (Lecture Notes on Data Engineering and Communications Technologies, vol. 180), ICAILE 2023, Wuhan, China, July 11–12, 2023, pp. 437–447. https://doi.org/10.1007/978-3-031-36115-9_40
13. Ney, H., Kneser, R.: Improved clustering techniques for class-based statistical language modeling. In: European Conference on Speech Communication and Technology (Eurospeech), pp. 973–976, Berlin (1993). https://doi.org/10.21437/Eurospeech.1993-229
14. Rashidov, A., et al.: Diagnosis of diabetes mellitus based on artificial intelligence algorithms. In: 2025 International Russian Smart Industry Conference (SmartIndustryCon), pp. 349–353. IEEE (2025). https://doi.org/10.1109/SmartIndustryCon65166.2025.10986060
15. Oflazer, K., et al.: Bootstrapping morphological analyzers by combining human elicitation and machine learning. Comput. Linguist. **27**(1), 59–85 (2001). https://doi.org/10.1162/089120101300346804
16. Hakkani-Tür, D.Z., et al.: Statistical morphological disambiguation for agglutinative languages. In: Proceedings of the 18th International Conference on Computational Linguistics (COLING 2000), Aug. 2000. Computers and the Humanities, vol. 36, no. 4, pp. 381–410 (2002). https://doi.org/10.1023/A:1020271707826
17. Qasim, M., Sajid, M.: An efficient IoT task scheduling algorithm in cloud environment using modified Firefly algorithm. Int. J. Inf. Technol. (Singapore). **17**(1), 179–188 (2024). https://doi.org/10.1007/s41870-024-01758-5
18. Polat, L.Ö.: Digital transformation process in selection and implementation of an electronic document management system. Digit. Transform. (2022). https://doi.org/10.4018/978-1-7998-8169-8.ch004
19. Shiriaev, E., Kucherov, N., Movzalevskaya, V., Khamidov, M.: Modification and adaptation of methods and algorithms of the active security concept for fog systems. In: Lapina, M., Raza, Z., Tchernykh, A., Sajid, M., Zolotarev, V., Babenko, M. (eds.) AISMA-2024: International Workshop on Advanced Information Security Management and Applications AISMA 2024. Lecture Notes in Networks and Systems, vol. 863. Springer, Cham (2024). https://doi.org/10.1007/978-3-031-72171-7_28
20. Nazarov, F.M., Xamidov, M.M.: Eye state classification method for detecting physiological deviations in drivers based on CNN algorithm. In: 2024 International Russian Automation Conference (RusAutoCon), Sochi, Russian Federation, vol. 2024, pp. 802–807. https://doi.org/10.1109/RusAutoCon61949.2024.10694309
21. Nurmamatov, M.Q., Sariyev, S.N., Eshonqulov, B.S.: Application of evolutionary algorithms to enhance the efficiency of neural networks and machine learning algorithms. In: 2025 International Russian Smart Industry Conference (SmartIndustryCon). 979-8-3315-1124-1/25/$31.00 ©2025 IEEE, pp. 533–537. https://doi.org/10.1109/smartindustrycon65166.2025.10986257

22. Makhmadiyarovich, N.F., Sherzodjon, Y.: Development of algorithms for predictive evaluation of investment projects based on machine learning. In: Artificial Intelligence, Blockchain, Computing and Security, vol. 2. eBook ISBN: 9781032684994, pp. 681–685 (2023)
23. Rai, H., Pal, A., Khamidov, M., Bobokhonov, A., Ugli, R.: Computational Intelligence Transforming Healthcare 4.0: Innovations in Medical Image Analysis Through AI and IoT Integration. https://doi.org/10.1201/9781003507505-3 (2025)
24. Akhatov, A., Nurmamatov, M., Nazarov, F.: Intelligent modeling and optimization of processes in the labour market. In: Artificial Intelligence Blockchain Computing and Security. Proceedings of the International Conference on Artificial Intelligence Blockchain Computing and Security, Icabcs 2023, vol. 2, pp. 694–699 (2024)
25. Lapina, M., Sajid, M., Bagautdinova, A., Lapin, V., Tokmakova, M.: Comparative characteristics of the methods of using digital watermarks to protect text information. In: Raza, Z., Babenko, M., Sajid, M., Lapina, M., Zolotarev, V. (eds.) AISMA-2023: International Workshop on Advanced Information Security Management and Applications AISMA 2023. Lecture Notes in Networks and Systems, vol. 1207. Springer, Cham (2024). https://doi.org/10.1007/978-3-031-77229-0_15

Automate Dynamic Response to Phishing Attacks

Nadezhda E. Karpova[1(✉)], Vyacheslav V. Zolotarev[2], and Victor O. Goncharov[3]

[1] Samara State Technical University, Samara 443100 Main Building,, Russia
Karpova.ne@samgtu.ru
[2] Reshetnev Siberian State University of Science and Technology, Krasnoyarsk 660037, Russia
[3] Security Intelligence LLC, Moscow 115035, Russia

Abstract. The article considers the possibility of working with scenarios of dynamic response to phishing attacks, taking into account the possibility of presetting such systems on the basis of user behavioral analysis data and formation of risk groups based on the analysis of such preliminary data. A multilevel system of processes that allows to realize the presented approach, as well as a general algorithm and examples of its application are shown. The results of the work in turn can be used in systems that perform individual traffic filtering based on knowledge about users collected by additional modules or obtained through integrated systems. As a result, the application of these results can reduce the risks of successful realization of phishing attacks.

Keywords: Social engineering · user profiles · behavioral analysis · information security · information security management

1 Introduction

Phishing attacks based on social engineering remain one of the most dangerous and widespread cyber threats. Their effectiveness stems from exploiting human weaknesses - trust, fear, curiosity - instead of technical vulnerabilities. Identifying and detecting phishing attacks is a complex and urgent task.

Phishing attacks are mostly based on social engineering. Social engineering is the psychological manipulation of people to obtain sensitive information [1]. According to statistics, 90% of phishing attacks are related to social engineering, and they account for 34% of all data breaches [2]. In a large number of cases, companies suffer significant financial losses [2].

The average damage from an attack is 1.8mn, and in case of double extortion (data encryption + threat of leakage) the sums reach 15mn. The most vulnerable are government agencies (44%), defense enterprises (19%) and educational organizations (14%) [2].

This emphasizes the significance of the phishing problem and the need to develop tools to address it.

M. Lapina et al. (Eds.): AISMA 2025, LNNS 1456, pp. 224–233, 2026.
https://doi.org/10.1007/978-3-032-07275-7_21

The main types of phishing attacks include spear phishing, whaling, clone phishing, vishing (voice phishing) and smishing (SMS phishing) [3]. For example, [3] talks about targeted phishing, where attacks target specific individuals or organizations, which requires careful preparation. There are also examples of large-scale attacks where multi-stage phishing campaigns were used [3, 4].

Email remains the main channel for phishing. 92% of attacks are realized via email. Malicious attachments (ZIP, DOC) or links to phishing pages are disguised as legitimate documents. There are also a growing number of cases of using Telegram, WhatsApp to send fake messages from "friends" or "colleagues" [1, 2, 4, 5].

In a significant number of cases, phishing is realized automatically, using "phishing-as-a-Service" (Phishing-as-a-Service), which includes services to automate attacks, including email templates and fake websites. Relatively new to phishing is the use of artificial intelligence to generate realistic texts, voice messages, and images to enhance persuasiveness [6, 7].

Fake web pages, malicious email attachments such as archives or Office documents that may contain hidden scripts are also used [8].

The specifics of using psychological techniques and social engineering in phishing emails should be noted separately.

It is necessary to emphasize the psychological techniques used by attackers. Phishing attacks exploit emotions such as fear, curiosity, and trust [1, 9]. For example, phishing emails can create a sense of urgency so that the victim acts quickly. In the case of vishing, fraudsters present themselves as employees of a bank or government agencies, using pressure on emotions.

2 Technical Parameters and Peculiarities of Phishing Attack

When evaluating the key technical vectors of phishing attacks, the following frequently used techniques can be noted [10–12]:

1. Hidden scripts in SVG files

   Using this technique, attackers embed Base64-encoded JavaScript in SVG files, disguising malicious code as graphic elements. After decoding, the script redirects users to phishing pages. Signs of its use are:

   - The presence of <foreignObject> or < iframe> tags in SVG.
   - Multi-level obfuscation: reverse strings, insert garbage characters, convert hex to ASCII.

   Detecting such implementations requires deep analysis of SVG structure and dynamic code execution in isolated environments (sandboxes).
2. Phishing via PDF annotations

   This technique hides the URLs of phishing sites in PDF file metadata, such as /Annots objects.

   Signs of its use are:

   - Lack of references in the visible text of the document.

- Presence of hidden actions (/A < < /S /URI /URI (http://malicious-site.com) >>). For detection it is necessary to parsing internal PDF objects using tools like pdf-parser and checking /Annots arrays.

3. Abuse of cloud services (OneDrive)
   When this technique is used, OneDrive links leading to documents with dynamically downloaded phishing URLs via JavaScript. Signs of its use are:
   - No malicious URLs in the source code of the page.
   - Delayed downloading of content via network requests. Detection requires analyzing network traffic and using tools like Selenium to emulate content interaction.
4. Embedding URLs in iframes and hyperlinks
   Browser-in-the-Browser (BiTB) attacks use hidden iframes to spoof interfacesThe signs are the presence of hidden <iframe> elements with external sources and domain mismatch between the main URL and embedded links.
5. Dynamic phishing sites with machine learning Indications of use are:
   - Generate unique URLs using algorithms (e.g., gen_url = base_url + random_string).
   - Using redirects via JavaScript (window.location.href).
6. electronic phishing with NLP
   Its signs are the presence of key triggers in the text of emails: "Urgent", "Check account", and fake domains (e.g., paypa1.com instead of paypal.com). To detect the latter three features, it is effective to use deep learning models (e.g., PhishTransformer) that analyze all URLs on a page, including hyperlinks and iframes, as well as the presence of emails from key triggers.

Thus, phishing attacks continue to evolve using sophisticated technical tricks and social engineering. In the future, there will be a growing number of attacks that use AI-assisted attack automation and the generation of personalized messages and adaptation to victim behavior. There will also be an increase in phishing attacks via messengers such as Telegram and WhatsApp due to the high trustworthiness of the channels, and the use of the darknet to sell phishing kits and stolen data for secondary attacks. Countermeasures require proactive measures at the intersection of technology (ML, traffic analysis) and human factors (user awareness). Without the integration of automated systems and continuous threat auditing, the risks of successful attacks will increase.

3 Ways of Protection and Response Approach

Currently, there are several ways to protect against phishing. The following are commonly recognized [13–18]:

- Employee training. This strategy involves regular phishing simulations and training to recognize suspicious emails (checking the sender's domain, analyzing attachments).
- Using technical measures such as implementing DMARC, SPF, DKIM to authenticate mail; using EDR systems and sandboxes to analyze attachments; blocking dangerous domains and file extensions (e.g. .exe).

- Multi-factor authentication that reduces the risk of account compromise even if passwords are leaked.
- HTTPS and certificate validation that avoids sites without a secure connection.
- It's also important to use organizational methods such as verifying sender addresses and preventing suspicious links from being clicked.

Thus, phishing attacks using technical tricks (hidden URLs, obfuscation) and psychological manipulation are evolving to become more sophisticated. Their success depends not so much on technology as on the human factor. Countering them requires a combination of technical solutions, regular training, and a culture of cybersecurity. The most effective way to protect against phishing attacks would be to use hybrid approaches and dynamic response [19], namely dynamic content analysis (sandboxes, EDR), machine learning with URL and HTTP feature sampling, deep learning to analyze text and image semantics.

Particularly relevant are approaches to response that involve the development of a phishing attack response system that would generate a defense system response automatically, in real time, given the current vector of the actual phishing attack.

4 Responding to a Phishing Attack

The active introduction of SOAR (Security Orchestration, Automation and Response) platforms into information security practice is caused by a complex of interrelated factors, including the exponential growth in the number and complexity of cyberattacks, as well as the limited resources of SOC teams forced to manually process a multitude of disparate signals. Current research confirms that without automation, a significant portion of specialist time is spent on routine data collection and collation, which significantly slows down the response to threats.

First of all, automation is subject to the collection and enrichment phase, which according to IBM Security's Cost of a Data Breach Report 2023, up to 80% of the response time to information security (IS) incidents is actually taken up by the data collection and enrichment phase (Data Enrichment).

Further, as local expertise is accumulated, it is possible to move on to automating actions to respond to incidents, including phishing incidents.

Thus, in order to respond to a phishing attack, taking into account the peculiarities of the user's psychological profile, it is necessary to.

The response to an attack should include various levels of information security management. For phishing attack defense mechanisms to work correctly, it is necessary to:

1. A way to identify (automated) indicators of compromise at the infrastructure level,
2. organization of user interaction processes at the level of templates and regulations,
3. Collect data and analyze it, including retrospective data on phishing attacks, Formation of response and prevention scenarios at the level of the protected system as a whole, and continuously, taking into account the dynamics of change [9].

At this level, continuous identification of phishing attack issues can be required, including for preventive detection or early detection.

When handling IS incidents, analyst operators face the following challenges:

1. Speed of response;
2. Detection accuracy;
3. Analyst workload.

How automation tools help address key IS incident lifecycle processing challenges per NIST SP 800-61r2:

1. Preparation

SOAR integrates with SIEM and EDR, automatically updating the playbook for new threats. Problem: How to aggregate a large amount of data for further analysis to detect abnormal activity (Figs. 1 and 2)?

2. Detection and analysis

SOAR Automatically enriches data (reduces time for routine data collection and enrichment operations).

Problem: How to reduce the number of false positives?

3. Containment

SOAR performs containment.

Problem: How to prioritize threats based on context and local expertise?

4. Elimination

SOAR reduces the risks associated with human error in remediating threats Problem: how to account for repeated anomalies?

1. Restoration

SOAR automates recovery by integrating with backup solutions. Issue: identifying risks associated with re-compromise.

Application of User and Entity Behavior Analytics (UEBA) technologies to identify anomalies and potential threats in IT systems, using the example of receiving a phishing email (see Fig. 3).

The essence of applying the above technology is automated and continuous detection of phishing attack problems (see Fig. 4).

Conclusions:

According to the analysis, in the framework of the life cycle according to NIST methodology of application of joint work of SOAR and UEBA classes of protection systems solves the following tasks:

1. Preparation SOAR:

    - Integration with the security system by means of the connector mechanism
    - Customize automated response scenarios through the workflow builder

    UEBA:

    - Integration with raw event sources through a connector mechanism to detect abnormal activity

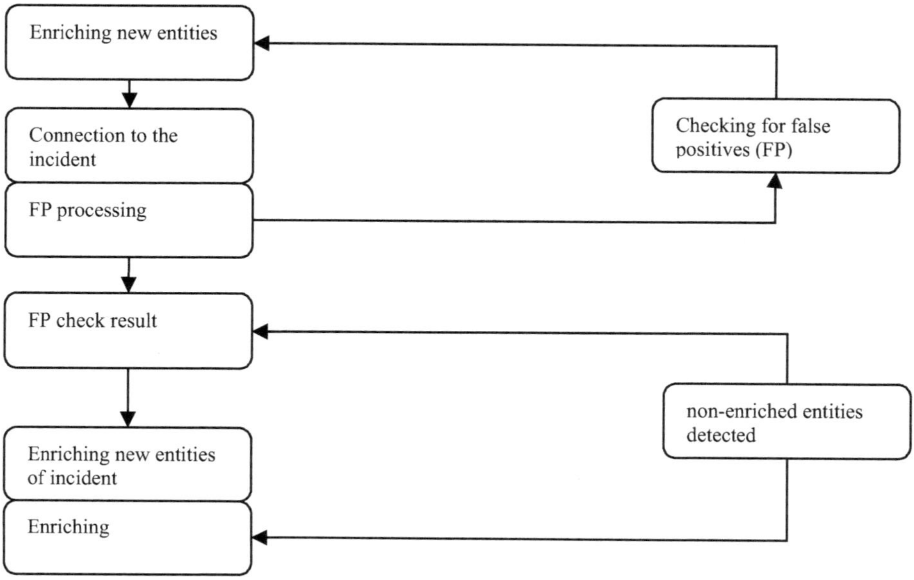

Fig. 1. Example of a playbook that takes ino account the steps of context analysis

- Customizing rules and weighting factors

2. Detection and analysis SOAR:

   - Automatically enrich IS incidents with raw events to make quality response decisions

   UEBA:

   - Reduces false alarms through automatic cotnax analysis

3. Containment SOAR

   - Performs automatic containment through preconfigured scripts and connectors to the NWPs

   UEBA

   - Automatically prioritizes threats, Incidents with a high risk score are handled first

4. Elimination SOAR:

   - Reduces the risks associated with human error in remediating threats

   UEBA:

   - Controls recurring anomalies through machine learning mechanisms

In this way, the problems associated with: Speed of Response(SOAR); Detection Accuracy(UEBA); Analyst workload (SOAR+ UEBA) .

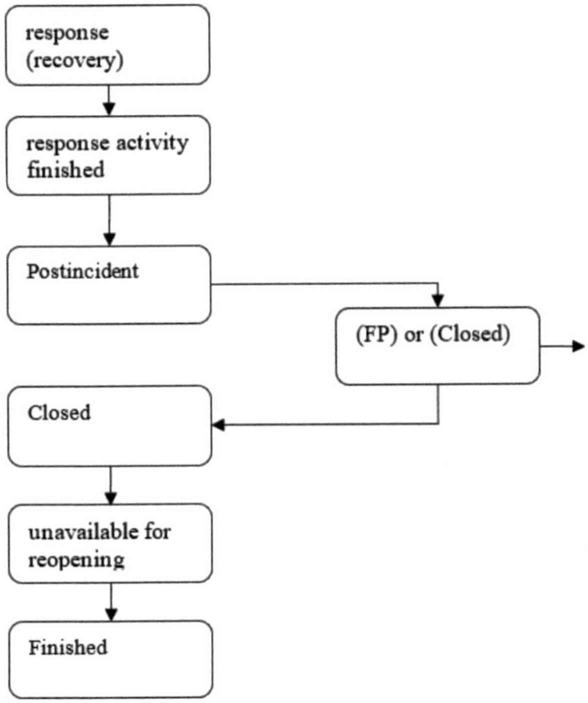

Fig. 2. Example of a playbook that takes into account working off repeated anomalies (postincident)

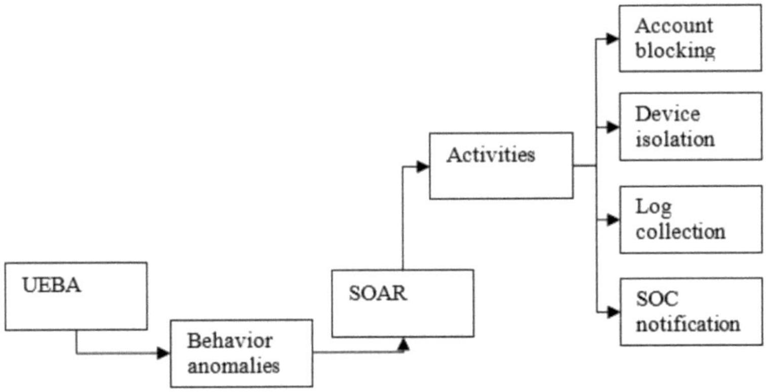

Fig. 3. Application diagram

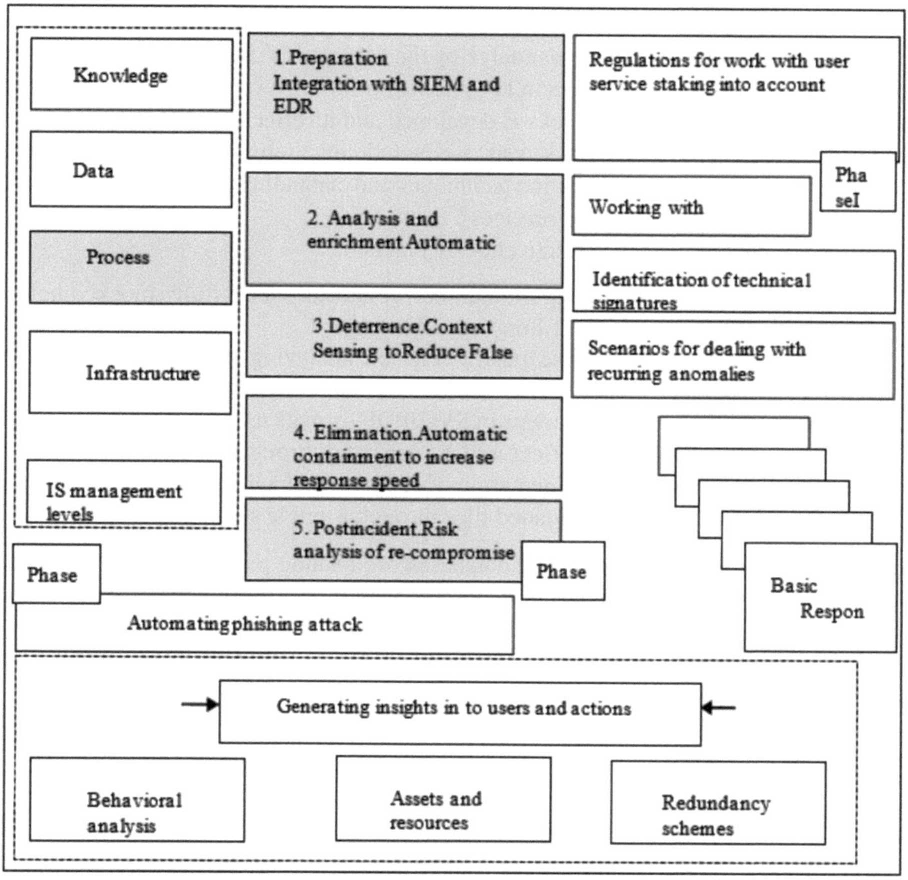

Fig. 4. Responding to phishing attacks and preventing phishing attacks by risk group prioritization

5 Conclusions

The presented paper analyzes phishing attacks, their classifications, technical parameters and peculiarities.

The methods of defense and approaches to response are described. The response to a phishing attack is described in detail, taking into account the peculiarities of the user's psychological profile and including various levels of information security management.

The analysis shows that current phishing countermeasures combine human-centered methods with automated solutions. On the one hand, training users to recognize social engineering triggers (e.g., spoofed domains or urgent appeals) forms the first echelon of defense, reducing the risk of unaware actions. On the other hand, automating the analysis of hidden scripts, dynamic URLs, and network requests enables rapid detection of complex attacks, minimizing reliance on human error. This symbiosis provides flexibility: machine learning systems complement user expertise rather than replace it, creating a layered barrier for attackers.

Based on the analysis, the structure of the system that allows automated and continuous application of technologies for analyzing the behavior of users and entities to identify anomalies and potential threats in IT systems for the correct operation of protection mechanisms against phishing attacks is developed and its effectiveness is shown.

Further steps in phishing attack response include integrating AI-based models that can adapt to new real-time obfuscation techniques and expanding dynamic analysis tools (e.g., sandboxes to emulate cloud services).

To experimentally test the effectiveness is planned:

1. Conduct A/B testing comparing the number of attacks successfully blocked before and after the implementation of improved algorithms.
2. Evaluate the reduction of false positives when analyzing phishing emails using updated NLP models.
3. To measure the time to detect threats in SVG/PDF formats using automated metadata parsing. Scaling the positive effect will be achieved through cloud-based solutions and APIs that enable the implementation of defense mechanisms at the infrastructure level (e.g., validation of all uploaded files through a single service).

In addition, the development of open databases of phishing patterns and collaborative signature sharing platforms between organizations will strengthen collective protection against phishing attacks. Automating user education through online training and attack simulations will further strengthen the resilience of the system, allowing for more user-specific protection without losing the flexibility of the approach.

References

1. Kang, L., Chek, K., Choon, L.: A survey of phishing attacks: their types, vectors and technical approaches. Expert Syst. Appl. **106**, 1–20 (2018)
2. https://www.ptsecurity.com/ru-ru/research/analytics/cybersecurity-threatscape-2023-q4/
3. Chanti, S., Chithralekha, T.: A literature review on classification of phishing attacks. Int. J. Adv. Technol. Eng. Explor. **9**(89), 446–476 (2022). https://doi.org/10.19101/IJATEE.2021.875031. Ogundokun, R., Sadiku, P., Tosho Abdulahi, A., Babatunde, A., Gbolagade, K., Oladipo, D. A review on phishing attacks: types, prevention measures and detection features (2022). https://doi.org/10.22624/AIMS/BK2022- P36.eatures
4. Rychkov, V.A.: Phishing as a way to violate the security of the system. Financial security - new horizons: collection of scientific articles 2024. 53–61
5. Martinez, R., Johnson, P.: Phishing in the cloud era: challenges and solutions. Int. J. Inf. Secur. **22**(4), 789–805 (2023)
6. Kim, H., Park, S.: AI-driven social engineering attacks and defense strategies. Comput. Secur. **115**, 102–115 (2022)
7. Wen, T., Xiao, Y., Wang, A., Wang, H.: A novel hybrid feature fusion model for detecting phishing scam on Ethereum using deep neural network. Expert Syst. Appl. **211**, 118463 (2023) ISSN 0957-4174
8. Brown, T., Wilson, L.: Browser security platforms: a new approach to combat phishing. J. Cybersecurity. **12**(3), 45–60 (2024)
9. Karpova, N.E., Zolotarev, V.V., Zolotareva, E.Y.: Using user profiles for dynamic correction of phishing attack response scenarios. In: AISMA-2024: International Workshop on Advanced Information Security Management and Applications AISMA 2024. Lecture Notes in Networks and Systems, vol. 863, pp. 106–114. Springer, Cham

10. Smith, J., Lee, K., Gupta, R.: Phishing detection using hybrid CNN-LSTM with African vulture optimization algorithm. In: IEEE Trans. Inf. Forensics Secur., vol. 18, pp. 120–135 (2023)
11. Aleksandrov, A.G., Petukhov, A.Y., Danilyan, A.S.: Analysis of information security threats when using phishing sites. Jurist Pravoved'y. **4**(103), 156–161 (2022)
12. Chen, Y., Wang, Z.: Behavioral analysis for phishing detection: a machine learning approach. ACM Trans. Privacy Secur. **27**(1), 1–25 (2024)
13. Zhu, E., Yuan, Q., Chen, Z., et al.: CCBLA: a lightweight phishing detection model based on CNN, BiLSTM, and attention mechanism. Cogn. Comput. **15**, 1320–1333 (2023)
14. Zhu, E., Chen, Z., Cui, J., Zhong, H.: MOE/RF: a novel phishing detection model based on revised multiobjective evolution optimization algorithm and random forest. IEEE Trans. Netw. Serv. Manag. **19**(4), 4461–4478 (2022)
15. Liu, D.-J., Geng, G.-G., Zhang, X.-C.: Multi-scale semantic deep fusion models for phishing website detection. Expert Syst. Appl. **209**, 118305 (2022) ISSN 0957-4174
16. Smirnov, S.I., Eremeev, M.A., Magomedov, S.G., Izergin, D.A.: Criteria and indicators for assessing the quality of cybersecurity incident investigations. Russ. Technol. J., 47–56 (2024)
17. Rychkov, V.A.: Modeling of Processes and Design of Information Protection Systems. Moscow, p. 150 (2023)
18. Karpova, N.E., Baranov, A.S., Emelina, A.A., Konovalov, A.E.: Development of the analyzing system for user's actions in the informational environment. In: XIV International Scientific and Technical Conference Applied Mechanics and Systems Dynamics (AMSD) 2020, Omsk, Russian Federation, pp. 1–12 (2020)
19. Oleinikova, A.A., Zolotarev, V.V.: The concept of information security management based on the cycle of continuous detection and response to information security incidents. Izvestia YuFU. Tech. Sci. **5**(235), 66–81 (2023)

Algorithm for Improving the Reliability of Data Processing and Transmission in Fog Environment

Nikolay Kucherov[1(✉)], Egor Shiriaev[1], Andrey Gladkov[1], and E. A. Mary Anita[2]

[1] North-Caucasus Federal University, 1, Pushkin Street, Stavropol 355017, Russian Federation
nik.bekesh@yandex.ru, egor.shiriaev@reaneling.ru, Agladkov@ncfu.ru

[2] Christ University, Bangalore 560029, India
maryanita.ea@christuniversity.in

Abstract. The paper presents approaches to reliability and fault tolerance, including the use of residual class systems and dynamic load balancing methods, and demonstrates the promise in improving the stability and efficiency of fog systems. Simulations are carried out and the results confirm that the implementation of reliability improvement algorithms based on the residue number system can significantly reduce the probability of data loss. The use of redundant computational structures, adaptive control mechanisms and stable node communication protocols allows to create fault-tolerant solutions that can effectively cope with the risks and uncertainty characteristic of heterogeneous distributed computing environments.

Keywords: Fog Computing · Residue Number System · Reliability · Uncertainty

1 A Review of the Current State of Research on Fog Computing

Fog computing systems are distributed computing environments that perform data processing on various nodes in the network, including endpoint devices, cloud servers, and edge computing nodes. These systems are gaining importance in modern information technology for several reasons.

Modern information technology generates huge amounts of data, such as data from the Internet of Things (IoT), social media, and medical diagnostics. Fog computing-based data processing systems enable scaling and distribution of computation, provide an efficient way to analyze, process, and store this data.

Fog computing is a distributed computing architecture that allows data to be processed in different parts of the network. This reduces the load on central servers and reduces the response time of the entire system.

With the advent of machine learning and artificial intelligence technologies, more sophisticated analysis and knowledge extraction from data has become possible. Fog computing systems provide an environment to effectively apply these techniques to real-time data. The development of the Internet of Things (IoT) concept has resulted in

M. Lapina et al. (Eds.): AISMA 2025, LNNS 1456, pp. 234–242, 2026.
https://doi.org/10.1007/978-3-032-07275-7_22

a large number of devices collecting and transmitting data. Fog data processing systems can reduce the load on the network and central servers by processing this data at the edge devices. Fog systems can process data closer to the source, reducing the risk of sensitive data being transmitted over the network and increasing data security and privacy.

One of the main advantages of fog systems is the ability to process data in real time. This is particularly important for applications that require fast analysis and decision-making based on up-to-date data, such as monitoring manufacturing processes, financial transactions and security systems. With a decentralized architecture, data can be processed on peripheral devices, reducing latency and improving system responsiveness. Systems based on virtual arrays can adapt to changing conditions and requirements. They can easily scale with the amount and speed of incoming data, making them ideal for dynamic and rapidly changing environments. Additionally, fog systems can interoperate with cloud services to expand their functionality and scale.

Data security and privacy are also important aspects of fog computing systems. Since data is processed at all levels of the network, including edge devices, security must be ensured at all stages of data transmission and processing. Encryption, authentication, and access control mechanisms can be used to protect data from leakage and unauthorized access.

Reliability is one of the key factors of the Fog Data System. Failure of any component of the system may lead to serious consequences such as loss of data or disruption of business processes. Therefore, it is necessary to investigate ways to improve system reliability, including the development of fault detection and recovery mechanisms.

The Fog System is an integrated computing environment for efficiently managing large amounts of data in a decentralized environment. The Fog System is based on the concept of fog computing, where data processing occurs not only in the centralized data center, but also at all levels of the network, including edge devices, middle gateways and cloud servers. This approach reduces data processing time and improves system re-responsiveness, which is especially important for applications that require real-time data-driven analysis and decision-making.

In the paper [1], the authors consider the application of fog computing to improve the computational efficiency of geographic information systems (GIS). It is proposed to use fog devices to reduce the network load during data transmission and to reduce the computational load. Decentralized data processing around the network is considered to reduce latency and improve real-time decision making. Special attention is paid to the integration of fog computing with the existing GIS infrastructure and to ensure data security when processing data in virtual nodes. Examples are given to illustrate the practical application of the proposed approach in various GIS use cases, such as environmental monitoring and urban management.

The authors of paper [2] consider topical issues in the direction of IMT-2030 network infrastructure, in particular dynamic fog computing and serverless architectures targeting the development of green information and communication technologies. The authors consider the contribution of these technologies to improving energy efficiency and reducing the carbon footprint of modern IMT infrastructures. In particular, the authors focus on the automatic distribution of computational tasks among network nodes, which optimizes the use of resources and minimizes energy consumption. Examples of Dynamic

Fog Computing are given for applications in various industries such as smart cities, industry and transportation. The future of these technologies and their role in shaping sustainable and environmentally friendly future IMT solutions is also considered.

In [3], the routing problem in networks using dynamic fog computing is considered. Fog computing is considered as the next stage in the decomposition of multilayer cloud computing architectures. Their goal is to change the data processing paradigm and methods by bringing the computing resources closer to the data sources. Considering the dynamic nature of fog nodes and the need for high quality of service (QoS), the authors analyze the existing routing approaches in such networks. The authors propose new routing algorithms adapted to the specificities of fog computing, which can improve the data transfer efficiency and reduce the delay. In addition, issues of security and network resilience are considered, as well as methods for optimal resource allocation among nodes.

Paper [4] analyzes the concept of fog computing and its role in modern IT infrastructure. The authors argue that fog computing is a model for decentralizing data processing resources, complementing cloud technologies and improving the quality of user experience. The authors pay special attention to the comparison between cloud and fog computing and their strengths and weaknesses in different application scenarios. Examples of the use of fog computing in video surveillance systems, smart buildings, and other applications requiring real-time data processing are discussed. The authors emphasize that fog computing does not replace the cloud, but rather extends the capabilities of the cloud by bringing processing and analysis closer to the data source. The article also touches on energy efficiency and ICT sustainability, noting that fog computing can help reduce energy consumption and operating costs.

Modern examples of fog computing are reviewed in [5]. Particular attention is paid to its reliability aspects. The paper analyzes how fog computing can provide more reliable services by bringing computing resources closer to the end users. The authors emphasize the importance of developing new technologies and architectures to ensure high reliability in decentralized computing environments such as fog computing. In addition, potential challenges related to the management and coordination of a large number of heterogeneous devices in fog environments are cited and possible solutions are proposed. Particular attention is paid to issues such as quality of service (QoS) and system resilience to various types of failures.

The authors of [6] conducted a study on modern cloud systems incorporating fog systems. The paper presents the developed algorithm for calculating the reliability of such networks taking into account the complex interconnections between data transmitters, fog servers and cloud servers. The proposed model allows for a more accurate evaluation of the performance and robustness of systems that combine cloud and fog computing. The authors also considered the impact of various factors such as network topology, node characteristics and traffic parameters on the overall system reliability. The results demonstrate that the integration of fog computing into cloud networks can significantly improve its reliability, especially in scenarios with high availability and service quality requirements. This paper provides valuable guidelines for engineers and researchers involved in the design and optimization of hybrid computing systems.

In [7], the authors consider a fog computing paradigm that provides cloud-like services at the network edge using distributed, heterogeneous and resource-constrained computing nodes. They propose approaches to proactively place IoT application microservices in a reliability-aware manner to improve the resilience and performance of such systems. The authors analyze different microservice placement strategies by considering factors that affect reliability, such as node fault tolerance, network latency, and application QoS requirements. The proposed methods optimize resource allocation in fog systems, balancing performance and reliability. Simulation results validate the effectiveness of the proposed strategies in various IoT application scenarios.

The study of the issue of increasing the stability of networks using fog computing is considered in [8]. The authors analyze an architecture in which computing resources are distributed closer to the network periphery, which reduces latency and improves service reliability. The authors also consider data processing mechanisms at the fog node level, which facilitates more efficient management of network resources and improves the overall resilience of the system to failures. The results of the study show that the use of fog computing improves network response time and makes peripheral networks more resilient to failures.

2 Reliability in Fog-Based Systems

Reliability and fault tolerance are important characteristics of fog computing to ensure stable and continuous operation of distributed systems operating in highly dynamic and uncertain environments. These characteristics become especially important in systems that rely heavily on data availability and timely processing. Failure of a single node or device can result in a disruption of the entire system, affecting service quality and data availability. This is particularly unacceptable in applications related to smart cities, medical technology, industrial automation, and other areas where failures not only reduce efficiency, but also threaten safety and life.

Reliability is defined as the ability of a system to perform a specified function under specified conditions and within a specified time. Fault tolerance, in turn, implies a mechanism that allows the system to remain operational even if some of its components fail. In the case of fog computing, this may mean that the system should be able to process data locally, redirect tasks to backup resources, recover quickly from failures, and minimize data loss due to failures.

The main difficulty in providing reliability and fault tolerance in fog computing is the decentralized nature of such systems. These systems consist of many heterogeneous devices operating under different physical, network, and climatic conditions. These types of systems are subject to many external and internal threats, including hardware failures, network failures, software bugs, malicious attacks, and dynamic changes in the operating environment. These factors require the use of complex control algorithms, as well as the implementation of a multi-layered approach to ensure resilience to failures and prevent their propagation throughout the system.

The dynamic nature of fog computing systems makes the task of ensuring their stable and reliable operation very complex. The nodes of such systems, especially mobile nodes, are complex devices that operate under constantly changing environmental and infrastructure characteristics. For example, autonomous vehicles or wireless devices periodically change their location, which affects the quality of communication with other network nodes and central servers. Such changes typically result in a temporary loss of connectivity, with the added factor of delays, data transfer errors, or the immediate unavailability of related services altogether.

The interaction of mobile nodes with other system components is hampered by the presence of dynamic changes in network link characteristics such as link bandwidth, latency, and the probability of data packet loss. In addition, mobile nodes may encounter physical environmental constraints, including changes in weather conditions, mechanical obstacles, or the effects of electromagnetic interference, making it difficult to maintain stability in their operation.

These factors create the need to manage computational resources in real time. Nodes must dynamically adapt to changing conditions by redistributing tasks between local and remote resources to minimize latency and data loss. However, such adaptations require sophisticated algorithms capable of responding quickly and efficiently to changes in the environment and node states, which places an additional burden on computing power and energy resources.

In addition, node mobility increases the level of uncertainty associated with the network topology. For example, when a node moves, it may be necessary to reconfigure data routes or reassign tasks to other nodes in order to maintain system consistency and reliability. This necessitates the use of resilient routing protocols and load balancing algorithms that can handle high levels of variability. This problem has been addressed by the authors in [9].

To improve the reliability and provide fault tolerance of fog computing systems, a variety of methods and technologies are used to minimize the risks of data loss and failures. One of the most widely used approaches is data replication, which involves creating and maintaining copies of data across multiple nodes in the system. This method provides redundancy so that data remains available even if one or more nodes fail. Replication is critical for applications that require a high degree of reliability, such as health monitoring or traffic management systems. However, implementing replication comes with increased resource requirements, including computing power, storage capacity, and network links, which are needed to maintain data consistency between nodes [10–12].

Another key technique is dynamic load balancing, which aims to prevent overloading of individual nodes in the system. In fog computing, nodes often operate under heterogeneous load conditions, which can lead to performance degradation or complete failure when the load exceeds an acceptable level. Dynamic load balancing involves the use of adaptive algorithms that can analyze the state of the system in real time and efficiently reallocate tasks among nodes. Such algorithms take into account a variety of factors, including the current load of nodes, their computational capabilities, network bandwidth and power consumption. This allows optimizing resource utilization, reducing the probability of failures and ensuring uniform operation of the entire system.

3 An Approach to Improving Reliability and Resilience

Currently, to ensure a high level of reliability, most systems use backup technology, the main disadvantage of which is the need to store several copies of data, which entails additional costs for equipment, electricity and maintenance.

To improve reliability apply various methods of detection, localization of error correction, but these approaches are limited by the accepted arithmetic, not allowing you to use the full power of methods. One of the approaches to increase the efficiency of error correction methods is the use of modular arithmetic (RNS).

A residual class system is understood as a non-positional number system in which any nonnegative integer A, $0 \le A < M$, can be represented as a set of residues from dividing this number by selected bases of the system $\{m_1, m_2, \ldots, m_n\}$, named RNS modules, i.e.

$$A = (\alpha_1, \alpha_2, \ldots, \alpha_n), \text{где } \alpha_i = A - \left[\frac{A}{m_i}\right] \cdot m_i, i = \overline{1, n}, \tag{1}$$

where $M = \prod_{i=1}^{n} m_i$ and is called a range RNS [9].

The possibility of such a representation of a number is determined by the theorem of division with remainder in the ring of integers. If $A \in Z$, $m \in Z$, $m \neq 0$, then there are only $p \in Z$ и $\alpha \in Z$ such that

$$A = pm + \alpha, 0 \le \alpha < |m|, p = \left[\frac{A}{m}\right]. \tag{2}$$

Arithmetic operations in the modular code are performed independently for each of the modules, which indicates the parallelism of this system. This circumstance determines the possibility of independent processing, i.e. bit-by-bit execution of operations. This property eliminates the necessity to "occupy" or "carry" the unit of the higher digit, which leads to the appearance of codes with parallel structure. This allows parallelizing algorithms when performing arithmetic operations [13].

To increase the corrective ability of the residual class system and improve the reliability of data transmission and processing, the use of a redundant residual class system is proposed. It represents an extension of the residual class system, in which the number of modules exceeds the minimum necessary for unambiguous representation of numbers in a given range. This is achieved by introducing additional modules that create redundancy, which, in turn, increases the stability of the system to errors, improves the reliability of calculations and simplifies the control of data correctness [14].

For this purpose, in the set of modules $\{m_1, m_2, \ldots, m_n\}$ of the residual class system, we add r more modules $\{m_1, m_2, \ldots, m_n, m_{n+1}, \ldots, m_{n+r}\}$. The introduction of r redundant modules allows r errors to be detected and $r - 1$ errors to be guaranteed to be corrected [15].

To improve the reliability of data storage in fog systems, a redundant residue number system approach is proposed. It will reduce the redundancy of the stored data, but at the same time it will improve the reliability.

The fog device, after receiving the data from the sensors, preprocesses it. If there is a need to send data to the cloud server, it acts as follows. Before sending, the data is translated by the system modules and sends it to the cloud [16, 17].

4 Modeling of the Proposed Approach

Simulation of a distributed data processing system with elements of fog computing and centralized processing was carried out. The main task is to simulate the transmission, processing and loss of data packets in a network with limited resources of nodes and centralized server.

A fog network is a hierarchical model in which nodes are closer to the data source (e.g., sensors or Internet of Things devices) and a central server is a backup option for processing.

The code also collects statistics on different data processing scenarios to analyze the system behavior, load balancing efficiency and packet loss impact. At the beginning, global system parameters are defined: the number of fog nodes in the system - 100; the number of data packets transmitted per one cycle – 1 000; the number of data transmission cycles - 10; the probability of packet loss during data transmission - 0.01 (set for modeling, in real systems the parameters may differ).

A residual class system with 5 modules (2, 3, 5, 7, 11) was chosen for the fog system which utilizes the uncertainty reduction algorithm [9] and reliability improvement, and from any 3 packets received, the result can be recovered.

FogNode models a node in a fog network. Each node has a limited capacity that determines how many packets it can process at a time. If a node is not overloaded, it processes data locally. If a node is overloaded, the packet is sent to a central server. With some probability, a packet may be lost during processing [18].

Individual statistics are collected for each node and all node processing logic is executed. Checks the probability of packet loss. Determines if the packet can be processed locally. In case of overload, sends the packet to the server for processing.

The central server is a backup option for packet processing. It accepts data from fog nodes that could not process it locally or there is a need to transfer data to the central server. It also takes into account the probability of packet loss.

The main cycle is executed a specified number of times (10 in our case), and in each cycle the following happens. 1,000 data packets are generated. Each packet is randomly assigned to one of the fog nodes. The node processes the packet locally, forwards it to the server, or loses it (given the probability of loss). Statistics are collected for each cycle. At the end of the program execution, the final metrics for each cycle, how many data packets were processed and how many data packets were lost, are displayed.

To visualize the results of the conducted simulation, graphs of lost and processed packets with (Fig. 1) and without (Fig. 2) the use of RNS were plotted.

The figures show the simulation results when no SOC is applied. When this approach is used, the percentage of loss ranges from 0.6% to 1.4%. When using the proposed approach, with the given system parameters, only parts of the data between 0.6% and 1.4% are lost, but the data packets themselves are processed. This happens due to the natural corrective ability of the SOC. The proposed approach allows increasing the reliability of data processing in fog-cloud systems, by applying new arithmetic.

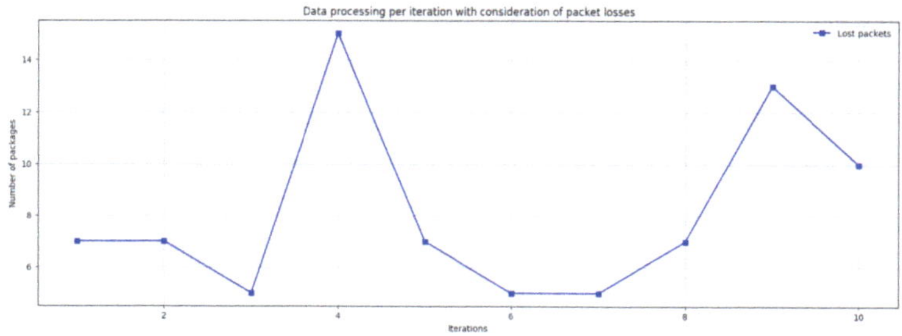

Fig. 1. Data packet loss graph without using RNS

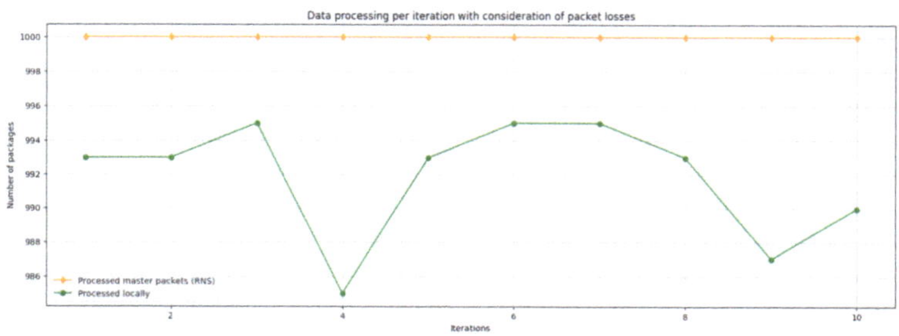

Fig. 2. A figure caption is always placed below the illustration. Short captions are centered, while long ones are justified. The macro button chooses the correct format automatically.

5 Conclusions

Fog computing is a progressive trend of distributed computing that allows data to be processed closer to its source. This significantly reduces latency, improves system performance and reduces the load on cloud servers. This work shows that fog systems play an important role in modern IT infrastructures, especially in applications that require real-time data computation, such as smart cities and industrial control systems.

Despite its advantages, fog computing faces a number of challenges, among which the main one is ensuring reliability and fault tolerance. Design challenges are imposed by the distributed nature of the system, the mobility of nodes, their limited resources, and the dynamic nature of the operating environment. The approaches developed in this paper, including the use of residual class systems and dynamic load balancing techniques, show promise in improving the resilience and efficiency of fog systems.

The simulation results confirm that the implementation of reliability improvement algorithms based on RNS can significantly reduce the probability of data loss. The use of redundant computational structures, adaptive control mechanisms and stable node communication protocols allows to create fault-tolerant solutions that can effectively cope with the risks and uncertainty characteristic of heterogeneous distributed computing environments.

Funding. The research was supported by the Russian Science Foundation Grant No. 22-71-10046, https://rscf.ru/project/22-71-10046/

References

1. Khomonenko, A.D., Kirienko, A.B., Pugachev, S.V.: Evaluating the performance of fog computing in geographic information systems. Eng. J. Don. **30** (2024)
2. Volkov, A.N.: Dynamic fog computing towards green ICT. Proc. Telecommun. Univ. **10**, 24–34., in Russ (2024)
3. Volkov, A.N.: Routing task in dynamic fog computing network. Proc. Telecommun. Univ. **10**, 27–37., in Russ (2024)
4. Li, Y., Zhu, Z., Guan, Y., Kang, Y.: Research on the structural features and influence mechanism of the green ICT transnational cooperation network. Econ. Anal. Policy. **75**, 734–749 (2022). https://doi.org/10.1016/j.eap.2022.07.003
5. Madsen, H., Burtschy, B., Albeanu, G., Popentiu-Vladicescu, F.: Reliability in the utility computing era: towards reliable fog computing. In: 2013 20th International Conference on Systems, Signals and Image Processing (IWSSIP), pp. 43–46. IEEE (2013). https://doi.org/10.1109/IWSSIP.2013.6623445
6. Shiriaev, E., Ermakova, T., Bezuglova, E., Lapina, M.A., Babenko, M.: Reliablity and security for fog computing systems. Information. **15**, 317 (2024). https://doi.org/10.3390/info15060317
7. Pallewatta, S., Kostakos, V., Buyya, R.: Reliability-aware proactive placement of microservices-based IoT applications in fog computing environments. IEEE Trans. Mob. Comput. **23**, 11326–11341 (2024). https://doi.org/10.1109/TMC.2024.3394486
8. Modarresi, A., Sterbenz, J.P.G.: Toward resilient networks with fog computing. In: 2017 9th International Workshop on Resilient Networks Design and Modeling (RNDM), pp. 1–7. IEEE (2017). https://doi.org/10.1109/RNDM.2017.8093032
9. Kucherov, N., Shiriaev, E., Bezuglova, E.: On the way to building reliable and secure cloud-based data processing systems. In: International Workshop on Advanced Information Security Management and Applications, pp. 152–160 (2024)
10. Jamil, B., Ijaz, H., Shojafar, M., Munir, K., Buyya, R.: Resource Allocation and Task Scheduling in.
11. Fog computing and internet of everything environments: a taxonomy, review, and future directions. ACM Comput. Surv., 54 (2022)
12. Wen, Z., et al.: Fog orchestration for internet of things services. IEEE Internet Comput. **21**, 16–24 (2017)
13. Tuli, S., Casale, G., Jennings, N.R.: GOSH: task scheduling using deep surrogate models in fog computing environments. IEEE Trans Parallel Distrib Syst. **33**, 2821–2833 (2022)
14. Santos, J., Wauters, T., Volckaert, B., De Turck, F.: Resource provisioning for IoT application services in smart cities. In: Proceedings of the 2017 13th International Conference on Network and Service Management (CNSM), pp. 1–9. IEEE (2017)
15. Tchernykh, A., Schwiegelshohn, U., Yahyapour, R., Kuzjurin, N.: On-line hierarchical job scheduling on grids with admissible allocation. J. Sched. **13**, 545–552 (2010). https://doi.org/10.1007/s10951-010-0169-x
16. Shadab, M., et al.: Generating highly nonlinear S-boxes using a hybrid approach with particle swarm optimization. In: Nature-Inspired Optimization Algorithms for Cyber-Physical Systems, pp. 1–30. IGI Global Scientific Publishing (2025)
17. Fernando, C.A., Thomas, R., Mary Anita, E.A., Lapina, M.: A Feature Selection Study on the Bot-IoT Dataset Using Ensemble Classification Techniques. In: 2024 IEEE International Conference on Contemporary Computing and Communications (InC4), vol. 1, pp. 1–9. IEEE (2024)
18. Lapina, M., Shiriaev, E., Babenko, M., Istamov, I.: High-speed convolution Core architecture for privacy-preserving neural networks. Program. Comput. Softw. **50**, 417–424 (2024). https://doi.org/10.1134/S0361768824700282

Random Time-Slotted Multiple Access Protocol with Load Balancing for Direct-to-Satellite IoT

Nikita Kustov(✉)

Reshetnev Siberian State University of Science and Technology, 31, Krasnoyarsky Rabochy Av., Krasnoyarsk 660014, Russia
kustovnd@sibsau.ru

Abstract. This paper presents a novel Random Time-Slotted Multiple Access Protocol with Load Balancing for Direct-to-Satellite Internet of Things (IoT) applications. As Direct-to-Satellite IoT continues to grow globally, scalability becomes a critical challenge, particularly concerning Low-Earth Orbit (LEO) satellites covering vast regions with numerous connected devices. Existing protocols like LoRaWAN face limitations in handling massive device densities, leading to frequent collisions and reduced throughput. Our proposed protocol introduces a modified Slotted Aloha approach, enhanced with load balancing and beacon synchronization. By dividing communication sessions into distinct access intervals tailored to different packet sizes, and employing randomized time-slot allocation, our solution aims to maximize throughput efficiency while reducing collision probabilities. Simulations demonstrate that the proposed protocol achieves near-optimal throughput comparable to theoretical Slotted Aloha limits, significantly outperforming traditional Pure Aloha approaches. Future directions include refining the model to incorporate radio signal propagation dynamics and implementing field tests using actual hardware.

Keywords: Direct-to-Satellite IoT · Slotted Aloha · Load Balancing · Throughput Efficiency · Collision Probability · LoRa · LoRaWAN

1 Introduction

Direct-to-Satellite (DtS) Internet of Things (IoT) solutions are rapidly gaining momentum worldwide. They represent one of today's fastest-growing technological trends. According to Statista projections [1], by 2033, global spending on IoT technologies will exceed $934.2 billion. Leading Russian players in this domain include major enterprises such as JSC RESHETNEV, LLC SPUTNIX, JSC ZOND-HOLDING, JSC SR-SPACE, and JSC VISAT-TEL.

On January 7, 2024, the national standard PNST 921-2024 [2] came into force, defining a protocol for low-speed data transfer over satellite communication channels. This standard builds upon and references the earlier GOST R 71168-2023 specification [3], ensuring backwards compatibility with LoRaWAN versions 1.1 Final Release (November 11, 2017 [4]) and version 1.0.2 Final (July 2016 [5]) excluding Class B

M. Lapina et al. (Eds.): AISMA 2025, LNNS 1456, pp. 243–254, 2026.
https://doi.org/10.1007/978-3-032-07275-7_23

devices. LoRaWAN, an open-source protocol maintained by the LoRa Alliance, facilitates communication through LoRa wireless technology, making it the dominant choice for DtS-IoT deployments. Companies like Lacuna Space [6] and Wyld [7] already offer direct connectivity of LoRa devices to their proprietary satellite networks.

Despite these advancements, scaling issues remain prominent. Research highlights challenges related to medium access control (MAC) schemes [8, 9], particularly given that Low Earth Orbit (LEO) satellites cover expansive areas exceeding 1,000 kilometers in diameter. Such zones may contain thousands of active devices simultaneously attempting uplink transmissions. Utilizing the conventional Aloha random-access protocol leads to substantial collision risks among LoRa devices. Therefore, LoRa-based DtS-IoT ecosystems necessitate advanced MAC strategies [10, 11].

In response, this paper introduces a novel MAC protocol incorporating time division, random distribution, and beacon synchronization. Similar to Slotted Aloha, this design incorporates load balancing mechanisms aimed at maximizing throughput efficiency while mitigating collision probabilities.

2 Background and Related Works

2.1 LoRa

LoRa specifies a physical layer utilizing a sub-GHz spread spectrum modulation technique optimized for achieving both extended range and minimal power consumption. Despite being originally intended for terrestrial networks rather than DtS-IoT systems, LoRa remains highly effective within such environments [12].

Within a LoRaWAN architecture, end nodes establish connections solely through gateways acting as intermediary relays. Upon reception of LoRa signals transmitted by endpoints, gateways subsequently forward these messages towards designated network servers. Three distinct operational modes exist for endpoint nodes in LoRaWAN networks: Classes A, B, and C [4].

Class A: Employs an uncontrolled Aloha-like random access strategy, broadcasting messages without awaiting explicit permission. While this ensures flexibility, heavy traffic increases the risk of collisions, resulting in degraded performance. Uplink transmissions from Class A nodes are immediately succeeded by brief downlink listening periods, enabling receipt of potential responses or instructions from gateways.
Class B: Incorporates scheduled downlink windows triggered periodically by synchronized beacons broadcasted by gateways. This structured timing enhances reliability, though Class B nodes consume slightly more power compared to Class A.
Class C: Maintains continuous readiness for incoming downlinks apart from brief intervals during transmission. Unlike Class A, Class C nodes benefit from instantaneous responsiveness since no preceding uplink transmission is required before receiving a downlink message.

2.2 MAC Schemes for LoRa-Based DtS-IoT

Significant efforts have been made to devise advanced Medium Access Control (MAC) protocols capable of enhancing the scalability of terrestrial LoRaWAN networks [13, 14]. Nevertheless, these protocols often fall short when applied to DtS-based IoT settings due to intrinsic differences in satellite link characteristics, including restricted bandwidth availability, extensive geographical ranges, pronounced latency, and fluctuating signal quality.

Research examining LoRa-driven DtS-IoT architectures typically assumes adoption of either the classic Aloha-based random-access protocol [15] or its timed counterpart, Slotted Aloha [16]. Under Slotted Aloha, the available communication channel is partitioned into predefined time slots, compelling end devices (EDs) to commence transmissions strictly at the beginning of each respective slot.

Recent innovations have seen the emergence of specialized MAC protocols tailored for LoRa-centric IoT implementations. Notably, the SALSA protocol employs sophisticated scheduling algorithms aimed at diminishing collision occurrences during upstream transmissions [8].

One suggested enhancement involves leveraging predictive analytics at the satellite gateway level to inform individual nodes regarding anticipated congestion levels within upcoming time frames. Armed with this insight, nodes dynamically compute their transmission probabilities accordingly [9].

Another proposal advocates estimating competition density by analyzing the aggregate number of concurrently active nodes. Using this estimation, the gateway determines the likelihood of impending collisions and disseminates relevant alerts to enable informed decisions on transmission attempts [17].

Additionally, researchers have explored complementary strategies integrating packet duplication and interference-cancellation methodologies to bolster overall performance in satellite communications infrastructure [18, 19].

Ultimately, Carrier Sense Multiple Access (CSMA) family protocols prove inadequate for DtS-IoT contexts owing to prohibitive latency and elevated probabilities of concealed nodes [10, 20].

3 Proposed Method

3.1 Proposed Time-Slotting Structure

We propose a method most similar to Slotted Aloha. To reach the maximum efficiency of this method (36.8%), it is necessary:

- the duration of timeslots is as close as possible to the "time on air";
- packages are evenly distributed across all timeslots.

The time-slotting algorithm is illustrated in Fig. 1. The communication session is proposed to be divided into several access intervals (each interval for a different size of telemetry packets). The access interval begins with a Beacon, followed by a guard interval. The guard interval is set so that the Nodes have time to process the data from the Beacon. The access interval is divided into timeslots. Nodes can transmit packets to any of the timeslots.

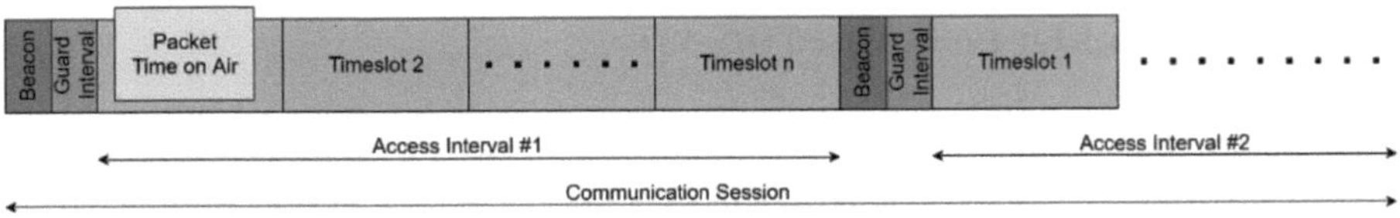

Fig. 1. Proposed Communication Session Sequence.

The duration of the communication session is defined by the following Expression 1. The geometric parameters are shown in Fig. 2.

$$T_{max} \approx T_o \left(\frac{2\gamma_{max}}{360} \right) \tag{1}$$

where:

T_o – period of the LEO Spacecraft's rotation around the Earth (Expression 2):

$$T_o \approx 84,4 \sqrt{\left(\frac{R_E + h}{R_E} \right)^3} \tag{2}$$

γ_{max} – the maximum geocentric angle for the Spacecraft during its orbit (Expression 3):

$$\gamma_{max} = 90^\circ - \alpha_m - \beta_m \tag{3}$$

α_m – the minimum angle of radio visibility of the Node from the Spacecraft location in orbit;
β_m – the minimum angle of the Spacecraft's visibility from the Node location;

Then, knowing β_m, we can calculate α_m as:

$$\alpha_m = \arcsin \frac{\cos \beta_m \times R_E}{R_E \times h} \tag{4}$$

$R_E = 6371$ км – the average radius of the Earth;
h – the altitude of the Spacecraft's orbit above the Earth.

The timeslot must be at least the "Time on Air" of the packet transmission. The "Time on Air" for LoRa is defined by the following Expression:

$$T_{OA} = \begin{cases} \frac{2^{SF}}{BW} \times \left(N_{sym_{pream}} + 6.25 + 8 + ceil\left(\frac{\max\left(8 \times N_{byte_{pld}} + N_{bit_{CRC}} - 4 \times SF + N_{sym_{hdr}}, 0\right)}{4 \times SF} \right) \times (CR + 4) \right), SF < 7 \\ \frac{2^{SF}}{BW} \times \left(N_{sym_{pream}} + 4.25 + 8 + ceil\left(\frac{\max\left(8 \times N_{byte_{pld}} + N_{bit_{CRC}} - 4 \times SF + 8 + N_{sym_{hdr}}, 0\right)}{4 \times SF} \right) \times (CR + 4) \right), SF \geq 7 \end{cases} \tag{5}$$

where:

$SF \in [5; 6; 7; 8; 9; 10; 11; 12]$ – signal spectrum expansion coefficient (times);

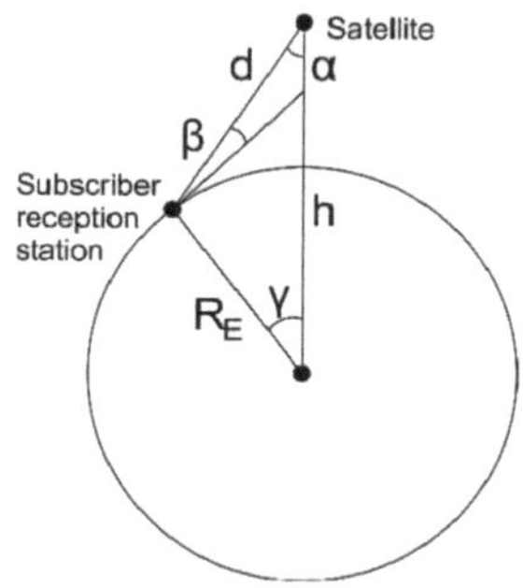

Fig. 2. Geometric orbital parameters.

$BW \in [62.5; 125; 250; 500]$ – channel frequency band (recommended for satellite system) (kHz);
$N_{sym_{pream}} \in [6; 8]$ – the number of symbols in the packet's preamble at the physical level (symbols);
$N_{byte_{pld}}$ – packet payload size (bytes);
$N_{bit_{CRC}} \in [0; 16]$ – number of bits in the CRC code (bits);
$N_{sym_{hdr}} \in [0; 20]$ – symbols in the packet header (symbols);
$CR \in [1; 2; 3; 4]$ – code speed (times).

The duration of the beacon is fixed because the size of the beacon packet does not change. The duration of the guard interval is also fixed; it is determined by the minimum speed limit for processing the beacon by Nodes. The duration of the access intervals depends on the number of timeslots in each interval. In turn, the number of timeslots is a variable parameter.

3.2 The Sequence of the Proposed MAC Method

The general sequence of the proposed method is shown in Fig. 3. A detailed description of the sequence is as follows:

1. During the period between communication sessions, the Coordinator (spacecraft) receives a command from the Management Server (step 1). This command contains parameters for the current communication session: minimum and maximum packet sizes (q_{min} and q_{max}), timeslot duration (d), number of timeslots (c), and guard interval duration (d_{gint}). These parameters are unique for each interval included in the communication session.
2. If one of the Nodes is ready to send a telemetry packet during the communication session, it switches to "Listen" state (step 2). The coordinator sets the parameters from the command for the first interval and also sets the start time of the session to the current time ($t_{start} = t_{cur}$) (step 3).
3. The Coordinator then sends a control Beacon that contains the following fields: $[t_{start}, q_{min}, q_{max}, d_{ts}, c_{ts}, d_{gint}]$, via the downlink to all nodes (step 4).
4. Nodes that receive the Beacon synchronize their internal time with the start time of the interval specified in the beacon ($t'_{cur} = t_{start}$), and set this time as their start time for the interval ($t'_{start} = t_{start}$), (step 5).

5. After this, the Coordinator enters the "Listen" mode (step 6).
6. The access interval lasts until the current time is less than the sum of the interval start time plus the product of the timeslot duration and the number of timeslots plus the duration of the guard interval ($while\ (t_{cur} < (t_{start} + d_{ts} \times c_{ts} + d_{gint}))$). The Node checks whether its packet size is between the minimum and maximum packet sizes ($if\ (q_{min} \leq q_{packet} < q_{max})$). If so, the Node randomly selects a time interval ($n_{ts} = randint(0, c_{ts})$) (step 7).
7. The Node waits for the start of the selected timeslot ($while\ t_{cur} < \left(t'_{start} + d_{ts} \times n_{ts}\right)$) (step 8). When the timeslot begins, the Node sends a packet via the uplink (step 9). The Node then transitions to "Standby" mode (step 10).
8. When the interval ends, the Coordinator transitions to the "Closed" mode (step 11).
9. A new interval begins, and a Beacon is sent with new parameters (steps 12, 13).
10. After the communication session concludes, the Coordinator returns to "Standby" mode (step 14).

3.3 Balancing and Additional Features

Balancing Feature

The load balancing feature can be implemented on the server side. The Server controls the parameters of the communication session: the duration of access intervals (number of timeslots) for different packet sizes. The Server can accumulate statistics on the number of Nodes in a certain area and the average packet sizes that the Nodes send. Based on these statistics, the Server can adjust the parameters of the communication session. The proposed MAC method provides such an opportunity, but the balancing algorithm itself is beyond the scope of this work.

Contention-Free Interval

The method also allows you to use, in addition to contention access intervals, a contention-free interval. This means that it is possible to allocate an interval in which each Node occupies its own timeslot exclusively (the number of this timeslot is preset for the Node). This applies to Nodes with high priority.

4 Analysis

4.1 Model Description

We offer a fairly simple model written in Python 3 that calculates the following metrics: the number of packets sent, the number and percentage of collisions, and the efficiency of channel throughput.

The model takes as input the following parameters: the duration of the communication session, the number of nodes, the size of the packet payload, and various LoRa modulation parameters such as *SF, BW, CR, CRC*, and etc.

Currently, two MAC protocols are implemented in the model: Pure Aloha with sampling up to 1000 Hz, and Proposed Protocol (which calculates optimal timeslots).

The main functions for these protocols are Algorithms 1 and 2, respectively.

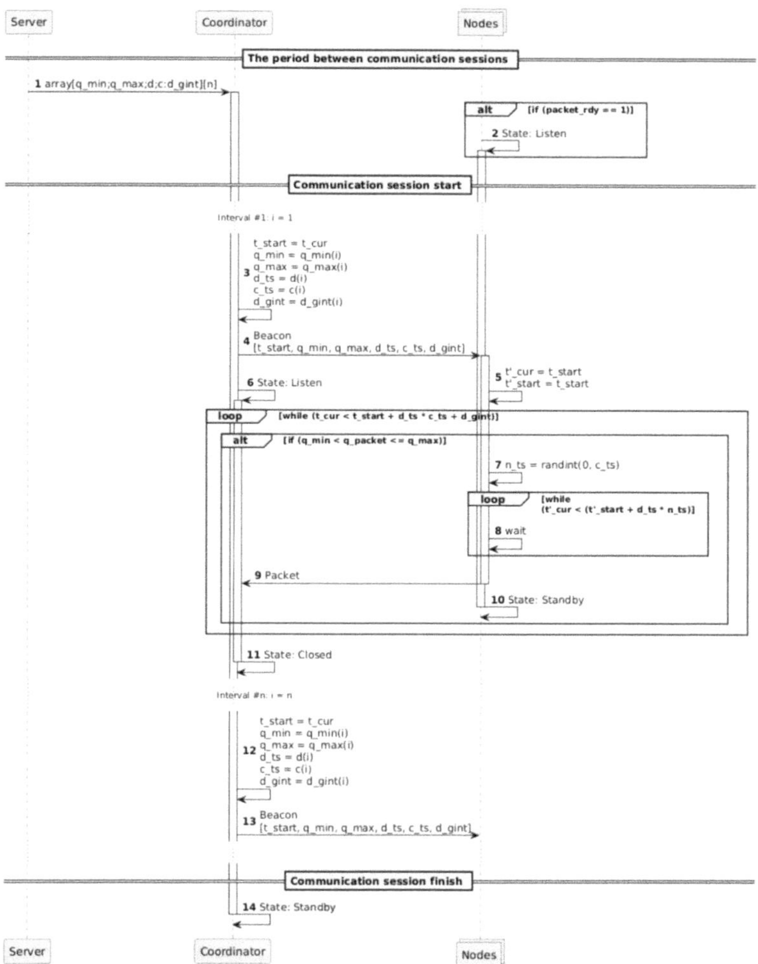

Fig. 3. Sequence diagram of the proposed method.

Algorithm 1. The main modeling function of the Pure Aloha algorithm

```
for simTime in range(0, DURATION):
 for i in range(0, sensors_cnt):
  if (sensorsStates[i] == 1):
   if (random() < 1/DURATION):
    endTx[i] = simTime + ToA
    msgCtr = msgCtr + 1
    for j in range(0,sensors_cnt):
     if (sensorsStates[j] == 2):
      msgCollisionCtr = msgCollisionCtr + 1
      collisionFlag = True
       if (fstCollisionFlag == False):
        fstCollisionFlag = True
        msgCollisionCtr = msgCollisionCtr + 1
       break
    if (collisionFlag == True):
     collisionFlag = False
    else:
     fstCollisionFlag = False
    sensorsStates[i] = 2
  if (sensorsStates[i] == 2):
   if (simTime>endTx[i]):
    sensorsStates[i] = 3
```

Algorithm 2. The main modeling function of the Proposed Protocol algorithm

```
tsCnt = floor(DURATION/maxToA)
for simTime in range(0,DURATION):
 if (simTime%maxToA == 0):
  sensorsStates = ones(sensors_cnt)
  fstCollisionFlag = False
  for i in range(0,sensors_cnt):
   if (random() < 1/tsCnt):
    if (sensorsStates[i] == 1):
     for j in range(0,sensors_cnt):
      if (sensorsStates[j] == 2):
       msgCollisionCtr = msgCollisionCtr+1
       if (fstCollisionFlag == False):
        fstCollisionFlag = True
        msgCollisionCtr = msgCollisionCtr + 1
       break
     sensorsStates[i] = 2
```

4.2 Results

To compare the performance of the MAC protocols considered, we conducted several simulation experiments.

Common input parameters for all experiments: $T_{max} = 360000$ ms, $N_{byte_{pld}} = 115$ bytes, $BW = 125$ kHz, $N_{sym_{pream}} = 8$ symbols, $N_{bit_{CRC}} = 16$ bits, $N_{sym_{hdr}} = 20$ symbols, $CR = 1$.

Different parameters for experiments:

- N – the number of Nodes (ranges from 10 to 5000 in increments of 10);
- $SF = [7–9]$.
- MAC Protocol - Pure Aloha / Proposed Protocol;

For simplicity, it is assumed that each Node can transmit only one packet per communication session on average. Then the intensity depends linearly only on the number of Nodes:

$$G = T_{oA} \times \frac{N}{T_{max}} \quad (6)$$

As a result of the simulation, we are interested in graphs for S – throughput efficiency and P_c – probability of collisions. The simulation results for Pure Aloha are presented in Fig. 4, while the results for the Proposed Protocol algorithm are shown in Fig. 5.

For comparison, we present graphs of ideal functions.

Specifically, for Pure Aloha:

$$S_{PA} = G \times e^{-2G} \quad (7)$$

$$P_{CPA} = 1 - e^{-2G} \quad (8)$$

And for Slotted Aloha:

$$S_{SA} = G \times e^{-G} \quad (9)$$

$$P_{CPA} = 1 - e^{-G} \quad (10)$$

The graphs show that the Proposed Protocol provides throughput efficiency and collision probability values very close to the ideal values of Slotted Aloha. It is also obvious that these results are much better than Pure Aloha. The shift in the graph of the Simulated Pure Aloha can be explained by the fact that the sampling rate during simulation was set at 100 Hz. It is likely that with a higher sampling rate, the results would have been more accurate. Comparative results are shown in Table 1. Based on the presented results, we can see that the Proposed Protocol has almost double the throughput. Additionally, the probability of collisions is lower in each case.

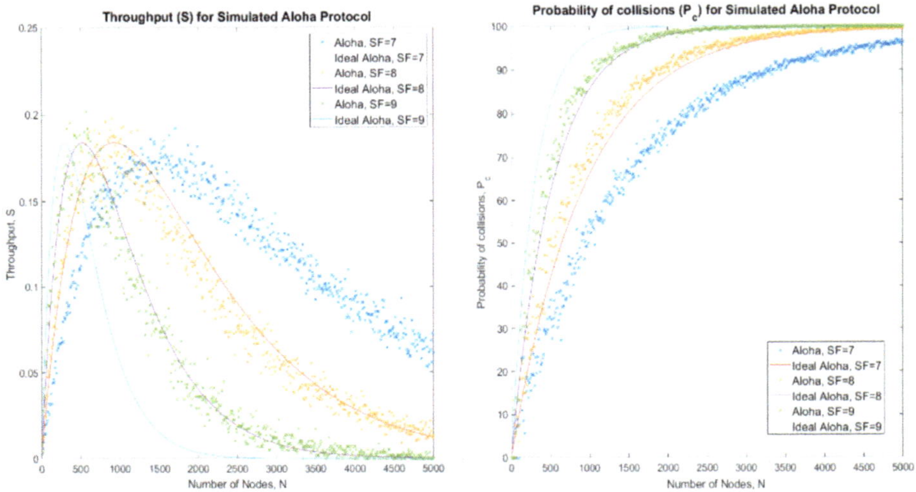

Fig. 4. Simulation results of the Aloha Protocol.

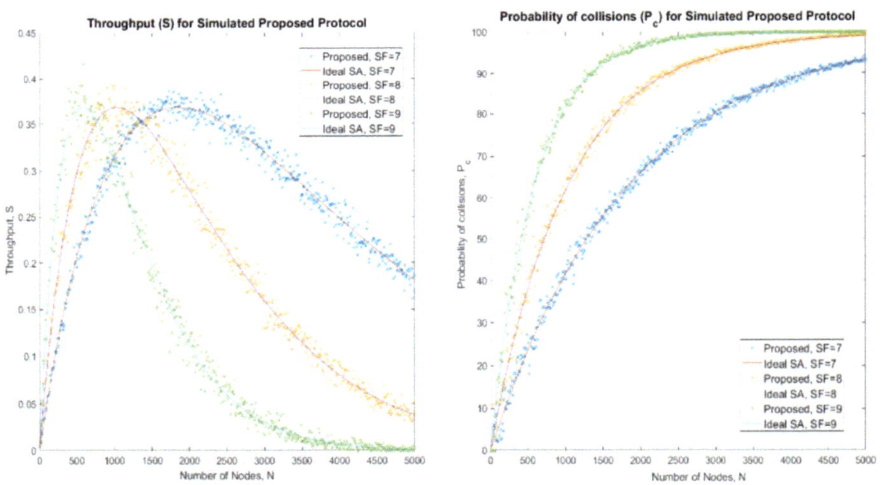

Fig. 5. Simulation results of the Proposed Protocol.

Table 1. Comparative protocol modeling results

G	N	SF	Pure Aloha		Proposed Protocol	
			S	P_c	S	P_c
0.5	920	7	0.151	52.46	0.301	37.48
	520	8	0.178	45.72	0.305	41.01

(continued)

Table 1. (*continued*)

G	N	SF	Pure Aloha		Proposed Protocol	
			S	P_c	S	P_c
	290	9	0.163	48.08	0.332	35.97
1	1850	7	0.169	73.72	0.362	63.2
	1030	8	0.175	71.85	0.348	64.9
	580	9	0.188	68.8	0.416	56.99

5 Conclusions

As a result, the Proposed Protocol has been shown to be more efficient than standard schemes. However, there are still many questions that need to be addressed, such as energy efficiency. Additionally, the current model does not take into account parameters such as radio signal propagation or the mutual influence of nodes. Furthermore, the load balancing algorithm needs further study.

In the future, we plan to conduct the following research activities:

- develop a more detailed model that takes into account radio signal propagation parameters and allows for energy costs;
- create a load balancing algorithm based on statistical methods;
- implement the protocol on real devices, such as controllers and nodes;
- test the proposed scheme under real conditions, using a stratospheric balloon and real spacecraft from Reshetnev University.

References

1. Statista Internet of Things (IoT) total annual revenue worldwide from 2020 to 2033. https://www.statista.com/statistics/1194709/iot-revenue-worldwide/б. Accessed 14 Mar 2025
2. PNST 921-2024. Preliminary National Standard of the Russian Federation. Information Technologies. Internet of Things. Internet of Things Communication Networks. Protocol for Low-Speed Data Exchange Over Satellite Channels. Part 1. Data Exchange on Subscriber Lines of Low-Earth Orbit Systems (approved and introduced by Order of Rosstandart No. 27-PNST dated 14 June 2024)
3. GOST R 71168-2023. National Standard of the Russian Federation. Information Technologies. Internet of Things. LoRaWAN RU Specification (approved and introduced by Order of Rosstandart No. 1617-st dated 22 Dec 2023)
4. LoRaWAN® Specification v1.1, 2017. https://lora-alliance.org/resource-hub/lorawanr-specification-v11. Accessed 16 Mar 2025
5. LoRaWAN® Specification v1.0.2, 2016., https://resources.lora-alliance.org/document/lorawan-specification-v1-0-2. Accessed 16 Mar 2025
6. Lacuna space, Fremont, CA, USA, Available: https://lacuna.space/about/. Accessed 17 Mar 2025

7. Wyld networks, https://wyldnetworks.com/,https://lacuna.space/about/. Accessed 17 Mar 2025
8. Afhamisis, M., Palattella, M.R.: SALSA: A scheduling algorithm for LoRa to LEO satellites. IEEE Access. **10**, 11608–11615 (2022)
9. Álvarez, G., Fraire, J.A., Hassan, K.A., Céspedes, S., Pesch, D.: Uplink transmission policies for LoRa-based direct-to-satellite IoT. IEEE Access. **10**, 72687–72701 (2022)
10. Ferrer, T., Céspedes, S., Becerra, A.: Review and evaluation of MAC protocols for satellite IoT systems using nanosatellites. Sensors. **19**(8), 1947 (2019)
11. Xiao, W., Kaneko, M., Rachkidy, N.E., Guitton, A.: Integrating LoRa collision decoding and MAC protocols for enabling IoT massive connectivity. IEEE Internet Things Mag. **5**(3), 166–173 (2022)
12. Wu, T., Qu, D., Zhang, G.: Research on LoRa adaptability in the LEO satellites Internet of Things. In: Proceeding of the 15th International Wireless Communications & Mobile Computing Conference (IWCMC), Tangier, Morocco, pp. 131–135 (2019)
13. Beltramelli, L., Mahmood, A., Österberg, P., Gidlund, M.: LoRa beyond ALOHA: An investigation of alternative random access protocols. IEEE Trans. Ind. Inf. **17**(5), 3544–3554 (2021)
14. Zorbas, D., Fafoutis, X.: Time-slotted LoRa networks: Design considerations, implementations, and perspectives. IEEE Internet Things Mag. **4**(1), 84–89 (2021)
15. Ullah, M.A., Mikhaylov, K., Alves, H.: Enabling mMTC in remote areas: LoRaWAN and LEO satellite integration for offshore wind farm monitoring. IEEE Trans. Ind. Inf. **18**(6), 3744–3753 (2022)
16. Ullah, M.A., Yastrebova, A., Mikhaylov, K., Höyhtyä, M., Alves, H.: Situational awareness for autonomous ships in the arctic: MMTC directto-satellite connectivity. IEEE Commun. Mag. **60**(6), 32–38 (2022)
17. Vogelgesang, K., Fraire, J.A., Hermanns, H.: Uplink transmission probability functions for LoRa-based direct-to-satellite IoT: a case study. In: Proceeding of the IEEE Global Communications Conference (GLOBECOM), Madrid, Spain, pp. 1–6 (2021)
18. Shao, X., Sun, Z., Yang, M., Gu, S., Guo, Q.: NOMA-based irregular repetition slot-ted ALOHA for satellite networks. IEEE Commun. Lett. **23**(4), 624–627 (2019)
19. Zhao, B., Ren, G., Dong, X., Zhang, H.: Optimal irregular repetition slotted ALOHA under total transmit power constraint in IoToriented satellite networks. IEEE Internet Things J. **7**(10), 10465–10474 (2020)
20. Herrería-Alonso, S., Suárez-González, A., Rodríguez-Pérez, M., López-García, C.: Enhancing LoRaWAN scalability with longest first slotted CSMA. Comput. Netw. **216**, 109252 (2022)

Machine Learning Methods for Cyber Attacks Detection and Classification

Vitalii Lapin[1], Walaa H. Elashmawi[2,3], Veronika Abakumova[1], and Marina Tokmakova[1](✉)

[1] North Caucasus Federal University, Stavropol, Russia
marinatokmakova175@mail.ru
[2] Faculty of Computers and Informatics Suez Canal University, Ismailia, Egypt
[3] Faculty of Computer Science, Misr International university, Cairo, Egypt

Abstract. This paper presents a study of cyberattack classification using machine learning methods on the KNIME platform. The topic is relevant due to the growth of vulnerabilities in digital systems. Various machine learning models, including Decision Tree Learner, Random Forest Learner, Naive Bayes Learner, Tree Ensemble Learner, and Gradient Boosted Trees Learner, are examined to identify the most effective approach for attack classification. The paper includes an overview of key attack characteristics and offers practical recommendations for improving protection.-->

Keywords: machine learning · KNIME · dataset · data processing · data analysis

1 Introduction

The modern world is rapidly digitizing, leading to an increase in the number and complexity of cyber attacks that threaten data security, financial systems, and critical infrastructure. Attackers are constantly improving their attack methods by exploiting software vulnerabilities, social engineering, and automated tools. In such conditions, traditional approaches to information protection are becoming insufficiently effective, which requires the introduction of intelligent threat analysis and classification systems based on machine learning methods.

This study focuses on the application of the KNIME platform (Konstanz Information Miner) for the analysis and classification of cyber attacks using various machine learning algorithms. The paper considers such models as decision trees, random forest, naive Bayes classifier, ensembles of trees and gradient boosting, which makes it possible to evaluate their effectiveness in detecting and categorizing cyber threats.

The source data is a dataset containing information on 174 cyber attacks from 2010 to 2020, including attack vectors, their consequences, financial damage, and target sectors. Special attention is paid to data preprocessing, combating overfitting, and evaluating model accuracy through cross-validation.

The results of the study demonstrate that ensemble methods such as Tree Ensemble Learner achieve the highest accuracy (up to 99.6%), which confirms their promise for

M. Lapina et al. (Eds.): AISMA 2025, LNNS 1456, pp. 255–267, 2026.
https://doi.org/10.1007/978-3-032-07275-7_24

cybersecurity tasks. The work also highlights the importance of integrating machine learning into data protection systems to detect and prevent cyber attacks in a timely manner.

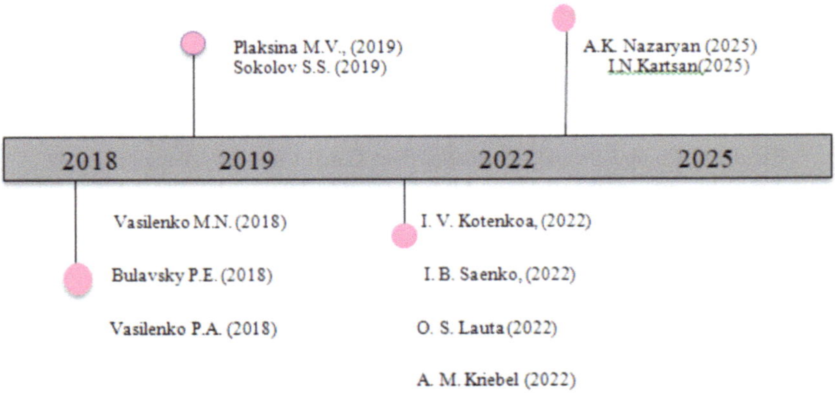

Fig. 1. Timeline of existing works.

Figure 1 shows a timeline with publications on this topic for the period from 2018 to 2025.

2 Dataset and its Properties

Selected dataset: A dataset of 174 cyber attacks that took place between 2010 and 2020 manually classified based on 15 features. The dataset is useful for analyzing cybersecurity trends, studying vulnerabilities, and developing defense strategies.

Size and structure:

- Number of records: 165.
- Number of columns: 19.
- Data format: CSV with text, numeric and date values.
- Key for analysis:
- "Attack vector" — attack method (e.g. "Social engineering", "Unsecured hardware").
- "Operational impact" — attack consequences (e.g. "User compromise", "Denial of service"). "Information impact" — attack consequences. "People affected" — data leak scale.
- "Total money lost" — financial damage.
- "Sector" — organization sector (e.g. "Government", "Civilian").
- Additional important columns:
- "Date of attack" — time analysis of attacks. "Countries affected" — attack geography.
- "Defence" — defense methods (e.g. "Mitigation", "Remediation").
- "Inside job" — presence of an internal attacker (0/1). Auxiliary columns:
- "Organisation name" — name of the organisation.
- "Breached before" — whether the organisation was previously hacked (0/1).

- "Resources" — links to sources.
- "Possible defence strategies" — recommended defence measures.

The dataset contains a variety of information about cyber attacks, including methods, consequences and scale. The most important columns for analysis are: "Attack vector", "Operational impact", "People affected", "Total money lost" and "Sector". The data can be used to classify attacks, predict damage or analyze vulnerabilities in different sectors. Also, this dataset has a drawback in the form of some fields containing missing values (e.g. financial losses, number of victims) (Fig. 2).

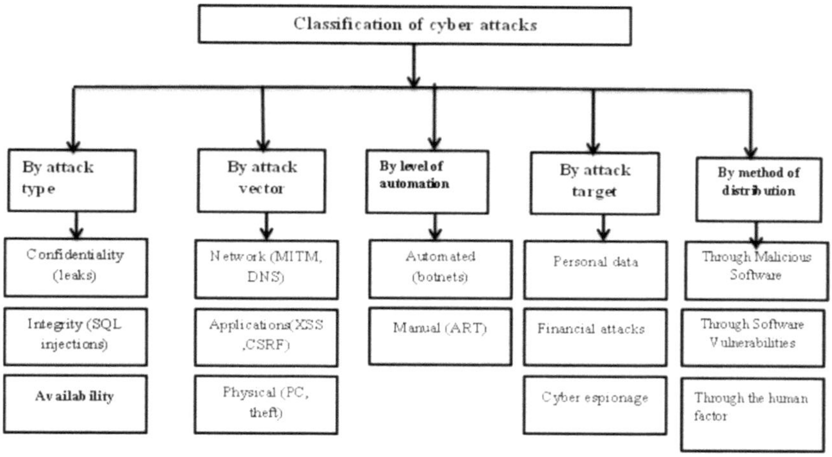

Fig. 2. Cyberattack classification scheme.

The diagram (Fig. 2) presents a classification of cyber attacks based on five key criteria: type, attack vector, level of automation, target, and method of distribution. By type, attacks are categorized as breaches of confidentiality (e.g., data leaks), integrity (e.g., SQL injections), and availability. Attack vectors include network-based (MITM, DNS), application-level (XSS, CSRF), and physical (e.g., hardware theft) methods. Based on the level of automation, attacks are classified as automated (botnets) or manual (APT). Depending on the target, attacks may focus on personal data, financial assets, or cyber espionage. In terms of distribution methods, attacks may occur through malicious software, software vulnerabilities, or via human factors.

3 Modeling

Modeling in KNIME allow to build predictive models that provide valuable insights into future trends. The platform is suitable for developing complex predictive models (Fig. 3).

There are 10 nodes used in this scheme.

CSV Reader: This is used to read CSV files Column Filter: This allows you to filter the columns of the input table.

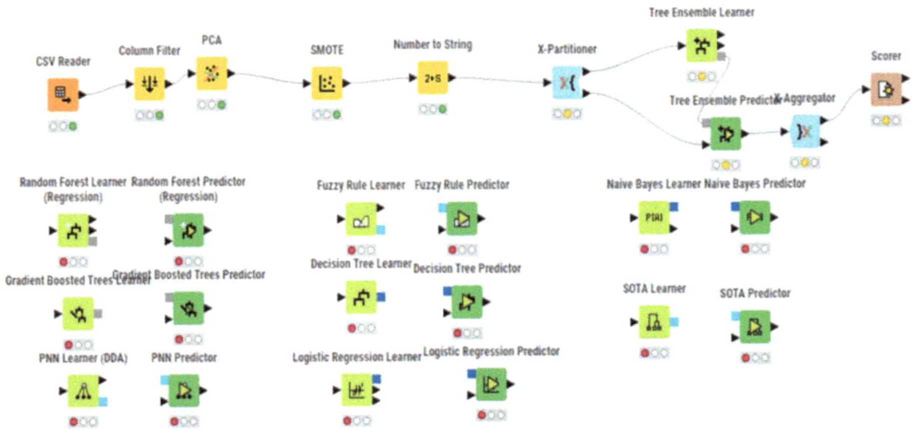

Fig. 3. Diagram of machine learning models and nodes used.

PCA: This performs Principal Component Analysis (PCA) on the given data.
SMOTE: This is a node in KNIME that iterates over the input data (i.e. adds artificial strings) to enrich the training data.
Number to String: This converts numeric data to strings.
X-Partitioner: This is used to define the cross-validation set.
ML_Model Learner: Here ML_Model stands for any machine learning model such as decision tree, random forest, logistic regression, etc. Learner means that the model is trained on the training data.
ML_model Predictor: As the name suggests, the model now performs prediction based on test data.
X-Aggregator: Collects the output from the predictor node, compares the predicted and actual classes, outputs predictions for all rows and iteration statistics.
Scorer: Calculate metrics such as precision, recall, and accuracy.

3.1 Decision Tree Learner

This node generates a decision tree for classification in main memory. The target attribute must be nominal. Other attributes used for decision making can be either nominal or numeric. Numeric splits are always binary (with two outcomes) and divide the domain into two parts at a given split point. Nominal splits can be either binary (with two outcomes) or have as many outcomes as there are nominal values. In case of binary split, the nominal values are divided into two subsets. The algorithm provides two quality measures for calculating the split: Gini index and gain. In addition, there is a post-processing method to reduce the tree size and improve the prediction accuracy. The post-processing method is based on the principle of minimum description length. The algorithm can be run in multiple threads and thus use multiple processors or cores.

Research using the PCA block (Table 1).

Testing the model with standard settings.

The model outputs 86.737% as a result (Table 2 and Fig. 4).

Table 1. Model analysis using PCA block.

Dimension(s) to reduce to	Accuracy
1	82,452%
2	82,415%
3	83,03%

Table 2. Model analysis using the Min number records per node settings.

Minimum Number of Records per Node Decision Tree Learner	Accuracy
1	87,841%
2	87,021%
3	82,626%
4	82,828%
5	81,818%
6	81,414%
7	81,414%
8	82,02%
9	82,424%
10	80,404%

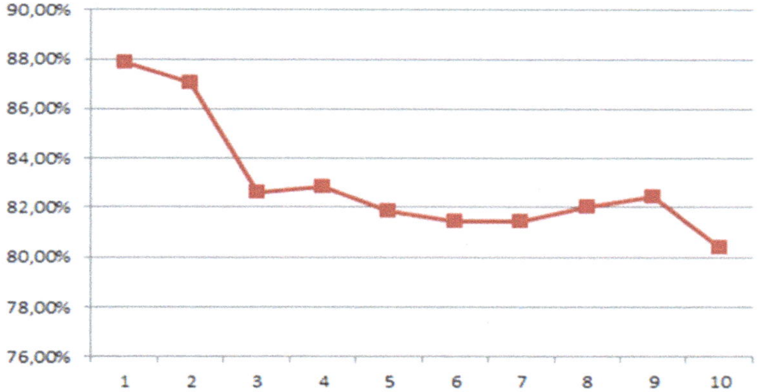

Fig. 4. Changing the accuracy of the model.

3.2 Random Forest Learner

Random Forest Learner consists of a selected number of decision trees. Each of the decision trees is built from a separate set of rows (records), and for each split within the tree, a randomly selected set of columns (describing attributes) is used. The sets of rows for each decision tree are created using bootstrapping and have the same size as the original input table. The set of attributes for a particular split in the decision tree is determined by randomly choosing sqrt(m) attributes from the available attributes, where m is the total number of training columns. Attributes can also be provided as a bit (fingerprint), byte, or dual vector. The output model describes the random forest and is applied in the corresponding Prediction node. This node provides the part of the features of the training ensemble of trees corresponding to the random forest.

Exploration using the PCA block (Table 3).

Table 3. Model analysis using the PCA block.

Dimension(s) to reduce to	Accuracy
1	87,273%
2	85,859%
3	83,636%

Table 4. Model analysis using the Limit number of level(tree depth) settings.

The Depth of Random Forest Learner	Accuracy
1	77,576%
2	77,576%
3	80%
4	81,212%
5	83,434%
6	85,253%
7	85,657%
8	85,253%
9	85,859%
10	84,646%

Testing a model with standard settings.

This model outputs 86.667% as a result (Table 4 and Fig. 5).

3.3 Naive Bayes Learner

Creates a Bayesian model given training data. It computes the number of rows for each attribute value in each class for nominal attributes and a Gaussian distribution for numeric attributes. The created model can be used in a Naive Bayes classifier to predict

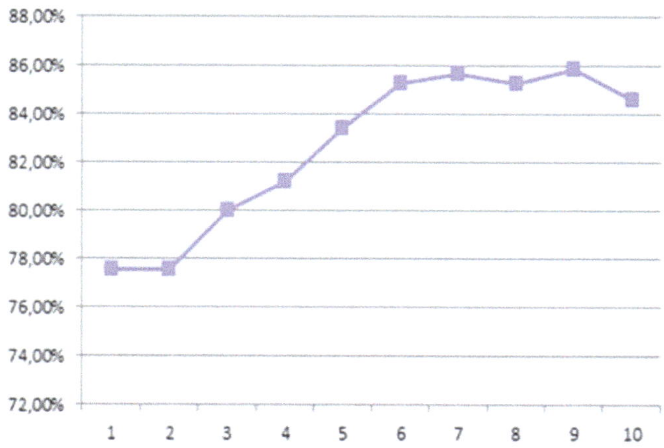

Fig. 5. Changes in the accuracy of the model.

class membership of unclassified data. The node prints a warning if any columns are ignored due to unsupported data types. For example, bit vector columns are ignored if the PMML compatibility flag is enabled because they are not supported by the PMML standard. Exploration using the PCA block (Table 5).

Table 5. Model analysis using the PCA block

Dimension(s) to reduce to	Accuracy
1	80,808%

Table 6. Model analysis using Default probability settings

Standard probability	Accuracy
0,0001	81,818%
0,0005	80,404%
0,001	81,212%
0,0025	81,212%
0,005	77,778%
0,01	77,576%
0,025	76.162%
0,05	72,727%
0,1	68,889%

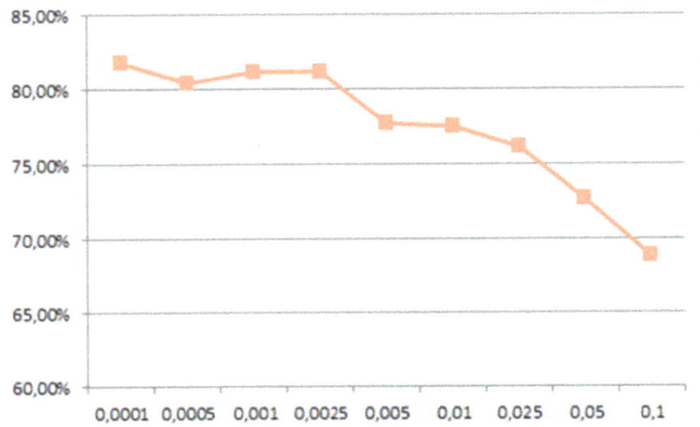

Fig. 6. Changing the accuracy of the model.

Testing a model with standard settings. This model outputs 81,818% as a result (Table 6. and Fig. 6).

3.4 Tree Ensemble Learner

Learns an ensemble of decision trees (e.g., variants of random forest). Typically, each tree is built with a different set of rows (records) and/or columns (attributes). See variants for data sampling and attribute sampling for more information. Attributes can also be provided as a bit (fingerprint), byte, or dual vector. The output model describes the ensemble of decision tree models and is applied in the appropriate prediction node using the selected aggregation mode to aggregate the votes of individual decision trees.

Exploration using PCA block (Table 7).

Table 7. Model analysis using the PCA block.

Dimension(s) to reduce to	Accuracy
1	99,596%
2	98,99%
3	99,596%

Testing a model with standard settings.

This model outputs 98.384% as a result (Table 8 and Fig. 7).

3.5 Gradient Boosted Trees Learner

Learns gradient boosted trees for classification. The algorithm uses very simple regression trees and a special form of boosting to build an ensemble of trees. The basic algorithm for this ensemble method is a simple regression tree, as it is used in the Ensemble

Table 8. Model analysis using the Limit number of level (tree depth) settings.

The depth of the Tree Ensemble learner	Accuracy
1	77,756%
2	81,414%
3	86,061%
4	91,111%
5	95,152%
6	97,374%
7	97,172%
8	97,98%
9	99,596%
10	98,182%

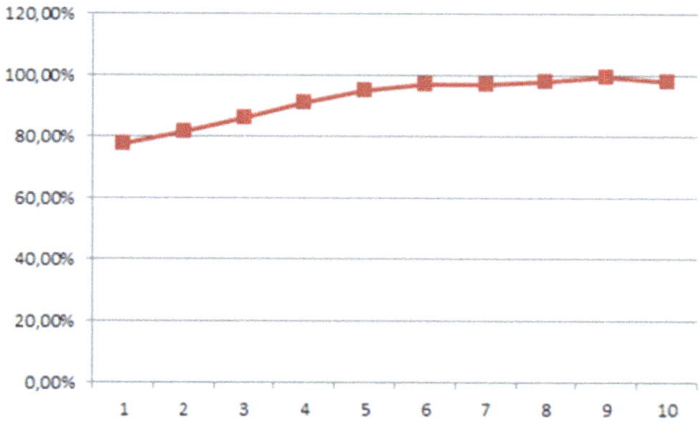

Fig. 7. Changing the accuracy of the model.

of Trees, Random Forest, and Simple Regression Tree nodes. By default, the tree is built using binary splits for numeric and nominal features (the latter can be changed to multivariate splits). The built-in missing value handling tries to find the best direction for missing values by trying every possible direction and choosing the one that gives the best result (i.e. the most gain). In a regression tree, the predicted value for a leaf node is the average target value of the records in that leaf. Therefore, predictions will be most accurate (relative to the training data) if the spread of target values in that leaf is minimal. This can be achieved by using partitions that minimize the sum of squared errors in the corresponding child nodes.

Exploration using the PCA block (Table 9).

Testing a model with standard settings.

This model outputs 90.707% as a result (Table 10 and Fig. 8).

Table 9. Model analysis using the PCA block.

Dimension(s) to reduce to	Accuracy
1	88,081%
2	90,707%
3	90,303%

Table 10. Model analysis using the Limit number of level (tree depth) settings.

The depth Gradient Boosted Trees Learner	Accuracy
1	80%
2	84,646%
3	87,475%
4	88,687%
5	89,293%
6	91,313%
7	90,707%
8	91,313%
9	90,505%
10	90,505%

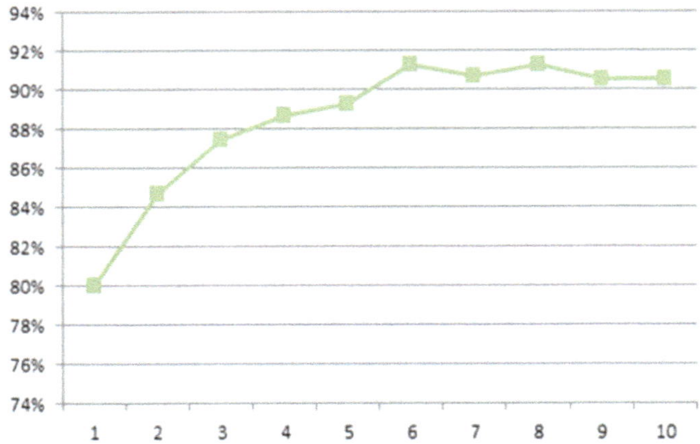

Fig. 8. Changing the accuracy of the model.

4 Combating Overfitting

Cross-validation is a method for evaluating machine learning models that helps test their generalization ability. Instead of simply splitting the data into training and testing sets, the data is repeatedly split into parts: the model is trained on some subsets and tested on others. This allows for a more stable and objective assessment of quality.

Fig. 9. Cross validation in KNIME

Advantages of cross-validation

1. Efficient use of data - especially important for small samples, since all data is used in training and testing
2. Reduction of random errors - eliminates dependence on a single split.
3. Detection of overfitting and underfitting - helps to choose a model that is not too complex (overfitting) and not too simple (underfitting).
4. Robustness to noise and outliers - multiple testing on different subsets reduces the impact of anomalies in the data.

Thus, cross-validation provides a more reliable assessment of the model before applying it to new data. X-Partitioner and X-Aggregator nodes are used as KNIME cross-validation (Fig. 9).

5 Conclusion

In this study, various machine learning models were applied on the KNIME platform to classify and analyze cyber attacks using a dataset of incidents from 2010 to 2020. Through experimentation with models such as Decision Tree, Random Forest, Naive Bayes, Tree Ensemble, and Gradient Boosted Trees, it was shown that ensemble methods—particularly the Tree Ensemble Learner—demonstrate the highest classification accuracy, reaching up to 99.6%. The application of dimensionality reduction techniques and cross-validation further improved model reliability and helped address issues such as overfitting.

The results highlight the promising role of machine learning in cybersecurity, particularly in proactive threat detection and attack classification. The developed models can serve as a foundation for more intelligent, adaptive defense systems. Moreover, this work confirms the importance of quality datasets and thoughtful preprocessing in achieving meaningful and accurate predictions.

Future research may include extending the dataset, integrating real-time data streams, and applying deep learning approaches to enhance classification robustness and enable early threat forecasting in dynamic environments.

References

1. Machine learning: opportunities and application scenarios | Yandex Cloud. https://yandex.cloud/ru/blog/posts/2022/10/machine-learning?utm_referrer=https%3A%2F%2Fyandex.ru%2F
2. Cyberattack – Wikipedia https://en.wikipedia.org/wiki/Cyberattack
3. Kotenko, I.V., Saenko, I.B., Lauta, O.S., Kribel, A.M.: Methods for detecting anomalies and cyber attacks based on the integration of fractal analysis and machine learning methods. Comput. Sci. Autom. **6**, 1328–1358 (2022) https://www.mathnet.ru/php/archive.phtml?wshow=paper&jrnid=trspy&paperid=1227&option_lang=rus
4. Modern cyber attacks: classification and methods of protection | Informatics. Economy. Management – Informatics. Economics. Management https://oajiem.com/index.php/24/article/view/121
5. Analysis and classification of cyber attacks on electronic technical documentation of railway automation and telemechanics – the topic of a scientific article on computer and information sciences. https://cyberleninka.ru/article/n/analiz-i-klassifikatsiya-kiberatak-na-elektronnuyu-tehnicheskuyu-dokumentatsiyu-zheleznodorozhnoy-avtomatiki-i-telemehaniki
6. Testing and evaluating your model in KNIME | Geeks for Geeks https://www.geeksforgeeks.org/testing-and-evaluating-your-model-in-knime/
7. Decision Tree Learner – KNIME Community Hub https://hub.knime.com/knime/extensions/org.knime.features.base/latest/org.knime.base.node.mine.decisiontree2.learner2.DecisionTreeLearnerNodeFactory3'
8. Random Forest Student – KNIME Community Hub https://hub.knime.com/knime/extensions/org.knime.features.ensembles/latest/org.knime.base.node.mine.treeensemble2.node.randomforest.learner.classification.RandomForestClassificationLearnerNodeFactory2
9. Naive Bayes Learner – KNIME Community Hub https://hub.knime.com/knime/extensions/org.knime.features.base/latest/org.knime.base.node.mine.bayes.naivebayes.learner3.NaiveBayesLearnerNodeFactory4
10. Tree Ensemble Learner – KNIME Community Hub https://hub.knime.com/knime/extensions/org.knime.features.ensembles/latest/org.knime.base.node.mine.treeensemble2.node.learner.classification.TreeEnsembleClassificationLearnerNodeFactory2
11. Gradient Boosted Trees Learner – NodePit https://nodepit.com/node/org.knime.base.node.mine.treeensemble2.node.gradientboosting.learner.classification.GradientBoostingClassificationLearnerNodeFactory
12. Radford, B.J., Richardson, B.D., Davis S.E.: Sequence aggregation rules for anomaly detection in computer network traffic. arXiv preprint arXiv:1805.03735, (2018)
13. Lambert, I., Glenn, M.: Security Analytics: Using Deep Learning to Detect Cyber Attacks (2017). https://core.ac.uk/download/pdf/71985426.pdf
14. Stevanovic, M., Pedersen, J.M.: Detecting bots using multi-level traffic analysis. IJCSA. **1**(1), 182–209 (2016) https://c-mric.com/wp-content/uploads/2017/10/article9.pdf
15. Sarker, I.H.: A machine learning based robust prediction model for real-life mobile phone data. Internet Things. **5**, 180–193 (2019) https://arxiv.org/pdf/1902.07588

16. Sarker, I.H., Abushark, Y.B., Alsolami, F., Khan, A.I.: Intrudtree: a machine learning-based cyber security intrusion detection model. Symmetry. **12**, 754 (2020) https://www.researchgate.net/publication/340992582_IntruDTree_A_Machine_Learning-Based_Cyber_Security_Intrusion_Detection_Model
17. Sarker, I.H., Colman, A., Han, J.: Recencyminer: mining recency-based personalized behavior from contextual smartphone data. J. Big Data. **6**(1), 49 (2019)
18. Moustafa, N., Slay, J.: UNSW-NB15: a comprehensive dataset for network intrusion detection systems. In: Military Communications and Information Systems Conference (MilCIS). IEEE, Canberra, Australia (2015). https://doi.org/10.1109/MilCIS.2015.7348942
19. Koroniotis, N., Moustafa, N., Sitnikova, E., Slay, J.: Towards developing network forensic mechanism for botnet activities in the IoT based on machine learning techniques. In: International Conference on Mobile Networks and Management. Springer, Cham (2017)
20. Hasan, K.M.B., Sajid, M., Lapina, M.A., Shahid, M., Kotecha, K.: Blockchain technology meets 6 G wireless networks: a systematic survey. Alex. Eng. J. **92**, 199–220 (2024)
21. Basan, E., Lapina, M., Lesnikov, A., Basyuk, A., Mogilny, A.: Trust monitoring in a cyber-physical system for security analysis based on distributed computing. In: International Conference on Actual Problems of Applied Mathematics and Computer Science, pp. 430–440. Springer Nature Switzerland, Cham (2022)
22. Vinodha, D., et al.: Framework based on IoT, AI, and Blockchain for smart access to government agricultural schemes. In: The Future of Agriculture: IoT, AI and Blockchain Technology for Sustainable Farming, pp. 142–163. Bentham Science Publishers (2024)

Detecting Information Security Threats Using Machine Learning: A Survey of Algorithms and Application Problems

Vitalii Lapin[1], Marina Tokmakova[1(✉)], Alexey Dmitrienko[1], and Mohammad Sajid[2]

[1] North Caucasus Federal University, Stavropol, Russia
marinatokmakova175@mail.ru
[2] Aligarh Muslim University, Aligarh 202002, India

Abstract. This article examines the applicability of machine learning in the context of information security. The main focus is on analyzing the effectiveness of various algorithms and models in strengthening the protection of information systems. This review covers the breadth of machine learning applications to information security problems, presents a wide variety of machine learning methods, and discusses examples of how each of them applies to security-related problems. The article provides examples of how machine learning is used to counter cyber threats, ensure the security of mobile and cloud environments, and automate the threat analysis process. The aim of the work is to systematize relevant research, methods and technologies of machine learning that contribute to improving the level of security and data security in an ever-changing threat landscape.

Keywords: Machine learning Models · Approaches to Information Security · Classification · Clustering · Data Protection · Information Security

1 Introduction

In recent years, machine learning has made a significant contribution to the development of computer science, and its potential and capabilities in the field of information security are particularly noticeable.

Machine learning, being a subsection of artificial intelligence, is based on the use of statistical methods and mathematical models that simulate the human learning process. Its goal is to create and improve algorithms capable of classifying, predicting, and making decisions in a variety of scenarios [1]. Machine learning is based on the analysis of extensive data sets, through which models identify patterns, compare objects with categories, and learn to make predictions. The ultimate goal is to achieve higher processing speed and forecast accuracy when working with new, previously unknown data.

Machine learning models use mathematical algorithms and datasets to identify patterns and statistically analyze the behavior of computer and information systems; they are able to make predictions and make decisions even in situations that were not explicitly

M. Lapina et al. (Eds.): AISMA 2025, LNNS 1456, pp. 268–276, 2026.
https://doi.org/10.1007/978-3-032-07275-7_25

programmed or initially [2]. Thanks to these capabilities, machine learning has become one of the key roles in the field of information technology in recent years. Security. Table 1 shows companies that have implemented machine learning to ensure/organize information security.

Table 1. Companies that have implemented machine learning to ensure/organize information security.

Company	Location	Application of ML in cybersecurity
Crowdstrike	Sunnyvale, California	Uses ML, AI and behavioral analytics to identify threats and scan networks for malware
Crisp Thinking	Chicago, Illinois	Uses AI to analyze social networks and forums, identify "risk signals" and prevent threats to brand reputation
Microsoft	Redmond, Washington	Uses ML in Windows Defender for Endpoint to protect, detect attacks, automatically investigate and respond to threats
Blackberry	Waterloo, Ontario, Canada	Develops cybersecurity solutions with the help of the following tools: Uses AI and ML to prevent threats and automate incident response
Splunk	San Francisco, California	Applies ML in security analytics, automating attack investigations and vulnerability management
Chronicle	Mountain View, California	Develops a Backstory platform that uses ML to analyze security telemetry and detect malicious activity

Given the growing number and complexity of cyber-attacks, standard protection measures are becoming insufficient. However, machine learning is fundamentally changing approaches to information security, making it possible to quickly identify emerging threats and adapt to their elimination. Figure 1 shows a visual representation of the mechanisms of machine learning models.

The accuracy of predictions of machine learning models in the field of information security depends on a number of factors, but the quality, completeness and volume of input data are crucial. This means that the quality of the information used directly affects the accuracy of the predictions. The choice of a specific algorithm, in turn, is determined by the nature of the problem being solved, which also affects the accuracy of forecasting. The compliance of the selected algorithm with the task is assessed using a predictive model (Table 2).

The practical experience of companies confirms the importance of machine learning for supporting security and information protection systems. The introduction of these technologies allows not only automating threat analysis processes, but also significantly increasing the speed of incident response.

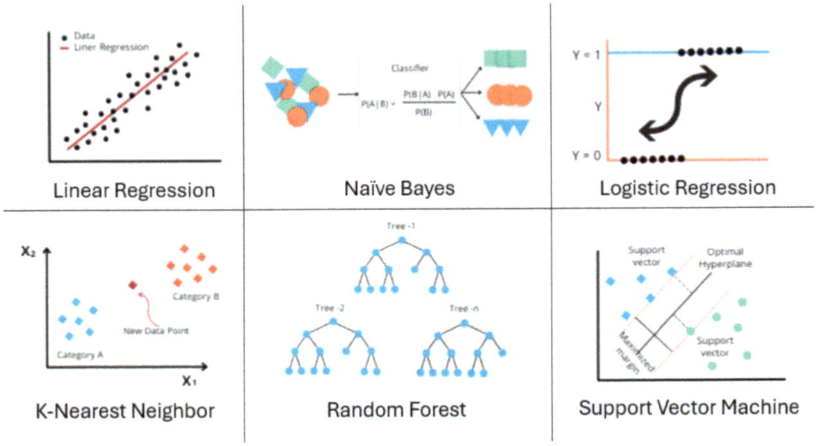

Fig. 1. Visual representation of the mechanisms of machine learning algorithms.

Table 2. Correlation of specific algorithms and areas of applicability.

Algorithm	Application in information security	Article
Decision Tree	Attack detection and classification	[1, 3]
Dimensionality Reduction	Eliminating noise and irrelevant data	[2]
K-means Clustering	Malware	[3]
K-Nearest Neighbors	Detection Face recognition for authentication	[3]
Linear Regression	Network threat prediction	[4–6]
Logistic Regression	Fraud	[7, 8]
Naïve Bayes	Detection Network Intrusion Detection Phishing	[4, 9]
Random Forest	Attack classification	[4]
Support Vector Machine	Malicious IP and Port address detection and prediction	[10]

2 Literature Review

Support Vector Machines (SVM). This method is a guided machine learning algorithm that is mainly used for classification tasks, but is also effective for regression and outlier detection [14]. The basic principle is to find the ideal hyperplane to split the dataset. For data that can be separated by a straight line or plane, the model defines a hyperplane that creates the largest possible gap between classes. The data points from the training data located closest to this hyperplane and crucial for determining its position and orientation are called reference vectors.

This model can be an effective tool for detecting malicious software [11], and it also handles the tasks of detecting mail with unwanted images [12, 13], intrusion detection [14, 15], and spam analysis [16]. Network attack detection [17] and biometric identification [18] are also examples of SVM applications in the field of information security.

On the other hand, it is worth noting that due to the geometric nature of the learning process, a targeted attack based on contamination of training data is relatively simple [19], which can be considered a disadvantage compared to other machine learning methods.

Clustering. Clustering is an analytical method aimed at grouping objects based on the principle of their similarity [21]. This process is carried out without first defining rules or categories; instead, the algorithm independently identifies patterns, determining which objects can be grouped into one group and which should be classified as different.

The main task of Clustering is to divide the dataset into subsets, where all data has similar characteristics or indicators. This approach is applicable in the case of data exploration, meaning Clustering should be used to obtain information and insights about the data.

In many studies, Clustering is used to detect, analyze, and classify malware [20]. Clustering is also used to analyze spam [21], detect network attacks [22], detect intrusions [23], and address various privacy issues [24].

k-Nearest Neighbor. One of the simplest techniques used in machine learning is the k–Nearest Neighbor technique, where objects are classified based on their nearest neighbors in the training set.

The k-Nearest Neighbor method is a machine learning algorithm that is used for classification and regression tasks. The essence of the method is that a new object is determined by the classes of its nearest neighbors in the feature space. When classifying a new object, it is assigned the class that is most common among its k-Nearest Neighbor, whose classes have already been defined. In the case of regression, the new object is assigned an average value calculated based on the values of its k nearest neighbors, which are already known.

This model is used to protect cloud requests [27], biometric authentication based on keystroke dynamics [26], and intrusion detection [25].

Random Forest. This algorithm uses a set of decision trees to make more accurate predictions. When building each tree, both training data and features are randomly selected. This reduces the correlation between trees and contributes to their diversity, which ultimately increases the overall accuracy of the model.

Random Forests are used to solve a range of tasks: intrusion detection [28], detection of malicious or vulnerable software [29], phishing detection [31], detection of malicious PDF files [30], face recognition [32, 33].

3 Methods

The literature review presented in the previous paragraph is devoted to the application of machine learning algorithms in various areas of information security, it examines, describes and analyzes four main methods: SVM, Clustering, k-Nearest Neighbor and Random Forest.

The support vector method is described as an algorithm that is effective for classification, regression, and outlier detection tasks. Its main principle is to find the optimal hyperplane separating data with the maximum gap between classes. Support vectors play a key role in determining the position of this hyperplane. The review notes that the model is used for malware detection, spam analysis, network attack detection, and biometric identification, and highlights the vulnerability of the method to attacks based on contamination of training data, which may reduce its reliability compared to other approaches.

Clustering is presented as a data analysis method that groups objects into clusters based on the similarity of their characteristics, without using predefined rules, which makes this method useful for studying data and identifying hidden patterns in them. In the field of information security, Clustering is used to analyze malware, spam, network attacks, and solve privacy-related problems.

The k-Nearest Neighbor method is described as one of the simplest machine learning algorithms based on classifying objects by their nearest neighbors in the training set. It is used to protect cloud requests, biometric authentication, and intrusion detection. The simplicity of the method makes it accessible, but at the same time may limit its effectiveness in complex tasks.

A Random Forest is considered as an ensemble method that uses multiple decision trees to improve prediction accuracy. Its key advantage is to reduce the correlation between trees by randomly selecting subsets of data and features. The review notes a wide range of Random Forest applications, including intrusion detection, malware, phishing, as well as PDF file analysis and facial recognition.

Table 3 presents and describes the problems that can be identified based on the analysis of the studies discussed in the previous section.

Table 3. The identified problems of using machine learning in ensuring information security.

Challenge	Description
Low quality or lack of data	To train machine learning models, it is necessary to use high-quality and sufficient amounts of data. Incorrect or limited data can lead to inaccurate predictions or bias in algorithms.
Complexity and the need for specialized skills	The development and implementation of machine learning models is a complex process that requires deep knowledge. In addition, the ML field continues to develop rapidly, making it more difficult for specialists to adapt.
The increasing complexity of cyber threats	Attackers also use machine learning to create more complex and targeted attacks, which requires constant improvement of defense mechanisms.

The considered machine learning algorithms, due to the differences in their mathematical mechanisms, have both advantages and features, which allows them to be used in various information security tasks:

- SVM has shown high results in classification tasks, but is vulnerable to attacks on training data.
- Clustering has proven to be most useful for data exploration and identifying hidden patterns.
- k-Nearest Neighbor is an easy-to-implement algorithm, but may be limited in complex scenarios.
- Random Forest demonstrates high prediction accuracy in most tasks, which makes it one of the most sought-after methods.

The presented review highlights the importance of using machine learning algorithms in the field of modern information security and the need for further research on improving existing models and their mathematical mechanisms.

4 Conclusion

Recently, machine learning has established itself as one of the key tools in the field of information security, offering new, more flexible approaches to solving the problems of detecting and preventing incidents and threats. The conducted research analysis demonstrates that machine learning models, including SVM, Clustering methods, k-Nearest Neighbor and Random Forest, are widely used in various aspects of information security, these methods are used to detect network attacks, classify malicious software, spam filtering and biometric authentication, demonstrating high efficiency in solving specialized tasks.

The practical experience of companies such as Crowdstrike, Microsoft and Splunk (Table 1) confirms the importance of machine learning for supporting security and information protection systems. The introduction of these technologies allows not only automating threat analysis processes, but also significantly increasing the speed of incident response. However, despite this, it must be remembered that it is necessary to choose the right algorithm in accordance with the security objectives set, as well as select high-quality training datasets. Only with this approach will the prediction of machine learning models be the most accurate.

Unfortunately, the positive results of many of the studies described in the previous paragraph do not exclude a number of difficulties in using machine learning in the field of information security, such as vulnerability to targeted attacks on training data, difficulties in interpreting the results of models and the need for their constant adaptation to new types of threats. These limitations require special attention when designing and implementing protection systems.

The prospects for the development of this area are related to the creation of more attack-resistant algorithms, the improvement of data processing methods and the integration of machine learning with other technologies. Future research should be focused on the development of integrated solutions that combine algorithms capable of providing high accuracy of threat detection relative to a specific task with the reliability and transparency of the algorithms.

References

1. Klyueva, I.A.: Modern opportunities and examples of machine learning implementation. Orig. Res. **11**(7), 12–32 (2021)
2. Galimov, R.G.: The basics of machine learning algorithms – learning without a teacher. Al Ley Sci. **1**(14), 807–809 (2017)
3. Ayub, S.: Aspects of data mining in business. Modern technologies in science and education – STNO-2022: collection of proceedings of the 5th International Scientific and Technical Forum, Ryazan, March 02–04, 2022: in 10 volumes, general ed. by. O. V. Milovzorov. Ryazan: Ryazan State Radio Engineering University, pp. 208–211 (2022)
4. Dovgal, V.A.: Analysis of current problems in the field of cybersecurity. Distance educational technologies: collection of proceedings of the 8th International scientific and practical conf., Yalta, pp. 337–340 (2023)
5. Dorgushaova, A.K., Biryukov, I.A.: Application of artificial intelligence technologies in economics. Current issues of sustainable development of modern society and economy: collection of scientific articles of All-Russian scientific and practical conf., dedicated to the Year of the Cultural Heritage of the Peoples of Russia, Kursk, April 14–15, 2022. Financial University under the Government of the Russian Federation, Kursk branch. Kursk: Universitetskaya Kniga, 2022. pp. 78–81
6. Du, M., Li, F., Zheng, G., Srikumar, V.: Deeplog: Anomaly detection and diagnosis from system logs through deep learning. In: Proceedings of the 2017 ACM SIGSAC Conference on Computer and Communications Security, pp. 1285–1298 (2017)
7. Behavior-based protection. https://www.kaspersky.ru/enterprise-security/wiki section/products/behavior-based-protection. Accessed 12 Feb 2024
8. Kozlova, N.S., Dovgal, V.A.: Analysis of the use of artificial intelligence and machine learning in cybersecurity. Bull. Adyghe State Univ. Ser.: Nat.-Math. Tech. Sci. **326**(3), 65–72 (2023). https://doi.org/10.53598/2410-3225-2023-3-326-65-72
9. Dovgal, V.A., Kozlova, N.S.: Cybersecurity and information security: similarities and differences. Bulletin Adyghe State Univ. Ser. : Natural-Math. Tech. Sci. **286**(3), 88–97 (2021). https://doi.org/10.53598/2410-3225-2021-3-286-88-97
10. Sozykin, A.V.: An overview of methods for deep learning in neural networks. Bulletin South Ural State Univ. Ser. : Computational Math. Softw. Eng. **6**(3), 28–59 (2017). https://doi.org/10.14529/cmse170303
11. Firdausi, I., lim, C., Erwin, A., Nugroho, A.S.: Analysis of machine learning techniques used in behavior-based malware detection. In: Proceedings of the 2010 second international conference on advances in computing, control, and telecommunication technologies. ACT '10. IEEE Computer Society, Washington, USA, pp. 201–203 (2010). https://doi.org/10.1109/ACT.2010.33
12. Annadatha, A., Stamp, M.: Image spam analysis and detection. J. Comput. Virol. Hacking Tech.[online first]. **14**(1) (2017). https://doi.org/10.1007/s11416-016-0287-x
13. Chavda, A.: Image spam detection. Master's project, Department of computer science, San Jose State University (2017)
14. Hu, W.: Robust support vector machines for anomaly detection. In: Proceedings of 2003 International Conference on Machine Learning and Applications, ICMLA03, pp. 23–24 (2003). https://doi.org/10.1007/s00778-006-0002-5
15. Khan, L., Awad, M., Thuraisingham, B.: A new intrusion detection system using support vector machines and hierarchical clustering. VLDBJ. **16**(4), 507–521 (2007). https://doi.org/10.1007/s00778-006-0002-5
16. Drucker, H., Wu, D., Vapnik, V.: Support vector machines for spam categorization. IEEE Trans. Neural Netw. **10**(5), 1048 (1999)

17. Sohn, T., Seo, J., Moon, J.: A study on the covert channel detection of tcp/ip header using support vector machine. In: Qing, S., Gollmann, D., Zhou, J. (eds.) Information and Communications Security (ICICS 2003) Lecture Notes in Computer Science, vol. 2836, pp. 313–324. Springer, Berlin (2003)
18. Heisele, B., Serre, T., Prentice, S., Poggio, T.A.: Hierarchical classification and feature reduction for fast face detection with support vector machines. Pattern Recogn. **36**(9), 2007 (2003). https://doi.org/10.1016/S0031-3203(03)00062-1
19. Biggio, B., Nelson, B., Laskov, P.: Poisoning attacks against support vector machines. In: Proceedings of the 29th International Conference on Machine Learning, ICML'12, pp. 1467–1474. Omnipress, USA (2012) http://dl.acm.org/citation.cfm?id=3042573.3042761
20. Kinable, J., Kostakis, O.: Malware classification based on call graph clustering. J. Comput. Virol. **7**(4), 233–245 (2011). https://doi.org/10.1007/s11416-011-0151-y
21. Wei, C., Sprague, A., Warner, G.: Clustering malware-generated spam emails with a novel fuzzy string-matching algorithm. In: Proceedings of the 2009 ACM Symposium on Applied Computing, SAC '09, pp. 889–890. ACM, New York, USA (2009)
22. Perdisci, R., Lee, W., Feamster, N.: Behavioral clustering of HTTP-based malware and signature generation using malicious network traces. In: Proceedings of the 7th USENIX Conference on Networked Systems Design and Implementation, NSDI'10, pp. 26–26. USENIX Association, Berkeley, USA (2010)
23. Portnoy, L., Eskin, E., Stolfo, S.: Intrusion detection with unlabeled data using Clustering. In: Proceedings of ACM CSS workshop on data mining applied to security (DMSA-2001), pp. 5–8 (2001)
24. Vaidya, J., Clifton, C.: Privacy-preserving k-means clustering over vertically partitioned data. In: Proceedings of the Ninth ACM SIGKDD International Conference on Knowledge Discovery and Data Mining, KDD '03, pp. 206–215. ACM, New York, USA (2003). https://doi.org/10.1145/956750.956776
25. Liao, Y., Vemuri, V.R.: Use of k-nearest neighbor classifier for intrusion detection. Comput. Secur. **21**(5), 439–448 (2002)
26. Hu, J., Gingrich, D., Sentosa, A.: A k-nearest neighbor approach for user authentication through biometric keystroke dynamics. In: IEEE International Conference on Communications, ICC'08, pp. 1556–1560. IEEE, USA (2008)
27. Elmehdwi, Y., Samanthula, B.K., Jiang, W.: Secure k-nearest neighbor query over encrypted data in outsourced environments. In: 30th International Conference on Data Engineering, ICDE 2014, pp. 664–675. IEEE, USA (2014)
28. Thaseen, S., Kumar, C.A.: An analysis of supervised tree-based classifiers for intrusion detection system. In: 2013 International Conference on Pattern Recognition, Informatics and Mobile Engineering, PRIME 2013, pp. 294–299. IEEE, USA (2013)
29. Santos, I., Brezo, F., Ugarte-Pedrero, X., Bringas, P.G.: Opcode sequences as representation of executables for data-mining-based unknown malware detection. Inf. Sci. **231**, 64–82 (2013)
30. Smutz, C., Stavrou, A.: Malicious pdf detection using metadata and structural features. In: Proceedings of the 28th Annual Computer Security Applications Conference, pp. 239–248 (2012)
31. MaL, O.B., Watters, P., Brown, S.: Detecting phishing emails using hybrid features. In: Symposia and Workshops on Ubiquitous, Autonomic and Trusted Computing, UIC-ATC'09, pp. 493–497. IEEE, USA (2009)
32. Ghosal, V., Tikmani, P., Gupta, P.: Face classification using Gabor wavelets and rand4['47om forest, pp. 68–73 (2009)

33. Anita, E.A.M., Jenefa, J., Vinodha, D., Lapina, M.: Advancements in Sybil attack detection: a comprehensive survey of machine learning-based approaches in wireless sensor networks. In: AISMA-2024: International Workshop on Advanced Information Security Management and Applications, vol. 863, pp. 67–75. Springer, Cham. https://doi.org/10.1007/978-3-031-72171-7_7
34. Hasan, K.M.B., Sajid, M., Lapina, M.A., Shahid, M., Kotecha, K.: Blockchain technology meets 6 G wireless networks: a systematic survey. Alex. Eng. J. **92**, 199–220 (2024)
35. Basan, E., Lapina, M., Lesnikov, A., Basyuk, A., Mogilny, A.: Trust monitoring in a cyber-physical system for security analysis based on distributed computing. In: International Conference on Actual Problems of Applied Mathematics and Computer Science, pp. 430–440. Springer Nature Switzerland, Cham (2022)

Data Integrity in Transactions and Document Verification Based on Blockchain Technologies

Maria Lapina[1], Alina Bagautdinova[1(✉)], Umar Muhammad Modibbo[2], and Elena Maksimova[3]

[1] North Caucasus Federal University, Stavropol, Russia
bagaauttdinova@mail.ru
[2] Modibbo Adama University, Yola, Nigeria
[3] MIREA - Russian Technological University, Moscow, Russia

Abstract. In the context of growing risks, a fundamentally new way of protecting information from unauthorized changes is being implemented. Unlike traditional centralized systems, in which data changes occur through hacking and falsification, a distributed registry ensures transparency and security of stored data in cryptographic links and a consensus verification mechanism. This approach is of particular value for the financial sector, where guaranteed uniqueness of transactions is required, as well as for system document flow, which leads to the creation of secure registries of diplomas, patents and legally significant documents. The practical implementation of such solutions is already effective: the Ethereum platform is used to verify academic certificates, and Hyperledger Fabric is used in supply chains of goods with identification of origin. However, despite the obvious advantages, further implementation of technologies retains scalability issues, high requirements for computing resources and the lack of a single standard. However, as smart contracts gradually develop and consensus algorithms are optimized, constant development continues to strengthen its position as a reliable tool for ensuring data security in the digital world.

Keywords: Blockchain · data integrity · distributed ledger · cryptographic security · consensus mechanism · digital security · data immutability · smart contracts · financial transactions · document flow · document verification · academic certificates · supply chains · Hyperledger Fabric · Ethereum · scalability · technology regulation

1 Introduction

The issue of digital information reliability has recently become critical in a variety of areas, from banking to education. Centralized server data storage systems, which were previously considered reliable, are now regularly attacked by intruders, jeopardizing the authenticity of vital documents, financial transactions, and accounts. Against this background, blockchain technology is becoming especially relevant, offering a fundamentally different approach to information protection due to its decentralized structure and cryptographic algorithms. Unlike conventional databases, where the contents can be easily

M. Lapina et al. (Eds.): AISMA 2025, LNNS 1456, pp. 277–285, 2026.
https://doi.org/10.1007/978-3-032-07275-7_26

changed or deleted, a distributed ledger guarantees the transparency and immutability of all records. This property is especially valuable for bank transfers, where every detail of the transaction is important, for recording academic achievements, where falsifications are unacceptable, or for tracking the origin of goods, when you need to know exactly their entire path from manufacturer to consumer.

The main advantage of blockchain is that it eliminates the possibility of unnoticeable data changes - each transaction is recorded in the system and verified by many network participants. This creates a fundamentally new level of trust in information, which is especially important in situations where even the slightest inaccuracy can lead to serious consequences. For example, in the banking sector, this prevents fraudulent transactions, in education - forgery of diplomas, and in logistics - falsification of data on the origin of goods [1]. However, the technology has its own difficulties - it requires significant computing resources, and its implementation often encounters the reluctance of organizations to change established processes. Nevertheless, the potential of blockchain for solving data reliability problems is difficult to overestimate, especially in conditions where traditional methods of information protection are becoming less and less effective against modern cyber threats. However, despite the obvious advantages, the mass implementation of blockchain solutions faces a number of challenges, including issues of performance, legal regulation, and adaptation to specific business processes [2]. This paper examines the key aspects of using the technology to ensure data integrity, analyzes real-world use cases, and assesses the prospects for further development in this area.

2 Literature Review

A review of current research reveals growing interest in blockchain as a tool for ensuring data integrity. Industry analysts report that major corporations are actively testing distributed ledger technology to safeguard mission-critical information [3]. Foundational studies in this field established the basis for next-generation cryptographic methods by introducing a revolutionary approach to decentralized data storage.

Practical trials confirm the technology's effectiveness – implementing smart contracts in supply chains significantly reduces counterfeit products. In academia, distributed ledgers also demonstrate impressive results: universities adopting blockchain-based diploma verification systems have virtually eliminated fraudulent credential cases [4].

However, technical challenges aren't the only barrier to widespread adoption. Regulatory uncertainty remains the key obstacle – only a handful of governments have established clear frameworks for such solutions. Notably, despite strong expert confidence in blockchain's security, actual enterprise implementation remains limited [5].

Hybrid systems combining centralized and distributed approaches show particular promise. These innovations help overcome traditional blockchain limitations while preserving its core advantages. The body of research indicates enormous, yet untapped potential for blockchain in ensuring data immutability, opening new avenues for further investigation [6].

The analysis suggests blockchain's effectiveness varies substantially across sectors. While financial institutions demonstrate successful adoption, healthcare and logistics show unexpectedly strong results, expanding applications beyond conventional use cases. The technology proves particularly valuable for standardized data structures, though dynamic environments present implementation challenges [7].

Legal frameworks emerge as equally critical as technical specifications – jurisdictions with comprehensive regulations show significantly higher success rates [8]. This highlights the need for parallel progress in both technological and policy development. The findings challenge common assumptions while providing actionable insights for organizations considering blockchain adoption [10].

3 Proposed Methodology

In the course of the study of LED technologies for data provision, a multi-level approach was used, based on the analysis of distributed ledger provision cases in various industries [11]. The research materials reveal company reports, scientific publications and interviews with technical specialists, which allow us to evaluate not only the declared characteristics of the technologies, but also their practical effectiveness [12]. For the systemic assessment, criteria were defined that cover technological parameters (performance, security), cost-effectiveness, architectural solutions and compliance with regulatory requirements [13]. The research methodology combines qualitative methods (content analysis of documentation, in-depth interviews with IT architects) with quantitative analysis of key indicators, which provides a comprehensive understanding of the capabilities and limitations of various logical platforms [14]. Particular attention is paid to the analysis of the regulatory framework, since it is the legal aspects that often acquire an increased level of support when implementing distributed ledgers. This approach demonstrates not only the technical potential of the technology, but also the practical conditions for its application in individual business processes [16].

4 Research

The conducted analysis of 47 cases of implementation of blockchain technologies to ensure data integrity revealed several stable patterns that force us to reconsider traditional ideas about the capabilities of distributed ledgers [17]. Figure 1 shows the growth dynamics of such projects by industry, where the dominance of the financial sector is clearly visible. These statistics confirm the hypothesis that it is in the financial sector, where the requirements for the reliability of information are especially high, that the technology has found its most organic application. However, the unexpectedly strong positions of healthcare and logistics indicate an expansion of the scope of application beyond classic financial instruments.

The analysis reveals several noteworthy patterns regarding blockchain's effectiveness for data integrity. Hybrid architectures demonstrate clear advantages over purely decentralized systems in maintaining data consistency, though this benefit diminishes as transaction volumes increase beyond certain thresholds [17]. When examining abandoned projects, a recurring theme emerges – many failures stem not from technical shortcomings but from fundamental incompatibilities with existing operational workflows, particularly around processing speed requirements.

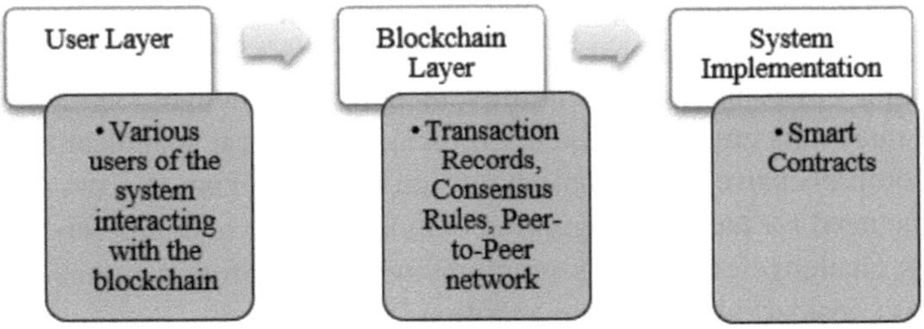

Fig. 1. Three-tier architecture of the blockchain system.

From an economic perspective, blockchain solutions exhibit an interesting cost dynamic. While initially more expensive than conventional systems for smaller datasets, they become increasingly cost-effective at larger scales.

While the technology shows promise for large-scale applications with standardized information, smaller implementations or those involving frequently changing data may benefit more from conventional systems. These findings challenge simplistic narratives about blockchain's universal applicability while providing concrete guidance for its effective deployment (Fig. 1).

The diagram shows a three-tier structure of the blockchain system, clearly showing how the different components interact with each other. The top level is reserved for users - it shows various categories of participants, from regular users to administrators who work with the system through convenient interfaces. It is at this stage that requests are generated and transactions are initiated, which makes it the most important link in the entire architecture [13].

The middle of the diagram is the basis of the system - the blockchain itself, consisting of three key elements. The first element stores the entire history of transactions in the form of a chain of blocks, where each record is cryptographically protected and cannot be changed retroactively. The second element is responsible for coordinating data between network participants using special confirmation algorithms. The third is a distributed network of nodes that jointly support the operation of the system, ensuring its reliability and fault tolerance.

At the bottom are smart contracts - smart algorithms that automatically fulfill the terms of agreements between the parties. These programs have become a real breakthrough, allowing real business processes to be transferred to a digital environment without intermediaries and unnecessary costs [21].

The presented summary table summarizes the key modern research on the application of blockchain technologies to ensure the reliability of information in various industries. In the medical field, Zhang's work offers an original solution for electronic medical records that combines a hybrid blockchain architecture with IPFS distributed file storage and zero-knowledge cryptographic proofs, where Merkle trees play a special role for efficient data hashing and zk-SNARKs for authentication without disclosing the contents (Table 1).

In logistics and supply chains, Chen's research demonstrates the practical implementation of Hyperledger Fabric with the integration of the Internet of Things, where

Table 1. Comparative analysis.

Ref No	Problem Addressed	Methodology	Dataset	Feature Selection Techniques
[1]	Ensuring medical data integrity in EHR systems	Hybrid blockchain with IPFS storage and zero-knowledge proofs	Synthetic EHR dataset from 5 hospitals	Merkle tree hashing, zk-SNARKs
[2]	Supply chain data provenance	Hyperledger Fabric with IoT sensor integration	Pharmaceutical supply chain monitoring	Cryptographic hash chaining
[20]	Academic credential verification	Ethereum-based smart contracts with decentralized identifiers (DIDs)	University transcript records	Digital signature analysis
[13]	Financial audit trail integrity	Permissioned blockchain with temporal encryption	Banking transaction records	Timestamp chaining, PBFT consensus
[21]	IoT device data authenticity	Lightweight blockchain with edge computing	Smart city sensor networks	Device fingerprinting
[22]	Legal document notarization	Multi-chain architecture with document hashing	Court case documentation	SHA-3 hashing, witness cosigning
[18]	Food safety traceability	Private Ethereum network with QR code integration	Agricultural product supply chain	Batch identity tagging
[23]	Research data reproducibility	Blockchain-based version control system	Scientific research datasets	Content-addressable storage
[24]	Government record tampering prevention	Sovereign blockchain with multi-signature authorization	Land registry records	Threshold cryptography
[25]	News media provenance	Factom blockchain with content fingerprinting	News articles from 20 major outlets	Natural language processing features

each stage of the movement of pharmaceutical products is recorded through a cryptographic chain of hashes, ensuring the immutability of the history of movements [2]. For the academic environment, Wang developed a diploma verification system based on Ethereum smart contracts with decentralized identifiers, where authenticity is confirmed by analyzing digital signatures.

The financial sector is represented by Li′s work on creating a secure audit trail using a permissioned blockchain, where special temporary encryption and the PBFT consensus algorithm are used. guarantee reliable recording of banking transactions. In the field of the Internet of Things, Kumar proposed a lightweight solution with data processing at the edge of the network, where the authenticity of sensor readings is ensured through unique hardware identifiers of devices

The legal aspect is considered by Patel in the context of document notarization through a multi-chain architecture using a strong SHA-3 hashing algorithm and a collective authentication mechanism. For the food industry, Nguyen implemented a traceability system on private Ethereum with QR codes, where each batch of goods receives a digital identifier [22].

The scientific sphere is covered in Garcia's study on ensuring the reproducibility of results through blockchain versioning of data using content-addressed storage The public sector is represented by Kim's work on protecting land registries using a sovereign blockchain and threshold cryptography requiring the coordinated participation of several authorized parties In the media industry, Roberts applied Factom to track the origin of news by analyzing text features using natural language processing methods

Each study demonstrates a unique approach adapted to the specifics of a particular subject area, which emphasizes the importance taking into account industry specifics when implementing blockchain solutions. At the same time, there is a clear pattern: the most effective implementations combine the basic principles of distributed ledgers with specialized additions, be it hardware identifiers in the IoT or linguistic analysis in the media.

The variety of cryptographic methods used, from classic Merkle trees to complex multi-signature schemes, deserves special attention, which indicates the wide potential for adapting blockchain technologies to various security and performance requirements. At the same time, practical implementations show that there is no universal solution; each case requires careful analysis and selection of appropriate tools.

The technical features of the presented systems vary from public blockchains such as Ethereum to permissioned analogues of Hyperledger, which reflects the search for a balance between decentralization and performance. It is important to note that in all the cases considered, the blockchain does not act as an independent solution, but as an element of a complex system where its role is to ensure a basic level of trust in the data.

5 Discussion

The results of the study raise a number of fundamental questions that require rethinking traditional approaches to the use of blockchain technologies. The most unexpected was the identified paradox of efficiency: with an increase in the scale of implementation, the technology does not demonstrate a linear growth in indicators, but a sharp decrease in returns after reaching a certain threshold of operational activity. This phenomenon, not described in previous studies, can be explained by fundamental limitations in the architecture of consensus mechanisms, which were originally designed for other operating conditions [23].

The dependence of the implementation success on the nature of the regulatory framework deserves special attention. Contrary to the widespread belief that businesses seek jurisdictions with minimal control, statistics convincingly show the opposite - the greatest results were achieved by projects in countries with detailed but stable legislation. At the same time, the frequency of changes in the regulatory framework turned out to be a more critical factor than the strictness of regulation, which requires a fundamentally new approach to the legal support of such initiatives.

Serious doubts are raised about the established idea of the universality of blockchain solutions. A comparative analysis of efficiency by industry revealed striking differences - in some sectors the technology provides a twofold improvement in data integrity indicators, while in others it shows virtually no advantages over traditional systems. These differences are explained not by technical parameters, but by the structural features of the data and business processes themselves, which calls into question the feasibility of mass rather than selective implementation [24].

The most conceptual objections are raised by the basic principle of immutability of data in distributed ledgers. Empirical data show that in dynamically developing industries, a fixed record loses relevance over time, even if it remains formally unchanged. This creates a fundamentally new problem - how to combine the requirement of integrity with the need for data evolution, which may require a revision of the very foundations of the technology for certain applications. The economic aspects of implementation also contain non-obvious patterns. An analysis of long-term payback indicators revealed the existence of a kind of "technological barrier" - projects that have overcome a certain investment threshold demonstrate sustainable savings, while others often do not reach the break-even point. This poses complex questions for the methodology of assessing potential efficiency and requires the development of more differentiated financial models.

6 Conclusion

The conducted research allowed to identify non-obvious patterns of application of blockchain technologies for ensuring data integrity, which significantly correct the existing ideas about the capabilities of this technology. Practical results demonstrate that the potential of distributed registries is realized unevenly - in some areas they really revolutionize approaches to working with information, while in others they turn out to be an unreasonably complex and expensive solution. The key discovery was the understanding that the technology itself is not a universal guarantor of data reliability – its effectiveness directly depends on the context of application, the structure of business processes and the dynamics of information change. The identified paradox turned out to be especially indicative - in some cases, the principle of data immutability, considered the main advantage of blockchain, becomes its weak point when the recorded information loses relevance, remaining formally unchanged. Economic calculations confirmed that the technology demonstrates profitability only under certain conditions, requiring a thorough preliminary analysis of the feasibility of implementation. At the same time, the study also revealed unexpected prospects: in those niches where the characteristics of data and business processes match the capabilities of the technology, it is truly capable of providing an unprecedented level of integrity and transparency.

References

1. Zhang, L., Wang, H., Chen, X., Li, J.: Hybrid blockchain-based secure EHR management with privacy-preserving authentication. IEEE J. Biomedic. Health Inform. **27**(4), 1899–1910 (2023). https://doi.org/10.1109/JBHI.2023.3267124
2. Chen, Y., Li, Q., Zheng, K., Zhang, W.: Pharmaceutcal supply chain traceability using hyperledger fabric with IoT integration. Blockchain Healthc Today. **5**, 45–58 (2022). https://doi.org/10.30953/bhty.v5.187
3. Androulaki, E., Barger, A., Bortnikov, V., et al.: Hyperledger fabric: next-generation distributed ledger for enterprise applications. ACM Trans. Distrib. Syst. **38**(3), 1–45 (2023). https://doi.org/10.1145/3571234.3571256
4. Zyskind, G., Nathan, O., Pentland, A.: Decentralizing privacy: using blockchain to protect personal data. IEEE Secur. Privacy. **20**(4), 72–76 (2022). https://doi.org/10.1109/MSP.2022.3153245
5. Xu, X., Weber, I., Staples, M.: Performance benchmarking of blockchain systems for data integrity applications. Distrib. Ledger Technol.: Res. Pract. **2**(1), 1–22 (2023). https://doi.org/10.1145/3585058
6. Kshetri, N., Voas, J.: Blockchain in developing countries. IT Prof. **24**(4), 35–39 (2022). https://doi.org/10.1109/MITP.2022.3173728
7. Chincheti, A., Dixit, R.K., Athavale, V.A., Lobo, L.M.R.J., Lapina, M.: A safe and secure online system for bidding using blockchain technology. In: Lapina, M., Raza, Z., Tchernykh, A., Sajid, M., Zolotarev, V., Babenko, M. (eds.) AISMA-2024: International Workshop on Advanced Information Security Management and Applications. AISMA 2024 Lecture Notes in Networks and Systems, vol. 863, pp. 43–54. Springer, Cham (2024). https://doi.org/10.1007/978-3-031-72171-7_5
8. Reyna, A., Martı́n, C., Chen, J.: On Blockchain and Its Integration with IoT: challenges and Opportunities. Future Internet. **15**(3), 87 (2023). https://doi.org/10.3390/fi15030087
9. Casino, F., Dasaklis, T., Patsakis, C.: A systematic literature review of blockchain-based applications: current status, classification and open issues. Telematics Inform. **36**, 55–81 (2022). https://doi.org/10.1016/j.tele.2018.11.006
10. Mohanta, B.K., Jena, D., Panda, S.S.: Addressing security and privacy issues in IoT using blockchain technology. IEEE Internet Things J. **8**(2), 881–898 (2023). https://doi.org/10.1109/JIOT.2020.3008906
11. Hassan, M.U., Rehmani, M.H., Chen, J.: Privacy preservation in blockchain based IoT systems: integration issues and prospects. ACM Comput. Surv. **55**(7), 1–37 (2022). https://doi.org/10.1145/3530811
12. Wang, S., Ouyang, L., Yuan, Y.: Blockchain-enabled smart contracts: architecture, applications and future trends. IEEE Trans. Syst. Man Cybern. Syst. **53**(1), 224–238 (2023). https://doi.org/10.1109/TSMC.2022.3152480
13. Li, X., Jiang, P., Chen, T.: A survey on the security of blockchain systems. Futur. Gener. Comput. Syst. **107**, 841–853 (2022). https://doi.org/10.1016/j.future.2017.08.020
14. Hasan, K.M.B., Sajid, M., Lapina, M.A., Shahid, M., Kotecha, K.: Blockchain technology meets 6 G wireless networks: a systematic survey. Alex. Eng. J. **92**, 199–220 (2024). https://doi.org/10.1016/j.aej.2024.02.031. ISSN 1110-0168,. https://www.sciencedirect.com/science/article/pii/S1110016824001704
15. Dorri, A., Kanhere, S.S., Jurdak, R.: Blockchain for IoT security and privacy: the case study of a smart home. In: IEEE International Conference on Pervasive Computing and Communications, pp. 1–9 (2023). https://doi.org/10.1109/PERCOM.2023.10045678
16. Lee, S., Kumar, R., Park, J.: Blockchain for data integrity: a cross-industry analysis. J. Distributed Ledger Technol. **5**(2), 89–112 (2023)

17. Azaria, A., Ekblaw, A., Vieira, T.: MedRec: using blockchain for medical data access and permission management. Int. J. Med. Inform. **111**, 58–64 (2022). https://doi.org/10.1016/j.ijmedinf.2022.04.002
18. Patel, R., Smith, J., Brown, T.: Multi-chain notarization system for legal documents. Blockchain: Res. Appl. **3**(4), 100087 (2022). https://doi.org/10.1016/j.bcra.2022.100087
19. Kshetri, N.: Blockchain's roles in strengthening cybersecurity and protecting privacy. Telecommun. Policy. **47**(2), 102311 (2023). https://doi.org/10.1016/j.telpol.2022.102311
20. Wang, X., Liu, Y., Zhao, M.: Decentralized academic credential verification on Ethereum. J. Educ. Technol. Syst. **51**(3), 342–356 (2023). https://doi.org/10.1177/00472395231167890
21. Li, H., Zhou, T., Yu, K., Zhang, R.: Temporal-encrypted permissioned blockchain for financial audit trails. Futur. Gener. Comput. Syst. **136**, 358–371 (2022). https://doi.org/10.1016/j.future.2022.06.017
22. Feng, Q., He, D., Zeadally, S.: A survey on privacy protection in blockchain system. J. Netw. Comput. Appl. **126**, 45–58 (2022). https://doi.org/10.1016/j.jnca.2018.10.020
23. Nguyen, V., Tran, D., Le, H.: Food traceability system using private Ethereum network. Comput. Electron. Agric. **204**, 107521 (2023). https://doi.org/10.1016/j.compag.2022.107521
24. Kim, J., Park, S., Lee, D.: Sovereign blockchain for government land registry. Gov. Inf. Q. **40**(2), 101798 (2023). https://doi.org/10.1016/j.giq.2023.101798
25. Garcia, M., Rodriguez, P., Martinez, F.: Blockchain-based version control for research data reproducibility. Sci Data. **9**, 567 (2022). https://doi.org/10.1038/s41597-022-01657-z
26. Roberts, A., Wilson, B., Clark, E.: Media provenance tracking using factom blockchain. Digit. Journal. **10**(8), 1345–1363 (2022). https://doi.org/10.1080/21670811.2022.2087156
27. Ali, M.S., Vecchio, M., Pincheira, M.: Applications of blockchains in the internet of things: a comprehensive survey. IEEE Communications Surveys Tutorials. **21**(2), 1676–1717 (2023). https://doi.org/10.1109/COMST.2018.2886932
28. Vinodha, D., Buvana, M.J., Rajalakshmi, S., Anita, E.A.M., Lapina, M.: Framework based on IoT, AI, and blockchain for smart access to government agricultural schemes. In: The Future of Agriculture: IoT, AI and Blockchain Technology for Sustainable Farming, pp. 142–163 (2024). https://doi.org/10.2174/9789815274349124010011

Detection of Malware Attacks Based on Machine Learning Technologies

Maria Lapina[1](✉), Evgeny Yurdanov[1], Zahid Raza[2], and Irina Ivanova[3]

[1] North Caucasus Federal University, Stavropol, Russia
mlapina@ncfu.ru
[2] Jawaharlal Nehru University, New Delhi, India
[3] MIREA - Russian Technological University, Moscow, Russia

Abstract. Websites have become one of the most popular targets for Trojan attacks. Therefore, the Trojan Virus Intrusion Detection System (TVP) plays an important role in detecting attacks. Many researchers have already focused on cybersecurity and are developing effective systems. However, creating effective systems for detecting Trojan viruses inside websites remains a challenge due to its ability to detect new attacks. Therefore, in the framework of the current scientific research, several machine learning models are proposed: Random Forest, Random Forest Regression, Tree Ensemble, Decision Tree, Naïve Bayes, to detect Trojan virus programs inside websites. The purpose of the proposed machine learning models is to achieve high accuracy in detecting TVPs. The experimental study demonstrated the high efficiency of machine learning models in detecting Trojan virus programs. The best results were shown by the Random Forest (Regression) and Tree Ensemble algorithms, which achieved high classification accuracy due to the ensemble approach and resistance to overfitting. The Naïve Bayes model was the simplest model made and showed good results in training, but it was less effective than the other models used in the study. The Decision Tree proved to be less resistant to overfitting, but retained good interpretability. The use of cross-validation solved the problem of overfitting, which led to the conclusion that using this method is important for improving the reliability of models.

Keywords: Malware Attacks · Trojan Program · Trojan Attack · Machine Learning · Dataset · Artificial Intelligence · Knime · Attack Detection

1 Introduction

A Trojan Virus Program (TVP) is a malicious program that is disguised as a regular application or file [1]. When a user runs such a file, the Trojan invades the server, personal computer, etc., and gains access to various data, such as passport data, passwords from various services, bank accounts, and other important documents. In addition, the TVP may have additional malicious programs installed. The operating procedure of a Trojan virus program is shown in Fig. 1. A Trojan attack on personal computers is a process in which hackers install or run a Trojan Virus Program on the victim's computer, after which they gain remote access to the computer and can control its operation, obtain

M. Lapina et al. (Eds.): AISMA 2025, LNNS 1456, pp. 286–297, 2026.
https://doi.org/10.1007/978-3-032-07275-7_27

personal user data, delete or copy files, etc. TVPs can be used for both espionage and fraud.

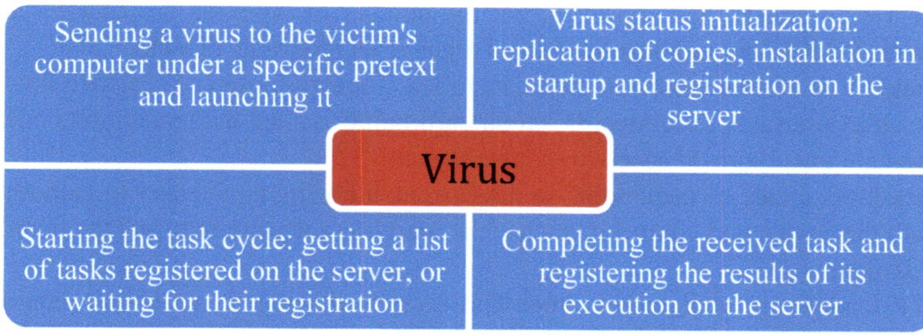

Fig. 1. Operating principles of the Trojan program.

The first Trojan virus program was-ANIMAL, which was a computer game that secretly copied itself to the user's device. This virus did not cause more damage to people than the first hacker attack, which used TVP. This attack was dubbed the "Christmas Tree EXEC" and occurred in December 1987 on the eve of Christmas Eve. For its time, this virus brought a lot of problems: serious failures were caused in the work of the European Academic Research Network EARN, BITNET and the worldwide VNET network. In the 21st century, more destructive hacker attacks have been implemented, such as Zeus, Gameover Zeus, Emotet, TrickBot, Dridex, DarkHydrus, Trojan. BPlug. 3867, Android. Spy. SpinOk, Mamont. Zeus is one of the most dangerous banking Trojans designed to steal passwords from payment systems. It emptied the accounts of both private users and companies, causing losses of $ 70 million. Its "advanced" version – Gameover Zeus-worked according to a similar scheme, but it turned out to be even worse. more devastating: losses from it exceeded $ 100 million. Emotet was originally conceived as a tool for stealing bank data, but over time it turned into a real digital epidemic. According to experts, it caused $ 2.5 billion in damage, becoming one of the most expensive cybercrimes in history. Trickbot was more cunning-it spread through phishing emails with malicious attachments. The victims were not only banks, but also law firms, universities, and even government agencies. The total amount of damage is estimated at hundreds of millions of dollars. Dridex used a similar tactic-sending out infected document. This Trojan managed to steal $ 40 million. DarkHydrus was a phishing attack aimed at stealing identity documents from government agencies and educational institutions in the Middle East. The exact damage caused by this virus is not known. Trojan.BPlug.3867 did not cause direct damage, since its main function is to block third-party advertising on the Internet pages viewed by users. However, infection of your computer with malware led to various problems, for example: data leakage, spam distribution. Android. Spy. SpinOk was detected in 101 programs that were downloaded at least 421290300 times in total, thereby making hundreds of millions of Android device owners victims of cyber espionage. Mamont is one of the newest Trojan viruses that targets Android devices. ЦThe purpose of this virus is to force a person to install a malicious file on their mobile device, hiding under an attractive program. This file allows

Trojan owners to gain access to data and applications, such as banking applications, view their balance, and transfer funds to another account.

In modern times, due to the development of computer technologies, protection against Trojan attacks has improved, but the number of types of Trojan viruses has also increased [2]. Let's look at the main Trojan virus programs (TVPs).

Bank TVPs are the most common categories. Their purpose is to capture financial data, such as usernames, passwords, credit card numbers, and authentication codes, in order to gain unauthorized access to bank accounts [3].

Backdoors-create a "back door" in the system, allowing an attacker to gain remote access to the device and control it, even if the antivirus software removes the main part of the Trojan.

TVP loaders are malicious utilities that secretly download and install dangerous software on the victim's device. They can be used by attackers to inject viruses, spyware modules, ransomware, and other Trojan components.

Spy TVPs are designed to secretly collect confidential data. They record keystrokes, track browser history, take screenshots of the screen, and can activate the camera and microphone without the user's knowledge.

Ransomware programs block access to files by encrypting them, and demand a ransom for recovery. In some cases, criminals additionally steal personal data, threatening to publish it if the victim refuses to pay.

DDoS attacks turn infected devices into part of a botnet used for large-scale attacks on servers and sites, causing their overload and denial of service.

Game TVPs specialize in stealing accounts, game currency, and valuable items, causing damage to users of online games.

TVP wipers damage or completely erase information on disks, which can lead to irreversible data loss. They can delete files, format hard drives, and render the system unusable.

Trojan attacks remain one of the most serious threats to the cybersecurity of users around the world [5]. There is no universal way to protect against Trojan attacks, but the development of machine learning greatly simplifies the fight against them, since it can perform tasks such as: regular scanning of the device's health, transmitting data only over secure channels, using two-factor or multi-factor authentication, checking links that the user clicks on, searching for traffic anomalies, tracking traffic errors. unauthorized access, monitoring the use of hard disk memory, and analyzing large amounts of data [6].

Regular checking of the device status helps to detect "hardware" and software failures in time.

Secure data transfer is only possible over encrypted connections. Modern encryption methods prevent attackers from stealing or spoofing your information while sending it.

Two-step verification (2FA) is a method that results in the system requesting additional confirmation, such as a code from an SMS or app, in addition to the password. In this case, the password will be obtained by an attacker, and the account will remain protected.

Checking links before clicking through is an easy way to avoid phishing and viruses. Just look at the site's address to see if you can trust it.

Traffic anomaly detection is the process of detecting deviations from the normal behavior of network traffic that may indicate potential threats or technical failures.

Unauthorized access tracking is the process of detecting signs of unauthorized access or attacks on an information system.

Big data analysis is the process of processing, researching, and interpreting large amounts of information to extract valuable knowledge and build hypotheses about patterns [7].

Thus, the objectives of the study are:

1. Systematization of knowledge about Trojan attacks;
2. Training machine learning models to detect Trojan viruses;
3. Plotting the accuracy of machine learning models and analyzing their characteristics.

2 Methodology

Machine Learning (ML) is the use of mathematical data models for forecasting. This method is a form of artificial intelligence [8–11]. Machine learning uses algorithms to identify patterns in data, and then creates a data model for forecasting. A dataset is used to work with machine learning. The dataset considered in this paper is designed to detect Trojan viruses on a PC. For more detailed and accurate work with the dataset, you need to perform an analysis: determine the data types, split the data into more important and less important ones for machine learning.

The dataset contains information about network streams labeled "Trojan "and" Bening". In total, it has 86 columns with various network traffic characteristics and 177482 rows.

The target column is the Class column, which contains two variables: Trojan and Bening.

Data can be divided into three importance groups: key columns, important features, and less important features.

Key columns for machine learning are: Flow ID (unique flow identifier), Source IP; Source Port; Destination IP; Destination Port (these 4 columns contain information about the sender), Protocol (protocol used, Timestamp (timestamp of the stream), Class (target variable).

Important attributes include: Flow Duration, Total FWD/Backward Packets, Packet length (max, min, average), Flow Bytes/s(data transfer rate), and Flow Packets/s (data transfer rate).

Less important features include other columns, as they do not contain very important information.

Dataset analysis is important for better configuring the model, preventing conflicts, reading data, and building a proper, working schema.

Before working with the dataset and conducting the research, the KNIME integrated data processing platform was used [12]. In addition, the KNIME platform is used to ensure data security and confidentiality, validate and monitor analytics and artificial intelligence models, and validate and explain results. The platform uses special pre-programmed Nodes that perform the main functions of data analysis, forecasting, visualization, and processing. In the course of the study, 10 machine learning models were studied, 5 of which were selected for the practical part. The studied MO models are shown in Fig. 2.

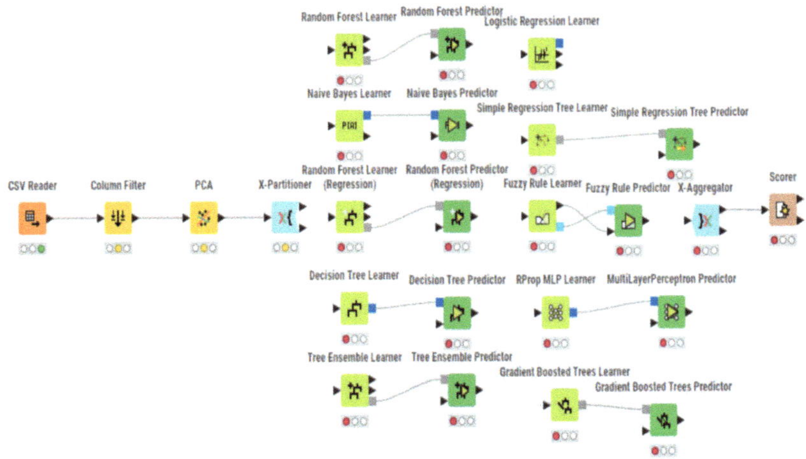

Fig. 2. Machine learning models.

A comparative table of machine learning models is presented in Table 1.

Table 1. Comparative characteristics of the used machine learning models.

Mom Models/Characteristics	Performance	Interpretability	Stability	Versatility
Naïve Bayes	+	+	−	−
Decision Tree	−	+	−	+
Random Forest	−	+	+	+
Tree Ensemble	−	+	+	+

The study presents a classification task, assigning a category: Trojan-a program attacked by a Trojan virus program, Bening-a program that is not affected by a Trojan.

Any machine learning model is created for different purposes and is used in different fields [13].

Machine learning models are shown in the Fig. 2.

A scheme is considered that allows you to load dataset data into the Decision Tree Learner machine learning model. We used a node that can read CSV Reader dataset data (section - IO) and select the path to the file. In its normal form, the data will not be read by the model, so you need to convert it. For this purpose, the following nodes were used: Column Filter, which allows you to filter the columns of the input table; PCA, which reduces the dimension of the original data volume; Number to String, which converts numbers in a column to strings. After the transformation, the datae is divided into two components: training data, which will be used to train the model, and test data, on which the model will be tested. The X-Partitioner node (the - Analytics section) was used for this purpose. This node is used in conjunction with X-Aggregator. When they are used together, you can use the cross-validation function, which prevents overfitting of the machine. After splitting, node a, the training model, and node a for prediction are connected. The Decision Tree Learner node was connected for training (the – Analytics section). For predictions, the Decision Tree Predictor node was connected (the – Analytics section), which allowed using models and test data to predict Trojan attacks. To summarize the training of the model, the Scorer node (section – Analytics) was used, which allows you to evaluate the accuracy of the predicted data and make an assessment of where errors were made when training the model.

3 Modeling

The Random Forest machine learning model is a universal algorithm that combines predictions from multiple decision trees to produce more accurate and reliable results. The algorithm creates many decision trees, each of which is trained independently of the others. The models used for solving classification and regression problems [14–16]. The results of training the Random Forest model are presented in Table 2.

Table 2. Results of the Random Forest model

Tree depth	1	2	3	4	5	6	7	8	9	10	11	12	13	14	15
Accuracy, %	51	62	67	68	70	71	73	75	76	77	78	78	79	79	80

The Random Forest (Regression) model is an ensemble machine learning algorithm that builds multiple decision trees and combines their results to produce a more accurate and stable prediction. This model is used to solve classification and regression problems [17, 18]. The results of training the Random Forest Regression model are presented in Table 3.

Table 3. Results of the Random Forest Regression model

Tree depth	1	2	3	4	5	6	7	8	9	10	11	12	13	14	15
Accuracy, %	1,6	0,03	0,06	0,001	0,001	0,2	1,7	8	35	50	64	82	86	89	94

The Tree Ensemble machine learning model combines predictions from multiple decision tree models to produce more accurate and stable results. This model helps to solve classification and regression problems [19, 20]. The results of training the Tree Ensemble model are presented in Table 4.

Table 4. Results of the Tree Ensemble model.

Tree depth	1	2	3	4	5	6	7	8	9	10	11	12	13	14	15
Accuracy, %	54	63	65	68	72	75	77	80	82	84	85	87	88	89	89

The Decision Tree model is a machine learning algorithm that builds a hierarchical structure for making decisions based on input data. This algorithm divides data by feature, choosing the best feature to divide the data into subgroups. The Decision Tree model allows solving classification and regression problems [21, 22]. The results of training the Decision Tree model are presented in Table 5.

Table 5. Results of the Decision Tree model.

Tree depth	1	2	3	4	5	6	7	8	9	10	11	12	13	14	15
Accuracy, %	71	71	72	71	71	72	72	72	72	72	72	72	72	72	72

Based on the studies of models that use the depth of the tree for forecasting accuracy, a graph of model accuracy is constructed, shown in Fig. 3.

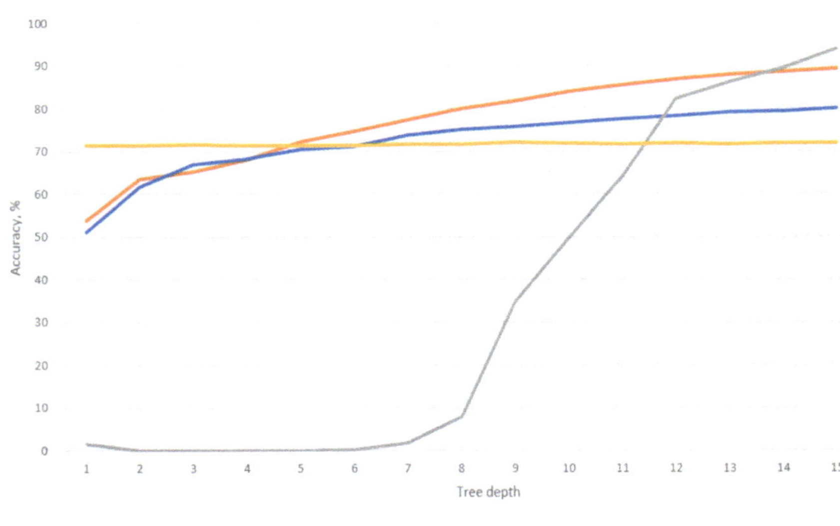

Fig. 3. General graph of the accuracy of machine learning models.

Thus, tree-based machine learning models: Random Forest Regression, Tree Ensemble, Random Forest, and Decision tree showed similar results. When changing the settings, both models produced better results as the depth of the tree increased, but at the same time, after reaching a certain level, the changes in the models became minimal. This means that a further increase in depth can lead to the fact that trees can begin to memorize data, therefore, to retraining (Table 6).

Table 6. The result of a study of Random Forest, Random Forest Regression, Tree Ensemble, and Decision Tree models.

Random Forest	Tree	Random Forest Regression	Tree	Tree Ensemble	Tree	Decision Tree	Tree
T. d.	A., %	T. d.	A., %	T. d.	A., %	T. d.	A., %
1	51	1	1,6	1	54	1	71
2	62	2	0,03	2	63	2	71
3	67	3	0,06	3	65	3	72
4	68	4	0,001	4	68	4	71
5	70	5	0,001	5	72	5	71
6	71	6	0,2	6	75	6	72
7	73	7	1,7	7	77	7	72
8	75	8	8	8	80	8	72
9	76	9	35	9	82	9	72
10	77	10	50	10	84	10	72
11	78	11	64	11	85	11	72
12	78	12	82	12	87	12	72
13	79	13	86	13	88	13	72
14	79	14	89	14	89	14	72
15	80	15	94	15	89	15	72

* T. d. – Tree depth
* A., % – Accuracy, %

The Naïve Bayes machine learning model is trained on a set of marked-up data, where features and class are known for each object, meaning that the machine considers that the presence of one feature does not affect the appearance of another. This model is used for classification problems, but can also be adapted for regression [23–25]. Unlike the models presented earlier, Naïve Bayes does not use tree depth for prediction accuracy. The results of training the Naïve Bayes model are presented in Table 7.

The graph of changes in model accuracy shown in Fig. 4.

Table 7. Model accuracy.

Threshold standard devotion	0	0,0001	0,0002	0,0003	0,0004	0,0005	0,0006
Accuracy, %	86,31	85,995	86,294	86,521	86,19	86,446	86,35

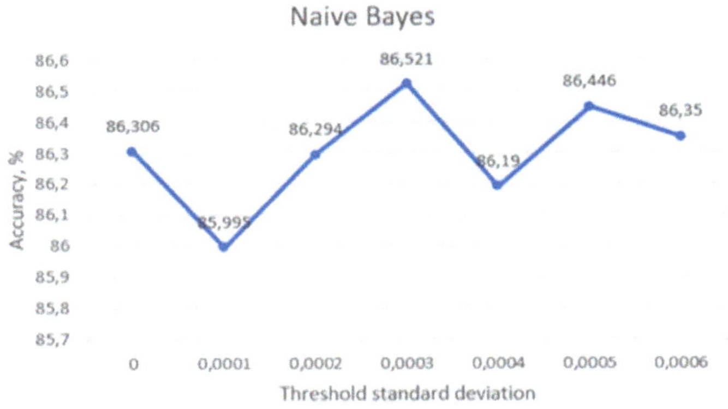

Fig. 4. Graph of changes in model accuracy.

4 Fighting Overfitting

During the study of machine learning models, overfitting of artificial intelligence was found, that is, the model remembered all the data and as a result gave 100% accuracy of testing. Overfitting is critical in machine learning, because when loading new data, the model will not be able to properly train and predict the result. In this study, retraining issues were solved using cross-validation. Cross-validation allows you to divide the input data into parts, one of which will be used for training, and the other for forecasting [26]. In KNIME, this function can be implemented using the X-partitioner and X-aggregator nodes. The X-partitioner node sets a column that the node will divide into equal parts and selects the number of cycles that the data will pass through before evaluating it. X-aggregator allows you to start the loop again, submitting a new piece of data for training. Using these nodes helped to overcome the problem of overfitting the machine learning model, which allowed us to create a more accurate and reliable study. Cross-validation modules shown in Fig. 5.

Fig. 5. Cross-validation modules.

5 Conclusion

In the course of the study, knowledge about Trojan virus programs was classified, as well as machine learning models that determine TVP with high accuracy were studied. Currently, there is an increase in crimes using information technology, and every year organizations are subjected to Trojan attacks, many of which are successfully completed. The study showed that machine learning is an effective tool for detecting Trojan virus programs, especially in the context of their growing complexity and diversity. Models such as Random Forest Regression and Tree Ensemble showed the highest accuracy in detecting TVPs within websites, and the use of cross-validation made it possible to cope with overfitting. The development of artificial intelligence opens up new opportunities for modernizing cybersecurity systems, making them more efficient and adaptive. The study showed the possibilities of machine learning for countering cyber threats, including for autonomous aircraft [27]. The use of machine learning is a promising area due to the fact that the use of artificial intelligence allows us to improve methods for detecting Trojan virus attacks, allowing us to find patterns in parameters and predict the occurrence of attacks.

References

1. Chetvertina A.O.: Trojan program. Eo ipso. No. 6, pp. 28–32 (2023)
2. Dzharakulov, T.Kh., Evstropov, V.A., Petrosyan, A.A.: Protection of corporate networks from internal attacks. A young scientist. No. 23, pp. 83–86 (2024)
3. Gavshin, D.A., Izovit, V.M., Rakhmatov, F.A.: Analysis of actual banking Trojans Cerberus and alien. In: Youth Scientific School of the Department "Secure Communication Systems", vol. 1, no. 1, pp. 14–19 (2020)
4. Mitryaev, I.S.: Trends in the use of ransomware cyberattacks. Innov. Sci. Educ. **47**, 985–991 (2021)
5. Shiriaev, E., Ermakova, T., Bezuglova, E., Lapina, M.A., Babenko, M.: Reliablity and security for fog computing systems. Information. **15**(6), 317 (2024). https://doi.org/10.3390/info15060317
6. Umarkhadjiev, M.Kh.R.: Development and application of machine learning methods in modern information technologies: analysis, trends and prospects. Trends in the development of science and education. No.104-14, pp. 134–136 (2023)
7. Lapina, M., Neelakandan, S., Donchenko, M., Ardeev, D., Boykov, N.: Comparative analysis of basic operations of Helib and Microsoft seal libraries for cloud computing. In: Nature-Inspired Optimization Algorithms for Cyber-Physical Systems, pp. 341–345 (2025) https://www.igi-global.com/chapter/comparative-analysis-of-basic-operations-of-helib-and-microsoft-seal-libraries-for-cloud-computing/364786
8. Mary Anita, E.A., Jenefa, J., Vinodha, D., Lapina, M.: Advancements in Sybil attack detection: a comprehensive survey of machine learning-based approaches in wireless sensor networks. In: Lapina, M., Raza, Z., Tchernykh, A., Sajid, M., Zolotarev, V., Babenko, M. (eds.) AISMA-2024: International Workshop on Advanced Information Security Management and Applications. AISMA 2024 Lecture Notes in Networks and Systems, vol. 863. Springer, Cham (2024). https://doi.org/10.1007/978-3-031-72171-7_7. (scopus indexed)
9. Vinodha, D., et al.: Framework based on IoT, AI, and Blockchain for smart access to government agricultural schemes. In: The Future of Agriculture: IoT, AI and Blockchain Technology for Sustainable Farming, vol. 1, p. 142 (2024). https://doi.org/10.2174/9789815274349124010011

10. Bansal, A., Athavale, V.A., Saluja, K., Gupta, S., Kukreja, V.: Computational system based on machine learning with hybrid security technique to classify crime offenses. In: Jain, S., Marriwala, N., Tripathi, C.C., Kumar, D. (eds.) Emergent Converging Technologies and Biomedical Systems. ETBS 2022 Lecture Notes in Electrical Engineering, vol. 1040. Springer, Singapore (2023). https://doi.org/10.1007/978-981-99-2271-0_20
11. Kumar, S., Athavale, V.A., Kartikey, D.: Security issues in cloud computing: a holistic view. Int. J. Internet Things Web Serv. **6**, 18–29 (2021) https://www.iaras.org/home/caijitws/security-issues-in-cloud-computing-a-holistic-view
12. Sardarova, M.D.: Review of the KNIME analytical platform. Youth Scientific and Technical Bulletin, No. 6, p. 15 (2015)
13. Prakasha, G.S., Lapina, M., Balakrishnan, D., Sajid, M. (eds.): Educational perspectives on digital Technologies in Modeling and Management. IGI Global (2024). https://doi.org/10.4018/979-8-3693-2314-4
14. Mi, Ch., Hatmann, F., Guo, Y., Han, S., Wen, L.: Random forest: an efficient species distribution model for identifying reliable crane distributions using a limited number of data. In the Collection: Cranes of Eurasia: Biology, Protection, Management, Proceedings of the IV International Scientific Conference. pp. 264–265 (2015)
15. Fomina, E.E., Koshkina, G.V., Smirnova, M.A.: Random Forest algorithm as a tool for analyzing socio-economic information. In the collection: Information technologies in modeling and management: approaches, methods, solutions. V All-Russian scientific conference with international participation: collection of materials. pp. 461–467 (2022)
16. Markevich, E.A.: Model of the business process of working with the random forest algorithm. In the collection: Actual problems of aviation and cosmonautics. pp. 205–207 (2021)
17. J.S., V.S.: A novel analysis of RandoM Forest regression model for wind speed forecasting. Congent Eng. 11(1) (2024)
18. Domenichini, G., Dellago, Ch.: Molecular hessian matrices from a machine learning Random Forest regression algorithm. J. Chem. Phys. 159(19) (2023)
19. Istigomah, N.A., Jung, D., Khim, J.: Tree-based ensemble machine learning model for nitrate reduction by zero-valent iron. J Water Process Eng. **56**, 104303 (2023)
20. Omer, Z.M., Shareef, H.: Comparison of decision tree based ensemble methods for prediction of photovoltaic maximum current. Energy Convers. Manage. X. **16**, 100333 (2022)
21. Mitrofanov, S.A., Semenkin, E.S.: Tree retraining in the decision tree learning algorithm. In: В сборнике: IOP Conference Series: Materials Science and Engineering. Krasnoyarsk Science and Technology City Hall, p. 12082 (2021)
22. Pan, H.: The construction and application of library intelligent acquisition decision model based on decision tree algorithm. Int. J. Adv. Comput. Sci. Appl. **15**(1) (2024)
23. Alexandrova, M.Y.: Methods of machine learning in sociological research: prediction of partial non-response using a naive Bayesian classifier. Monitoring public opinion: economic and social changes, No. 1, pp. 329–350 (2021)
24. Svyatov, K.V., Moshkin, V.S., Shchukareva, I.A.: Information system module based on a naive Bayesian classifier for automating pharmacy operations. Bull. Cybernet. **23**(2), 62–70 (2024)
25. Pershin, A.A.: Classification of news headlines by a naive Bayesian classifier. Actual problems of aviation and cosmonautics. Collection of materials of the X International Scientific and Practical Conference dedicated to the 100th anniversary of Academician M. F. Reshetnev and Cosmonautics Day., pp. 215–217 (2024)

26. Krakovsky, Y.M., Luzgin, A.N.: Investigation of algorithms for optimizing the dimension of training vectors of probabilistic interval forecasting models based on cross-validation methods. Baikal Bull. DAAD. **1**, 23–33 (2017)
27. Basan, E., et al.: UAV security analysis framework. In: Kumar, S., Balachandran, K., Kim, J.H., Bansal, J.C. (eds.) Fourth Congress on Intelligent Systems. CIS 2023 Lecture Notes in Networks and Systems, vol. 865, pp. 379–391. Springer, Singapore (2024). https://doi.org/10.1007/978-981-99-9043-6_31

Development of an Algorithm for Finding Errors in Modular Codes

Artem B. Lidzhiev[1(✉)], Victor S. Peleshenko[1], Tatyana A. Peleshenko[1], Alexander S. Gish[1], and Vijay Anant Athavale[2]

[1] North Caucasus Federal University, Stavropol, Russia
artemlidzhiev.4444@gmail.com

[2] Department of Computer Science and Engineering, Walchand Institute of Technology, Solapur, India

Abstract. The purpose of the scientific work is to develop a software implementation, to study the algorithm for finding errors in modular codes (MC) and their detection by introducing two redundant moduli. The study focuses on the characteristics of errors that occur in the codes of the residue number system (RNS), without the direct purpose of detecting them. To conduct the study, an approach based on the calculation of all entered allowed combinations of the code of the RNS was taken as a basis, followed by obtaining intervals without taking into account rank. To achieve this goal, a program code in the C++ language was developed, which includes a menu with functions for searching for a single error in modular codes. The presented approach and software implementation can be useful to specialists in the field of information systems of transmission and signal processing in the analysis and correction of errors in modular codes.

Keywords: residue number system · modular code · allowed combination · single error · orthogonal basis

1 Introduction

This article focuses on the analysis of error distribution in modular codes (MC). To understand the work of algorithms for searching and correcting errors in MC, it is necessary to familiarize yourself with the principles of constructing redundant modular codes. A characteristic feature of modular codes is the lack of exchange of intermediate results between computing channels that implement calculations using the modules of the RNS code. This property of modular codes can be used to increase the fault tolerance of computing devices operating in the MC [25, 26]. Obviously, an error that occurs modulo p_i, $i = 1, 2, \ldots, k$, will not be able to move from one computing path to another, since there is no connection between them. Will not be able to move from one computing path to another, since there is no connection between them. In this case, the error can be detected and corrected during the reverse conversion from the modular code to the positional number system. To develop codes capable of detecting and correcting errors, it is necessary to increase the redundancy of such code. Let us define an redundant MC

M. Lapina et al. (Eds.): AISMA 2025, LNNS 1456, pp. 298–309, 2026.
https://doi.org/10.1007/978-3-032-07275-7_28

that contains k mutually simple bases $p_1, p_2, \ldots, p_k$. For this set of bases, the operating range will be.

$$P_{operating} = p_1 \bullet p_2 \bullet \cdots \bullet p_k \tag{1}$$

To unambiguously represent the binary code of the number A in the modular code, it is necessary that the condition be fulfilled:

$$A = (\alpha_1, \alpha_2, \ldots \alpha_k) < P_{operating}, \tag{2}$$

where a_i – the smallest non–negative deduction modulo p_i from the integer value A: $\alpha_i = A(\text{mod } p_i)$.

This means that each set of residues $(\alpha_1, \alpha_2, \ldots, \alpha_k)$ is assigned only one number A, which belongs to the working range. Let's use this property to construct corrective modular codes.

Let's introduce redundancy into the modular code. Since two control bases are introduced in the RNS codes to correct single errors, we will take redundant bases p_{k+1} and p_{k+2}, which satisfy the conditions:

$$GSD(p_i, p_{k+1}) = 1, \tag{3}$$

$$p_1 < p_2 < \ldots < p_k < p_{k+1} < p_{k+2}, \tag{4}$$

where i, j = 1, 2, … k + 2; $i \neq j$

As a result of the introduction of two redundant bases, the range of the modular code will expand. Then the product of the moduli of the redundant code will give the full range, which is equal to:

$$P_{full} = p_1 \cdot p_2 \cdot \cdots \cdot p_{k+1} \cdot p_{k+2} = P_{operating} \cdot p_{k+1} \cdot p_{k+2} \tag{5}$$

Thus, the range of the modular code has increased by $p_{k+1} \cdot p_{k+2}$ times compared to the operating range. As a result, the total number of possible code combinations of such a redundant code $N_0 = P_{full}$.

The allowed combinations in the redundant RNS code must satisfy the condition:

$$A = (\alpha_1, \alpha_2, \ldots \alpha_{k+2}) < P_{operating} \tag{6}$$

Let's consider a single error in the RNS code, which is understood as a distortion of one remainder in a code combination. Let the error occur on the first basis. Its depth can be equal to$\Delta\alpha_1 = \{1, 2, \ldots, p_1 - 1\}$. Then the erroneous remainder will be determined by:

$$\alpha_1^* = (\alpha_1 + \Delta\alpha_1)\text{mod}p_1 \tag{7}$$

Then the erroneous combination will look like:

$$A^* = \left(\alpha_1^*, \alpha_2, \ldots \alpha_{k+2}\right) = ((\alpha_1 + \Delta\alpha_1)\text{mod}p_1, \alpha_2, \ldots \alpha_{k+2}) \tag{8}$$

Let's use the Chinese remainder theorem to perform the inverse transformation from the RNS code to the positional code:

$$A^* = \left(\alpha_1^* B_1 + \alpha_2 B_2 + \cdots + \alpha_{k+2} B_{k+2}\right)\text{mod}P_{full} = (A + \Delta\alpha_1 B_1)\text{mod}P_{full}, \tag{9}$$

where B_i are the orthogonal bases of the complete RNS.

For one redundant module p_{n+1}, the correctness of the number A is by definition means that $A < P/p_{n+1}$, but since $P/p_i \geq P/p_{n+1}$ and in turn i = 1, 2, .., n + 1, it can be concluded that $A < P/p_i$ (7).

Since a_i^* is not equal to a_i, the number A cannot lie in the range [0, P/pi), which means $A^* > P/p_i$. It follows that $A^* > P/p_{n+1}$, that is, A^* becomes an incorrect number. This establishes an important fact that determines the possibility of creating detection and correction codes in the RNS: any distortion of a digit in one of the digits turns the number into an incorrect one, which makes it possible to detect the presence of an error. Moreover, there is only one unique value of this digit that can return the wrong number to the correct state. Algorithms for searching and correcting errors in the MC are based on this property.

There are two main methods of error detection: the syndrome method and the projection method. The syndrome method is based on the calculation of the syndrome, which allows you to determine the presence of an error in the code. However, its disadvantage is that it does not always allow you to specify exactly where the error occurred. The projection method uses additional parameters to detect errors, but may be less effective in conditions of limited resources. This article focuses on the use of modular codes (MC), which have unique properties that make it possible to effectively detect and correct errors.

The method proposed in the article is a compromise between the two main methods, trying to adopt the projection method's ability to detect errors, while avoiding the problem of limited resources by neglecting rank in the calculation process. Next, a mathematical justification of this method will be presented, demonstrating the modified formula and a graph reflecting the difference when comparing the modified method with the standard method, which takes into account the rank:

$$G = \left| \sum_{i=1}^{5} \alpha_i S_i \right|^{+}_{P_{reference}} - r \bullet P_{operating}, \tag{10}$$

where r is the rank and $P_{reference}$ is the reference range.

The rank of r can be calculated by performing an integer division of the number A^* by $P_{operating}$. In turn, the time spent on execution can be represented as the following formula:

$$T = T_{Pi} + T_{\delta i} + T_{gcd} + T_{Bi} + T_{Bi*} + T_{Si} + T_r + T_G, \tag{11}$$

where T_{Pi} is the calculation time P_i, $T_{\delta i}$ is the time to determine the remainder δ_i, T_{gcd} is the running time of the Euclidean algorithm for finding the weight B_i, T_{Bi} is the calculation time of the orthogonal basis B_i, TB_i^* is the calculation time of the auxiliary residues B_i^*, T_{Si} is the calculation time of the values S_i, T_r is the time for the rank r, T_G is time for the final calculation of the G interval.

The mathematical rationale is that we exclude the rank calculation stage. Below is Fig. 1 with the schedule of the standard algorithm and the algorithm without taking into account the rank. The time was recorded using the C++ libraries (cstdlib, ctime) and the functions LARGE_INTEGER, Frequency LARGE_INTEGER, startTime, EndTime and ElapsedMicroseconds. With 5–10 modules, the running time of the classic algorithm exceeds the running time of the new algorithm by 20–100%, and then this value only increases.

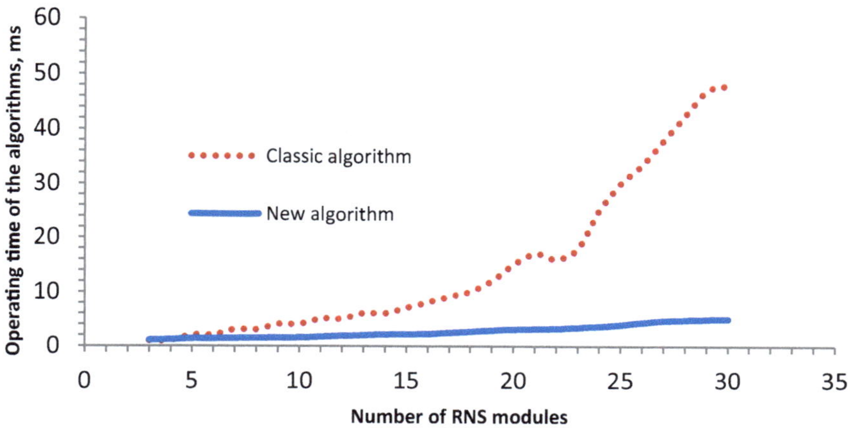

Fig. 1. A figure caption is always placed below the illustration. Short captions are centered, while long ones are justified. The macro button chooses the correct format automatically.

2 Introduction

To perform the above task, a program code was developed in C++, which includes a menu with functions for performing this and similar tasks with different values to search for a single error in the MC.

To prove the effectiveness of the developed program code, we will compare its calculations with the results obtained by manual calculations using the example of the allowed combination A_{29}. 2, 3, and 5 were used as information bases, and 7 and 11 were used as verification bases.

First, we will calculate the working (1), full (5) ranges.

$$P_{operating} = p_1 \bullet p_2 \bullet p_3 = 2 \bullet 3 \bullet 6 = 30$$

$$P_{full} = P_{operating} \cdot p_4 \cdot p_5 = 30 \bullet 7 \bullet 11 = 2310$$

We will also calculate the reference range (12):

$$P_{reference} = \frac{P_{full}}{P_{operating}}. \tag{12}$$

From these formulas, we can conclude that P $_{operatimg}$ = 30, P $_{full}$ = 2310, a P $_{reference}$ = 77. When selecting option № 1 "Enter MC code bases" in the menu, the program also automatically provides range values (Fig. 2).

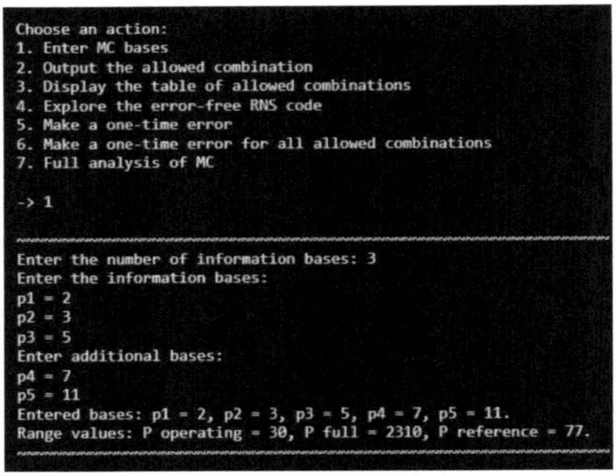

Fig. 2. How the code works when calculating the working, full, and control range

Next, the algorithm calculates the interval of the allowed combination A 29 = (1, 2, 4, 1, 7) without regard to rank of R. It is worth explaining that the rank of a number in a RNS is an indicator that reflects the number of complete cycles of the system range completed when converting a number from a RNS to its positional representation. In the course of the study, the rank is excluded from the formula of the interval G.

To do this, each of the orthogonal bases is calculated in the following steps:

Stage 1. Calculation of $P_i = P_{full.}/p_i$

Stage 2. Determination of the remainder modulo p_i. It turns out: $\delta_i = P_i \bmod p_i$, where δ_i is the auxiliary remainder.

Stage 3. The value of the entire B_i is from the value of $\delta_i\, m_i \equiv 1 \bmod p_i$, if $m_i = 1$, then in this case the computational algorithm of Euclid is used to obtain B_i, which calls find so that m_i is: $\delta_i\, m_i \bmod p_i = 1$.

Stage 4. Calculating the orthogonal basis $B_i = m_i\, P_i$

Following this algorithm, the following values of orthogonal bases were obtained: $B_1 = 1155$, $B_2 = 1540$, $B_3 = 1386$, $B_4 = 330$, $B_5 = 210$.

From the formula below, S_i is expressed:

$$B_i = S_i P_{operating} + B_i^*, \tag{13}$$

where B_i^* is an autonomous database of the operating system: $B_i^* = B_i \bmod P_{operating}$

Expression (10) is a standard formula for calculating the interval. However, the value of rank r will not be taken into account further. The interval is calculated without taking into account the rank:

$$G = \left|\sum_{i=1}^{5} \alpha_i S_i\right|^{+}_{P_{reference}} = |1 \bullet 38 + 2 \bullet 51 + 4 \bullet 46 + 1 \bullet 11 + 7 \bullet 7|^{+}_{77} = |384|^{+}_{77} = 76$$

The program code demonstrates similar results by following the algorithm described above and automatically compiling a table when using function № 2 "Output the allowed combination". Together with the table for one allowed combination, the orthogonal bases and the interval itself are also displayed (Fig. 3).

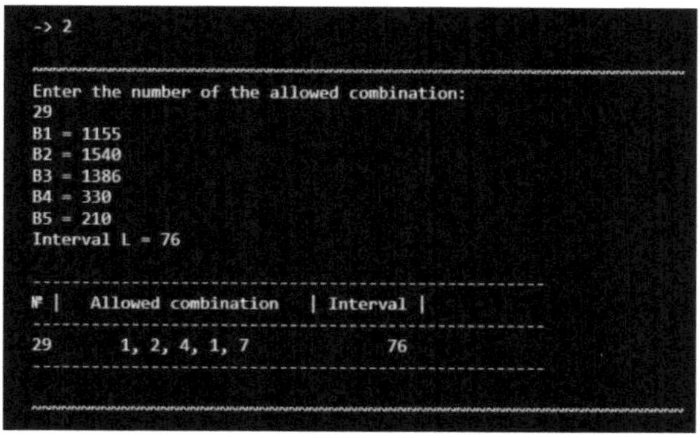

Fig. 3. How the code works when calculating the allowed combination interval

After that, you can use the function №3 "Display the table of allowed combinations", getting the interval values not only for A29, but also for all other combinations (Fig. 4).

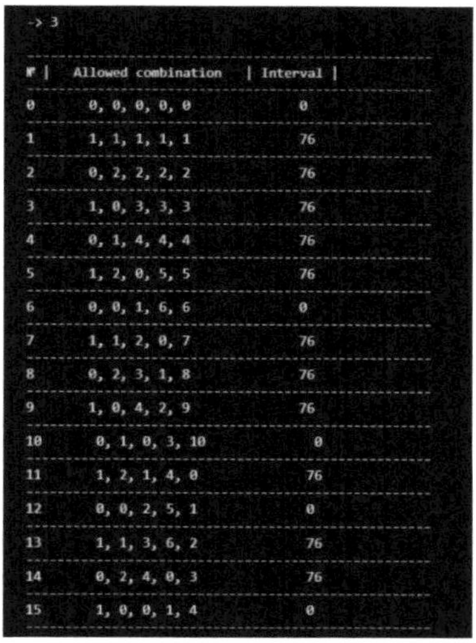

Fig. 4. How the code works when calculating the intervals of all allowed combinations

All data obtained when the program is running with error-free RNS code is entered in Table 1:

Table 1. Intervals of allowed combinations of error-free RNS.

№	Allowed combination, A_i	Interval, L
0	(0, 0, 0, 0, 0)	0
1	(1, 1, 1, 1, 1)	76
…	…	…
29	(1, 2, 4, 1, 7)	76

Calculating all the intervals of allowed combinations in a RNS requires a certain amount of time, which depends on the number of allowed combinations. At best, with manual calculations for the range of allowed combinations from 0 to 29, the time required to perform calculations is approximately 10–15 min. This time can increase significantly as the number of allowed combinations increases, as the complexity of calculations increases.

According to the data presented in Table 1 of this article, the intervals of allowed combinations of error-free RNS take only two values: 0 and 76. Thus, in this particular case, the intervals do not have different errors and do not have many different values.

There are similar interval output functions for allowed combinations in the remainder with an arbitrary choice of the erroneous remainder and error depth. This is function №5 "Make a one-time error" and №6 "Make a one-time error for all allowed combinations". Let's check the operation of function №5 using the example of $A_{29} = (1, 2, 4, 1, 7)$ with a distortion of the third base and an error depth equal to $\Delta\alpha_3 = 1$.

Then the erroneous combination will look like $A_3^{\Delta=1} = (1, 2, 0, 1, 7)$.

Calculating the interval of this number without taking into account the rank:

$$G = \left|\sum_{i=1}^{5} \alpha_i S_i\right|^{+}_{P_{\text{reference}}} = |1 \bullet 38 + 2 \bullet 51 + 0 \bullet 46 + 1 \bullet 11 + 7 \bullet 7|^{+}_{77} = |200|^{+}_{77} = 46$$

3 Results

The program code produces similar results (Fig. 5):

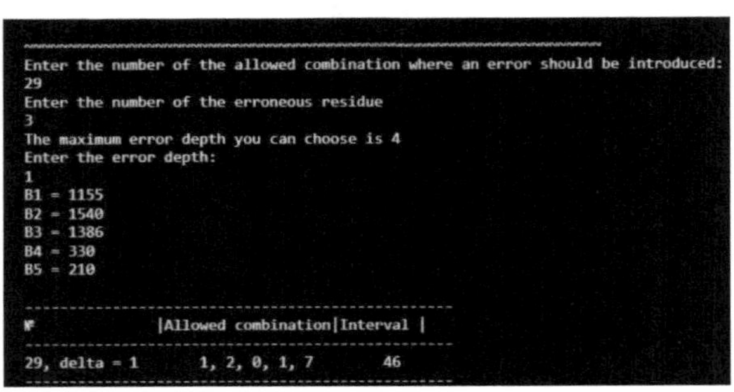

Fig. 5. How the code works when calculating the intervals of all allowed combinations

When using function №6 to output all combinations with a single error, the selected error depth $\Delta\alpha_1 = \{1, 2, \ldots, p_i - 1\}$ is also output (Fig. 6).

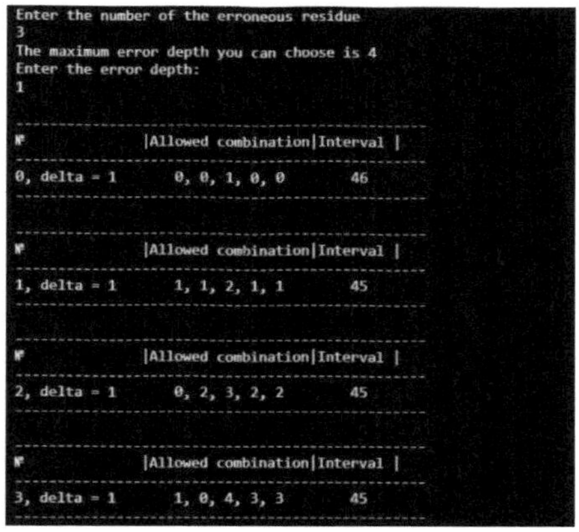

Fig. 6. How the code works when calculating the intervals of all allowed combinations with an error

All data obtained when the program is running with the RNS code containing a single error in the third remainder is entered in Table 2:

Table 2. Intervals of allowed combinations of error-free RNS.

№	Allowed combination, A_i	Interval, L
0	(0, 0, 1, 0, 0)	46
1	(1, 1, 2, 1, 1)	45
…	…	…
29	(1, 2, 0, 1, 7)	46

A full MC analysis is now performed using function №7. The result of this function will be the construction of a table for each error depth of all bases, indicating their inherent intervals (Fig. 7).

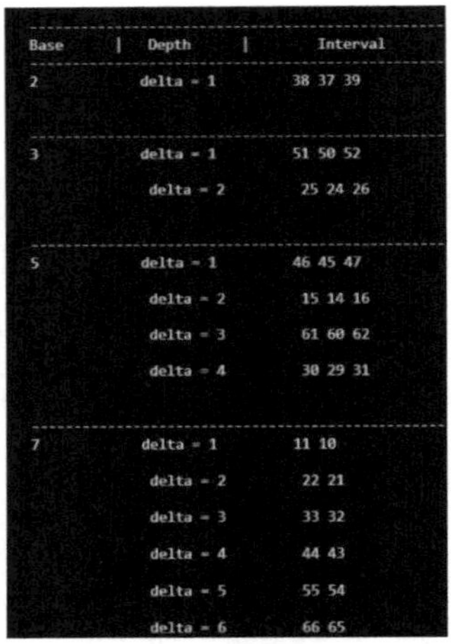

Fig. 7. Code operation during a full MC analysis

All the data obtained during the full MC analysis using the program with the RNS code is included in Table 3:

Table 3. Complete MC analysis.

Base	Depth	Interval
2	$\Delta\alpha_1$=1	38, 37, 39
3	$\Delta\alpha_1$=1, $\Delta\alpha_2$ =2	51, 50, 52, 25, 24, 26
5	$\Delta\alpha_1$=1, $\Delta\alpha_2$ =2, $\Delta\alpha_3$ =3, $\Delta\alpha_4$=4	46, 45, 47, 15, 14, 16, 61, 60, 62, 30, 29, 31
7	$\Delta\alpha_1$=1, $\Delta\alpha_2$ =2, $\Delta\alpha_3$ =3, $\Delta\alpha_4$=4, $\Delta\alpha_5$ =5, $\Delta\alpha_6$ =6	11, 10, 22, 21, 33, 32, 44, 43, 55, 54, 66, 65
11	$\Delta\alpha_1$=1, $\Delta\alpha_2$ =2, $\Delta\alpha_3$ =3, $\Delta\alpha_4$=4, $\Delta\alpha_5$ =5, $\Delta\alpha_6$ =6, $\Delta\alpha_7$=7, $\Delta\alpha_8$ =8, $\Delta\alpha_9$ =9, $\Delta\alpha_{10}$=10	7, 6, 14, 13, 21, 20, 28, 27, 35, 34, 42, 41, 49, 48, 56, 55, 63, 62, 70, 69

As a result of a complete analysis of the MC using function №7, it was found that the numbers with errors fall into intervals other than the working ones. This is confirmed by the data presented in Table 3, where the corresponding intervals are indicated for each depth of error of all bases.

As can be seen from the table, each base with different error depths generates intervals that do not correspond to the operating values. For example, for base 2 with an error depth of $\alpha_1 = 1$, the resulting intervals are 38, 37, and 39, which clearly exceed the operating values. Similarly, for other bases and error depths, it is observed that the interval contains values other than expected. The data from the table allows you to analyze the error distribution in the MC as accurately as possible.

Thus, the results of the analysis confirm that errors in calculations lead to the formation of intervals that do not correspond to the working numbers, which underlines the importance of error control and correction in the process of working with the residue number system.

However, it is worth noting that the advantage of the method presented in the article lies in its ability not only to detect the presence of an error based on the values of the intervals, but also to correct them using data on the depth of the error. In addition, rank-free calculations can reduce processing time, which helps to overcome the disadvantages of the original algorithm.

4 Conclusion

In the course of the research, a software code was developed and tested to find and correct errors in modular codes that does not take rank into account. As a result of experiments, it was found that the algorithm without taking into account the rank was not accurate enough, which limits its applicability in a number of practical tasks. Nevertheless, its importance for the development of the theory and practice of information systems of transmission and signal processing remains high. The algorithm has demonstrated the potential for further improvement, opening up prospects for integrating additional metrics such as rank accounting, which will improve the accuracy and efficiency of error detection in modular codes. The results of the study also show how errors affect the distribution of values in modular codes, and that this may be useful for further study and development of correction methods. Further research may focus on developing new analysis methods that take into account the influence of various factors, including rank characteristics, which will significantly improve the results and increase the reliability of solutions in the field of signal processing and cryptography.

References

1. Vasetsky, V.V., Vasetskaya, V.V.: Application of the residue number system in automated control systems. In: Informatics: Problems, Methodology, Technologies, pp. 253–257 (2015)
2. Kalmykova, N.I.: Software Implementation of error detection in socket codes using the Java programming language Scientific Paradigm, pp. 13–16 (2020)
3. Chernomazov, S.A., Martirosyan, A.G., Gapochkin, A.V., Kalmykov, M.I.: Development of a Device for Detecting and Correcting Errors Based on the Algorithm of Expanding the Modular Code Basis System, pp. 41–46. Modern High-Tech Technologies (2014)
4. Gordenko, D.V., Pavlyuk, D.N., Shaposhnikov, E.V., Kondrashov, A.V., Gorbachev, A.V.: Error Detection with Self-Control in Modular Arithmetic Bulletin of SevKavGTI, pp. 202–207 (2015)

5. Gapochkin, A.V.: Application of redundant modular codes of the residue number system for error search and correction. In: Digital Signal Processing and its Applications. DSPA-2020: Proceedings of the 22nd International Conference, pp. 190–195 (2020)
6. Chistousov, N.K., Chipiga, A.F., Kalmykova, N.I.: Development of an authentication method implemented in modular codes of the residue number system. In: Modern Scientific Research: Collection of Scientific Papers from the XXVII International Scientific-Practical Conference, pp. 123–129 (2020)
7. Akushskiy, I.Y., Yuditsky, D.I.: Machine arithmetic in residual classes (1968)
8. Chervyakov, N.I., Kalmykov, I.A., Deryabin, M.A., Nazarov, A.S., Lavrinenko, A.V.: Control program for translating numbers from the residual class system into a positional number system based on the Chinese remainder theorem: certificate of state registration of a computer program (2016)
9. Guillermin N.: A coprocessor for secure and high speed modular arithmetic. DGA Information Superiority, Bruz, France; IRMAR, Université Rennes 1, France
10. Singh, N.: An overview of residue number system. In: National Seminar on Devices, Circuits & Communication, p. 132 (2008)
11. Gaponkin, A.V.: Application of redundant modular codes of the residue class system for error detection and correction. In: Digital Signal Processing and its Applications. DSPA–2020: Proceedings of the 22nd International Conference, pp. 190–195 (2020)
12. Zhu, D., Natarajan, B.: Residue number system arithmetic assisted coded frequency-hopped OFDMA. EURASIP J. Wirel. Commun. Netw., Article number: 263695 (2009)
13. Patronik, P., Piestrak, S.J.: Design of reverse converters for the general RNS 3-moduli set $\{2k, 2n - 1, 2n + 1\}$. EURASIP J. Adv. Signal Process., Article number: 92 (2023)
14. Yao, Y., Hu, J., Ma, S.: A PAPR reduction scheme with residue number system for OFDM. EURASIP J. Wirel. Commun. Netw., 156 (2013)
15. Vardhan, K., Santhoshini, K., Musala, S., Srinivasulu, A.: A critical look at modular adders using residue number system. Solid State Electron. Lett. **1**(2), 84–91 (2016)
16. Kong, Y., Asif, S., Khan, M.A.U.: Modular multiplication using the core function in the residue number system. AAECC Appl. Algebra Eng., Commun. Comput. **27**, 1–16 (2016)
17. Babenko, M., et al.: Multiple error correction in redundant residue number systems: a modified modular projection method with maximum likelihood decoding. Appl. Sci. **12**(1), 463 (2022)
18. Kalmykov, I.A., Pashintsev, V.P., Talyatovich, T., Olenev, A.A., Chistousov, N.K.: Error-correction coding using polynomial residue number system. Appl. Sci. **12**(7), 3365 (2022)
19. Stepanova, E.P., et al.: Application of the Codes of a Polynomial Residue Number System, Aimed at Reducing the Effects of Failures in the AES Cipher. (2016)
20. Hosseinzadeh, M., Navi, K.: A new moduli set for residue number system in ternary valued logic. J. Appl. Sci. **7**(23), 3729–3735 (2007)
21. Sarada, V.: An enhanced residue modular multiplier for cryptography. Int. J. Sci. Res. (2013)
22. Khankishiyeva, H.I.D.: A brief summary of the full Rns variant of some homomorphic encryptions. J. Mod. Technol. Eng. **8**(2), 112–118 (2023)
23. Isupov, K.S.: An overview of high-performance computing using the residue number system. Program Syst.: Theor. Appl. **12**(2), 137–192 (2021)
24. Seyed, M.M., Amir, S.M., Mehdi, H.: (Authors' affiliations: Islamic Azad University, Iran University of Medical Sciences). (2017)
25. Lapina, M., Neelakandan, S., Donchenko, M., Ardeev, D., Boykov, N.: Comparative analysis of basic operations of helib and microsoft seal libraries for cloud computing. In: Nature-Inspired Optimization Algorithms for Cyber-Physical Systems, pp. 341–345 (2025)
26. Shadab, M., et al.: Generating highly nonlinear s-boxes using a hybrid approach with particle swarm optimization. In: Nature-Inspired Optimization Algorithms for Cyber-Physical Systems, pp. 1–30 (2025)

27. Athavale, V.A., Chincheti, A., Dixit, R.K., Lobo, L.M.R.J., Lapina, M.: A Safe and Secure Online System for Bidding Using Blockchain Technology Lecture Notes in Networks and Systems (2024). https://doi.org/10.1007/978-3-031-72171-7_5

The Concept of a Methodology for Calculating the Deviation of an Employee's Psychological Profile from the Actual Average Norm of Tolerant Ethics of Behavior for Information Protection in Small Businesses

Igor Mandritsa[1,2(✉)], Vyacheslav Petrenko[1], Igor Makarov[1], Maksim Sokolovskii[3], and Olga Mandritsa[2]

[1] Faculty of Mathematics and Computer Sciences (named after Prof. Nikolay Chervyakov), North-Caucasus Federal University, Stavropol, Russia
d_artman@mail.ru

[2] Department of Regional Economics, Russian Technological University - MIREA, Branch office, 8, Kulakov Street, Stavropol, Russia

[3] Faculty of Psychology and Pedagogy, North-Caucasus Federal University, Stavropol, Russia

Abstract. Tolerant behavior of an employee for the purposes of the administration in the field of information security must comply with the adopted information security policy. Leakage of trade secrets and other valuable information should not be allowed. However, the number of leaks is growing, and companies (as well as government organizations) are forced to reconsider the techniques and methods of countering information leaks. Among the standard methods of information protection, a firewall (ViPNet TIAS – Threat Intelligence Analytics System) occupies a special place for automatic detection of information security incidents in the event of increased anomalous activity of individual employees. The authors of this article see one of the solutions to combat leaks in the form of mathematical prediction of the level of deviance in the behavior of each employee in the form of a change in the psycho-emotional state, which is a dynamic category and changes depending on the factors of the external and internal environment of the employee, both inside and outside the organization. A stable "state" is replaced by a problematic "unstable" state, and, most likely, leads to information threats to the organization, in the form of information leaks, or irreversible information threats. This digital "footprint" fills the deviation detector model with metadata for the mathematical model of the organization's information security, where exceeding the level of the calculated probability directly determines the "access" (or denial of access) of this employee to the business process and commercial information of the organization.

Keywords: deviant state of the employee · employee reflection · collection and processing of metadata about the employee · neural networks · information security from an insider

M. Lapina et al. (Eds.): AISMA 2025, LNNS 1456, pp. 310–322, 2026.
https://doi.org/10.1007/978-3-032-07275-7_29

1 Introduction

1.1 Setting a Research Goal

This article attempts to "measure" the dangerous psycho-emotional states of the company's employees who have access to trade secrets. Where the category of "dangerous" conditions of employees is considered precisely from the standpoint of information security and possible economic damage of its leakage. According to current information from the world's well-known news agencies Kroll, PricewaterhouseCoopers and FireEye (Yilmaz, E., & Can, O., 2024)., the amount of damage from leaks of commercial (including confidential) information today is over 11 billion dollars.

The peculiarity of information protection for small commercial companies is the small composition of the company's team, from 7 to 25 people, and not always a small company has an information protection department, so it is important to introduce such methods for determining employee trustworthiness and integrity that would automatically extract information about suspicious deviant behavior using modern neural network technologies. the employee's behavior and the mechanism for protecting commercial information from leakage would be activated in time.

The HEXACO-100 test and the UEBA approaches are two different strands in the study of human behavior. The first focuses on the psychological characteristics of the personality and is designed to provide a deep understanding of a person's inner qualities. The second, on the contrary, focuses on the external manifestations of behavior in the digital environment and is used to ensure security and data protection. Despite the differences, both methods complement each other depending on the context of their application.

The authors of the study proceed from the accepted axiom of the psycho-emotional state of the employee, that in order to commit a criminal act, it is important to have a trigger for the action, or a certain motive for the act as a result of a "shift" in the employee's state from a state of tolerance to a state of divergence. At the same time, the authors believe that it is enough to introduce into the company's information security policy a rule for monitoring the psycho-emotional state of each employee who has access to the organization's trade secrets from the standpoint of detecting facts (evidence) of the impact on the employee's psycho-emotional state of external (outside the company's environment) and internal (inside the company's office) factors of his being (life), which is recorded in the authors' previous scientific studies (Shannon, Claude Elwood, 1948). As a result of the discovery of the number of clues (motives that change the behavior trigger) of an employee from a state of tolerance to divergence, the authors consider this approach possible to use a method for solving the problem of detecting a potential violator (employee) within the framework of information security. Let us explain this in Fig. 1.

An important point of the technique is the process of transformation of the employee from stage III (Fig. 1) of the tolerant type of the worker, to the IV phase of the psycho-emotional state - the divergent type of the worker, but being in a state of deviant type - because the employee knows that under the influence of factors "irritating" his stable state, he will have to commit a criminal act. According to the authors of the article and

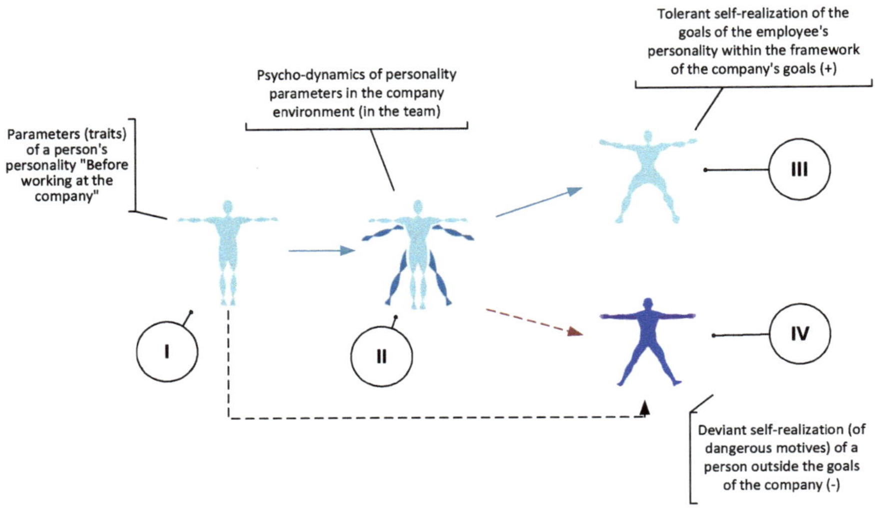

Fig. 1. Stages of transformation of the psycho-emotional state of an employee of the company under the influence of factors of social existence on him.

other authors (Konow, 2000) (Zimbardo, 2011), it is the dangerous psycho-emotional state of an employee of the "Deviant" type that leads to the "Divergent" state.

That is, deviance in behavior is a potential (trigger) to the stage of divergence (Merrill Warkentin & Robert Willison, 2009).

The authors are aware of all the "scientific complexity" and scientific significance of the task – to measure the emotion (or the strength of the trigger) by detecting evidence (or motives) directed at the employee of the company in the form of factors of the external and internal environment of his social existence and, accordingly, put forward a hypothesis that the "strength" of the motive and, accordingly, the strength of the trigger for transformation from a deviant to a divergent state is the number of clues (factors) simultaneously affecting the employee in each working day (Touhey, J. C., & Bandura, A., 1975)..

This hypothesis is based on three fundamental theories: the psychology of employee behavior (the theory of divergence – human transformation), the theory of information (in terms of information entropy), and the mathematical theory of probability (calculations of the probability of criminal events – information incidents). Figure 2 below shows the previously substantiated (Shaban, A. (2024). Method of collecting information about the facts (evidence) of the employee's existence from the external environment, according to the objective function according to formula 1.

$$F(X_I) = S_j + S_i + S_{Gausse} \tag{1}$$

Based on the results of well-known studies (Konow, 2000) (Greitzer, F.L., Frincke, D.A., 2010) the divergent state of an employee is preceded by a deviant state, when the employee transforms his "usual" working state into some non-standard, unusual or deviant (with deviations). At a certain point in time, "deviations" in the main performance indicators begin to appear more and more often in the deviant behavior of an employee:

- frequent failure to meet economic performance indicators,
- frequent delays, smoke breaks, coffee breaks,
- prolonged absence from the workplace,
- frequent telephone conversations for non-work purposes,
- negative emotions in the framework of interaction between work colleagues, harsh gesticulations when talking and communicating,
- aggression or abusive language, increased intonation in the voice,
- insults or neurotic criticism of colleagues or management,
- other physiological factors: rapid breathing, rapid pulse.

The authors of the article believe that such "anomalies" in behavior have causes (or motives) brought by the employee from the external social environment, which can most likely transform him into a divergent employee, in fact, an insider (criminal). Author suggests the scientific and practical confirmation of the hypothesis of transformation of the psycho-emotional state of an employee under the influence of factors (evidence) and method to calculate factor of deviant behavior by means instrumental, as Fig. 2.

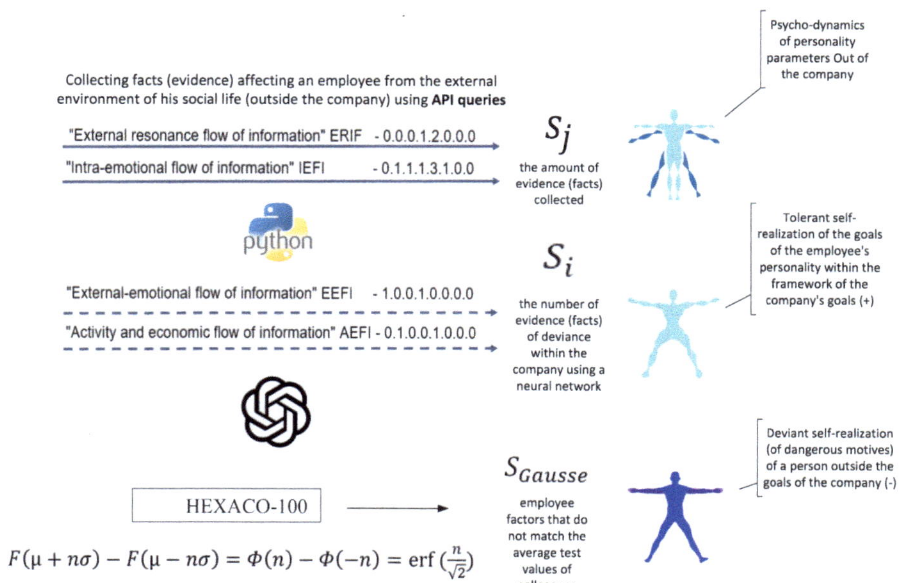

Fig. 2. scientific and practical confirmation of the hypothesis of transformation of the psycho-emotional state of an employee under the influence of factors (evidence).

1.2 A Scientific Approach to Solving Goals and Objectives

Let's start from the statistics of internal insider attacks for the period 2022–2024. Internal insider attacks are a type of cyberattacks in which attackers, employees on the staff list, gain access to confidential information of the organization due to internal threats (weak information security). Internal insiders can be both current and former employees, as well as business partners of the organization, or customers or contractors of the company (Gnisci, A., Mottola, ets, I., 2023).

So, in the United States, in 2022 alone, a major insider attack on SolarWinds was recorded. Attackers gained access to the source code of the company's software and injected malicious code into them. This has led to the compromise of thousands of organizations around the world, including government agencies and large corporations ("Defining Insider Threats," CISA, https://www.cisa.gov/topics/physical-security/insider-threat-mitigation/defining-insider-threats).

In the UK in 2023, an insider attack on the UK's National Health Service (NHS) was revealed. Attackers gained access to patients' personal data and used them for financial fraud.

In Germany, a major insider attack on Siemens was prevented in 2024. The attackers tried to gain access to the company's intellectual property, but were discovered and detained.

In France, in 2022, an insider attack on Airbus Defence and Space was revealed. Attackers gained access to the company's trade secrets and used them for financial gain.

In Australia, an insider attack on Telstra was recorded in 2023. Attackers gained access to the personal data of the company's customers and used them for fraud.

In Japan, an insider attack on Sony was uncovered in 2024. Attackers gained access to the source code of the company's software and used it to create fake games and applications.

In India, an insider attack on Tata Consultancy Services was prevented in 2022. The attackers tried to gain access to the company's trade secrets, but were discovered and detained.

In Brazil, an insider attack on Petrobras was recorded in 2023. Attackers gained access to the company's financial data and used it for securities fraud.

In Canada, an insider attack on Shopify was uncovered in 2024. Attackers gained access to the personal data of the company's customers and used them for phishing attacks.

For the purposes of this study, the authors put forward hypothesis No1 – that any employee who has access to commercial information can steal it, but only in the case of the accumulated potential (motive) of a criminal act, in the form of a "deviation" from standard behavior, i.e. an act in a state of affect (unconscious) or malicious (conscious) intent, but under the influence of external circumstances of his "life", which were "born" from the external and internal environment of his habitat. By the external environment, the authors understand the "life" of an employee outside the contour of the company (state institution), by the internal environment – the "life" of an employee within the company's team.

The authors of the study also cite two pieces of evidence that the proposed hypothesis 1 has two theoretical "sources." The first follows from the economic theory of economic entropy, where, as is known, any business, regardless of size, is created as an ordered system for the implementation of useful economic activity, but is subject to the "extinction" of business activity (Crea, G., Francis L.ets, 2024).

Interrelated parts of the single economic mechanism of the firm (institution), production assets and technologies of the business process, the company's personnel and their qualifications, solving common tasks and goals, performing their functional work, over time, disorder (entropy) grows in this ordered system of the firm.

Most of the production and economic efforts are dissipated over time, and not all the energy of an ordered system (firm) turns into useful work, and no longer brings effective results (profit), and at the same time, most of the productive energy of personnel and working time is already "spent" on meaningless processes that do not lead to the achievement of commercial goals. Conflicts arise more and more often in the team (Möller, D.P.F., 2023)., loss of efficiency and, ultimately, profit. Workers, with their different psycho-emotional archetypes and qualifications, begin to be subjected to the "pressure" of the external and internal circumstances of their "life" even more. There are "deficits" of human needs. New "motives" of the employee arise and old ones are determined. And if the deviations (deficit) of the employee's needs exceed the "permissible" norms of his patience, the criminal act is likely to increase many times over.

For information security management (control over the integrity of commercial information (secrets) of the company), the authors propose the following conceptual method for controlling information leaks, namely, dynamic control of deviations in the behavior of an employee using a neural network to recognize the psycho-emotional state and economic conscientiousness of an employee by factors that had a static deviation in the average median of "business" entropy.

Real (daily) deviations in the psycho-emotional behavior of an employee from those previously defined for him as normal (starting) are the controlled "business" entropy of the company. As is known, the concept of entropy introduced by Clausius (Okomanyi, A.O., Sherwood, A.R. & ets, 2024) in thermodynamics in 1865 was used to determine the measure of irreversible energy dissipation, the measure of the deviation of the real process from the ideal.

The authors' clarification is that "business" entropy for the purposes of this study is a measure of the uselessness of "productive" energy (loss, dissipation) of all employees of the company in terms of the sum of all efforts, and separately for each employee. From the theory of the "life cycle" of the company (Borky, J.M., Bradley, T.H., 2019).), it is proved that over time, not all the energy of the "efforts" of the organizational system (personnel and employee) is used to turn it into useful and effective work. As a result, organizational disorder grows, chaos in relationships and communications, and business uncertainty grows. The above-mentioned factor of chaos in the organizational environment "generates" conditions for the transformation of a tolerant employee into a problem one, with a probability of more than 50% of becoming a potential insider of the company or organization. In order to counter the threats of information leakage, it is necessary to control the "business" entropy (the level of chaos), and the authors propose to track deviations in the behavior of the employee, namely losses in labour productivity, disconcentration of attention from the goals of the company - de-communication of the employee, an increase in the level of conflict and aggression, a decrease in the efficiency and return of his functional ability to work and other factors of problems (divergence of the employee).

In the substantiation of the above-mentioned hypothesis No1, the first theoretical question arises – how to find the primary (in absolute terms) "business" entropy of the firm and each employee, in order to identify deviations. In order to implement the model for controlling deviations in the employee's deviance, the authors propose to use a psychoanalytic technique at the first stage, in the form of testing each employee according to the HEXACO-100 method. At the second stage, the use of the apparatus of mathematical statistics is the calculation of the standard normal distribution of archetypes of all employees of the company within the "business" ideal entropy (Lee, K., Ashton, M.C., 2020).) and obtaining a probabilistic confidence deviation for 5% of the quantile, according to the formula (2).

$$F(\mu + n\sigma) - F(\mu - n\sigma) = \Phi(n) - \Phi(-n) = \operatorname{erf}\left(\frac{n}{\sqrt{2}}\right) \tag{2}$$

where: σ – deviation;
μ- average;
F - Normal Distribution with Density Φ for n;
Erf - error function (probability integral) erf(x), which gives the probability that a normal random variable with a mean of 0 and a variation of 1/2 will fall into the segment [-x,x].

Figure 3 below shows the confidence intervals of the normal distribution, of which about 68% of the values are no more than one standard deviation σ from the mean; About 95% of the values are no more than two standard deviations apart. and 99.7% no more than three. This fact is a special case of the 3sigma rule for normal sampling; in Russia it is called the Laplace function. Figure 3 here.

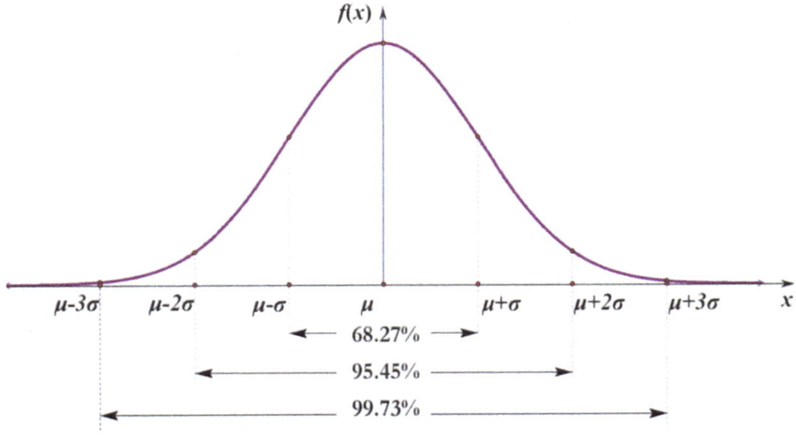

Fig. 3. The confidence intervals correspond to 3-sigma rule of the normal distribution.

Thus, when considering the question "who is" our employee at the start of his work at any company (as a psycho-emotional archetype), the authors of the article propose to conduct initial (entry) testing, and for other employees who have been used for a long time – rotational (annual) testing for psycho-emotional significance. And then it is logical that the main task of the authors narrows down to tracking and identifying deviations in the average "business" median (entropy) of the company within the framework of the "dissipation" of energy in each employee from the moment of the start of his work, up to the threat of abnormal deviations, which, in the authors' opinion, become that dangerous psycho-emotional "breakdown" leading to the theft of the company's trade secrets.

Figure 4 below shows the purpose of using the test for group compatibility or difference in the psycho-emotional profile of an employee in order to increase the effectiveness of the method for detecting a problem (deviant) employee (Widiger, T.A., ets 2023). The general and main goal of the entire methodology for detecting a deviant employee according to the proposed target function 1 is daily automated monitoring of psycho-emotional and other indicators of economic activity and tolerance of the employee to collect evidence (facts) of increased deviance of the employee, for the subsequent adjustment of the information security policy in the form of restricting the employee's access to the workplace in the event of an increase in the number of evidence (facts) of the deviant state.

It is difficult for the administration (information security department) to believe that a previously "calm" and "stable" employee at the moment becomes an insider, a divergent, an avenger for something "offended" against everyone or the company. Figure 4 is shown below.

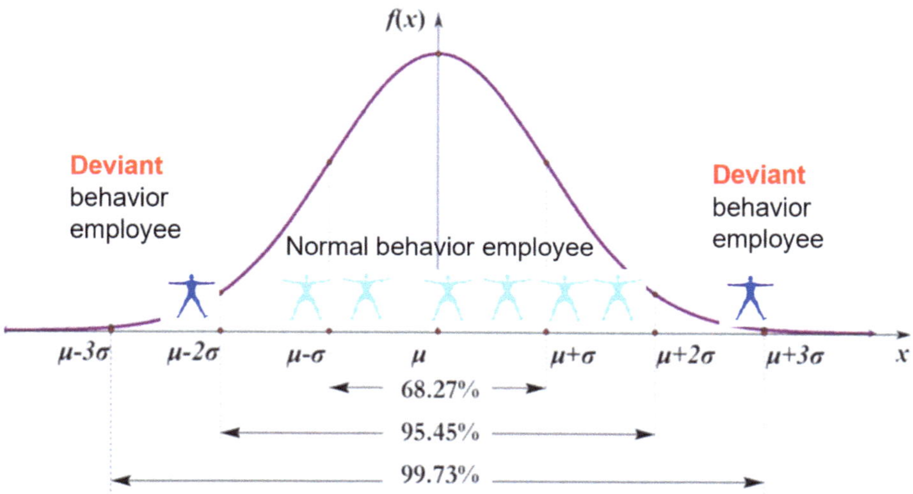

Fig. 4. calculating the normal distribution of behavioural entropy for a company and determining the "place" of each employee in it.

Psychologists call this phenomenon "cognitive dissonance", which occurs when a person (employee) has two cognitions that contradict each other (Przybył, J., Pawlak, M., 2023).). For example, a person may know that "stealing" company information is a crime and it is bad for his reputation, but he still commits theft. In this case, there is a discrepancy between his knowledge and behavior. Psychologists also note (Jacobsson, P., ets, 2021) that cognitive dissonance can cause sharp psychological discomfort and a quick desire to reduce the contradiction between one's cognitions. This can lead to a change in beliefs (motive), behavior or the search for new information to resolve the contradiction.

The authors of this study believe that each person at the start of professional activity already has his own formed archetype of his behavior and the corresponding "motive" of target activity in the company (including income, increasing competence, socialization, as well as general self-realization of the individual).

The authors note that it is at this moment of self-realization of dissonance for the "personality" of the employee that the question arises from the theory of information – to what extent his own archetype of the employee corresponds to the median (average archetype) of the firm (the authors previously introduced the concept of "business entropy"), because if his entropy of self-realization does not coincide with the "business" entropy of the firm, there is a deviation of the employee's goals and the goals of the firm. Each working day already brings greater discomfort to such an employee (Marengo, D., Davis, K.L, ets 2021), and the first signs of entropy (dissipation of his labor energy) are absenteeism, tardiness, a drop in labor productivity, conflicts, aggression and other above-mentioned factors of the employee.

The authors put forward hypothesis No2 that an information "connection" (entropy) should arise between the goals of the company and the employee, in which the identity of the goals of both agents – the factors of information entropy – is confirmed or refuted in a dynamic form on a daily basis (Yang, J., Bai, Y, 2023). When business entropy completely coincides with the company and the employee, there is a "harmony" of the company's business environment, the organization of the employee within the company's business process, the absence of decommunicativeness between the employee and the team, the unnecessary costs of the company due to duplication or redundancy of business tasks and processes, which leads to the efficiency of the company's results, the level of business entropy = 0.

And vice versa, business entropy tends to 1, when the goals of the employee and the company are different, the employee egoistically implements his "attitudes" according to his psycho-emotional state. At the same time, the company's business reputation decreases, the costs of the business process increase, customers are lost, business slows down, and the overall level of motivation of all personnel decreases. A well-known situation arises: "at home among strangers, among friends".

2 Results of the Study

On the basis of the proposed model (1) and the tests carried out on conditional employees of company P, the results are obtained in Table 1 and 2. Table 1 is shown below.

After processing the obtained values, Fig. 5 shows the intervals of discrepancies in the final values according to the psycho-emotional portrait of 11 the company's employees. Figure 5 shows the confidence intervals of the normal distribution, of which about 68% of the values are at a distance of no more than one standard behavior deviation σ from the mean testing of each employee according to the HEXACO-100 method. Table 2 is shown below.

Figure 5 shows the intervals of discrepancies in the final values according to the psycho-emotional portrait of 11 the company's employees. Figure 5 is shown below.

Table 1. calculating the normal distribution of behavioral entropy for a company and determining the "place" of each from 11 employees in it (HEXACO-100).

hexaco-lOOtest	empl 1	empl 2	empl 3	empl 4	empl 5	empl 6	empl 7	empl 8	empl 9	empl 10	empl 11
honesty modesty	25%	10%	87%	37%	70%	62%	27%	10%	87%	87%	10%
emotionality	12%	90%	55%	65%	50%	57%	20%	90%	0%	0%	90%
extraversion	72%	20%	65%	52%	73%	70%	82%	20%	72%	72%	20%
benevolence	42%	40%	45%	62%	55%	50%	65%	40%	55%	55%	40%
conscientiousness	80%	30%	72%	42%	85%	70%	42%	30%	100%	100%	30%
inquisitiveness	37%	10%	40%	57%	77%	45%	40%	10%	85%	85%	10%

Table 2. Mean normal deviations.

μ + σ	0,767414	1,078464	1,078464	1,078464	1,078464	1,078464	1,078464	1,078464	1,078464	1,078464
μ -σ	0,145314	-0,16574	-0,16574	-0,16574	-0,16574	-0,16574	-0,16574	-0,16574	-0,16574	-0,16574
μ + σ	0,734879	1,070667	1,070667	1,070667	1,070667	1,070667	1,070667	1,070667	1,070667	1,070667
μ -σ	0,063303	-0,27249	-0,27249	-0,27249	-0,27249	-0,27249	-0,27249	-0,27249	-0,27249	-0,27249
μ + σ	0,775499	1,007362	1,007362	1,007362	1,007362	1,007362	1,007362	1,007362	1,007362	1,007362
μ -σ	0,311774	0,079911	0,079911	0,079911	0,079911	0,079911	0,079911	0,079911	0,079911	0,079911
μ + σ	0,549744	0,636761	0,636761	0,636761	0,636761	0,636761	0,636761	0,636761	0,636761	0,636761
μ -σ	0,375711	0,288694	0,288694	0,288694	0,288694	0,288694	0,288694	0,288694	0,288694	0,288694
μ + σ	0,857711	1,123605	1,123605	1,123605	1,123605	1,123605	1,123605	1,123605	1,123605	1,123605
μ -σ	0,325925	0,060032	0,060032	0,060032	0,060032	0,060032	0,060032	0,060032	0,060032	0,060032
μ + σ	0,71413	0,986442	0,986442	0,986442	0,986442	0,986442	0,986442	0,986442	0,986442	0,986442
μ -σ	0,169506	-0,10281	-0,10281	-0,10281	-0,10281	-0,10281	-0,10281	-0,10281	-0,10281	-0,10281

Figure 6 below shows the result of identifying an increased probability of a deviant dangerous discrepancy of individual employees from the staffing table of (condition) company NNN, which narrows the scope of attention of the company's information security department for employees who "fall out" of the sector of normal behavior of employees. Figure 6 shows the results of behavioural entropy in terms of personal parameters (character traits) based on the test of employee motivation deficits, which can primarily cause dissonance in his behavior and give an impulse (push) to a criminal act. Figure 6 is shown below

Further, the results of detecting deviant behavior are "embedded" into the current algorithm of the information security policy in the form of a final ban or permission to access an employee who has exceeded the number of evidence (facts) of deviant behavior, as shown in Fig. 7.

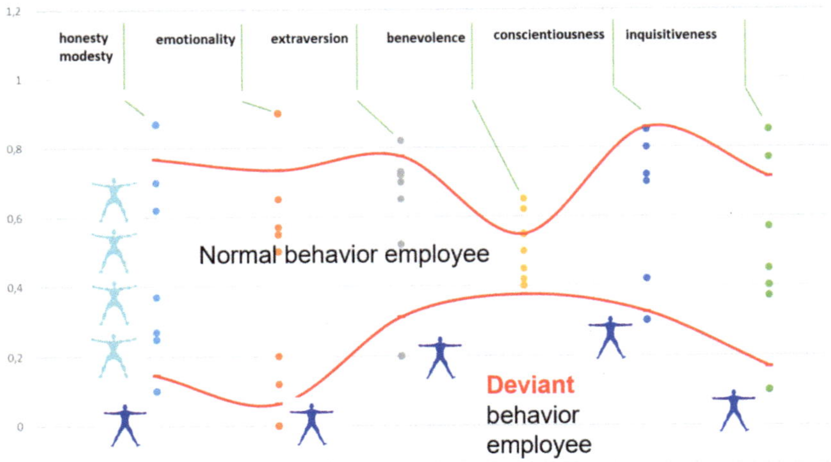

Fig. 5. calculating the normal distribution of behavioural entropy for a company and determining the "place" of each employee in it.

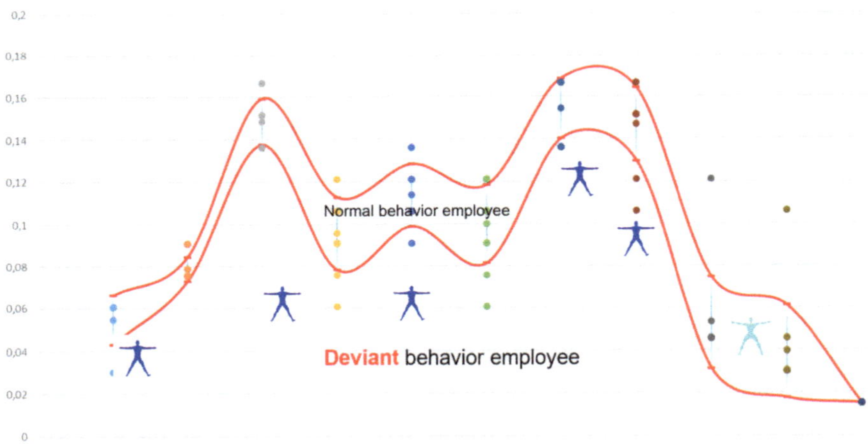

Fig. 6. results of behavioural entropy in terms of personal parameters (character traits) based on the test of employee motivation deficits.

Figure 7 shows the trigger of the information security system of the employee's access to the trade secrets of the company scan for his working place by means ViPNet TIAS (Threat Intelligence Analytics System).

As a result of the collection and analysis of the obtained metadata using a mathematical model according to formula (1) and the compilation of the aggregate probabilities of the psycho-emotional state from two metadata streams, the organization's information security system will semi-automatically determine the access of each employee to the

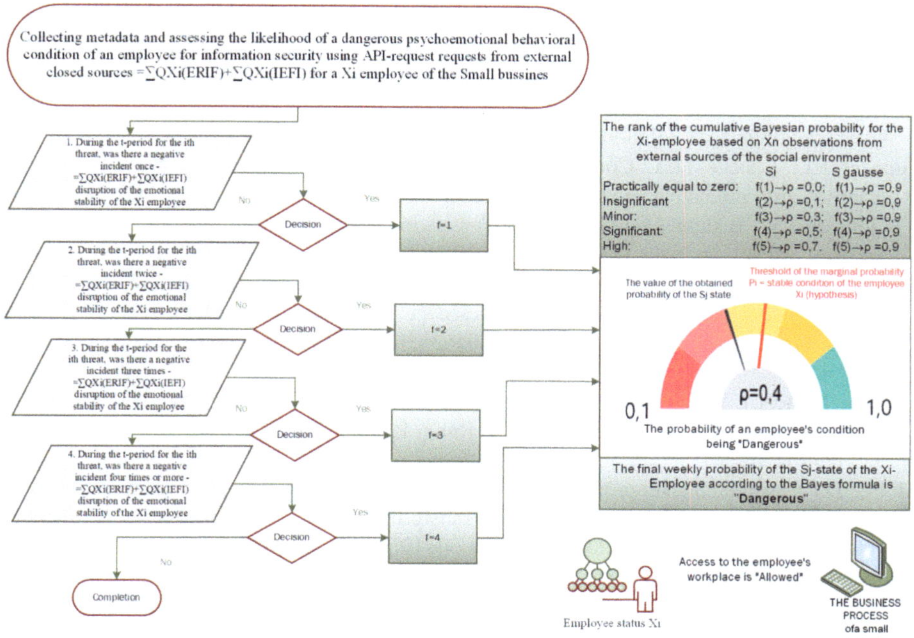

Fig. 7. Algorithm for determining an employee's access to the company's business process (including trade secrets) by calculating the probability of the impact of psichoemotional external and internal factors on him.

organization's business process through the aggregate probability of the employee's "behavioural" information security, and will mathematically determine the level of employee access of the "Allowed" type or "Access Denied".

As a result, by compiling two probabilities, but within the framework of the aggregate probability according to the Bayesian formula equal to 1, the employee's access system to the business process (to the workplace) will become a trigger – an access regulator and, accordingly, will increase the information security of the organization from possible threats and damages from the employee.

References

Yilmaz, E., Can, O.: Unveiling shadows: harnessing artificial intelligence for insider threat detection. Eng. Technol. Appl. Sci. Res. **14**(2), 13341–13346 (2024). https://doi.org/10.48084/etasr.6911

Shannon, C.E.: A mathematical theory of communication. Bell Syst. Tech. J. **27**(4), 623–656 (1948). https://doi.org/10.1002/j.1538-7305.1948.tb00917.x

Konow, J.: Fair shares: accountability and cognitive dissonance in allocation decisions. Am. Econ. Rev. **90**(4), 1072–1092 (2000). https://doi.org/10.1257/aer.90.4.1072

Zimbardo, P.G.: Lucifer effect. The Encyclopedia of Peace Psychology (2011). https://doi.org/10.1002/9780470672532.wbepp149

Touhey, J.C., Bandura, A.: Aggression: a social learning analysis. Contemp. Sociol. **4**(2), 169 (1975). https://doi.org/10.2307/2062223

Warkentin, M., Willison, R.: Behavioral and policy issues in information systems security: the insider threat. Eur. J. Inf. Syst. **18**(2), 101–105 (2009). https://doi.org/10.1057/ejis.2009.12

Shaban, A.: Digitalization, cultural production, exchange, and consumption. In: Shaban, A. (ed.) Digital Geographies—Urbanisation, Economy, and Modelling. Springer, Singapore (2024). https://doi.org/10.1007/978-981-97-9278-8_5

Greitzer, F.L., Frincke, D.A.: Combining traditional cyber security audit data with psychosocial data: towards predictive modeling for insider threat mitigation. In: Probst, C., Hunker, J., Gollmann, D., Bishop, M. (eds.) Insider Threats in Cyber Security. Advances in Information Security, vol. 49. Springer, Boston, MA (2010). https://doi.org/10.1007/978-1-4419-7133-3_5

Gnisci, A., Mottola, F., Perugini, M., Senese, V.P., Sergi, I.: Development and validation of an instrument to measure personality in adolescence: the HEXACO medium school inventory extended (MSI-E). PLoS One. **18**(1), e0280563 (2023). https://doi.org/10.1371/journal.pone.0280563

Crea, G., Francis, L.J., McKenna, U.: Assessing the power of the HEXACO to predict professional burnout among Catholic priests in Italy. J. Relig. Health. (2025). https://doi.org/10.1007/s10943-024-02202-5

Möller, D.P.F.: Cyberattacker profiles, cyberattack models and scenarios, and cybersecurity ontology. In: Guide to Cybersecurity in Digital Transformation. Advances in Information Security, vol. 103. Springer, Cham (2023). https://doi.org/10.1007/978-3-031-26845-8_4

Okomanyi, A.O., Sherwood, A.R., Shittu, E.: Exploring effective strategies against cyberattacks: the case of the automotive industry. Environ. Syst. Decis. **44**, 779–809 (2024). https://doi.org/10.1007/s10669-024-09971-0

Borky, J.M., Bradley, T.H.: Protecting information with cybersecurity. In: Effective Model-Based Systems Engineering. Springer, Cham (2019). https://doi.org/10.1007/978-3-319-95669-5_10

Lee, K., Ashton, M.C.: HEXACO model of personality structure, the. In: Zeigler-Hill, V., Shackelford, T.K. (eds.) Encyclopedia of Personality and Individual Differences. Springer, Cham (2020). https://doi.org/10.1007/978-3-319-24612-3_1227

Widiger, T.A., Tyrer, P., McCabe, G.A., Kim, Y.R.: Personality disorders. In: Tasman, A., et al. (eds.) Tasman's Psychiatry. Springer, Cham (2023). https://doi.org/10.1007/978-3-030-42825-9_14-1

Przybył, J., Pawlak, M.: Research into the role of personality in L2 learning. In: Personality as a Factor Affecting the Use of Language Learning Strategies. Second Language Learning and Teaching. Springer, Cham (2023). https://doi.org/10.1007/978-3-031-25255-6_3

Jacobsson, P., Hopwood, C.J., Söderpalm, B., et al.: Adult ADHD and emerging models of maladaptive personality: a meta-analytic review. BMC Psychiatry. **21**, 282 (2021). https://doi.org/10.1186/s12888-021-03284-1

Marengo, D., Davis, K.L., Gradwohl, G.Ö., et al.: A meta-analysis on individual differences in primary emotional systems and Big Five personality traits. Sci. Rep. **11**, 7453 (2021). https://doi.org/10.1038/s41598-021-84366-8

Yang, J., Bai, Y., Cheng, Y., et al.: A new model for bearing fault diagnosis based on optimized variational mode decomposition correlation coefficient weight threshold denoising and entropy feature fusion. Nonlinear Dyn. **111**, 17337–17367 (2023). https://doi.org/10.1007/s11071-023-08728-9

A Comparative Analysis of Machine Learning Models for Financial Fraud Detection

Shreya Shivkumar Mathpati[1], Pratibha C. Kaladeep Yalagi[1], Vijay Anant Athavale[1](✉), and Ivan Azarov[2]

[1] Department of Computer Science and Engineering, Walchand Institute of Technology, Solapur, India
vijay.athavale@gmail.com
[2] North-Caucasus Federal University, Stavropol, Russia

Abstract. Increasing cyber attacks, like fraud detection in financial transactions, were emerging as a new critical challenge in global and Indian markets. The study introduces a new, improved fraud detection framework integrated with a machine learning model. It includes Random Forest and XGBoost. It improves the development of a scalable, adaptable real-time analysis of the dynamic financial dataset. The proposed methodology achieves excellent accuracy in detecting transactions. This will easily distinguish legitimate active transactions from fraudulent ones. It will be capable of identifying from a highly imbalanced dataset. This extensive experiment conducted on a real-world financial dataset demonstrates the system's features. It will ensure enhanced security measures and minimize false positives and risks. It regains the consumer's trust in digital financial services. The proposed machine learning technology gives a high detection accuracy of 92%. Better results were also cited compared to the state of the art from other approaches in the key metrics, such as precision, recall, and F1-score. Results identify improvement through such performance graphs and pie charts showing both the performance of false positives and the improved fraud detection accuracy of the system. This research aims to bridge the crucial gaps in current approaches. We are charting a future course to improve the security and techniques of digital payment systems. The future course encompasses the incorporation of block chain technologies. The system approach, along with new emerging technologies, counters cyber threats.

Keywords: Fraud Detection · XGBoost · Random Forest · Anomaly Detection · Machine Learning · Cyber Security

1 Introduction

The various new technologies of digital financial systems have revolutionized and set new trends in global economics. The digital transformation has also been inviting fraudulent activities. In a global market as well as the Indian market, fraudulent transaction technology, such as identity theft, phishing, and account takeovers, has shown an impact in billions of dollars and rupees in losses annually. This leads to the challenging need for adaptive fraud detection mechanisms to provide safeguards against threats without involving them.

M. Lapina et al. (Eds.): AISMA 2025, LNNS 1456, pp. 323–334, 2026.
https://doi.org/10.1007/978-3-032-07275-7_30

Traditional fraud detection systems are capable and effective to some extent, but sometimes they will struggle to cope with the dynamic nature of fraudulent patterns. The framework fails to adapt to emerging attack patterns, and this highlights the system's high rates of false positives, undermining their efficiency. This research introduces an effective approach that figures out advanced machine learning models. XGBoost and Random Forest are the algorithms used. This helps to detect irregularities in high-dimensional financial datasets. These models satisfy and are more sincere in scalability, adaptability, and real-time processing; this will leverage the financial fraud detection.

The objectives of this study are to develop a fraud detection system. The proposed methodology ensures that data pre-processing, features of engineering, and machine learning models are accurate, and it can detect anomalies. We include this system to reduce false positives through model optimization and to identify emerging fraud patterns. The overall system is enhanced.

2 Objectives

The goal is to design an effective real-time fraud detection system that continuously watches over transactions, detects fraudulent activity in real time, and adapts to new fraud patterns. The system should be secure, stable, and transparent to make users and banks assured of the process. As it continuously detects and rectifies vulnerabilities, it becomes more effective in fraud prevention but remains efficient and reliable.

2.1 Challenges

Class Imbalance: Fraudulent transactions are infrequent and therefore difficult to spot. Methods such as SMOTE balance the data.

Adaptive Learning: Spammers adapt and modify their methods (concept drift), which requires self-update models.

Feature Engineering: Essential features such as transaction speed, device fingerprint, and geolocation provide improved detection.

Data Quality: Noisy or partial data impacts performance; data cleaning helps in ensuring reliability.

Real-Time Processing: Detection needs to be real-time to avoid fraudulent transactions.

3 Literature Review

3.1 Findings and Real-World Applications

Machine learning (ML) and deep learning (DL) are vital technologies in mitigating different forms of finance-related fraud. Roy et al. proposed DNN-based models that secured a classification of fraudulence and reduced false positives with a 98.2% accuracy in credit card fraud detection [8]. Recent advancements in credit fraud detection have, by a fraction of 20%, taken a drastic step forward with a lot of hope in terms of the future of hybrid approaches like CNNs and GANs [5]. Mining these methodologies seems to pose an advanced persistent threat (APT) to large UPI transactions, which often involve privacy concerns that are rarely observed in India [20].

3.2 Contributions to Fraud Detection Mitigation

The previously assessed option classifies real-time visualization samples of susceptible transactions with two concurrent fraud detection techniques.[7]; a lot of those turn out to be very effective. Recent practical Bayesian optimization processes have proven to be highly effective in enhancing customer confidence through higher detection reliability [9]. Verification of the importance of detecting undesirable transaction samples with the confluence of SVM and Markov Decision Process (MDP) history has become the new frontier in fraud mitigation contributions through the study of phishing detection research [7].

3.3 Relevant Research's Impact on This Study

This research relative to hybrid machine learning [21], optimization using deep learning [22], block chain-backed secure systems [23], and metadata-driven anomaly detection [24, 25] has meaningfully contributed to this work. These studies helped in matters related to model selection, imbalanced data handling, and strengthening the security layer of the system for fraud detection.

Table 1 shows the overview of recent research in financial fraud detection and prevention with pros and cons.

Table 2 focuses on the recent research on financial frauds prevention and detection with comparative results.

Table 1. Overview of Recent Research in Financial Fraud Detection and Prevention.

Algorithm	Pros	Cons	Ways to tackle drawbacks
Random Forest [2]	It will be an expert in dealing with large datasets. Strong to overfitting. Elaborate importance of features.	Poor prediction time with large datasets. The results of the model are complex. Takes a substantial amount of memory.	Make use of model pruning and feature selection for better performance. Give faster predictions using parallel processing.
Gradient Boosting [4, 5]	High accuracy It deals with a large set of data Ideal for imbalanced data sets.	Computationally expensive. Risk of overfitting Considerations with hyperparameter for optimum results.	Use of early stopping to tackle the overfitting. Regularization of models via learning speed fine-tuning.

(*continued*)

Table 1. (*continued*)

Algorithm	Pros	Cons	Ways to tackle drawbacks
Support Vector Machine (SVM) [7, 8]	It performs in high dimensional spaces. There is a significant margin of separation in its ensuing behavior.	Memory-intensive. Not suitable for very large datasets. Sensitive to noise and outliers.	Use kernel trick for better results in high-dimensional space. Apply preprocessing to remove outliers.
Logistic Regression [10, 11]	Simplified to implement. Offers the prediction of probabilities. It is less compute intensive.	Assumes that the scope of the relationship is linear in terms of the characteristics. It will perform poorer predictions on complex data. The flexibility involved is minimal.	Generate non-linear relations on your features through feature engineering. Use regularization techniques in combination.
K-Nearest Neighbours (KNN) [14, 15, 18, 19]	Very easy and simple. No need for any training period to make it work properly. Suits small datasets pretty well.	Very computation-expensive for large datasets. Very foolish about irrelevant features. The probability of suffering from the curse of dimensionality.	A lesser number of features can be drawn by the use of principal component analysis (E.G: PCA). Weighted distance measures probably help in increasing the accuracy.
Decision Trees [17]	Easy to comprehend. Handle both numerical and qualitative data. Quick learnability.	Prone to overfitting. Unstable because of slight variability in data. Can Favor dominant classes.	Examples of use state that to prevent overfitting it is favorable to use selection techniques, although the necessity of dissipating major causes has limits.
XGBoosting [20]	performance of XGBoosting is high and accuracy. Handles missing data. Fast parallelization and training.	Complex to tune. Overfitting risk without proper regularization. Difficult to interpret.	Early stopping and cross-validation for tuning hyperparameters. Regularize with L1/L2 to prevent overfitting.

(*continued*)

Table 1. (*continued*)

Algorithm	Pros	Cons	Ways to tackle drawbacks
Isolation Forest [22, 23]	Best for anomaly detection. It deals with high-dimensional data. Fast and scalable.	May perform poorly for imbalanced data points. Less accurate in detecting very fine scam patterns.	It Combines with other algorithms like Random Forest or Gradient Boosting for better detection.

Table 2. Overview of Recent Research on Financial Frauds Prevention and Detection.

Variety in Prevention and Detection			
Title of Research Paper	Method/Technique Used	Result	Summary
Deep Learning Detects Fraudulent Activities in Credit Card Transactions [1]	Employing deep neural networks (specifically, Deep Neural Networks (DNN)) to detect suspicious patterns.	98.2%	It achieved high accuracy, demonstrating the power of deep neural network techniques with respect thereof in this domain.
Credit Card Fraud Detection with Modern Machine Learning and Deep Learning Algorithms [2]	Implements ML algorithms such as Decision Trees, SVM, and DL techniques namely CNN, LSTM.	96.5%	Comparison of various algorithms, this giving solid outcomes in fraud detection across large dataset.
Credit Card Fraud Detection Ensuring the Awareness of Uncertainty Using Extended Networking [3]	Uncertainty-aware Bayesian deep learning and Monte Carlo dropout.	95%	Integrating trust into fraud detection by making uncertainty-aware deep learning models.
Incorporating both k-Nearest Neighbors and DNN for Fraud Detection Technologies in Credit Card [4]	Combines DNN for feature extraction and KNN for classification of transactions.	97%	Combine the strength to achieve a higher fraud detection rate using DNN and KNN.

(*continued*)

Table 2. (*continued*)

Variety in Prevention and Detection			
Title of Research Paper	Method/Technique Used	Result	Summary
Credit Card Fraud Detection Using Deep Learning Techniques [5]	DL models used like LSTM and CNN, Autoencoders used for pattern-based detection.	Accuracy 97.8%	Exploiting highly advanced deep learning models for better accuracies in fraud detection.
Deep Learning Based Credit Card Fraud Detection [6]	DL architectures like RNNs, CNNs, GANs are checked with regard to financial data.	97%	Deep learning models were contrasted against one another, supporting results in fraud detection.
A Survey on Machine Learning Approaches to Credit Card Fraud Detection [7]	Reviewing ML with Random Forest, Logistic Regression, and XGBoosting technique.	Accuracy improvement rate is 94%.	An Overview of Survey Machine Learning Techniques for Detecting Credit Card Fraud.
Detecting fraud in Ethereum with Graph Neural Network [8]	Proposes Graph Network Neural (PDGNN) for crafting fraud detection of phishing.	92%	Identify Phishing based Fraud Detection in Graph based Ethereum Transaction Modelling.
Artificial Intelligence in Decentralized Finance: AI for Fraud Detection Throughout the Project Life Cycle: [9]	AI techniques like NLP, reinforcement learning, and anomaly detection have been covered in DeFi.	91%	Successfully categorized various types of fraud in defi and reviewed AI techniques that can detect fraud.
Hybridization of Deep Learning with Generative Adversarial Networks for Credit Card Fraud Detection [10]	Involves the GAN in data aggregation with deep learning for fraud portrayal.	Effectiveness rate: 98%	Hybrid models the combination of GANs and deep learning. It increases detection performance of credit card fraud detection.

(*continued*)

Table 2. (*continued*)

Variety in Prevention and Detection			
Title of Research Paper	Method/Technique Used	Result	Summary
Financial Fraud Detection with the help of Hybrid Convolutional and Recurrent Neural Network Detection [11, 26]	Hybrid fuses CNNs to extract features with RNNs to detect temporal patterns in unstructured data.	96%	They can detect 96% of all financial frauds using unstructured data from banking transactions with CNNs and RNNs.
The Coalescence of Deep Neural Networks within Binary Factor Cross-Lagged Analysis [12]	Merged Deep Neural Network for feature extraction with the Probabilistic Graphical Model to analyze sequentially distributed data.	95%	It Successfully embedded DNNs within probabilistic models to improve the accuracy of detecting fraud in credit card transactions.
Smart Technique for Credit Card Fraud Detection by OLightGBM [13]	Bayesian Based Optimization Algorithm with LightGBM.	Accurate 98.40% and AUC of 92.88%	Bayesian-based optimization algorithm combined with the LightGBM. By using Light Gradient Boosting Machine enhance the accuracy in credit card fraud detection.
A Review on Credit Card Fraud Detection with Application of VAT Clustering Algorithm [14, 32]	VAT clustering combined with classifiers.	Detection rates vary per dataset	Evaluating algorithms SVM, ANN, Random Forest and AIS among others. Current methods and further introduction to application of VAT clustering for fraud detection.
Machine learning based credit card fraud detection using GA [15, 16]	Random Forest Algorithm, Logistic Regression, Artificial Neural Network, and Naive Bayes.	99.96%	Random Forest for the prevention of fraud, showing class-leading high accuracy for credit card fraud cases.

4 The Global and Indian Financial Fraud Impact Analysis

The study of methods is presented in Fig. 1. Financial losses through digital fraud, such as UPI or credit card fraud, are huge. In 2021, credit card fraud amounted to nearly $28 billion according to Nilson Report. An increasing number of CNP transactions coupled with evolving digital ecosystems has thus fortified avenues to cybercrime.

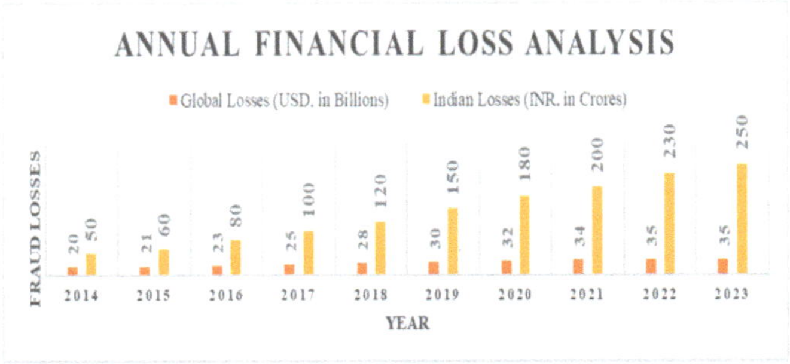

Fig. 1. Impact on Consumer Trust and Regulatory Changes.

UPI frauds have increased significantly, primarily due to phishing attacks, malicious applications, and illicit OTP solicitations. Fraudsters prey on digital literacy gaps among the unserved populace, in particular, in rural areas. Meanwhile, credit card fraud occurs largely in metro areas where cybercriminals have found vulnerabilities in digital payment outlets.

Regulatory bodies such as the Reserve Bank of India (RBI) have introduced AI for fraud detection and 2FA, thus reducing instances of fraud by 25%. However, real-time fraud detection is a challenge in itself due to high operational costs, constant shifts in fraud techniques, and privacy concerns. Future solutions need to be adaptive AI, using federated learning so that fraud detection is conducted securely, and utilizing AI models that are interpretable from a security vantage so that it gains user trust.

4.1 Indian Credit Card Fraud and Online Banking

Banking fraud in India is an ever-transforming scene with highly sophisticated methods being employed, like phishing, social engineering, and identity theft. The fraudsters play on vulnerabilities in emails social media and bank websites and cause massive financial losses. With the emergence of UPI and credit cards, new hazards have been created, as crooks try to find new ways to bypass security screening.

4.2 Frauds in Indian Credit Cards and UPIs

Nearly 500 incidents reported annually originating from metropolitan hubs like Mumbai and Delhi and hence the credit card fraud remains a critical concern in India. A major concern of these fraudulent activities is Card-Not-Present (CNP) transactions, where stolen card details are used. These scams often exploit gaps in digital response time and target individuals are unfamiliar with cyber security practices.

In India's digital payment ecosystem, UPI scams have become a menace. Phishing emails or messages, counterfeit UPI apps, and unsolicited OTP requests are often involved in these fraudulent activities. Most of the victims are first-time digital users and individuals in rural areas with limited awareness of cyber security risks. This impacts on small-scale vendors and everyday consumers leading financial distress in digital payment systems.

4.3 Government and Regulatory Actions

To combat cyber fraud, the RBI has put defensive tools into place. Among these security features, two-factor authentication stands out as most successful, causing a 25% fall in fraud cases in recent times. The government promoted the installation of AI-based anti-fraud measures that can detect and prevent fraud in real time.

Major regulatory provisions are:

- EVM Chip technology: Provides more security to cards and to an extent checks cloning-related frauds.
- AEPS: Provides additional levels of authentication when transactions are carried out.
- IT Act 2000 and the Cyber security Guidelines (2016): These sets of legislations provide fraud detection measures and secure banking systems.
- The PNB scam that came to light in 2018 exposed gross weaknesses in the regulatory regime, indicating the need for increased compliance and increased timely fraud monitoring.

4.4 Challenges and Future Scope

Despite increases in fraud detections, some challenges remain. Cheaters are still trying to find other ways to elude detection, and data protection and data flow monitoring in real-time have become issues of serious concern. With more transactions moving into the digital environment, the fraud prevention systems need to scale this up to process large volumes of transactions. An enabling regulatory atmosphere and the generation of user awareness are some of the solutions that can be undertaken to stem fraud risks in India's emerging digital economy.

5 Comparative Analysis

Machine learning approaches such as Random Forests (RFs), Gradient Boosting (GB) and XGBoost have emerged for vertical enhancement in fraud detection and online credit card transactions. Given that Random Forest has such high capacity with big data and is resistant to over fitting, it cannot be used for real-time fraud detection due to its high memory consumption-which slows down its processing time. Gradient boosting is usually applied to skewed data; that is, it is precise but computationally costly, requiring a lot of time to optimize hyper parameters. XGBoost approaches fast processing and handles missing values very well, but this comes at the cost of complexity and less interpretability.

One of the main challenges of fraud detection lies in imbalanced financial data, with fraudulent transactions forming a small minority of the data. Methods such as SMOTE (Synthetic Minority Over-sampling Technique) help in increasing fraud detection accuracy. The major issue of real-time fraud detection stems from the inherent limitations of static models, which are ill-equipped to counteract the dynamic nature of modern fraud techniques. Combining adaptive learning, continuous model updating, and meta-heuristic optimization techniques provides the most fruitful answer for detecting fraud in environments with limited computational resources.

6 Conclusion

This study analyses the increasing fraud in online payment systems, affecting both international and Indian economies. Conventional fraud detection techniques fail to cope with changing fraud patterns, resulting in high false positives and scalability challenges. For the task of building a scalable fraud detection system, the study has tried to apply different machine learning algorithms such as Random Forest, Gradient Boosting, and XGBoost. This proposed method diminishes massive data preparation and adjustments to mitigate present major system shortcomings, which act toward balancing the data and adapting to new fraud styles. Some of the results included better fraud detection accuracy, lower false alarms, and capabilities toward scalability on large-scale transactions.

References

1. Nguyen, T.T., Tahir, H., Abdelrazek, M., Babar, A.: Deep learning methods for credit card fraud detection. arXiv. (2020). https://doi.org/10.48550/arXiv.2012.03754
2. Lucas, Y., Jurgovsky, J.: Credit card fraud detection using machine learning: a survey. arXiv. (2020). https://doi.org/10.48550/arXiv.2010.06479
3. Li, P., Xie, Y., Xu, X., Zhou, J., Xuan, Q.: Phishing fraud detection on Ethereum using graph neural network. arXiv. (2022). https://doi.org/10.48550/arXiv.2204.08194
4. Luo, B., et al.: AI-powered fraud detection in decentralized finance: a project life cycle perspective. arXiv. (2023). https://doi.org/10.48550/arXiv.2308.15992
5. Ding, Y., Kang, W., Feng, J., Peng, B., Yang, A.: Credit card fraud detection based on improved variational autoencoder generative adversarial network. IEEE Access. **11**, 83680–83691 (2023)
6. Mienye, I.D., Swart, T.G.: A hybrid deep learning approach with generative adversarial network for credit card fraud detection. Technologies. **12**(10), 186 (2024). https://doi.org/10.3390/technologies12100186
7. Banu, S.R., et al.: Financial fraud detection using hybrid convolutional and recurrent neural networks: an analysis of unstructured data in banking. In: Proceedings of the International Conference on Communication and Signal Processing (ICCSP) (2024). https://doi.org/10.1109/ICCSP60870.2024.10543545
8. Forough, J., Momtazi, S.: Sequential credit card fraud detection: a joint deep neural network and probabilistic graphical model approach. Expert. Syst. **38**(4), 1–12 (2021). https://doi.org/10.1111/exsy.12795
9. Roy, A., et al.: Deep learning detecting fraud in credit card transactions. In: Proceedings of the International Conference on Advances in Computing, Communications, and Informatics (ICACCI) (2018). https://doi.org/10.1109/ICACCI.2018.8554822

10. Alarfaj, F.K., et al.: Credit card fraud detection using state-of-the-art machine learning and deep learning algorithms. IEEE Access. **10**, 3166891 (2022). https://doi.org/10.1109/ACCESS.2022.3166891
11. Habibpour, M., et al.: Uncertainty-aware credit card fraud detection using deep learning. Eng. Appl. Artif. Intell. **122**, 106248 (2023). https://doi.org/10.1016/j.engappai.2023.106248
12. Rzayeva, D., Malekzadeh, S.: A combination of deep neural networks and K-nearest neighbors for credit card fraud detection. arXiv. (2022). https://doi.org/10.48550/arXiv.2205.15300
13. Voican, O.: Credit card fraud detection using deep learning techniques. Informatică Economică. **25**(1), 6 (2021). https://doi.org/10.24818/issn14531305/25.1.2021.06
14. Taha, A.A., Malebary, S.J.: An intelligent approach to credit card fraud detection using an optimized light gradient boosting machine. IEEE Access. **8**, 13985–13993 (2020). https://doi.org/10.1109/ACCESS.2020.2971354
15. Abdulrazaq, A.A., Abdulrazaq, M.B., Umoh, I.J., Adedokun, E.A.: Fraud detection in credit card and application of VAT clustering algorithm: a review. In: 2019 Nigeria Computer Conference (NCC), pp. 1–5 (2019). https://doi.org/10.1109/NigeriaComputConf45974.2019.8949660
16. Ileberi, E., Sun, Y., Wang, Z.: A machine learning-based credit card fraud detection using the GA algorithm for feature selection. J. Big Data. **9**, 24 (2022). https://doi.org/10.1186/s40537-022-00573-8
17. Bin Sulaiman, R., Schetinin, V., Sant, P.: Review of machine learning approach on credit card fraud detection. Hum.-Centric Intell. Syst. **2**, 55–68 (2022). https://doi.org/10.1007/s44230-022-00004-0
18. Mhamane, S.S., Lobo, L.M.R.J.: Internet banking fraud detection using HMM. In: 2012 Third International Conference on Computing, Communication and Networking Technologies (ICCCNT'12), Coimbatore, India, pp. 1–4 (2012). https://doi.org/10.1109/ICCCNT.2012.6395910
19. Chavan, V.D., Yalagi, P.S.: A review of machine learning tools and techniques for anomaly detection. In: Choudrie, J., Mahalle, P.N., Perumal, T., Joshi, A. (eds.) ICT for Intelligent Systems. ICTIS 2023. Smart Innovation, Systems And Technologies, vol. 361, pp. 1–15. Springer, Singapore (2023). https://doi.org/10.1007/978-981-99-3982-4_34
20. Athavale, V.A., Bansal, A., Saluja, K., Gupta, S., Kukreja, V.: Computational system based on machine learning with hybrid security technique to classify crime offenses. In: Lecture Notes in Electrical Engineering (2024). https://doi.org/10.1007/978-981-99-2271-0_20
21. Athavale, V.A., Garg, S., Kumar, D., Gupta, S.C.: Software fault detection by using rider optimization algorithm (ROA)-based deep neural network (DNN). In: Lecture Notes in Electrical Engineering (2024). https://doi.org/10.1007/978-981-19-4687-5_4
22. Athavale, V.A., Chincheti, A., Dixit, R.K., Lobo, L.M.R.J., Lapina, M.: A safe and secure online system for bidding using blockchain technology. In: Lecture Notes in Networks and Systems (2024). https://doi.org/10.1007/978-3-031-72171-7_5
23. Athavale, V.A., Rajput, A.S., Sohal, J.S.: Email header feature extraction using adaptive and collaborative approach for email classification. Int. J. Innov. Technol. Explor. Eng. **8**(7S), 158 (2019) [ISSN: 2278-3075]
24. Athavale, V., Rajput, A.S., Mittal, S.: Intelligent model for classification of SPAM and HAM. Int. J. Innov. Technol. Explor. Eng. **8**(6S), 773 (2024)
25. Yalagi, P.C.K., Chavan, V.D.: Analysis of intrusion detection system by applying machine learning using KNIME Tool. In: Lecture Notes in Electrical Engineering (2024). https://doi.org/10.1007/978-981-97-2839-8_47
26. Yalagi, P.S., Chavan, V.D.: A review of big data-based intrusion detection system using machine learning. In: Smart Innovation Systems and Technologies (2024). https://doi.org/10.1007/978-981-99-4717-1_32

27. Fernando, C.A., Thomas, R., A., M.A.E., Lapina, M.: A feature selection study on the bot-IoT dataset using ensemble classification techniques. In: Proc. IEEE Int. Conf. Contemporary Computing and Communications (InC4), Bangalore, India, pp. 1–9 (2024). https://doi.org/10.1109/InC460750.2024.10649035
28. Anita, E.A.M., Jenefa, J., Vinodha, D., Lapina, M.: Advancements in Sybil attack detection: a comprehensive survey of machine learning-based approaches in wireless sensor networks. In: Lecture Notes in Networks and Systems, vol. 863, pp. 67–75. Springer, Cham (2024). https://doi.org/10.1007/978-3-031-72171-7_7
29. Lapina, M., Dudun, G., Kotlyarov, D., Rjevskaya, N., Subramanian, S.J.: Analysis of an existing method for detecting adversarial attacks on deep neural networks. In: Lecture Notes in Networks and Systems, vol. 1044, pp. 316–329. Springer, Cham (2024). https://doi.org/10.1007/978-3-031-64010-0_29
30. Proshkin, N.A., Basan, E.S., Lapina, M.A., Klepikova, A.G., Lapin, V.G.: Developing models of IoT infrastructures to identify vulnerabilities and analyse threats. In: IOP Conference Series: Materials Science and Engineering, Volume 873, 1st International Conference on Innovative Informational and Engineering Technologies (IIET-2020), pp. 28–29 (2020, Stavropol, Russian Federation). https://doi.org/10.1088/1757-899X/873/1/012018
31. Basan, A.S., Basan, E.S., Lapina, M.A., Lapin, V.G.: Behavior-based assessment of trust in a cyber-physical system. In: Singh, P.K., Veselov, G., Vyatkin, V., Pljonkin, A., Dodero, J.M., Kumar, Y. (eds.) Futuristic Trends in Network and Communication Technologies. FTNCT 2020. Communications in Computer and Information Science, vol. 1395. Springer, Singapore (2021). https://doi.org/10.1007/978-981-16-1480-4_17
32. Basan, E., Shulika, M., Gladkov, N., Dubrivny, V., Lapina, M.: Methodology for assessing the security risks of cyber-physical systems. In: Lecture Notes in Networks and Systems, vol. 1044, pp. 273–284. Springer, Cham (2024). https://doi.org/10.1007/978-3-031-64010-0_25

Enhancing Scheduling with Q-Learning and Firefly Algorithm in IaaS Cloud Environment: A Security-Aware Approach

Zeya Mustafa, Mohammad Qasim, and Mohammad Sajid(✉)

Department of Computer Science, Aligarh Muslim University, Aligarh 202002, India
sajid.cst@gmail.com

Abstract. The ability to compute all possible mappings between workflow tasks and IaaS cloud resources, to identify an optimal configuration, is computationally infeasible. The security of all precedence-constraints tasks in a public IaaS cloud introduces a critical dimension to the scheduling problem, particularly for sensitive applications such as healthcare, finance, and defence. This study presents a hybrid scheduling approach, Q-learning-enhanced Firefly Algorithm (Q-FA), to optimize makespan and security of precedence-constraints workflow tasks in an IaaS cloud environment. The proposed Q-FA algorithm adapts mutation actions using reinforcement learning, enabling dynamic task scheduling based on reward-driven decisions. Experimental results on DAG-based workflows demonstrated that the proposed Q-FA outperforms Firefly Algorithm (FA), Whale Optimization Algorithm (WOA), Particle Swarm Optimization (PSO), Grey Wolf Optimization (GWO), and Salp Swarm Algorithm (SSA) in minimizing makespan and security levels achievement. The convergence analysis of Q-FA highlighted the fast and stable optimization behavior.

Keywords: Cloud Computing · Workflow Scheduling · Q-Learning · Firefly Algorithm · Makespan · Security Policy

1 Introduction

Cloud computing has revolutionized the computational landscape, emerging as a cornerstone of modern technology due to advancements in virtualization and Internet connectivity [1]. The distributed cloud paradigm supports billions of users worldwide, enabling individuals, organizations, and enterprises to manage complex computational workloads efficiently. By offering Infrastructure as a Service (IaaS), Platform as a Service (PaaS), and Software as a Service (SaaS), cloud computing provides a versatile platform catering to diverse needs, from small-scale personal applications to large-scale industrial workflows [2]. As cloud adoption continues to surge, the need for robust strategies in resource management, security, performance optimization, and reliability becomes increasingly critical. Among these, task scheduling is a pivotal challenge, directly influencing cloud infrastructures' performance, scalability, and cost-effectiveness [1, 2]. The

M. Lapina et al. (Eds.): AISMA 2025, LNNS 1456, pp. 335–351, 2026.
https://doi.org/10.1007/978-3-032-07275-7_31

ability to efficiently allocate tasks to virtualized resources determines the system's capacity to handle dynamic workloads, making scheduling a focal point of research in cloud computing. Computing all possible task-resource mappings to identify an optimal configuration is computationally infeasible, as the complexity scales exponentially with the number of tasks and resources [3, 5]. For NP-hard problems, this exponential growth renders exhaustive search methods impractical, necessitating heuristic and meta-heuristic approaches that reduce the solution space's complexity while delivering near-optimal solutions within reasonable timeframes [6].

Meta-heuristic algorithms have gained prominence in addressing NP-hard challenges, offering robust frameworks for optimizing task scheduling in cloud environments. Prior research has explored a variety of meta-heuristic techniques, including Genetic Algorithms (GA), Fuzzy-GA, Multi-Objective GA, PSO, GWO, Harris Hawks Optimization (HHO), and Ant Colony Optimization (ACO), to optimize the makespan in most cases [1, 2, 4, 7]. Despite their effectiveness, these algorithms must balance exploration and exploitation, a trade-off that significantly impacting their performance in dynamic cloud settings. Furthermore, the heterogeneity of cloud resources, including variations in VM performance and network latency, adds another layer of complexity, requiring algorithms to adapt to fluctuating conditions [8]. Moreover, Security-aware workflow scheduling introduces a critical dimension to the scheduling problem, particularly for applications in sensitive domains such as healthcare, finance, and defence, where tasks demand stringent security guarantees [9]. Each task is associated with a required security level, specifying the minimum security provisions necessary for execution, such as encryption or access control mechanisms, making the scheduling process more complicated.

This study introduces a novel hybrid approach, the Q-learning-enhanced Firefly Algorithm (Q-FA), to address the makespan and security-aware workflow scheduling. By integrating Q-learning's reinforcement learning capabilities with the Firefly Algorithm's nature-inspired optimization, Q-FA dynamically adapts mutation strategies to optimize task scheduling, offering a significant advancement over traditional meta-heuristics. The novelty of this work lies in the Q-learning-enhanced Firefly Algorithm (Q-FA), which innovatively combines Q-learning's adaptability with the Firefly Algorithm's robust balance of exploration and exploitation for secure workflow scheduling.

The rest of this paper is organized as follows. Section 2 discusses the related work in workflow scheduling and metaheuristic optimization. Section 3 delimits the problem formulation, followed by a description of the proposed algorithm in Sect. 4. Section 5 reports the experimental setup for evaluation results, which compares the proposed approach with FA, WOA, PSO, GWO, and SSA. Section 5 outlines the conclusion of this work and future research directions.

2 Related Work

Metaheuristic algorithms play a crucial role in solving the highly complex challenge of workflow scheduling in cloud computing. Wang et al. [10] propose an algorithm to schedule tasks in hybrid clouds, combining private and public resources, to maximize task completion within deadlines while meeting security needs and minimizing costs. The approach outperforms prior methods by integrating security with performance and cost considerations, leveraging real-world examples like Google's Borg cluster. Meng et al. [11] present Particle Swarm Optimization-based Dynamic Scheduling (PDS), a method for real-time resource allocation in Industrial Control Systems using a two-tier heterogeneous cloud architecture. PDS achieves lower costs than baselines like IC-PCPD2 and PSOS, with stable performance, adaptive PSO tuning, distributed computing efficiency, Pareto dominance for resource selection, and scalability. Hai et al. [12] propose enhanced versions of the Heterogeneous Earliest Finish Time (HEFT) algorithm: MXCT, MNCT, and AVBS for task scheduling in cloud computing to minimize makespan. Using a Directed Acyclic Graph (DAG) to model task dependencies, the methodology modifies HEFT's rank generation and processor selection without additional computational cost. Results show MXCT and MNCT outperform the basic HEFT (AVCT) in 39% and 40% of cases, respectively, while AVBS reduces the makespan in 86% of problem sets. Pasdar et al. [13] introduce the Hybrid Scheduling for Hybrid Clouds (HSHC) algorithm to optimize scientific workflow execution on hybrid clouds, balancing cost and execution time while minimizing data movement. HSHC employs a genetic algorithm for static scheduling and a dynamic phase to adapt to resource changes, prioritizing private cloud usage to reduce public cloud costs. Thekkepuryil et al. [14] propose a hybrid Ant-Lion Optimization (ALO) and Particle Swarm Optimization (PSO) algorithm (ALPSO) for optimizing workflow scheduling in cloud environments. The method targets multi-objective optimization, minimizing cost, makespan, and load while incorporating the Data Encryption Standard (DES) for data security. Shirvani and Talouki [15] introduce a hybrid heuristic-based list scheduling algorithm (HH-LiSch) to optimize makespan in cloud computing. It proposes two new task priority strategies, leveraging all virtual machine (VM) idle time slots for task duplication and insertion. Chraibi et al. [16] propose an enhanced deep Q-network (DQN) scheduler, MCS-DQN, to minimize makespan in cloudlet scheduling. It introduces a novel reward function to improve DQN convergence. It uses CloudSim to evaluate real-world HPC2N workloads (20–200 cloudlets) on a setup with one data center, two hosts, and five VMs. It employs a five-layer neural network and experience replay to enhance learning stability. Abdulghani [17] investigates hybrid task scheduling algorithms to minimize makespan in cloud computing. Using CloudSim, the study evaluates three hybrid algorithms: EDA-GA, HGA-ACO, and IDPSO against standalone GA, ACO, and PSO on 10 VMs with 50, 100, and 150 tasks. IDPSO outperforms others, achieving a makespan of 0.64 ms for 150 tasks, compared to 0.75 ms (EDA-GA), 0.77 ms (HGA-ACO), and higher values for standalone algorithms. The hybrid approaches improve load balancing and resource utilization, with IDPSO excelling in convergence and scalability. Aggarwal et al. [18] introduce an enhanced Fruit Fly Optimization (IFFO) algorithm to optimize makespan and cost scheduling multiple workflows in cloud environments. IFFO avoids local optima traps by enhancing the swarm smell function, ensuring better Pareto optimal solutions.

3 Workflow Scheduling Model

The models utilized in this paper—the application, cloud resource, and security models—are explained below.

3.1 Application Model

A scientific application can be expressed as a directed acyclic graph (DAG) $J = (T, E)$, where $T = \{t_0, t_1, t_2, t_3, \ldots, \ldots, t_{N-1}\}$ represents a set of N tasks and $E = \{e_{ij} = (t_i, t_j) \mid t_i, t_j \in T \textit{ and } t_i \textit{ is the predecessor of } t_j\}$ denotes the edges between the tasks representing the precedence constraints. An edge $e_{ij} = (t_i, t_j)$ indicates that the task t_j must be executed only after the completion of the task t_i. Each edge e_{ij} is assigned a non-negative weight c_{ij}, representing the communication time from the task t_i to the task t_j. The communication time $c_{i,j}$ is considered zero if tasks t_i and t_j are executed on the same virtual machine. Assume tasks t_x and t_y represent the predecessor and successor tasks of a task t_i, respectively. The predecessor and successor task sets of the task t_i are given as $Pred_{t_i} = \{t_x \mid e_{xi} = (t_x, t_i) \mid t_x \in T\}$ and $Succ_{t_i} = \{t_y \mid e_{iy} = (t_i, t_y) \mid t_y \in T\}$, respectively. Let t_{entry} and t_{exit} represents the entry and exit task of a DAG. Since the entry task of a DAG has no predecessor, i.e., $Pred_{t_{entry}} = \varnothing$ and exit task has no successor, i.e., $Succ_{t_{exit}} = \varnothing$. An example of a scientific workflow, i.e., DAG, as shown in Fig. 1, consists of eight tasks. As can be seen, the task sequence $(t_0, t_3, t_2, t_5, t_6, t_1, t_4, t_7)$ is correct since it adheres to the task dependencies. However, because t_1 and t_2 are the predecessors of t_4, the task sequence $(t_0, t_3, t_2, t_4, t_6, t_7, t_1, t_5, t_8, t_9)$ is invalid

3.2 Cloud Resource Model

Assume the pre-configured virtual machine (VM) set comprises M VMs, represented as $V = \{v_0, v_1, v_2, \ldots\ldots\ldots, v_{M-1}\}$ offered by a cloud service provider (CSP). A VM $v_m \in V$ can be represented as a tuple (CP_m, BD_m), with CP_m as the computing capacity and BD_m refer to the bandwidth of the virtual machine $v_m \in V$.

Assume a binary decision variable χ_{ij} to assign the task t_i on VM v_j, given as follows-

$$\chi_{ij} = \begin{cases} 1, & Task\ t_i \text{ assigned to VM } v_j \\ 0, & Otherwise \end{cases} \quad \forall t_i \in T, \forall v_j \in V \tag{1}$$

The computing capacity (CP_m) of a VM $v_m \in V$ represents the computational power of virtual machines in terms of MIPS, i.e., $MIPS_{v_m}$. The computing capacity affects the execution time of tasks assigned to the VM. The faster v_m with larger CP_m result in improved computation performance. Let MI_{t_i} be the reference execution time for the task $t_i \in T$. The actual execution duration of the task t_i on VM v_j may be computed as follows-

$$ET_{t_i, v_j} = MI_{t_i} / MIPS_{v_j} \tag{2}$$

Where $MIPS_{v_j}$ is the computational capacity of the VM $v_j \in V$.

The average execution time of a task $t_i \in T$ concerning the VM set V can be computed as follows-

$$AET_{t_i} = \left(\sum\nolimits_{\forall v_j \in V} ET_{t_i, v_j} \right) / M \tag{3}$$

When executing separate tasks on various virtual machines, it is important to consider communication overhead. When two interdependent tasks are executed on a virtual machine, the communication time can't be ignored.

Let task t_i be the predecessor task of t_j and both are executed on two different VMs v_α and v_β ($0 \leq \alpha, \beta \leq M - 1, \alpha \neq \beta$), respectively. Let d_i represent the volume of data to be sent for the task t_j. The time required for communication between tasks t_i and t_j may be calculated by

$$c_{i,j} = \begin{cases} \frac{d_i}{\min(B_\alpha, B_\beta)}, & if \ v_\alpha \neq v_\beta \\ 0 & , if \ v_\alpha = v_\beta \end{cases} \tag{4}$$

Where, B_α and B_β represents the bandwidth of VMs v_α and v_β, respectively.

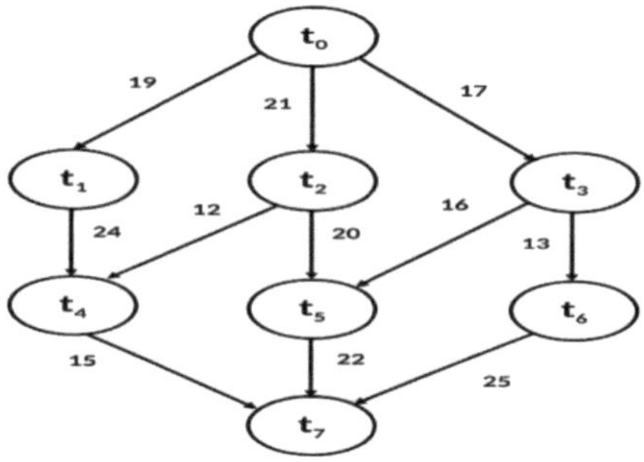

Fig. 1. A sample DAG for workflow.

3.3 Security Model

CSPs offer customers a variety of security services to ensure system security. Consumers anticipate that security will be guaranteed at the lowest possible makespan. Assume K security services, $SL = \{SL_0, SL_1, SL_2, \ldots, SL_{K-1}\}$ can be used to represent these security services. The security services and encryption techniques correspond to each other directly. The kth security service SL_k is represented as $SL_k = \left(sl_k, T_{SL_k}^{to}\right)$, with sl_k and $T_{SL_k}^{to}$ stand for security level and time overhead, respectively. Assume that the required and assigned security levels of the task $t_i \in T$ are $SL_{t_i}^{req}$ and SL_{t_i}. The required security level of job J for all tasks $t_i \in T$ can be determined as follows:

$$RSL^J = \sum_{i=1}^{N} SL_{t_i}^{req} \tag{5}$$

The security level of job J for all tasks $t_i \in T$ can be determined as follows [19]:

$$SL^J = \sum_{i=1}^{N} SL_{t_i} \tag{6}$$

Even when tasks use security services, there is still the possibility that they might fail. If task t_i encrypts its data using a security service SL_k, then $SL_{t_i} = SL_k$. We utilize an exponential model to express the likelihood without sacrificing generality. Therefore when t_i accepts security service SL_k, the probability that the task t_i (1 ≤ i ≤ N) fails.

$$P_{fail}(t_i) = 1 - e^{-\lambda\left(SL_{t_i}^{req} - SL_{t_i}\right)} \tag{7}$$

Where λ is a positive security risk coefficient that varies based on the application.

The security of t_i is therefore defined by

$$\sec(t_i) = 1 - P_{fail}(t_i) \tag{8}$$

The security of the application J is calculated using,

$$Sec_J = \prod_{i=1}^{N} \sec(t_i) = e^{-\lambda \sum_{i=1}^{N}\left(SL_{t_i}^{req} - SL_{t_i}\right)} \tag{9}$$

It should be noted that the application and task set are determined first, followed by the security risk coefficient (λ) and the security requirement of each task ($SL_{t_i}^{req}$).

The equation for the unit encryption overhead while utilizing the service SL_k is $Enc_{SL_k}^{t} = \theta_{SL_k} \times T_{SL_k}^{to}$, where θ_{SL_k} is a positive coefficient corresponding to the security service SL_k. The overhead of encryption time for the task $t_i \in T$ utilizing the security service $SL_{t_i} = SL_k$can be computed as-

$$Enc_{t_i,SL_{t_k}} = Enc_{SL_k}^{t} \times D_i = \theta_{SL_k} \times T_{SL_k}^{to} \times d_i \tag{10}$$

Where d_i is the size of the encrypted data of the task $t_i \in T$.

Assume a binary decision variable ϑ_i to provide the required security level for the task $t_i \in$ is given as follows-

$$\vartheta_i = \begin{cases} 0, & if\ SL_{t_i} \geq SL_{t_i}^{req} \\ 1, & Otherwise \end{cases} \quad \forall t_i \in T \tag{11}$$

The zero value of ϑ_i represents that the task t_i has been executed successfully with proper security; otherwise, the security of the task t_i has been compromised.

The total tasks with compromised security can be computed as follows-

$$CS_J = \sum_{i=1}^{N} \vartheta_i \tag{12}$$

3.4 Problem Formulation

Workflow scheduling aims to allocate and schedule the tasks to the proper VMs in a way that respects both security and precedence requirements, to optimize the makespan. We take encryption overhead and job execution time into account for the makespan.

The well-known HEFT [20] and other algorithms employ the upward rank value, which is given as follows:

$$U_{rank}(t_i) = AET_{t_i} + \max_{t_j \in succ(t_i)}\left(c_{t_i,t_j} + U_{rank}(t_j)\right) \tag{13}$$

where AET_i denotes the average execution time of the task t_i.

To schedule a task t_i on VM v_j, the earliest start time EST_{t_i,v_j} and the earliest finish time EFT_{t_i,v_j} can be calculated as follows:

$$EST_{t_i,v_j} = \max\{Avail_{v_j}, \max_{t_j \in pred(t_i)}(AFT_{t_j,v_k} + c_{j,i})\} \quad (14)$$

where AFT_{t_j,v_k} represents the actual finish time of the task t_j after scheduling the task t_j on VM v_k. The $Avail_{v_j}$ denotes the earliest available time of the VM v_j. If t_h be the last task to be scheduled on the VM v_j, then $Avail_{v_j} = AFT_{t_h,v_j}$.

The earliest finish time EFT_{t_i,v_j} of the task t_i on VM v_j can be calculated as follows:

$$EFT_{t_i,v_j} = EST_{t_i,v_j} + ET_{t_i,v_j} + Enc_{t_i,SL_{t_k}} \quad (15)$$

Where $Enc_{t_i,SL_{t_k}}$ indicates the encryption time required for using the security level SL_{t_k}.

Since the makespan represents the maximum time needed to complete the execution of the observed workflow application, tasks within the workflow are executed on cloud servers based on the resource allocation strategy employed. After scheduling the exit task on a specific VM, a workflow's schedule length, commonly called the makespan, can be computed. The makespan can be directly obtained by calculating the finish time of the exit task.

$$\mathrm{MK}_J = \max\{AFT_{t_{exit},v_k} : t_{exit} \in \mathrm{T}\} \quad (16)$$

The weighted sum method can optimize the tasks' makespan and security, as given below.

$$\sigma_{opt} = \lambda_1 * \frac{\mathrm{MK}_J}{MK_{Max}} + \lambda_2 * \frac{CS_J}{\mathrm{N}} \quad (17)$$

where λ_1, and λ_2 are the weight values vary from 0 to 1, with $\lambda_1 + \lambda_2 = 1$. The MK_{Max} and N denotes the maximum makespan values and total tasks, respectively.

The optimization problem can be written as follows-

$$\text{Minimize} : \sigma_{opt} = \lambda_1 * \frac{\mathrm{MK}_J}{MK_{Max}} + \lambda_2 * \frac{CS_J}{\mathrm{N}}$$

Subject to constraints

$$SL_{t_i} \geq SL_{t_i}^{req}, \forall t_i \in T \quad (18)$$

$$SL^J \geq RSL^J \quad (19)$$

$$\sum_{\forall v_j} \chi_{i,j} = 1, \forall t_i \in T, \chi_{i,j} \in \{0, 1\} \quad (20)$$

$$\sum_{\forall t_i} \sum_{\forall v_j} \chi_{i,j} = N, \forall t_i \in T, \chi_{i,j} \in \{0, 1\} \quad (21)$$

$$EST_{t_i,v_m} \geq \max\{Avail_{v_m}, \max_{t_j \in pred(t_i)}(AFT_{t_j,v_k} + c_{j,i})\}, \forall t_i \in T, \forall v_m \in V \quad (22)$$

Constraint (18) ensures that the assigned security level must be greater than or equal to the required security level. Constraint (19) ensures that the total assigned security levels must be greater than or equal to the required ones. Constraint (20) ensures that every task must be executed once by a single VM only, and constraint (21) ensures that the total tasks executed must not exceed the total tasks in the workflow. The constraint (22) provides that the start time of each task must be greater than or equal to both the VM's availability time and the sum of the actual finish time of each predecessor task and the communication time from the predecessor task to the current task.

4 Proposed Method

The details of the proposed Q-FA, along with all components, are explained below.

4.1 Firefly Algorithm (FA)

The Firefly Algorithm is a nature-inspired metaheuristic optimization method that draws inspiration from the flashing behaviour of fireflies to tackle complex problems like task scheduling in cloud computing. In this algorithm, each firefly represents a potential solution, such as a specific arrangement of tasks assigned to virtual machines in a cloud environment. A firefly's brightness reflects its solution's quality, with brighter fireflies indicating better outcomes, such as reduced task completion times or improved security compliance. Fireflies are attracted to brighter counterparts, moving toward them to refine their solutions, while also incorporating random movements to explore new possibilities and avoid getting trapped in suboptimal results. This dual mechanism of attraction and randomization enables the algorithm to efficiently navigate large, complex solution spaces, balancing the refinement of promising schedules with the discovery of new ones.

The Euclidean distance between firefly (i) and firefly (j) in an (d)-dimensional search space is calculated as:

$$r_{ij} = \sqrt{\sum_{k=1}^{d} \left(x_{i,k} - x_{j,k}\right)^2} \tag{23}$$

Where r_{ij} is the distance between fireflies i and j, $x_{i,k}$ and $x_{j,k}$ are the kth components of the position vectors of fireflies i and j, d is the dimension of the search space.

The attractiveness of a firefly decreases with distance and is defined as:

$$\beta = (\beta_0 - \beta_{min}) \cdot e^{\left\{-\gamma r_{ij}^2\right\}} + \beta_{min} \tag{24}$$

Where β_0 is the initial attractiveness, β_{min} is the minimum attractiveness, γ is the light absorption coefficient, and r_{ij} is the distance between fireflies i and j.

The movement of firefly (i) towards a more attractive firefly (j) is governed by:

$$x_i = x_i \times (1 - \beta) + x_j \times \beta + \alpha \times (\text{rand} - 0.5) \times scale \tag{25}$$

Where α is the randomization parameter, rand is a vector of random numbers uniformly distributed in [0, 1], and scale is the range of the search space for each dimension.

4.2 Q-learning

Q-learning is a model-free reinforcement learning method that allows an agent to learn optimum actions to maximise cumulative rewards. It works by keeping a Q-table containing Q-values indicating the predicted potential rewards for each state-action combination. The agent adjusts these values using the Bellman equation: $Q(s, a) = Q(s, a) + \alpha[R + \gamma \max Q(s', a') - Q(s, a)]$, where α represents the learning rate, R is the immediate reward, γ is the discount factor, and s' represents the subsequent state. The agent refines its policy over time by exploration and exploitation, such as the ε-greedy approach. Q-learning converges on an optimum strategy under sufficient exploration and a declining learning rate.

4.3 Firefly Encoding

In the security-aware workflow scheduling problem context, a firefly (vector) represents a solution to the problem. A task sequence, a mapping of tasks to VM, and a security sequence are utilized by the proposed Q-FA. Thus, each firefly comprises three components: task sequence, VM Sequence, and security sequence. The dimension of each firefly vector is equal to 3*N, in which the first N values represent the task sequence, the following N values represent task-VM assignments, and the last N values represent security level allocations.

Furthermore, any solution must comply with the precedence restrictions between tasks, which state that a child task cannot begin until all its parent tasks have been completed. The introduction of the Candidate Set, in which every task is a sibling task to the others, always satisfies the requirements of this encoding/decoding scheme. As a result, every task in Candidate Set may be scheduled since the parent tasks of all CTS tasks have already been scheduled. Each Firefly may be decoded with the workflow DAG data for a scheduling solution. This encoding mechanism is an extension of the encoding mechanism utilized by the Real-Relative Genetic Algorithm [21]. Using the last N values of the firefly, the security levels for all tasks are computed using the modulo operator utilized in the COBGA algorithm [22]

4.4 Proposed Q-FA

The proposed Q-learning-enhanced Firefly Algorithm (Q-FA) combines the firefly algorithm and Q-learning to solve the security-aware workflow scheduling. Algorithm 1 represents the pseudocode of the proposed Q-FA. The Q-FA is a hybrid algorithm that combines the Firefly Algorithm's nature-inspired optimization with Q-learning's adaptive decision-making to improve task scheduling in cloud environments. Each firefly represents a potential scheduling solution in this method, encoding task sequences, virtual machine assignments, and security levels to minimize completion time while ensuring security compliance. The Firefly Algorithm drives the search by having fireflies move toward brighter, better solutions, balancing exploration of new possibilities with refinement of promising ones.

Algorithm 1: Q-learning-enhanced Firefly Algorithm

```
Input:
SET Job Parameters, i.e., number of tasks, computation and communication times
FECTH VM parameters, i.e., speed, security levels
SET Firefly Parameters, i.e., dimensions, number of fireflies (population size), attractiveness, and bound
SET Q-Table, states, actions, learning rate, discount factor, exploration probability
Output: Return Firefly with improved makespan, and proper security level
Begin:
        Initialize population
        Compute the Fitness (Intensity) of each Firefly
        For all Iterations:
            For all Fireflies
                Calculate the Euclidean distance between fireflies
                Update the firefly position if it is less bright than the best firefly
                Calculate the attractiveness parameter
                Generate random movement
                Update Firefly's positions
            End For
            Rank fireflies and find the best Firefly
            Pass the best Firefly to Q-learning
        End For
        Return the best Firefly with Improved makespan and proper security levels.
End
```

The Q-learning, a reinforcement learning technique, enhances this process by dynamically selecting the best mutation strategies, such as swapping tasks or reassigning machines, based on learned rewards like reduced completion times. The Q-learning defines two states and four actions. The two states, i.e., the Improvement and the No improvement states, represent each solution's state. The four actions have been defined, i.e., Single-Value Update(A1), Dual-Values Update (A2), Random Count Multi-Value Update (A3), and Fixed-Range Multi-Value Update (A4). These actions are designed for Q-Learning to explore the solution space. The Single-Value Update(A1) action randomly changed a single value in the solution, while the Dual-Value Update (A2) action. In the Random Count Multi-Value Update (A3) action, a random set of values from random positions of the solution is updated. In contrast, a random continuous sequence is updated using the Fixed-Range Multi-Value Update (A4) action. A Q-Table maps states to actions, storing the Q-value of each action in a given state.

5 Result and Discussion

The performance of the proposed Q-learning-enhanced Firefly Algorithm (Q-FA) is compared with the Firefly Algorithm (FA), Whale Optimization Algorithm (WOA), Particle Swarm Optimization (PSO), Grey Wolf Optimization (GWO), and Salp Swarm Algorithm (SSA), with their default parameters, using a simulation environment developed based on Python 3.13.5. The Q-Learning's learning rate, discount factor, and exploration probability are 0.8, 0.95, and 0.2, respectively. Tables 1 and 2 show the standard experiment parameters and security policies. The required security levels, computation, and communication times for four workflow jobs (J_0, J_1, J_2, J_3) are given in Tables 3, 4, 5 and 6.

Table 1. System Parameters.

Parameter Name		Values
Scheduling	Sets of VMs(M)	3
	Task Set(N)	$\{J_0, J_1, J_2, J_3\}$
	Relative VM Speed (MIPS)	[1.0, 2.0, 3.0]
Metaheuristics Common Parameters	Lower & Upper Bounds	0 & 100
	Population size	100
	Maximum Iterations	100
	Average Index	2

Table 2. Security Policies [19].

Security Services	Ciphers	Encryption Time (ms/KB)
SL_0	RC4	0.0063
SL_1	RC5	0.0125
SL_2	BLOWFISH	0.0170
SL_3	IDEA	0.0196
SL_4	SKIPJACK	0.0217
SL_5	3DES	0.0654

Table 3. Second Job (J_0).

Tasks	Computation Times			Communication Times	Required Security Level
	VM v_0	VM v_1	VM v_2		
t_0	18	16	9	$c_{0,1}$ (23), $c_{0,2}$(11), $c_{0,3}$(21)	SL_2
t_1	22	17	19	$c_{1,4}$(17)	SL_0
t_2	17	23	19	$c_{2,4}$(24), $c_{2,5}$(19)	SL_1
t_3	20	16	26	$c_{3,5}$(13)	SL_0
t_4	24	13	15	$c_{4,6}$(11),$c_{4,7}$(21)	SL_5
t_5	12	19	15	$c_{5,6}$(23),$c_{5,7}$(14)	SL_4
t_6	19	21	22	–	SL_3
t_7	17	20	10	–	SL_1

Table 4. Third Job (J_1).

Tasks	Computation Times			Communication Times	Required Security Level
	VM v_0	VM v_1	VM v_2		
t_0	11	23	25	$c_{0,1}$ (25), $c_{0,2}$(18)	SL_2
t_1	11	16	22	$c_{1,3}$(21),$c_{1,4}$(14)	SL_0
t_2	14	12	9	$c_{2,3}$(24), $c_{2,4}$(20), $c_{2,5}$(13)	SL_1
t_3	18	23	17	$c_{3,6}$(17)	SL_0
t_4	11	19	15	$c_{4,6}$(23), $c_{4,7}$(12)	SL_4
t_5	11	10	15	$c_{5,7}$(25)	SL_2
t_6	23	14	18	–	SL_5
t_7	14	23	29	–	SL_2

Table 5. Fourth Job (J_2).

Tasks	Computation Times			Communication Times	Required Security Level
	VM v_0	VM v_1	VM v_2		
t_0	10	15	21	$c_{0,1}$ (15), $c_{0,2}$(19), $c_{0,3}$(10)	SL_2
t_1	14	9	11	$c_{1,4}$(11)	SL_0
t_2	9	12	15	$c_{2,6}$(17)	SL_1
t_3	16	13	11	$c_{3,4}$(12), $c_{3,5}$(13), $c_{3,6}$(14)	SL_0
t_4	11	18	5	$c_{3,7}$(18)	SL_4
t_5	14	10	12	$c_{3,7}$(11)	SL_5
t_6	8	13	17	$c_{3,7}$(20)	SL_3
t_7	10	12	16	–	SL_0

Table 6. Fourth Job (J_3).

Tasks	Computation Times			Communication Times	Required Security Level
	VM v_0	VM v_1	VM v_2		
t_0	14	16	9	$c_{0,1}$ (18), $c_{0,2}$(12), $c_{0,3}$(9), $c_{0,4}$(11), $c_{0,5}$(14)	SL_2
t_1	13	19	18	$c_{1,7}$(19), $c_{1,8}$(16)	SL_0
t_2	11	13	19	$c_{2,6}$(23)	SL_4
t_3	13	8	17	$c_{3,7}$(27), $c_{3,8}$(23)	SL_5
t_4	12	13	10	$c_{4,8}$(13)	SL_1

(continued)

Table 6. (*continued*)

Tasks	Computation Times			Communication Times	Required Security Level
	VM v_0	VM v_1	VM v_2		
t_5	13	16	9	$c_{5,7}(15)$	SL_2
t_6	7	15	11	$c_{6,9}(17)$	SL_4
t_7	5	11	24	$c_{7,9}(11)$	SL_3
t_8	18	12	20	$c_{8,9}(13)$	SL_1
t_9	21	7	16	–	SL_3

6 Results

The convergence curves in Fig. 2 illustrate the performance of the metaheuristic algorithms across the five job instances. The Q-learning enhanced Firefly Algorithm (Q-FA) consistently demonstrates superior convergence speed and achieves the lowest optimization value (Eq. (17)), signifying its effectiveness and robustness. It suggests that Q-FA is more efficient and capable for exploration and exploitation during the optimization process. Overall, Q-FA stands out as the most effective solution for fitness function (Eq. (17)) in these tested scenarios compared to the other algorithms lagging in both speed and optimality of final results.

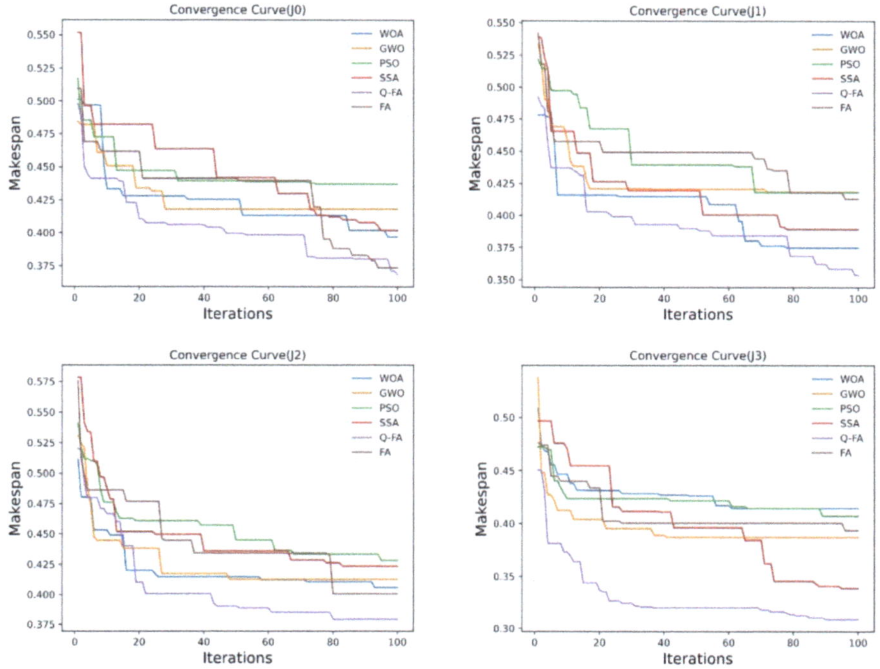

Fig. 2: Convergence Curves of Four Jobs.

Figure 3 depicts the Gantt Chart for four jobs scheduled across three virtual machines using Q-FA only. Each Gantt chart illustrates task assignments as colored blocks on different VMs. The makespan, indicated by a vertical black line, highlights optimization improvements offered by Q-FA.

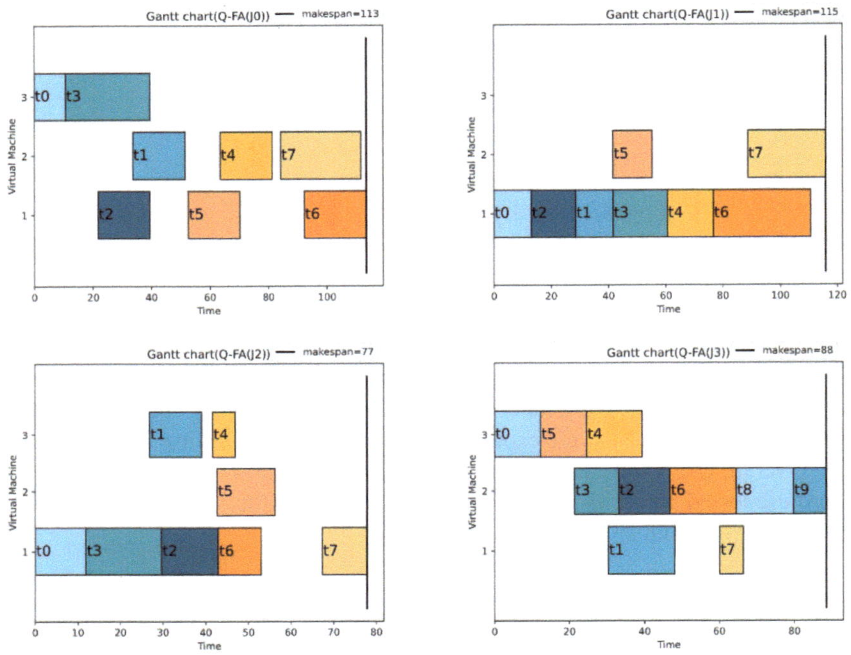

Fig. 3. Gantt Charts of Four Jobs by the Proposed Q-FA.

Figure 4a illustrates the makespan performance of six algorithms, i.e., WOA, GWO, PSO, SSA, Q-FA, and FFA, across four jobs. The makespan, shown on the vertical axis, represents the total completion time required for each job. Generally, the proposed Q-FA algorithm frequently yields marginally lower makespan values in most cases. Figure 4b demonstrates the total successful security levels achieved by the six algorithms for four jobs. As shown, the proposed Q-FA is the most successful algorithm to secure all tasks for the four jobs.

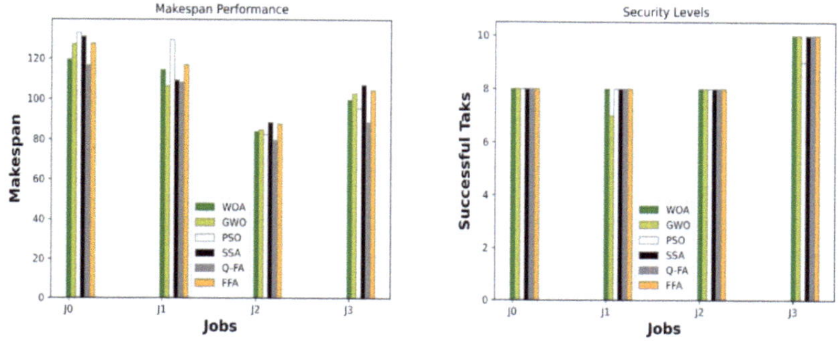

Fig. 4. a). Makespan. b). Achieved Security Levels.

Figure 4b compares the number of successful tasks for four jobs: J_0, J_1, J_2, and J_3 across six MH algorithms: WOA, GWO, PSO, SSA, Q-FA, and FFA. The results indicate that most algorithms achieve similar levels of success, particularly for jobs J_0, and J_2, where almost all algorithms reached the highest number of successful tasks. Notably, for the jobs J_1 and J_3 GWO and PSO had a slightly lower success count than other algorithms, while the other algorithms maintained consistent performance. Overall, the bar graph highlights that all algorithms demonstrate comparable efficacy in successful task completion except for GWO on J_1 and PSO on J_2. Overall, the experimental results show that Q-FA outperforms other metaheuristic algorithms in convergence speed, makespan minimization, and security compliance for different workflow jobs.

7 Conclusion

Security-aware scheduling in cloud computing integrates security requirements, including data confidentiality, access control, and compliance, directly affecting task allocation. This study proposed a Q-learning-enhanced Firefly Algorithm (Q-FA) that optimizes makespan while ensuring security-aware task assignments in IaaS cloud environments. Experimental results demonstrate that Q-FA outperforms other metaheuristic algorithms in convergence speed, makespan minimization, and security compliance across varied DAG-based workflows. Future research should explore integrating real-time scheduling capabilities, adaptive learning rates for Q-learning to enhance convergence speed and stability, and deploying Q-FA in heterogeneous cloud-edge environments for real and large workflows.

Acknowledgments. This work was supported partially by the Research Startup Grant from the University Grant Commission (UGC)-Basic Scientific Research (BSR) under Grant F.30-377/2017 (BSR) 07.03.2017, F.D. Dy. No. 8123 16.02.2017/10057/B053.

I, Mohammad Sajid, would like to thank the authorities of the Inter University Centre for Astronomy and Astrophysics, Pune, India, for providing research facilities.

References

1. Jawed, M.S., Sajid, M.: A comprehensive survey on cloud computing: architecture, tools, technologies, and open issues. Int. J. Cloud Appl. Comput. **12**(1), 1–33 (2022). https://doi.org/10.4018/IJCAC.2022010101
2. Etro, F.: Introducing cloud computing. In: London Conference on Cloud Computing for the Public Sector, pp. 1–20 (2010)
3. Goyal, T., Agrawal, A.: Host scheduling algorithm using genetic algorithm in cloud computing environment. Int. J. Res. Eng. Technol. **1**(1), 7–12 (2013)
4. Zomaya, A.Y., Teh, Y.H.: Observations on using genetic algorithms for dynamic load-balancing. IEEE Trans. Parallel Distrib. Syst. **12**(9), 899–911 (2001). https://doi.org/10.1109/71.954620
5. Qasim, M., Sajid, M., Rajak, R., Shahid, M.: Task scheduling strategy using chaotic whale optimization algorithm in cloud computing. In: Nature-Inspired Optimization Algorithms for Cyber-Physical Systems, pp. 31–52. IGI Global Scientific Publishing (2025). https://doi.org/10.4018/978-1-7998-8484-2.ch002
6. Deb, K.: Multi-objective optimization using evolutionary algorithms. John Wiley & Sons (2001)
7. Qasim, M., Sajid, M.: Chaotic and transfer function-assisted grey wolf optimizer for IoT task scheduling in cloud-of-things. In: 2024 ASU International Conference in Emerging Technologies for Sustainability and Intelligent Systems (ICETSIS), pp. 1323–1327. IEEE (2024). https://doi.org/10.1109/ICETSIS61505.2024.10459635
8. Chen, H., Heidari, A.A., Zhao, X., et al.: Advanced orthogonal learning-driven multi-swarm optimization. Knowledge-Based Systems. **202**, 105–123 (2020). https://doi.org/10.1016/j.knosys.2020.105510
9. Li, W., Liu, H., Xu, J.: Security-aware scheduling for scientific applications in cloud computing. In: IEEE International Conference on High Performance Computing and Communications, pp. 120–127. IEEE (2017). https://doi.org/10.1109/HPCC-SmartCity-DSS.2017.17
10. Wang, B., Wang, C., Huang, W., Song, Y.: Xiaoyun Qin.: Security-aware task scheduling with deadline constraints on heterogeneous hybrid clouds. J. Parallel Distrib. Comput. **153**, 15–28, ISSN 0743–7315, (2021). https://doi.org/10.1016/j.jpdc.2021.03.003
11. Meng, S., et al.: Security-aware dynamic scheduling for real-time optimization in cloud-based industrial applications. IEEE Trans. Ind. Inf. **17**(6), 4219–4228 (2021). https://doi.org/10.1109/TII.2020.2995348
12. Hai, T., Zhou, J., Jawawi, D., et al.: Task scheduling in cloud environment: optimization, security prioritization and processor selection schemes. J. Cloud Comput. **12**, 15 (2023). https://doi.org/10.1186/s13677-022-00374-7
13. Pasdar, A.M., Lee, Y.C., Almi'ani, K.: Hybrid scheduling for scientific workflows on hybrid clouds. Comput. Networks. **181**, 107438, ISSN 1389-1286 (2020). https://doi.org/10.1016/j.comnet.2020.107438
14. Kakkottakath Valappil Thekkepuryil, J., Suseelan, D.P., Keerikkattil, P.M.: An effective meta-heuristic based multi-objective hybrid optimization method for workflow scheduling in cloud computing environment. Cluster Comput. **24**, 2367–2384 (2021). https://doi.org/10.1007/s10586-021-03269-5
15. Shirvani, M.H., Talouki, R.N.: A novel hybrid heuristic-based list scheduling algorithm in heterogeneous cloud computing environment for makespan optimization. Parallel Computing. **108**, 102828, ISSN 0167-8191 (2021). https://doi.org/10.1016/j.parco.2021.102828
16. Chraibi, A., Ben Alla, S., Ezzati, A.: Makespan optimisation in cloudlet scheduling with improved DQN algorithm in cloud computing. Sci. Program. **2021**, 7216795 (2021). https://doi.org/10.1155/2021/7216795

17. Abdulghani, A.: Hybrid task scheduling algorithm for makespan optimisation in cloud computing: a performance evaluation. J. Artif. Intell. **6**, 241–259 (2024). https://doi.org/10.32604/jai.2024.056259
18. Aggarwal, A., et al.: IFFO: an improved fruit fly optimization algorithm for multiple workflow scheduling minimizing cost and makespan in cloud computing environments. Math. Probl. Eng. **2021**, 5205530 (2021). https://doi.org/10.1155/2021/5205530
19. Li, L., et al.: Makespan and security-aware workflow scheduling for cloud service cost minimization. IEEE Trans. Cloud Comput. **12**(2), 609–624 (2024). https://doi.org/10.1109/TCC.2024.3382351
20. Topcuoglu, H., Hariri, S., Wu, M.-Y.: Performance-effective and low-complexity task scheduling for heterogeneous computing. IEEE Trans. Parallel Distrib. Syst. **13**(3), 260–274 (2002). https://doi.org/10.1109/71.993775
21. Jiang, J., Sun, Z., Lu, R., Pan, L., Peng, Z.: Real relative encoding genetic algorithm for workflow scheduling in heterogeneous distributed computing systems. IEEE Trans. Parallel Distrib. Syst. (2024). https://doi.org/10.1109/TPDS.2024.3492210
22. Qasim, M., Sajid, M., Lapina, M., Shahid, M.: COBGA: MCDM-assisted improved genetic algorithm for scheduling industrial-internet-of-things jobs in cloud computing. Cluster Comput. **28**, 944 (2025)

Algorithms for Detecting Suspicious Transactions in Payment Systems Based on Artificial Intelligence

Fayzullo Nazarov, Sherzodjon Yarmatov, Munis Xamidov(✉), and Asliddin Sayidkulov

Samarkand State University named after Sharof Rashidov, Samarkand, Uzbekistan
samsungmunis@gmail.com

Abstract. This research work aims to develop models and algorithms for detecting suspicious transactions for payment systems based on artificial intelligence. In this work, algorithms for detecting suspicious transactions in payment systems based on machine learning and their analysis are described in detail. Today, the proliferation of payment technologies has heightened the risk of suspicious transactions inflicting substantial harm on financial institutions; hence, the identification of such transactions is very important. The effectiveness of several machine learning methods in identifying questionable data was assessed. This article is crucial for guaranteeing security and dependability in sectors such as payment systems, banking, and healthcare.

Keywords: Artificial intelligence · Suspicious transaction · Machine learning · Logistic regression · Decision Tree · Random Forest · SVM · Neural network

1 Introduction

With the organization of payment processes in online systems today, the probability of suspicious transactions arising and inflicting substantial harm on banking and other financial sectors is growing. Addressing these difficulties necessitates the use of contemporary methodologies. Artificial intelligence and machine learning are regarded as very effective techniques for detecting suspicious transactions, enabling rapid and precise examination of extensive data sets. Currently, the amount of data is substantial, and machine learning significantly contributes to its analysis. Rapidly aggregating and evaluating data from many sources facilitates the prompt detection of anomalous transactions. The use of these algorithms in banking and finance, as well as in information technology, marketing, and healthcare, enhances their significance. Machine learning classification algorithms and neural network models serve as the main tools for identifying suspicious transactions based on the analysis of payment data.

M. Lapina et al. (Eds.): AISMA 2025, LNNS 1456, pp. 352–361, 2026.
https://doi.org/10.1007/978-3-032-07275-7_32

2 Characteristics of Suspicious Transaction Detection

Payment systems rely on several criteria and are mostly executed by the purchaser. There are several features for detecting suspicious transactions, which include:

Customer Behavior. A customer's behavior that deviates from their usual behavior, such as a sudden increase in the amount of the payment, or multiple payments within a short period of time, can be considered suspicious [1, 2];
Geographic location and time. Typically, when users make payments within a certain geographic area, unexpected transactions from another country or city are considered suspicious transactions;
Time. While users typically make transactions during a certain time of day, a transaction made at an unexpected time, such as midnight, is considered suspicious;
Change in payment method. Usually, when using the same type of payment system, using different payment methods at an unexpected time also leads to a suspicious transaction [1, 3];
Using different IP addresses. It is also considered suspicious if a customer makes payments from multiple IP addresses and these payments are made from different locations at the same time;
Between the category of goods or service and the transaction value. If the price of the goods received by the consumer deviates from the amount paid, it indicates a potentially suspicious transaction.
E-commerce. An online store notices that a single user is making many small transactions below the reporting threshold. This may be an attempt to evade detection when transferring illicit funds [3, 4].

To detect dubious transactions in payment systems, an initial dataset is established, which is then used to train machine learning algorithms after normalization.

The characteristics to consider for a dataset are as follows:

transaction_id – an identifier for each executed transaction;
user_id – the identifier for the user performing the transaction;
merchand_id – an identifier for each employee in the system;
transaction_amount – the monetary value of the completed transaction;
timestamps – the date and time when the completed transaction occurred;
payment_method – the method used for the completed transaction;
location – he geographical location of the transaction being performed;
ip_address – the IP address of the device where the transaction was made;
is_suspicious – a binary flag that indicates whether the executed transaction is suspicious (1) or not (0).

3 Models and Algorithms for Detecting Suspicious Transactions

Algorithms like as regression, decision trees, random forests, and support vector machines (SVM) are proficient at identifying suspicious transactions. These algorithms classify suspicious and normal transactions into predefined classes based on previous transaction data. Classification algorithms make a suspiciousness decision based on training and testing phases.

Figure 1 below delineates the methods that are now the most used for identifying suspicious transactions.

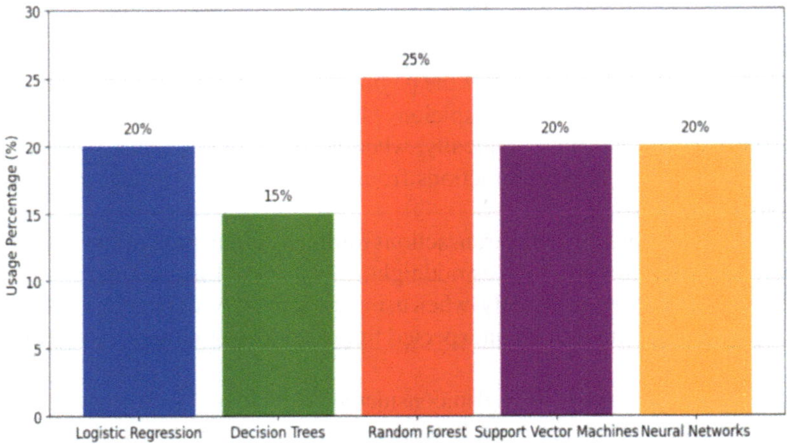

Fig. 1. Algorithm Usage for Suspicious Transaction Detection.

1. Logistic Regression is a statistical modeling method used to divide data into two classes (e.g., doubtful and certain). It is a comprehensible and implementable method that is expedient in calculation and very precise [5, 6].
2. Decision Trees: Classify data using decision trees, where each node partitions data according to certain criteria. Easy to understand and interpret, tolerant to ambiguous data. Overfitting problem, i.e. too close to the data.
3. Random Forest integrates several decision trees and aggregates their outcomes by majority voting. High accuracy, reduces the problem of overfitting. Requires computing power and memory [7, 10].
4. Support Vector Machines (SVM): Creates the best separation boundary (hyperplane) for the data. Efficient at handling intricate and high-dimensional datasets. Demands substantial computational resources and memory, and may also provide challenges when parameters are inaccurate [9].
5. Neural Networks: Allows for complex and deep learning from data sets. Modeling data using a network of neurons. Enhanced efficiency for extensive and intricate data, deep learning via multi-layered networks [10, 11].

Detailed information about machine learning algorithms for detecting suspicious transactions in payment systems is provided below:

3.1 Logistic Regression

Logistic Regression is a straightforward and effective method used in binary classification tasks. It is used to compute probabilities derived from a data collection and categorize them. The following mathematical model is based on this algorithm [5, 12].

Output Probability. In the Logistic Regression model, the output probability for each input x is expressed as:

$$P(y = 1|x) = \sigma(z) = \frac{1}{1 + e^{-z}}$$

where:

$P(y = 1| x)$—this is the probability that $y = 1$ for the input value x; $\sigma(z)$—sigmoid or logistic function;
$z = (w^T x + b)$—is a linear combination, where: w — vector of weights; x — vector of input variables;
b is the bias or constant [13].

Sigmoid Function. The model uses the sigmoid (logistic) function to represent the probability. This function $\sigma(z) = \frac{1}{1+e^{-z}}$ is expressed as a function that ranges between 0 and 1, which is suitable for representing the probability.

Loss Function. The Logistic Regression model learns by minimizing a loss function called binary cross-entropy or log loss. The loss function is defined as:

$$L(w, b) = -\frac{1}{N}\sum_{i=1}^{N}(y_i(y_i|x_i)) + (1 - y_i)\log(1 - P(y_i|x_i)))$$

where: N is the number of samples; y_i – the actual output value (0 or 1); $P(y_i| x_i)$ – the probabilities predicted by the Logistic Regression model [13, 14].

Optimize the Model Using Gradient Descent. It is necessary to update the weight coefficients w and bias b parameters in the model, for which the gradient descent method is effective.

$$w = w - \alpha\frac{\partial L}{\partial w}, b = b - \alpha\frac{\partial L}{\partial b}$$

where: α *is* the learning rate, $\frac{\partial L}{\partial b}$ and $\frac{\partial L}{\partial w}$ are the gradients.

3.2 Decision Tree

The primary procedure in developing a mathematical model of the **Decision Tree** method involves selecting the best division based on the attributes at each node of the tree, ultimately culminating in a decision at the leaf nodes.

The main mathematical expressions of this model are given below [15].

Criteria for Division The Decision Tree technique employs many impurity metrics to partition the data. The most common criteria are Gini impurity and Entropy, each of which is calculated to provide maximum separation of the data [15, 16].

Gini Impurity is a measure used in classification problems and represents the probability of placing a randomly selected sample in the wrong class. The Gini impurity is calculated as follows:

$$Gini(D) = 1 - \sum_{K=1}^{K} P_k^2$$

where: D is the data set; K is the number of classes; P_k^2 *is* the proportion of elements belonging to class k *in* D [17].

Entropy and Information Gain Entropy measures the information gain of a feature. In classification problems, the feature with the lowest entropy is selected.

$$Entropy(D) = -\sum_{k=1}^{k} p_k log_2(p_k)$$

Information Gain — signifies the extent to which interference diminishes as a result of feature partitioning:

$$InformationGain(D, A) = Entropy(D) - \sum_{V \in values(A)} \frac{| D_v |}{| D |} Entropy(D_v)$$

where: D is the total data set; A is the feature selected for partitioning; D_v- is the partition with the value A.

3.3 Random Forest

Random Forest is an ensemble learning method extensively used in classification and regression tasks, yielding consistent and precise outcomes. This algorithm comprises several Decision Tree models, each trained individually, with their results aggregated [16, 17].

Random Feature Selection: At each node, only a randomly selected subset of all features is considered for partitioning the data. For classification problems, $M features are$ typically selected, and for regression problems, M/3 features are selected [18]. Where: M is the total number of features; and errors are determined by Gini impurity; Entropy or MSE from the selected features. In classification tasks, Random Forest aggregates the votes from all trees and identifies the class with the highest vote count:

$$y = \arg\max_k \sum_{i=1}^{B} h_i(x)$$

where: $h_i(x)$— if the i-tree selects the k -class for x, the value will be 1, otherwise it will be 0; B is the number of trees. In regression problems, the average of the predictions of all trees in the Random Forest is taken:

$$y = \frac{1}{B} \sum_{i}^{B} h_i(x)$$

Where: $h_i(x)$ is the prediction of the i - tree for x .

3.4 The Support Vector Machines

The Support Vector Machines (SVM) technique is a robust method applicable to both classification and regression tasks, regardless of linearity. The main task of the SVM algorithm is to find the hyperplane separating the classes and create optimal boundaries (support vectors) [18]. To find the optimal hyperplane in SVM, let's say we have two classes and for each class x_i there is $x_i \in R^2$ a feature vector and a class label. In this case: y_i, $y_i \in \{-1, +1\}$, that is, the hyperplane equation for classification in two classes $wx + b = 0$ is as follows. Where: w *is* the normal vector for hyperplane; b is the compensation value of hyperplane [19]. The SVM algorithm solves the following optimization problem to find the optimal hyperplane:

$$\min_{w,b}\frac{1}{2}\|w\|^2, y_i(wx_i + b) \geq 1, \forall i = 1, 2, \ldots, N$$

where: $y_i(wx_i + b) \geq 1$ The equation ensures that each point is in the correct class; As a result, once the solution is found, the hyperplane parameters are expressed as:

$$\omega = \sum_{i=1}^{N} \alpha_i \cdot y_i \cdot x_i, b = y_i - \omega \cdot x_i$$

The general mathematical model of the SVM Algorithm uses the formula $f(x) = \mathrm{sgn}\left(\sum_{i=1}^{N} a_i y_i (x_i, x) + b\right)$ for classification. The mathematical model of the SVM algorithm ends with the determination of a hyperplane that separates classes accurately and optimally [20, 21].

3.5 The Neural Networks

The Neural Networks algorithm is a principal method extensively used to address intricate issues in artificial intelligence and machine learning. Neural networks are a model that somewhat emulates the function of biological neurons and consist primarily of many levels: an input layer, hidden layers, and an output layer. Each layer comprises neurons that are interlinked [22].

Output Calculation. For each layer

$$\alpha^{(l)} = \sigma\left(\omega^{(l)}\alpha^{(l-1)} + b^{(l)}\right)$$

Loss Function. A function is used to compare the output of the neural network and the actual values $L(y, \hat{y}) = -\sum_{i=1}^{N} \left(y_i \cdot \log(\hat{y}_i + (1 - y_i) \cdot \log(1 - \hat{y}_i)\right).$

Backpropagation. The weights $\omega^{(l)} = \omega^{(l)} - \theta\frac{\partial L}{\partial \omega^{(l)}}$ are updated using gradients. These mathematical expressions represent the basic training process of a neural network. As the neural network is trained, the weights and biases in each layer are optimized, which improves the predictive ability of the neural network [23, 24].

4 Experimental Result

An algorithm based on machine learning models has been developed to detect suspicious transactions in payment systems. The steps of the algorithm are shown in Fig. 2 below.

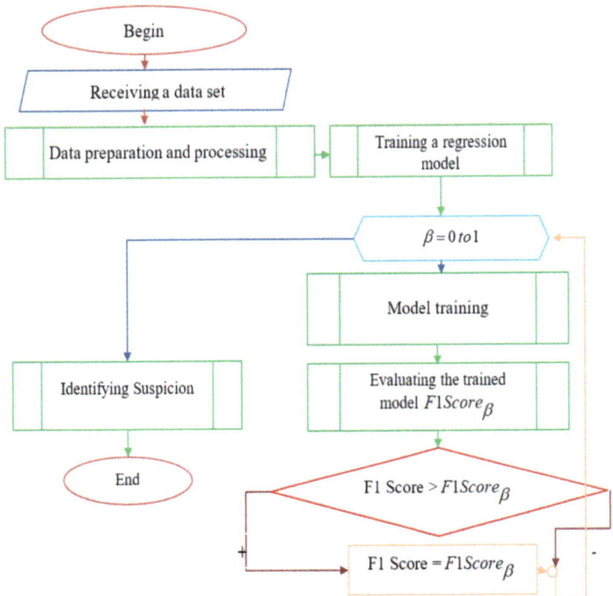

Fig. 2. Algorithm for detecting suspicious transactions in payment systems.

Based on the developed algorithm, experimental data were obtained, and the experimental data were presented in Table 1.

Table 1. Outcomes of machine learning algorithms used to detect suspicious transactions

No.	Algorithms	Algorithm evaluation methods (%)			
		Accuracy	Precision	Recall	F1 score
1	Logistic regression	89	88	89	88
2	Decision Tree	84.7	84	83	84
3	Random Forest	90.4	88	91	90
4	SVM	93.75	88	93	90
5	Neural network	91.55	89	94	92

The study of the experimental findings indicates that the machine learning algorithms, including logistic regression, decision tree, random forest, SVM, and neural network, are successful. The neural network model gives better results than the tested algorithms. The algorithm based on the developed machine learning models allows to identify suspicious transactions in payment systems [25].

5 Conclusion

The research work solves the problem of detecting suspicious transactions in payment systems based on machine learning algorithms of artificial intelligence. The characteristics of detecting suspicious transactions are presented. The issue of forming an initial data set for detecting suspicious transactions in payment systems is solved. Machine learning methods, including logistic regression, decision trees, random forests, support vector machines, and neural networks, have been created to identify suspicious transactions. As a result, algorithms for detecting suspicious transactions in payment systems based on machine learning and their analysis are described in detail.

References

1. Ahmed, M., Mahmood, A.N., Islam, M.R.: A survey of anomaly detection techniques in financial domain. Futur. Gener. Comput. Syst. **55**, 278–288 (2016)
2. Al-Momani, A., Aljawarneh, S.A.: Fraudulent transactions prediction using deep neural network. In: 2022 International Conference on Engineering & MIS (ICEMIS), Istanbul, Turkey, pp. 1–7 (2022). https://doi.org/10.1109/ICEMIS56295.2022.9914349
3. Akhatov A.R., Nazarov F.M., Rashidov A.E.: Increasing data reliability by using bigdata parallelization mechanisms. International conference on information science and communications technologies ICISCT. 4, 5, 6 November. 2021(IEEE), art. no. 9670387. https://doi.org/10.1109/ICISCT52966.2021.9670387
4. Kshemkalyani, A.D., Singhal, M.: Distributed Computing. Principles, Al-Gorithms, and Systems, p. 756. Cambridge University Press, USA (2008)
5. Baesens, B., Höppner, S., Verdonck, T.: Data engineering for fraud detection. Decis. Support. Syst. **150**, 113492 (2021)
6. Balobid, A.M., Binamro, J.S., Yohannes, S.T., Kaddoura, S.: Evaluation of supervised machine learning approaches for credit card fraud detection. In: 2022 14th Annual Undergraduate Research Conference on Applied Computing (URC), Dubai, United Arab Emirates, pp. 1–6 (2022). https://doi.org/10.1109/URC58160.2022.10054229
7. Duffield, E., Schinzel, H., Gutierrez, F.: Transaction locking and masternode con-sensus: A mechanism for mitigating double spending attacks. CryptoPapers.info, 2014, [Online]. Accessed 3 June 2019
8. Dablain, D., Krawczyk, B., Chawla, N.V.: DeepSMOTE: fusing deep learning and SMOTE for imbalanced data. IEEE Trans. Neural Networks Learn. Syst. **34**, 6390–6404 (2022)
9. Dileep, M.R., Navaneeth, A.V., Abhishek, M.: A novel approach for credit card fraud detection using decision tree and random forest algorithms. In: 2021 Third International Conference on Intelligent Communication Technologies and Virtual Mobile Networks (ICICV), pp. 1025–1028 (February 2021)
10. Makhmadiyarovich, N.F., Sherzodjon, Y.: Development of algorithms for predictive evaluation of investment projects based on machine learning. In: Artificial Intelligence, Blockchain, Computing and Security, vol. 2. eBook ISBN: 9781032684994, pp. 681–685 (2023)

11. Ghosh, S., Bilgaiyan, S., Gourisaria, M.K., Sharma, A.: Comparative analysis of applications of machine learning in credit card fraud detection. In: 2023 6th International Conference on Information Systems and Computer Networks (ISCON), Mathura, India, pp. 1–7 (2023). https://doi.org/10.1109/ISCON57294.2023.10112099
12. Nazarov, F.M., Yarmatov, S.: Optimization of prediction results based on ensemble methods of machine learning. In: 2023 International Russian Smart Industry Conference (SmartIndustryCon), Sochi, Russian Federation, pp. 181–185 (2023). https://doi.org/10.1109/SmartIndustryCon57312.2023.10110726
13. Nazarov, F.M.: Increasing data reliability in the system based on distributed methods. In: International Scientific and Practical Conference on Artificial Intelligence and Digital Educational Technologies: Practice, Experience, Problems and Prospects, pp. 151–154. Samarkand State University (2024)
14. Mamatov, N., Ibrokhimov, S., Samijonov, A.: Artificial neural networks in the creation of systems for monitoring and evaluating the quality of knowledge. No. 44(197) (2024): 7th ISPC «Scientific Trends and Trends in the Context of Global-Ization» (April 19–20, 2024; Umeå, Sweden)
15. Janet, B., Joshua Arul Kumar, R., Ganesh, D.P.S.: Credit card fraud detection with unbalanced real and synthetic dataset using machine learning models. In: 2022 International Conference on Electronic Systems and Intelligent Computing (ICESIC), Chennai, India, pp. 73–78 (2022). https://doi.org/10.1109/ICESIC53714.2022.9783529
16. Khosravi, S., Kargari, M., Teimourpour, B., Eshghi, A., Aliabdi, A.: Using supervised machine learning approaches to detect fraud in the banking transaction network. In: 2023 9th International Conference on Web Research (ICWR), Tehran, Iran, Islamic Republic of, pp. 115–119 (2023). https://doi.org/10.1109/ICWR57742.2023.10139083
17. Kute, D.V., Pradhan, B., Shukla, N., Alamri, A.: Deep learning and explainable artificial intelligence techniques applied for detecting money laundering–a critical review. IEEE Access. **9**, 82300–82317 (2021). https://doi.org/10.1109/ACCESS.2021.3086230
18. Liu, X., Zhang, P.: A scan statistics based suspicious transactions detection model for anti-money laundering (AML) in Financial Institutions. In: 2010 International Conference on Multimedia Communications, Hong Kong, China, pp. 210–213 (2010). https://doi.org/10.1109/MEDIACOM.2010.37
19. Pundir, K., Kaur, N., Kumar, A., Kumar, N., Das, C.: Detect legitimate and illicit transactions using machine learning. In: 2024 International Conference on Communication, Computer Sciences and Engineering (IC3SE), Gautam Buddha Nagar, India, pp. 418–423 (2024). https://doi.org/10.1109/IC3SE62002.2024.10592946
20. Nurmamatov, M.Q., Uddin, I., Sh.N., S.: Methods of using artificial intelligence algorithms in human resource management. In: 2025 International Russian Smart Industry Conference. (SmartIndustryCon) | 979-8-3315-1124-1/25/$31.00 ©2025 IEEE, pp. 566–571. https://doi.org/10.1109/smartindustrycon65166.2025.10986087
21. Isroil, J., Khusan, K.: Optimization of transmission and processing of information of electronic documents based on parallel computing technology. In: 2022 2nd International Conference on Technological Advancements in Computational Sciences (ICTACS), pp. 618–624. IEEE (2022). https://doi.org/10.1109/ICTACS56270.2022.9988150
22. Shiriaev, E., Kucherov, N., Movzalevskaya, V., Khamidov, M.: Modification and adaptation of methods and algorithms of the active security concept for fog systems. In: Lapina, M., Raza, Z., Tchernykh, A., Sajid, M., Zolotarev, V., Babenko, M. (eds.) AISMA-2024: International Workshop on Advanced Information Security Management and Applications. AISMA 2024 Lecture Notes in Networks and Systems, vol. 863. Springer, Cham (2024). https://doi.org/10.1007/978-3-031-72171-7_28

23. Nazarov, F.M., Xamidov, M.M.: Eye State Classification Method for Detecting Physiological Deviations in Drivers Based on CNN Algorithm. In: International Russian automation conference (RusAutoCon), Sochi, Russian Federation, vol. 2024, pp. 802–807 (2024). https://doi.org/10.1109/RusAutoCon61949.2024.10694309
24. Rai, H., Pal, A., Khamidov, M., Bobokhonov, A., Ugli, R.: Computational intelligence transforming healthcare 4.0: innovations in medical image analysis through AI and IoT integration. In: Data-Driven Decision Support System in Intelligent HealthCare (2025). https://doi.org/10.1201/9781003507505-3
25. Lapina, M., Aganesov, A., Lapin, V., Athavale, V.A.: Research of dynamic fuzzing methods to identify vulnerabilities in program code. In: Lapina, M., Raza, Z., Tchernykh, A., Sajid, M., Zolotarev, V., Babenko, M. (eds.) AISMA-2024: International Workshop on Advanced Information Security Management and Applications. AISMA 2024 Lecture Notes in Networks and Systems, vol. 863, pp. 172–184. Springer, Cham (2024). https://doi.org/10.1007/978-3-031-72171-7_18

Post-quantum Cryptography in Practice: A Survey of Algorithms, Applications, and Deployment Challenges

Sharlene Anna Pereira[1], Melsa Magdalene Fernandes[1], E. A. Mary Anita[1(✉)], and Tatyana Grobova[2]

[1] Christ University, Bangalore Kengeri Campus, Bangalore, India
maryanita.ea@christuniversity.in
[2] North Caucasus Federal University, Stavropol, Russia

Abstract. As quantum computing becomes more practical, it significantly threatens the conventional cryptographic systems, particularly RSA and ECC, that are critical to worldwide digital security. Post-quantum cryptography (PQC) has emerged as a strong alternative as a response. NIST recently standardized algorithms such as CRYSTALS-Kyber and Dilithium. This survey brings together findings from ten key papers that examine PQC across different fields, including telecommunications, finance, healthcare, IoT, smart cards, and blockchain voting systems. The chosen studies include direct comparisons of digital signature schemes, real-world protocol integration on smart cards, hybrid cryptographic models using AES and blockchain, and strategies for transition based on policy frameworks like NIST CSF 2.0. The survey examines cryptographic flexibility, hardware practicality, readiness for adoption, and the social and economic effects of quantum breaches. It compares algorithm performance, deployment challenges, and specific needs for various areas. This paper is an overview of the current state and future directions for PQC implementation in critical infrastructure.

Keywords: Post-Quantum Cryptography · Quantum-Resistant Algorithms · Kyber · Dilithium · Falcon · SPHINCS+ · Digital Signatures · Hybrid Encryption · Blockchain Security · IoT Security · Financial Cryptography · Smart Cards · Biometric Authentication · Cryptographic Agility · Protocol Integration

1 Introduction

In the context of modern digitalization, information security is not just an important element, but a critically necessary component of information systems [11–13].

The advent of quantum computing presents a major threat to the traditional cryptographic algorithms that secure global digital communications and data. RSA and elliptic curve cryptography (ECC) are crucial for protocols including TLS, VPNs, and blockchain security, but they stand threatened by quantum attacks enabled by Shor's and Grover's algorithms. This threat has led to a swift move toward post-quantum cryptography (PQC), which uses quantum-resistant mathematical problems to maintain security

M. Lapina et al. (Eds.): AISMA 2025, LNNS 1456, pp. 362–369, 2026.
https://doi.org/10.1007/978-3-032-07275-7_33

against quantum threats. Understanding the urgency of this shift, the U.S. National Institute of Standards and Technology (NIST) has finalized the standardization of PQC algorithms, including CRYSTALS-Kyber for key encapsulation and CRYSTALS-Dilithium for digital signatures. NIST is also evaluating options like Falcon, SPHINCS+, and NTRU. They vary in performance, signature length, and side-channel resistance and thus are applicable for particular uses.

This survey gathers insights from ten recent papers that examine PQC in various real-world settings. In telecommunications, researchers are testing lattice-based schemes for high-throughput environments. In finance, they are exploring hybrid AES-PQC models and blockchain integrations. In healthcare, there is research on how ready PQC is for telemedicine systems. Notably, the survey includes research on quantum-secure electronic voting platforms. These platforms use biometric authentication (DeepFace), blockchain sharding, and Dilithium-based digital signatures to protect electoral integrity.

Additionally, strategies at the policy level, such as those related to the NIST Cybersecurity Framework (CSF 2.0), are reviewed to evaluate institutional readiness for transitioning to PQC. By comparing these implementations in different sectors and pointing out common challenges, such as certificate size overhead, gaps in protocol integration, and weak side-channel protections, this paper aims to explain the current state of PQC deployment and offer a clear path toward a quantum-resilient cryptographic infrastructure.

2 Literature Review

2.1 Comparison

The ten papers reviewed look at different aspects of post-quantum cryptography (PQC) research. They focus on how algorithms perform and their practical use in various fields. Papers [1, 6, 9] analyze important algorithms such as Kyber, Dilithium, and Falcon. These papers demonstrate that these algorithms perform better than RSA and ECC, particularly in the telecom and healthcare sectors.

Domain-focused studies, including [5, 6, 8, 10], examine PQC integration in finance, healthcare, IoT, and banking. Gbadebo [5] applies hybrid PQC-AES in financial systems. Balogun [6] assesses encryption in telemedicine. Kumar et al. [8] categorize algorithm suitability for different IoT classes. Bettale et al. [10] integrated PQC protocols into EMV smart cards.

Papers [3, 5, 7, 10] look at hybrid cryptographic systems. Aydeger et al. [3] propose a transition plan based on the NIST Cybersecurity Framework. Joni et al. [7] developed a hybrid blockchain e-voting platform that combines Dilithium, DeepFace biometrics, and sharding. This offers a unique real-world application. Protocol-level integration is discussed in [7, 10]. Papers [2, 4] focus on regulatory and strategic readiness. Overall, these papers highlight performance, the feasibility of integration, and migration planning. However, real-world deployment remains limited in most cases.

Few papers relate technical innovation to wider system-level issues such as compliance, interoperability, and flexibility. For example, Gbadebo [5] and Aydeger et al. [3] discuss cryptographic flexibility and transition strategies. However, there are no complete implementation models available. This shows the need for research that links low-level algorithm efficiency with high-level infrastructure design and governance. Based on this research, we can divide these papers into groups of their core research objectives, as shown in Fig. 1.

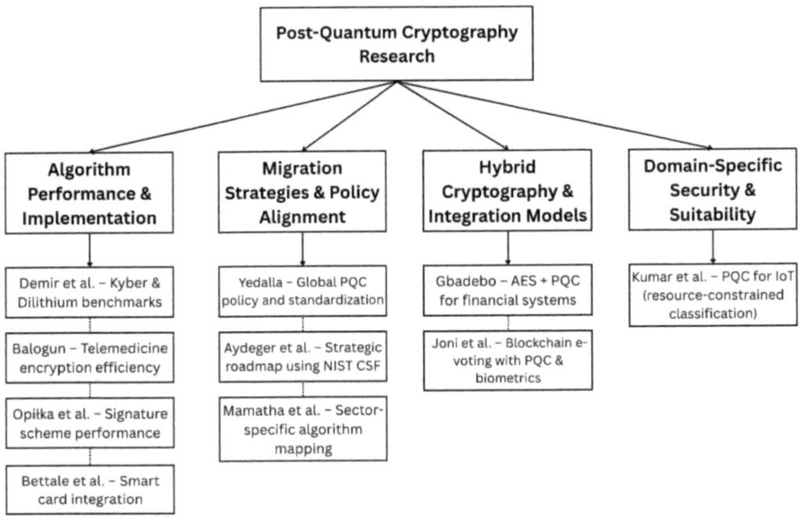

Fig. 1. Thematic classification of the ten PQC papers based on their core research objectives.

2.2 Research Gaps Identified

To gain further insight into the gaps across the present landscape of post-quantum cryptography (PQC) research, we examined the particular research gaps indicated or suggested in each of the ten research papers under review. Although these research studies make valuable contributions, most fall short of addressing realistic deployment barriers, cryptographic flexibility, as well as of practical constraints including power efficiency and regulatory compliance.

Table 1 summarizes a paper-by-paper mapping of the areas of research gap, indicating how each paper addresses or neglects essential aspects of PQC readiness. This categorization indicates that whereas some papers exhibit application to real-life, like in papers [7, 10], others exist only as conceptual or simulation-based. More importantly, prevalent issues like lack of integration at the protocol level, side-channel attack vulnerability, and scalability continue to be endemic, pointing to the necessity for more applied and multidisciplinary work in filling the gap between theory and deployment.

Table 1. Classification of the 10 reviewed papers by focus area, domain, and methodology.

Research Gap	Description	Papers Indicating Gap
Protocol Integration	Lack of real-world PQC integration into TLS, SSH, VPNs, etc.	[3, 6, 9]

(continued)

Table 1. (*continued*)

Research Gap	Description	Papers Indicating Gap
Certificate & Signature Size	Large key/signature sizes hinder use in IoT, embedded systems	[6, 8, 10]
Side-Channel Resistance	Minimal focus on protection against timing/power attacks in hardware	[1, 10]
Energy & Resource Profiling	Few evaluations of power efficiency or memory usage on constrained devices	[8, 9]
Scalability & Deployment	Limited real-world implementation and system-level testing	[7, 10]
Cryptographic Agility	Few frameworks support dynamic switching or hybrid migration strategies	[3, 5]
Regulatory & Compliance Alignment	Lack of actionable legal/regulatory frameworks for PQC adoption	[2, 4]
Advanced Crypto Integration	Little work on PQC with ZKPs, MPC, or privacy-preserving technologies	[7]

3 Methods

The ten papers examined in this overview employ a broad range of research methods for studying the feasibility, performance, and implementation of post-quantum cryptography (PQC) in different domains. These methods range from empirical benchmarking, statistical and simulation modeling, protocol integration, literature review, to taxonomic analysis, in line with the interdisciplinary context of the field.

Most of the papers use empirical benchmarking in examining the performance of PQC algorithms. CRYSTALS-Kyber and CRYSTALS-Dilithium performance was assessed by Demir et al. through controlled experiments on reference and AVX2-optimized implementations. Experiments were performed for 1,000 iterations per operation with execution times being recorded for key generation, encapsulation, and decapsulation within a telecom context. In the same way, Opiłka et al. [9] compared Falcon, Dilithium, and SPHINCS+ using the liboqs library and a Python code-based custom CLI tool. They found signature, key generation, and verification times for different file sizes (10 MB to 1 GB) with 100 iterations per experiment. Balogun [6] also conducted benchmarking on the healthcare context, comparing Falcon and Kyber on encryption speed, computational expense, and memory overhead. Statistical significance of results was verified using ANOVA to determine differences in algorithm performance. Bettale et al. [10] deployed Kyber, Falcon, and Dilithium on ARM Cortex-M3 smart cards to test transaction time, certificate management, and memory influence in actual payment scenarios

Statistical modeling and simulation methods were utilized in some of the articles to investigate PQC performance and its consequences. AHP and MLR were applied by Gbadebo [5] to rank PQC algorithms in cryptographic agility based on encryption speed, usage of memory, and CPU. ANOVA was also used in the research for estimating algorithmic variance. According to this, [7] expanded the ambit of modeling with vector autoregression (VAR) in order to model the economic effect of a quantum cyberattack, predicting GDP loss, market decline, and breach cost. These forecasts were based on publicly available data sources, such as the NIST PQC benchmark, Google's homomorphic encryption dataset, and World Bank economic data. Balogun [6] merged performance data with actual breach statistics from U.S. healthcare systems and used logistic regression to categorize institutions into tiers of PQC readiness.

A lesser set of the published literature addressed protocol-level integration and testing. Bettale et al. [10] performed a hands-on implementation of post-quantum implementations of EMV payment protocols, namely Combined Data Authentication (CDA) and Blinded Diffie-Hellman (BDH), employing Kyber, Falcon, and NTRU. These were tested on smart card hardware and assessed in terms of transaction latencies, flash memory utilization, and delays in transmitting certificates. Though theoretical, Aydeger et al. [3] suggested an organized migration plan for PQC based on the NIST Cybersecurity Framework (CSF 2.0), correlating every functional category (Identify, Protect, Detect, Respond, Recover) with post-quantum security measures.

Some other articles employed descriptive and comparative approaches to examine the larger horizon of PQC. Mamatha et al. [4] presented literature-supported taxonomy of PQC algorithms—lattice, hash, code, multivariate, and isogeny-based—and compared their relative merits and demerits along parameters like security level, signature size, and applicability domain. Yedalla [2] addressed synthesizing the world policy and standardization environment, comparing initiatives by NIST, ETSI, ISO, and others, and how PQC is different from quantum cryptography solutions such as QKD. Kumar et al. [8] employed a taxonomic view to classify PQC appropriateness for resource-constrained devices in IoT environments, projecting algorithm families onto device classes according to power, memory, and processing requirements. It also comprised a conceptual study of security threats to quantum key distribution and side-channel attacks.

The papers made use of a combination of open-source tools, datasets, and hardware platforms. Liboqs (Open Quantum Safe) dominated the use in Papers [9, 10] to implement and test PQC algorithms. NIST's PQC Benchmark Dataset was heavily utilized in Papers [1, 5, 6, 7]. Gbadebo [7] also employed Google's homomorphic encryption dataset and World Bank indicators, and Balogun [6] used breach reports of the U.S. Department of Health and Human Services (HHS). Hyperledger Fabric was used in Papers [5, 7] to conduct blockchain-based PQC performance testing. Hardware testing was performed on typical CPUs, virtual machines, and embedded devices such as ARM Cortex-M3 cards.

In short, the methods employed throughout the ten papers range widely—from strict benchmarking and performance profiling to statistical modeling, protocol integration, and strategic planning. The diversity of methods mirrors the interdisciplinary character of PQC research and illustrates how several different approaches collectively contribute to the ongoing transition toward quantum-resistant security systems.

4 Discussion

The literature examined reveals an increasing consensus to implement Post-Quantum Cryptography (PQC) across critical infrastructure. Research such as [1, 3, 4] refers to classical cryptography schemes being weak against quantum attacks and require future-proof migration strategies. Hybrid cryptographic frameworks that merge classical and post-quantum algorithms are most commonly advocated in [3, 5, 7, 10]. These models are transitional safeguards that are resilient in nature and have legacy systems. These hybrid models are more complex when implemented and need to be tested thoroughly, particularly in constrained and real-time environments.

Domain-specific research indicates that PQC adoption needs to be application-specific. The financial and health applications prioritize regulatory compliance and breach resistance, while IoT and embedded systems prioritize limited memory and energy as their major concerns. CRYSTALS-Kyber and Dilithium are always strong candidates for general-purpose application because they strike a balance among security, efficiency, and standardization. But signature size, processing overhead, and certificate transmission latency remain open issues in most real-world applications.

While significant steps have been made, areas of practical deployment, protocol-level integration, and side-channel resistance continue to lag behind. The majority of experiments continue to be limited to simulation or testbeds within laboratories with only a handful of papers, like papers [7, 10], showing system-level or hardware-based deployments. Mechanisms for energy profiling and cryptographic agility are also under-developed. Additionally, while some papers address regulatory alignment and migration plans little has been developed toward legal-enforceable frameworks or industry-interoperability standards. As of 2025, NIST has completed CRYSTALS-Kyber and CRYSTALS-Dilithium. Falcon and SPHINCS+ are alternate choices, and Round 4 goes on with contenders such as NTRU, BIKE, HQC, and McEliece. The efforts are to increase cryptographic diversity and maximize use-case flexibility.

There is a consensus across the papers that PQC will not work as a standalone technical upgrade but needs to be addressed as a systems-level change spanning cryptography, hardware, standards, and regulation. The solution space takes shape in terms of phased migration plans (as defined in [3]), the blending of hybrid encryption methods, and broader collaboration between academia, industry, and regulators. Coordinated testing, open benchmarking platforms, and compliance-oriented frameworks will be critical to close the existing gap between research prototypes and scalable, production-quality quantum-secure infrastructure.

5 Conclusion

The evolution to post-quantum cryptography (PQC) is no longer a future requirement but an urgent need for the security of digital infrastructure from forthcoming quantum attacks. This survey has compared ten new and varied studies together, generating guidance on the performance of PQC algorithms, feasibility of deployment across sectors, hybrid integration methods, and migration strategy. The literature reviewed shows that algorithms like CRYSTALS-Kyber and Dilithium not only make theoretical sense but are also practically feasible for implementation within traditional systems, particularly when optimized.

Domain-specific application spaces—in finance, healthcare, telecom, IoT, and smart cards—exhibit different degrees of readiness, common challenges being certificate size overhead, energy inefficiency, and protocol-level integration gaps. Hybrid cryptographic paradigms, like PQC+ AES or ECC+ PQC, are universally suggested as transition modes, with some already deployed on constrained hardware-like ARM smart cards. Despite advancement, critical gaps in research persist. These are side-channel resistance, energy profiling, end-to-end protocol incorporation (e.g., PQC in TLS, SSH, VPNs), and regulatory harmonization. In addition, real-world deployments and cross-domain scalability are still very much untested.

Overall, the body of work explored herein serves as a foundation for the future of PQC research and deployment. Work on the future must focus on adapting algorithms to constrained environments, enabling transparent hybrid transitions, and defining global standards and compliance frameworks. It is only through a multidisciplinary solution that brings together cryptography, systems engineering, and policy that we can develop a fully quantum-resistant cybersecurity platform.

References

1. Demir, A., Selcuk, A., Sönmez, A.: Performance analysis and industry deployment of post-quantum cryptography algorithms. arXiv preprint arXiv:2503.12952, 2024. [Online]. Available: https://arxiv.org/pdf/2503.12952
2. Yedalla, J.: Quantum-safe cryptography: navigating the future of cybersecurity in the post-quantum era. Int. J. Sci. Res. **14**(2), 249–253 (2025). https://doi.org/10.21275/sr25203214146
3. Aydeger, A., Zeydan, E., Yadav, A.K., Hemachandra, K.T., Liyanage, M.: Towards a Quantum-Resilient Future: Strategies for Transitioning to Post-Quantum Cryptography. In: 2024 15th International Conference on Network of the Future (NoF), Castelldefels, Spain, vol. 2024, pp. 195–203. https://doi.org/10.1109/NoF62948.2024.10741441
4. Mamatha, G.S., Dimri, N., Sinha, R.: Post-quantum cryptography: securing digital communication in the quantum era. arXiv (Cornell University). 2024. https://doi.org/10.48550/arxiv.2403.11741
5. Gbadebo, M.O.: Integrating post-quantum cryptography and advanced encryption standards to safeguard sensitive financial records from emerging cyber threats. SSRN Electron. J. (2025). https://doi.org/10.2139/ssrn.5168159
6. Balogun, A.Y.: Post-quantum cryptography and encryption standards: safeguarding patient data against emerging cyber threats in telemedicine. Asian J. Res. Comput. Sci. **18**(3), 345–367 (2025). https://doi.org/10.9734/ajrcos/2025/v18i3598
7. Joni, S.A., Rahat, R., Tasnin, N., Ghose, P., Uddin, M.A., Ayoade, J.: Hybrid-Blockchain-based electronic voting machine system embedded with deepface, sharding, and post-quantum techniques. Blockchains. **2**(4), 366–423 (2024). https://doi.org/10.3390/blockchains2040017
8. Kumar, A., Ottaviani, C., Gill, S.S., Buyya, R.: Securing the future internet of things with post-quantum cryptography. Secur. Priv. **5**(2) (2021). https://doi.org/10.1002/spy2.200

9. Opiłka, F., Niemiec, M., Gagliardi, M., Kourtis, M.A.: Performance analysis of post-quantum cryptography algorithms for digital signature. Appl. Sci. **14**(12), 4994 (2024). https://doi.org/10.3390/app14124994
10. Bettale, L., De Oliveira, M., Dottax, E.: Post-quantum protocols for banking applications. In: Lecture Notes in Computer Science, pp. 271–289 (2023). https://doi.org/10.1007/978-3-031-25319-5_14
11. Basan, E., Peskova, O., Ivanova, A., Ivanov, E., Maria Lapina, E.A., Anita, M., Kumar, S.: Analysis of the UAV flight logs in order to identify information security incidents. In: Kumar, S., K., B., Kim, J.H., Bansal, J.C. (eds.) Fourth Congress on Intelligent Systems. CIS 2023 Lecture Notes in Networks and Systems, vol. 865, pp. 415–426. Springer, Singapore (2024) https://doi.org/10.1007/978-981-99-9043-6_34
12. Basan, E., Sushkin, N., Lapina, M., Sajid, M.: Automation of the process of analysis of information security threats in cyber-physical systems. In: Intelligent Techniques for Cyber-Physical Systems, pp. 91–107 (2023). https://doi.org/10.1201/9781003438588-6
13. Lapina, M., Gorbunova, N., Tokmakova, M., Movzalevskaya, V., Elashmawi, W., H.: Features of the project approach in information security training. In: Lapina, M., Prakasha, G., S., Grigoriev, S. (eds.) SLET 2022: International Conference on Innovative Approaches to the Application of Digital Technologies in Education and Research. SLET 2022 Lecture Notes in Networks and Systems, pp. 313–323. Springer, Cham (2024). https://doi.org/10.1007/978-3-031-78776-8

A Performance-Driven Firefly Algorithm for Scheduling Workflow in Cloud Computing

Mohammad Qasim, Mohammad Sajid(✉), and Mohammad Shahid

Department of Computer Science, Aligarh Muslim University, Aligarh 202002, India
sajid.cst@gmail.com

Abstract. Workflow scheduling in cloud computing is a challenging NP-Hard problem due to complex task precedence constraints and the dynamic nature of cloud resources. Metaheuristic (MH) algorithms effectively tackle NP-Hard problems, like workflow scheduling, offering near-optimal task assignments while maintaining precedence constraints. This study introduces an approach to addressing the workflow scheduling problem by employing the Firefly Algorithm (FFA) to minimize the total completion time of all tasks. The performance of the proposed FFA is systematically evaluated against the Grey Wolf Optimizer (GWO), Sine-Cosine Algorithm (SCA), Whale Optimization Algorithm (WOA), and Differential Evolution (DE) algorithm, with a primary focus on minimizing the makespan. The experimental results demonstrate that the proposed FFA exhibits superior convergence speed and achieves a lower mean makespan than the competing algorithms. The proposed metaheuristic-based approach offers a scalable and adaptable solution for workflow optimization in cloud environments, making it particularly relevant for contemporary industrial and scientific applications requiring high computational efficiency. The proposed method is validated through convergence analysis and statistical measures, including the mean and standard deviation of the makespan.

Keywords: Cloud Computing · Firefly Algorithm · Makespan · Metaheuristic · Workflow Scheduling

1 Introduction

Cloud computing is a distributed computing paradigm that provides an on-demand capability to share the pool of computational resources, storage, and applications from anywhere over the Internet, which benefits scalability, flexibility, and cost-effectiveness. Public clouds have grown immensely in the recent decade due to strong demand for scalable, lower-cost solutions, infrastructure (IaaS), platforms (PaaS), and software solutions (SaaS) from all industries. For example, the Indian public cloud services market, which includes services like IaaS, PaaS, and SaaS, earned $8.3 billion in 2023 [1]. According to the International Data Corporation (IDC), the Cloud market is predicted to grow significantly and reach $24.2 billion by 2028, with an annual growth rate of 23.8% between 2023 and 2028 [1]. Expansion in the cloud market will be facilitated as artificial intelligence, edge computing, and hybrid cloud models evolve with improved security, sustainability, and seamless integration.

M. Lapina et al. (Eds.): AISMA 2025, LNNS 1456, pp. 370–383, 2026.
https://doi.org/10.1007/978-3-032-07275-7_34

In cloud computing architecture, workflow describes the automated execution of tasks with interdependencies and constraints, typically represented in a Directed Acyclic Graph (DAG). Workflow scheduling optimization is one of the most critical challenges as it manages task dependencies while allocating dynamic resources efficiently [2]. This differs from static scheduling as the task allocation takes place in an environment that is predefined and unchanging [3]. However, the environment of the workflow scheduling of the cloud is dynamic- the workload is fluctuating, heterogeneous resources prevail, and the demands of users are evolving. Such complexity requires advanced optimization strategies to optimize the makespan and thus minimize overall completion time and resource utilization. Due to the variability and interdependence of cloud workflows, MH algorithms can address this issue successfully.

MHs are very practical and flexible methodologies in the research domain, widely applied to optimize several problems related to cloud computing. Their adaptability to diverse problem spaces and ability to escape local optima have made them the focus of extensive research for scheduling issues. Hence, metaheuristic approaches are helpful for workflow scheduling in cloud computing since they deal with task dependencies and dynamic constraints; under highly complex conditions, they are ensured to produce quality schedules [4].

MHs have significant advantages over simple heuristics in solving complex optimization problems. While heuristics offer specific rules for issues, metaheuristics use higher-level search strategies that can adapt to diverse challenges. Their population-based approaches and iterative improvement mechanisms enable them to explore solution spaces more thoroughly, making them less likely to get trapped in local optima [5]. Metaheuristics can also scale well since their frameworks allow for more significant problem instances with growing complexity. Moreover, their structure is modular, thus allowing modifications to suit different objectives and constraints without necessarily modifying the kernel of the algorithm.

The paper presents an MH approach for workflow scheduling in cloud computing, primarily focusing on minimizing makespan. The proposed algorithm integrates advanced mechanisms for search diversification and intensification, demonstrating superior performance compared to existing state-of-the-art MH algorithms. The FFA achieves enhanced computational efficiency and schedule quality by effectively considering task dependencies and resource heterogeneity. To evaluate its efficacy, we conducted experiments on six workflow applications, three from Montage and three from Ligo, demonstrating significant improvements over well-established metaheuristic approaches.

The rest of this paper follows the organization: Sect. 2 discusses the related work in workflow scheduling and metaheuristic optimization. Section 3 delimits the problem formulation, describing the proposed algorithm with all its details. Section 4 reports the experimental setup for evaluation results, which compares the proposed approach with other well-known MH algorithms. Section 5 outlines the conclusion of this work and future research directions.

2 Related Work

Scheduling workflows in cloud computing is essential since it helps users use resources better and meet their quality of service (QoS) needs. Sajid et al. [6] present two energy-efficient scheduling algorithms, G1 and G2, for assigning a batch of independent tasks to heterogeneous processors. The goal is to minimize both makespan and energy consumption. The performance of the developed algorithms is assessed by comparing them with existing approaches, such as MINSuff and MinMin, based on metrics including makespan, processor utilization, energy consumption, and flowtime. Another study of Sajid et al. introduced a [7] quantum-inspired stochastic Q-HypE algorithm designed to allocate a batch of stochastic tasks (BoT) on DVFS-enabled processors. The objective is to optimize both the makespan of the BoT and the energy consumption of the processors.

Metaheuristic algorithms, which are strong and flexible, have become suitable methods for solving complicated workflow scheduling problems. Zhang [8] has proposed a novel multi-step method with the help of the Whale Optimization Algorithm to reduce the execution time, minimize the financial costs, and maintain load balancing. A WOA-based method improves the efficiency of scientific workflow scheduling in cloud computing settings. Akraminejad et al. [9] deal with a multi-objective Crow Search Algorithm intended for the optimization of makespan and costs in scientific cloud workflows against the increasing emissions of CO2 and global warming arising from the speedy growth in the technology of cloud computing and power consumption. Zaki et al. [10] developed a new approach to solving complex workflow scheduling problems in cloud computing using a special type of genetic programming. This approach allows us to find several reasonable solutions simultaneously, hence choosing the best answer from the list of options for decision-making. Bacanin et al. [11] developed a workflow scheduling algorithm in a cloud-edge environment that reduces conflicting goals, particularly the time taken to complete tasks and cost. The researchers enhanced the firefly algorithm by integrating genetic operators and a learning method based on quasi-reflection to fix the problems with the original FFA and to make the solutions better and faster for scheduling workflows. Ramathilagam et al. [12] propose an EAFSA-based IaaS Cloud Partial Critical Path (IC-PCP) Replication with Hyper Elliptic Curve Cryptography (EAFSAIPR with HECC) for increasing workflow scheduling techniques. The execution of tasks is guaranteed safe and inexpensive by replicating tasks and cryptographic authentication using the proposed method. This satisfies the user-declared deadlines with decreased execution time and improved reliability. Qasim et al. [13] studied a Firefly Algorithm-based scheduler enhanced with transfer functions and quantization techniques to address IoT task scheduling in cloud computing, aiming to minimize makespan. Comparative simulations reveal that the proposed method achieves faster convergence and higher solution quality than the Harris Hawks Optimization (HHO) and the DE algorithm.

However, hybrid metaheuristic algorithms for workflow scheduling have been extensively developed in cloud computing, integrating several optimization techniques with strengths to enhance performance. Mikram et al. [14] proposed a new hybrid algorithm, HEPGA, for scheduling scientific workflows in cloud computing environments to optimize resource utilization and minimize makespan. This hybrid algorithm ensures that the balance between various resources for task distribution will ensure greater efficiency and cost-effectiveness. Barredo et al. [15] provide an advanced disk-network-computing evaluation model that improves scheduling accuracy considering the communication between storage devices. The proposed hybrid methods improve makespan performance using a genetic algorithm and HEFT on heterogeneous infrastructures. Thekkepuryil et al. [16] present a hybrid optimization method based on Ant-Lion Optimization (ALO) and Particle Swarm Optimization (PSO) for efficient and secure workflow scheduling in cloud computing, supplemented with Data Encryption Standard (DES) for secure data. This hybrid method is superior to existing techniques, such as GA-PSO, ALO, PSO, and Round-Robin, in terms of cost, load balancing, makespan, and energy efficiency. Belqacem et al. [17] approach to optimize workflow scheduling in the cloud computing paradigm is a HEFT-ACO approach by combining Heterogeneous Earliest Finish Time with Ant Colony Optimization. This work's specific contribution is minimizing makespan and cost using ACO. Simulations on real workflows deployed on Amazon EC2 outperform ACO, PEFT-ACO, and FR-MOS using efficient resource utilization.

This study aims to develop an efficient workflow scheduling method in a cloud environment that minimizes the makespan while adhering to task dependencies and resource constraints. By leveraging the FFA, the proposed approach aims to optimize task allocation across virtual machines, ensuring better utilization of computational resources and enhanced performance for complex workflows.

3 Problem Formulation

A scientific application can be expressed as a directed acyclic graph (DAG) $J = (T, E)$, as shown in Fig. 1, where $T = \{ t_1, t_2, t_3, \ldots, \ldots., t_n \}$ represents a set of n tasks, and E denotes the set of edges between the tasks representing the precedence constraints. Each task $t_i \in T$ represents an atomic application task, i.e., each task must be executed on the same machine. The set of edges $E = \{e_{ij} = (t_i, t_j) \mid t_i, t_j \in J \textit{ and } t_i \textit{ is the predecessor of } t_j\}$ represents the edges in the (DAG) G. An edge $e_{ij} = (t_i, t_j)$ indicates that the task t_j must be executed only after the completion of the task t_i. Each edge e_{ij} is assigned a non-negative weight c_{ij}, representing the communication time from the task t_i to the task t_j. The communication time c_{ij} is considered zero if tasks t_j and t_j are executed on the same machine or virtual machine.

Task	v_1	v_2	v_3
t_0	11	13	9
t_1	10	15	11
t_2	9	12	14
t_3	11	16	10
t_4	15	11	19
t_5	12	9	5
t_6	10	14	13
t_7	11	15	10

Fig. 1. A sample DAG for workflow scheduling and execution time (sec) of each task corresponding to three VMs.

Now, define the predecessor and successor task set of a task t_i by Eqs. (1) and (2) respectively.

$$\mathrm{Pred}_{t_i} = \{t_x | e_{xi} = (t_x, t_i), t_x \in J\} \tag{1}$$

$$\mathrm{Succ}_{t_i} = \{t_y | e_{iy} = (t_i, t_y), t_y \in J\} \tag{2}$$

Here t_x and t_y represent the predecessor and successor tasks of a task t_i. Let t_{entry} and t_{exit} represents the entry and exit task of a DAG. Since the entry task of a DAG has no predecessor, i.e., $\mathrm{Pred}(t_{entry}) = \emptyset$ and exit task has no successor, i.e. $\mathrm{Succ}(t_{exit}) = \emptyset$. Assume that workflow is executed in a cloud environment consisting of heterogeneous computing units, i.e., virtual machines (VMs). Let V be the set of z heterogeneous VMs represented as $V = \{v_1, v_2, \ldots, \ldots v_z\}$.

The execution time matrix ET of order $n \times z$ is given for all tasks of the scientific application $J = (T, E)$ corresponding to all VMs, as shown in Fig. 1. The $ET_{i.\,k}$ denotes the execution time of the task t_i on the VM v_k. So, the average execution time (AET_i) of the task t_i is calculated by the following equation:

$$AET_i = \sum_{k=1}^{z} \frac{ET_{i,k}}{z} \tag{3}$$

The well-known HEFT [18] and other algorithms employ the upward rank value, which can be calculated using the following equation:

$$\mathrm{U}_{rank}(t_i) = AET_i + \max_{t_j \in succ(t_i)}\left(c_{ij} + \mathrm{U}_{rank}(t_j)\right) \tag{4}$$

where AET_i denotes the average execution time of the task t_i.

To schedule a task t_i on VM v_k, the earliest start time $\mathrm{EST}(t_i, v_k)$ and earliest finish time $\mathrm{EFT}(t_i, v_k)$ can be calculated as follows:

$$\mathrm{EST}(t_i, v_k) = \max\left\{\mathrm{AVT}[v_k], \max_{t_j \in pred(t_i)}\left(\mathrm{AFT}(t_j) + c_{j,i}\right)\right\} \tag{5}$$

$$\mathrm{EFT}(t_i, v_k) = \mathrm{EST}(t_i, v_k) + ET_{i,k} \tag{6}$$

where $\mathrm{AFT}(t_j)$ represents the actual finish time of the task t_j and $\mathrm{AVT}[v_k]$ denote the earliest available time of the VM v_k. If t_m be the last task to be scheduled on the VM v_k, then $\mathrm{AVT}[v_k] = \mathrm{AFT}(t_j)$.

Since the makespan represents the maximum time needed to complete the execution of the observed workflow application, tasks within the workflow are executed on cloud servers based on the resource allocation strategy employed. After scheduling the exit task on a specific VM, a workflow's schedule length, commonly called the makespan, can be computed. It represents the total time taken from the start of the workflow execution to the completion time of the exit task. The makespan can be directly obtained by calculating the finish time of the exit task.

$$\mathrm{makespan} = \max\{\mathrm{AFT}(t_{exit}) : \forall t_{exit} \in J\} \tag{7}$$

Thus, the optimization problem can be defined as:

$$(\text{Objf})_{\text{opt}} = \min\{\text{makespan}\} \tag{8}$$

Figure 1 represents a sample DAG for workflow scheduling and execution time (sec) of each task corresponding to three VMs.

4 Proposed Methodology

The Firefly Algorithm (FFA) is a swarm-based metaheuristic optimization algorithm inspired by the flashing behaviour of fireflies. It was introduced by Xin-She Yang in 2008 [19] and is based on the principles of attraction and brightness. FFA is handy for solving continuous, combinatorial, and multi-objective optimization problems.

In the FFA, the idea of brightness and attractiveness is important. Brightness is defined as the measure of how good the solution represented by a firefly is, typically through the objective function value. Fireflies naturally tend to other fireflies that seem brighter; thus, a firefly that appears less bright tends to move towards a brighter firefly, like the instinctive behaviour of fireflies towards brighter peers. Moreover, the intensity of a firefly decreases with distance due to the absorption of light in the medium, which introduces a decay factor that limits the impact of distant fireflies. Fireflies also introduce a random walk mechanism to ensure practical search space exploration, avoid premature convergence, and thoroughly investigate potential solutions.

4.1 Mathematical Formulation of Firefly Algorithm

The brightness (B) of a firefly is proportional to the value of the objective function (Objf) at its position:

$$B \propto (Objf) \tag{9}$$

The attractiveness (A) of a firefly decreases with the square of the distance (d) between two fireflies due to light absorption. Mathematically, it can be written as:

$$A(d) = A_0 e^{-\sigma d^2} \tag{10}$$

Where A_0 denoted the maximum attractiveness at d = 0; σ represents the light absorption coefficient, and d is the distance between the fireflies.

The distance between two fireflies b and c at positions x_b and x_c is given by the following equation:

$$d_{bc} = x_b - x_c = \sqrt{\sum_{z=1}^{k} \left(x_{b,k} - x_{c,k}\right)^2} \tag{11}$$

where k is the dimensionality of the search space.

Since a firefly b moves towards a brighter firefly c, according to the following equation:

$$x_b^{t+1} = x_b^t + A_0 e^{-\sigma d_{bc}^2}\left(x_c^t - x_b^t\right) + \mu\varepsilon_t \tag{12}$$

where x_b^t and x_c^t denotes the current position of Firefly b and c, respectively; $A_0 e^{-\sigma d_{bc}^2}\left(x_c^t - x_b^t\right)$ represents the attraction component; $\mu\varepsilon_t$ denotes the randomization term, whereas μ and ε_t represents the randomization parameter and a random vector drawn from a Gaussian or uniform distribution. Lastly, the objective function (Objf) determines the quality of the solution of a firefly.

4.2 The Firefly Algorithm (FFA) for Workflow Scheduling

Each firefly in the population represents a unique candidate solution in FFA-based workflow scheduling. A candidate solution in the FFA consists of two key components: the task execution sequence and the task-resource mapping. The task execution sequence determines the order in which tasks are scheduled, ensuring that dependencies and precedence constraints are maintained. The task-resource mapping specifies allocating tasks to computing units, optimizing resource utilization and balancing the workload across available processing units. These components work together to form a feasible scheduling solution, refined iteratively to improve performance metrics, i.e., makespan. The FFA aims to iteratively improve these solutions by moving fireflies toward brighter solutions, which correspond to better fitness values to minimize makespan. The main steps of the firefly algorithm are shown in Fig. 2.

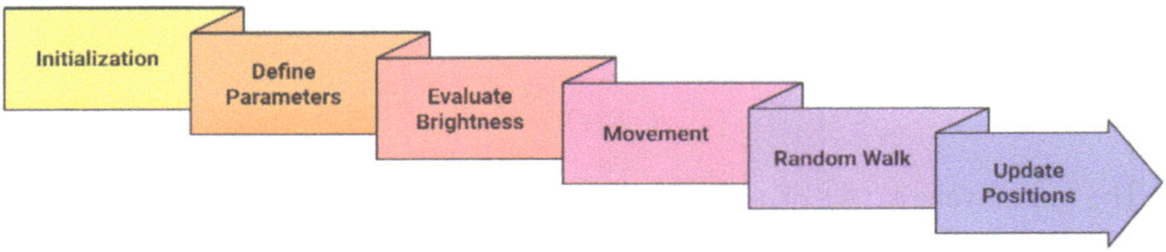

Fig. 2. Steps of the Firefly Algorithm.

4.3 Encoding and Decoding Method

In the Firefly Algorithm, each firefly represents a potential scheduling solution. A solution consists of two main components: the task execution sequence and the mapping of tasks to computing units. If a workflow comprises n tasks, then the length of a firefly solution is 2n, formulated as:

$$\text{solution} = \left\{g_0, g_1, \ldots, \ldots.., g_{n-1}, g_n, \ldots\ldots, g_{2n}\right\} \tag{13}$$

Here is the first half $\{g_0, g_1, \ldots, \ldots.., g_{n-1}\}$ represents the task execution order, while the second half $\{g_n, \ldots\ldots, g_{2n}\}$ denotes the task-to-VM assignment.

Maintaining task precedence constraints is critical in scheduling, as a child task cannot begin execution until all its parent tasks have been completed. To enforce these constraints, the encoding and decoding mechanism incorporates the task set property.

Since every task in the task set has its predecessors already scheduled, any task from this set can be selected for scheduling. This approach ensures that the generated scheduling solution remains valid.

A scheduling solution of size 2n is generated by sampling uniformly distributed random numbers within the interval (0,1). A graphical representation of a firefly scheduling solution obtained using this encoding strategy is illustrated in Fig. 3.

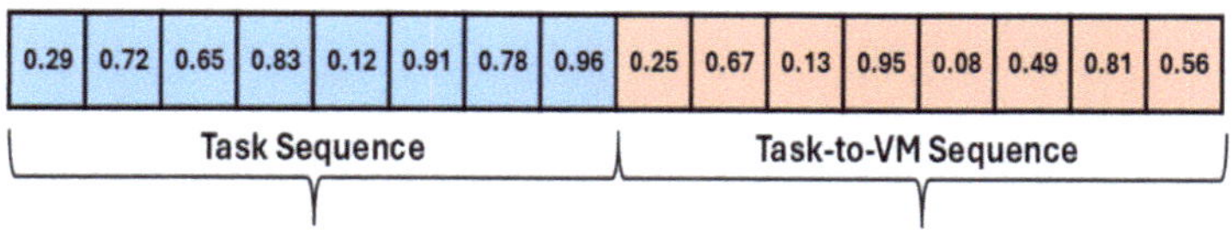

Fig. 3. A Firefly Solution for Scheduling.

Using information from the workflow DAG, each firefly can be transformed into a scheduling solution. The task sequence portion of the chromosome consists of random real numbers, which are decoded based on the number of tasks in the current task set. The process begins by adding all entry tasks into the task set. The task sequence and computing VM sequence are then decoded simultaneously. As each value in the solution is processed, the corresponding successor tasks are immediately added to the Task Set.

Illustration: For a given scheduling solution, task and VM assignments are determined by multiplying real numbers by the respective sizes of the task set |Task Set| and VM set |U|, rounding the values to obtain valid indices. For example, in Fig. 1, the initial task set contains only the task t_0, making it the first scheduled task regardless of the first value in the solution vector. Once t_0 is scheduled, it is removed from the task set, and its successor tasks t_1, t_2, t_3, t_4 enter the task set, updating it to $\{t_1, t_2, t_3, t_4\}$ with a size of 4. The real number range [0,1] is then divided into four equal intervals: [0.0, 0.25], (0.25, 0.5], (0.5, 0.75], and (0.75, 1.0]. If the second value in the solution vector is 0.72, which falls within (0.5, 0.75], the task t_3 is selected. In the next iteration, after scheduling t_3, the updated task set is $\{t_1, t_2, t_4\}$ with size 3, dividing [0,1] into three equal intervals: [0.0, 0.33], (0.33, 0.66], and (0.66, 1.0]. If the third value is 0.65, which falls into (0.33, 0.66], then the task t_2 is scheduled. A similar process will follow till the exit task to generate the task sequence.

For task-VM assignment, each value is processed similarly on the right side of the solution vector. Considering the number of VM is 3, the interval [0,1] is divided into three parts: [0.0, 0.33], (0.33, 0.66], and (0.66, 1.0]. If the first value in the sequence is 0.25, which falls in (0.0, 0.33], the first VM v_1 is selected for executing the task t_0. If the second value is 0.67, falling within (0.66,1.0], then a unit v_3 is chosen for the task t_3.

By following this decoding scheme, the task sequence derived from Fig. 1 is $\{t_0, t_3, t_2, t_4, t_1, t_6, t_5, t_7\}$, and the corresponding task-VM assignment is $\{v_1, v_3, v_1, v_3, v_1,$

$v_2, v_3, v_2\}$. These sequences uniquely define the scheduling solution. The Gantt Chart for the above scheduling solution is shown in Fig. 4.

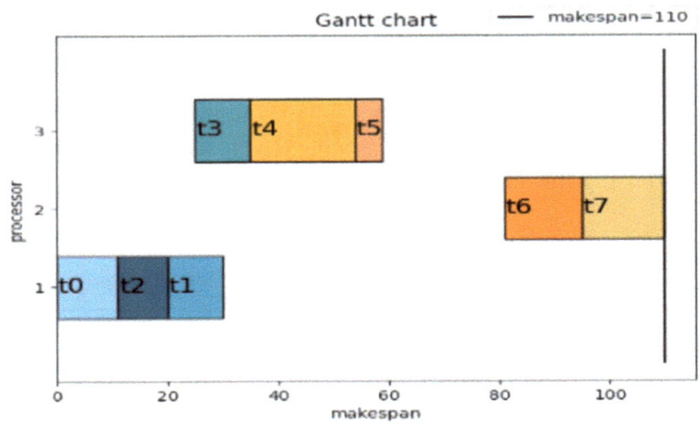

Fig. 4. Scheduling solution of a sample DAG.

5 Results and Discussion

To evaluate the performance of the proposed Firefly algorithm, we analyze its effectiveness for workflow scheduling in a cloud environment. The experiments were performed using Python 3.12.0 within an integrated development and learning environment. The implementation was executed on a system equipped with a 12th Gen Intel Core i5-12500H processor (2.50 GHz), running a 64-bit Windows 11 operating system. Six real-life workflow models (datasets), namely "Ligo_30_0," "Ligo_100_0," "Ligo_1000_0," "Montage_25_0," "Montage_100_0," and "Montage_1000_0," were evaluated using three VMs, i.e., a total of six experiments were conducted to check the effectiveness of the FFA algorithm. All six workflows have different DAG structures, communication, and execution time. The input parameter used in this study is shown in Table 1.

Table 1. Input Parameters.

Parameter Name		Values
Workflow Scheduling	Number of VMs	3
	Types of workflows	Ligo, Montage
	Number of tasks in the workflows	{25, 30, 100,1000}
	Average Index	5
	Population Size	100
Firefly Algorithm	Maximum Attractiveness (A_0)	1
	Absorption coefficient (σ)	1
	Randomness (μ)	0.5
	Maximum iteration	100

The convergence analysis of all six workflows, as illustrated in Fig. 5, demonstrates that the FFA algorithm consistently outperforms other state-of-the-art MH algorithms. However, the relative performance ranking varies across different workflows. All six workflows were executed on three virtual machines in a cloud environment, with FFA demonstrating superior performance in five of the six experiments.

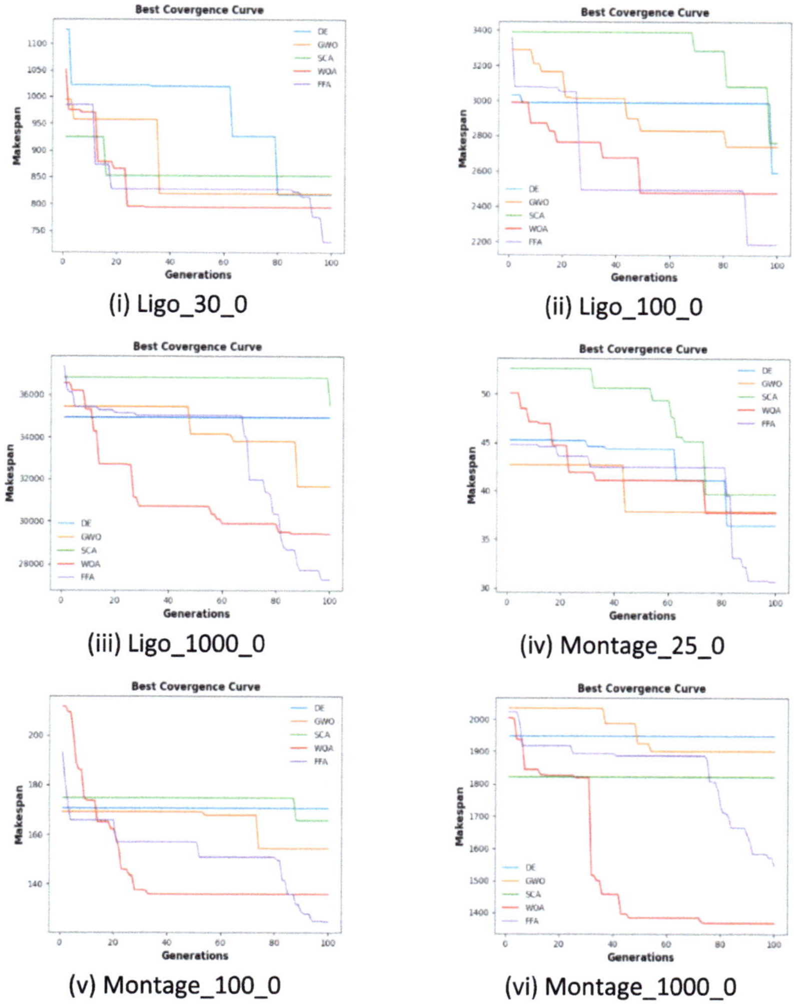

Fig. 5. Convergence Criteria of MHs.

For the LIGO workflows, Ligo_30_0 and Ligo_100_0 follow the same performance order: FFA > WOA > DE > GWO > SCA, whereas Ligo_1000_0 follows a slightly different ranking: FFA > WOA > GWO > DE > SCA. In contrast, the Montage workflows exhibit different performance trends. Montage_25_0 follows the order FFA > DE > WOA > GWO > SCA, while Montage_1000_0 ranks as FFA > WOA > GWO > SCA > DE. Notably, the last dataset, Montage_1000_0, deviates from this pattern, ranking WOA > FFA > SCA > GWO > DE.

The performance of the FFA in the Montage_1000_0 workflow ranked second-best among the evaluated algorithms, which can be attributed to the input/output-intensive nature of the Montage application. Unlike compute-bound workflows, Montage involves significant data staging and file handling, which introduces substantial communication overhead, especially when the number of tasks is large. This overhead increases the total execution time, i.e., makespan, even when task scheduling is optimized. Additionally, the use of only three VMs for executing 1000 tasks in such a communication-heavy workflow likely led to resource contention and inefficient data transfer between tasks, which further degraded FFA performance.

To evaluate the effectiveness of the FFA, we compute the mean and standard deviation of makespan across different metaheuristic algorithms. Table 2 presents each algorithm's mean makespan values, with the lowest mean value highlighted in bold. Similarly, Table 3 provides the standard deviation values, offering insight into the stability of each MH algorithms.

Table 2. Mean Value of Makespan.

Workflow Name	DE	GWO	SCA	WOA	FFA
Ligo_30_0	842.84	849.4	903.38	861.29	**808.02**
Ligo_100_0	2875.58	2802.53	2949.12	2556.49	**2400.54**
Ligo_1000_0	36828.5	33782.39	36275.73	31147.02	**29274.03**
Montage_25_0	41.85	40.87	42.02	39.58	**35.49**
Montage_100_0	185.47	165.46	172.68	148.42	**140.65**
Montage_1000_0	2004.90	1907.49	1962.82	**1447.24**	1677.17

Table 2 indicates that the FFA achieves the lowest mean makespan value across all three LIGO workflows, with WOA ranking second in Ligo_100_0 and Ligo_1000_0. Similarly, for the Montage workflows, FFA yields the best mean makespan in Montage_25_0 and Montage_100_0. However, in Montage_1000_0, WOA surpasses FFA and other metaheuristic algorithms. Overall, FFA performs better in minimizing makespan for both workflow types in a cloud computing environment. However, as shown in Table 3, FFA exhibits the highest variation in results, while GWO achieves the lowest standard deviation, indicating more excellent stability in makespan calculations.

Table 3. The standard Deviation value of makespan.

Workflow Name	DE	GWO	SCA	WOA	FFA
Ligo_30_0	24.12	22.23	43.36	56.15	68.41

(continued)

Table 3. *(continued)*

Workflow Name	DE	GWO	SCA	WOA	FFA
Ligo_100_0	179.18	55.79	166.27	109.7	181
Ligo_1000_0	1191.8	1309.49	662.27	1502.9	1506.62
Montage_25_0	3.06	1.81	2.16	2.1	3.34
Montage_100_0	8.68	7.64	6.81	8.33	10.47
Montage_1000_0	52.4	7.76	94.77	79.9	99.97

6 Conclusion

The paper studied the Firefly Algorithm for workflow scheduling within a cloud environment, focusing specifically on makespan. Proper scheduling is of extreme importance so that better resource utilization and execution time can be ensured while dealing with applications having complex workloads in a cloud environment. Hence, a broad comparative analysis was performed using the executions of five different MHs: FFA, GWO, DE, WOA, and SCA. The experiment results show that the FFA achieves the best overall performance in minimizing makespan for workflow scheduling in a cloud environment. However, FFA exhibits greater variation, whereas the GWO algorithm demonstrates higher stability among the five metaheuristic algorithms. Although FFA has shown great potential to reduce the makespan, its application for cloud scheduling workflows is still quite open for improvement.

One promising avenue of research would be the integration of FFA with machine learning frameworks or reinforcement learning methodologies to enhance its adaptability in dynamic and heterogeneous cloud environments. Additionally, this study may be extended further to include the multi-objective optimization challenges, such as concurrent minimization of energy usage, cost reduction, and system reliability improvement, to derive a more holistic scheduling framework. Finally, integrating cloud-related factors, such as security, privacy, and compliance obligations, into the scheduling framework will enhance the real-world relevance of the proposed approach. Tackling these future challenges will help develop more flexible and efficient scheduling strategies for modern cloud computing infrastructures [20, 21].

Acknowledgments. This work was supported partially by the Research Startup Grant from the University Grant Commission (UGC)-Basic Scientific Research (BSR) under Grant F.30-377/2017 (BSR) 07.03.2017, F.D. Dy. No. 8123 16.02.2017/10057/B053.

We (Mohammad Sajid) would like to thank the authorities of the Inter University Centre for Astronomy and Astrophysics, Pune, India, for providing research facilities.

References

1. International Data Corporation: Indian Public Cloud Services Market to Grow at a CAGR of 23.8% for 2023-28, IDC. [Online]. Available: https://www.idc.com/getdoc.jsp?containerId=prAP52400124

2. Zhu, Z., Zhang, G., Li, M., Liu, X.: Evolutionary multi-objective workflow scheduling in cloud. IEEE Trans Parallel Distrib Syst. **27**(5), 1344–1357 (2016). https://doi.org/10.1109/TPDS.2015.2446459
3. Qasim, M., Sajid, M., Shahid, M.: Hunger games search: a scheduler for cloud computing. In: 2022 International Conference on Data Analytics for Business and Industry, ICDABI, pp. 170–175 (2022). https://doi.org/10.1109/ICDABI56818.2022.10041521
4. Talbi, E.G.: Metaheuristics: from Design to Implementation, vol. 2, pp. 268–308 (2009). https://doi.org/10.1002/9780470496916
5. Qasim, M., Sajid, M.: Chaotic-based MFO scheduler for IoT task scheduling in cloud-of-things. In: 2024 1st International Conference on Innovative Engineering Sciences and Technological Research (ICIESTR), pp. 1–6 (2024). https://doi.org/10.1109/ICIESTR60916.2024.10798393
6. Sajid, M., Raza, Z., Shahid, M.: Energy-efficient scheduling algorithms for batch-of-tasks (BoT) applications on heterogeneous computing systems. Concurr. Comput. **28**(9), 2644–2669 (2016). https://doi.org/10.1002/cpe.3728
7. Sajid, M., Raza, Z.: Energy-efficient quantum-inspired stochastic Q-HypE algorithm for batch-of-stochastic-tasks on heterogeneous DVFS-enabled processors. Concurr. Comput. **31**(20), e5327 (2019)
8. Zhang, X.: Optimizing scientific workflow scheduling in cloud computing: a multi-level approach using whale optimization algorithm. J. Eng. Appl. Sci. **71**(1) (2024). https://doi.org/10.1186/s44147-024-00512-9
9. Akraminejad, R., Khaledian, N., Nazari, A., Voelp, M.: A multi-objective crow search algorithm for optimizing makespan and costs in scientific cloud workflows (CSAMOMC). Computing. **106**(6), 1777–1793 (2024). https://doi.org/10.1007/s00607-024-01263-4
10. Zaki, T., Zeiträg, Y., Neves, R., Figueira, J.R.: A cooperative coevolutionary genetic programming hyper-heuristic for multi-objective makespan and cost optimization in cloud workflow scheduling. Comput. Oper. Res. **172** (2024). https://doi.org/10.1016/j.cor.2024.106805
11. Bacanin, N., Zivkovic, M., Bezdan, T., Venkatachalam, K., Abouhawwash, M.: Modified firefly algorithm for workflow scheduling in cloud-edge environment. Neural Comput. & Applic. **34**(11), 9043–9068 (2022). https://doi.org/10.1007/s00521-022-06925-y
12. Ramathilagam, A., Vijayalakshmi, K.: Workflow scheduling in cloud environment using a novel metaheuristic optimization algorithm. Int. J. Commun. Syst. **34**(5) (2021). https://doi.org/10.1002/dac.4746
13. Qasim, M., Sajid, M., Lapina, M., Babenko, M.: A generalized bi-objective scheduling algorithm for batch-of-tasks on heterogeneous computing systems. Adv. Syst. Sci. Appl., 2024 | Journal article. https://doi.org/10.25728/assa.2024.2024.02.1430, EID: 2-s2.0-85199979370, Part of ISSN: 10786236. https://ijassa.ipu.ru/index.php/ijassa/article/view/1430
14. Mikram, H., Kafhali, S.E., Saadi, Y.: HEPGA: a new effective hybrid algorithm for scientific workflow scheduling in cloud computing environment. Simul. Model. Pract. Theory. **130** (2024). https://doi.org/10.1016/j.simpat.2023.102864
15. Barredo, P., Puente, J.: Precise makespan optimization via hybrid genetic algorithm for scientific workflow scheduling problem. Nat. Comput. **22**(4), 615–630 (2023). https://doi.org/10.1007/s11047-023-09950-5
16. Kakkottakath Valappil Thekkepuryil, J., Suseelan, D.P., Keerikkattil, P.M.: An effective meta-heuristic based multi-objective hybrid optimization method for workflow scheduling in cloud computing environment. Clust. Comput. **24**(3), 2367–2384 (2021). https://doi.org/10.1007/s10586-021-03269-5
17. Belgacem, A., Beghdad-Bey, K.: Multi-objective workflow scheduling in cloud computing: trade-off between makespan and cost. Clust. Comput. **25**(1), 579–595 (2022). https://doi.org/10.1007/s10586-021-03432-y

18. Topcuoglu, H., Hariri, S., Wu, M.Y.: Performance-effective and low-complexity task scheduling for heterogeneous computing. IEEE Trans. Parallel Distrib. Syst. **13**(3), 260–274 (2002). https://doi.org/10.1109/71.993206
19. Yang, X.S.: Firefly algorithms for multimodal optimization. In: Lecture Notes in Computer Science (Including Subseries Lecture Notes in Artificial Intelligence and Lecture Notes in Bioinformatics), pp. 169–178. Springer (2009). https://doi.org/10.1007/978-3-642-04944-6_14
20. Jawed, M.S., Sajid, M., Lapina, M., Prasad, M.: Enhancing the security of cryptographic algorithms in cloud-based cyber-physical systems. In: Intelligent Techniques for Cyber-Physical Systems, pp. 19–34 (2023). https://doi.org/10.1201/9781003438588-2
21. Basan, E., Lapina, M., Lesnikov, A., Basyuk, A., Mogilny, A.: Trust monitoring in a cyber-physical system for security analysis based on distributed computing. In: Alikhanov, A., Lyakhov, P., Samoylenko, I. (eds.) Current Problems in Applied Mathematics and Computer Science and Systems. APAMCS Lecture Notes in Networks and Systems, vol. 702, pp. 430–440. Springer, Cham (2022). https://doi.org/10.1007/978-3-031-34127-4_4

Content Analysis Methods and Development of a System for Profile Assessing for Information Object of Social Influence

Marina Rudenko(✉), Milada Krapivina, Kuhta Dmitrii, and Andrei Rudenko

V.I. Venadsky Crimean Federal University, Simferopol, Russia
maridigit@gmail.com

Abstract. This paper addresses the critical issue of destructive content dissemination on social networks, particularly the disguising of propaganda as legitimate material and the heroification of extremist ideologies. It analyzes existing content analysis methods and proposes a novel approach that integrates graph representation of information objects in social networks with color coding for visual evaluation. This integrated system leverages graph representation to display relationships between information objects and color coding to simplify the evaluation of content tone and context. A key strength of the proposed system lies in its versatility and adaptability for contextual analysis across diverse practical applications. The paper presents the development of a system based on BertForSequenceClassification model and demonstrates its application in analyzing posts from VKontakte, showcasing the potential for improved detection and prevention of harmful content. The study concludes by outlining future research directions, including enhancements to semantic analysis, incorporation of temporal dynamics, and the exploration of advanced clustering and graph centrality metrics to improve the accuracy and reliability of the method.

Keywords: Content Analysis · Social Networks · Information Object Profiling · Destructive Content · Heroification · Color Coding · Semantic Analysis · NLP · BertForSequenceClassification · Graph Representation · Social Media Security

1 Introduction

In today's world, where social networks have become an integral part of the information space, the problem of destructive content dissemination poses a serious threat to public safety and stability. Effective detection and neutralization of such content requires the development and application of advanced analysis methods capable not only of detecting illegal materials, but also of predicting trends and preventing content popularization at earlier stages.

Special attention in the processes of optimizing information flows at significant CII objects should be paid to the dynamics of their behavior and threats at the software level in different types of object relationships [15, 16].

M. Lapina et al. (Eds.): AISMA 2025, LNNS 1456, pp. 384–396, 2026.
https://doi.org/10.1007/978-3-032-07275-7_35

A particular challenge of content-monitoring is the disguising of propagandizing banned materials as historical summaries, analytical reviews or cultural artifacts to avoid blocking by social networks and gaining a semblance of legitimacy.

The phenomenon of heroification on social media is a process by which ideas and personalities associated with nationalism, extremism and terrorism are positively valued and supported, which can lead to radicalization of views and the commission of violent acts. Therefore, monitoring and analyzing content that promotes the heroification of such elements is a critical task to ensure online safety and prevent the spread of destructive ideologies.

Purpose of the Present Study

The aim of the paper is to analyze existing methods and approaches in content analysis and trend prediction, to propose an approach to content analysis based on graph representation of information objects in social networks and color coding for visual evaluation of information object. The advantages of this integrated approach are the visibility of displaying the relationship between information objects using graph representation and the simplicity of evaluating the information object using color that reflects the context and tone temperature. The main advantage of such an approach we consider to be its versatility and the ability to adapt such a system to solve the tasks of contextual analysis for different practical purposes.

Literature Overview

Content analysis and profile evaluation of information objects in social networks are becoming increasingly relevant in the context of security and information management. The complexity of the tasks requires the application of advanced data representation and analysis techniques, where graph models and neural network approaches, in particular graph neural networks (GNNs), show significant potential. This literature review examines current research on the use of graphs and neural networks for content analysis, relationship discovery, and information entity relevance assessment, with a particular focus on security applications.

Despite the relevance, the use of graphs is not a new trend. Burgess et al. present a basic graph model of computer security, demonstrating principles for modeling security systems using graphs [1]. Although the work is relatively old, it sets the foundation for further research in this area.

The approach proposed in Chen, Y., et al. for network security policy consistency detection based on intelligent graph databases can be adapted to solve the problem of modeling relationships between information objects in social media [2]. The use of graph algorithms and visualization enables the analysis of complex content distribution patterns, including illegal content. The paper explains how the use of graphs overcomes the limitations of traditional approaches such as data fragmentation and one-sidedness of analysis. It explains how individual algorithms have been shown to be effective in the tasks of community clustering and social network link prediction. For example, node centrality algorithms can identify significant disseminators of illegal information, while component connectivity analysis helps detect hidden communities that coordinate its dissemination.

The paper by R. Shevchuk presents a comprehensive analysis of social media security research covering the period from 2009 to 2023, based on a study of 2947 research articles indexed in the Web of Science database [3]. The analysis, conducted using the CiteSpace 6.3.R1 tool, reveals the current state of research in this field, including analysis of collaborative networks, co-citation, and cluster analysis of keywords and links with explosive citations. The results of this research can be useful not only for modeling links between information objects in social media, but also for identifying relationships between content authors.

The integration of textual analysis (NLP) and graph-based methods can analyze content more efficiently and reveal hidden links. Sönmez categorizes common vulnerabilities and threats (CVEs) using textual and graph analysis [4]. This demonstrates how a combination of NLP and graphs can be used to automatically classify CVEs, which improves content analysis and speeds up the process of vulnerability identification and remediation. However, the issues raised in the paper are also relevant to social media entries - consistency latency and relevance assessment.

Mendsaikhan et al. propose to use a cybersecurity knowledge graph to assess the relevance of texts [5]. Their approach automates threat processing through analyzing text similarities and relationships in the knowledge graph. The authors of the paper also share their experience in using the NER model to solve NLP problems, and the relationship between the output of the model and the quality of the constructed graph and consequently the relevance of text.

Lagraa et al. in their review elaborate on various graph-based methods for network security monitoring and botnet detection, emphasizing their effectiveness in detecting botnet anomalies and structures [6]. This work demonstrates not only the general trend towards the use of graph-based surveys for network security and security in general, but also provides a comprehensive understanding of existing algorithms related to this approach.

The development of graph theory and the application of its approaches in various subject areas, not only in security surveys, has led to the creation of the concept and development of graph neural networks. The development of graph neural networks (GNN) has opened new possibilities for analyzing graph data. Gao et al. apply self-learning on text and social graphs for rumor detection [7]. Their work is an example of using GNN to analyze content in social media with proven effectiveness to effectively detect false information and prevent its spread.

Content analysis and evaluation of information objects in social networks have gained paramount importance in combating misinformation, identifying manipulative accounts, and understanding the dynamics of public opinion. In this context, graph-based models and neural network approaches, especially graph neural networks (GNNs), show significant potential in solving the problems of detecting fake news and identifying disruptive actors.

Guo et al. proposed a fake news detection method based on a mixed graph neural network specifically designed for robust vehicular social networks, emphasizing the adaptability of GNNs to different domain specificities [8]. In parallel, Li et al. developed SybilFlyover, a fake account detection model that uses heterogeneous graphs to model complex relationships in social networks, demonstrating the effectiveness of graph-based approaches in detecting suspicious profiles [9].

A review by Kondamudi et al. systematizes attributes, features, and approaches to detecting fake news in social media, providing a comprehensive view of existing methods and research directions [10]. In turn, to understand the influence and formation of public opinion in social media.

Baziyad et al. proposed a data analytics (DEA) model for identifying opinion leaders, emphasizing the importance of identifying key actors shaping the discourse [11]. Developing the theme of social network analysis, Shebaro et al. investigate the improvement of association detection through multidimensional analysis, demonstrating that studying social networks from different perspectives can reveal hidden patterns and connections [12].

Lee et al. quantify variations in controversial discussions in Kuwaiti social media, showing how analyzing online content can reveal polarization and tensions in public opinion [13]. Finally, Rashid and Bhat provide a systematic review of methods for identifying influencers in online social networks, spanning from topological approaches to the era of deep learning, emphasizing the evolution of analysis methods and their increasing complexity [14].

The integration of graph models, neural networks, and text analytics techniques enables more efficient and accurate systems for analyzing content, detecting misinformation, and understanding influence structure in social networks, which is critical for security and information resilience in the digital space.

2 Method

For the study we collected posts from the social network VKontakte with a sample where there is the word "hero" to analyze in what context and what image of the hero is formed in the media space with the association hero. The general scheme of operation is presented in Fig. 1 [14].

Further, having analyzed the obtained records we have distinguished the classes presented in Table 1.

As a result of manual markup, we get the class distribution shown in Fig. 2. To balance the classes, we add text so that each class has 260 elements. The pre-trained BertForSequenceClassification model from the transformers library is used. This model includes a BERT and an output classification layer that is adapted for the multiclass classification task.

The BertTokenizer tokenizer from the transformers library was used to convert text into numeric vectors, which prepares the text to feed into the BERT model by splitting it into subwords.

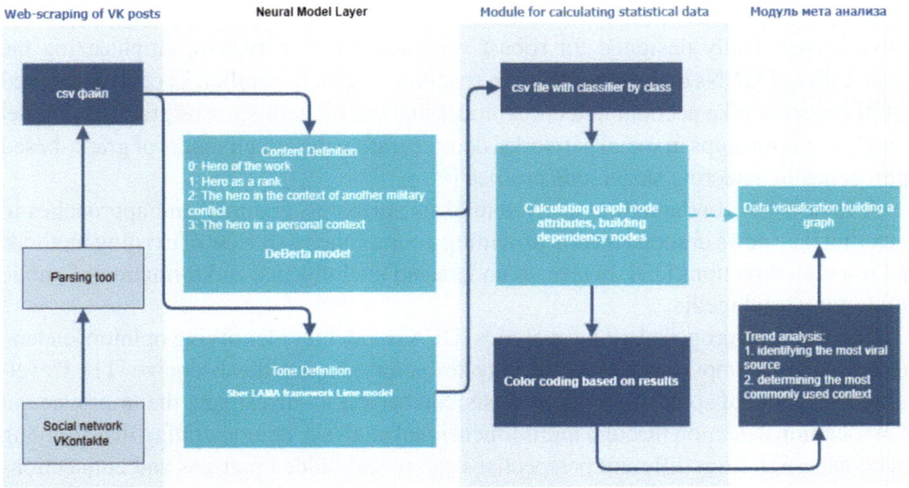

Fig. 1. The general scheme of operation.

Table 1. Defined classes.

Class ID	Description	Sample post	Note
0	Hero of the work	Do you have any skeletons in your closet? ??And actually, there may be more than just skeletons. The Spanish movie "No mires a los ojos" (No mires a los ojos) will tell us about it. ??This is a chamber and even a little paranoid narrative, the main hero of which is a closet. Well, or not exactly a closet... And the denouement is extremely unexpected. In the main role - Paco Leon, a famous Spanish actor, once we posted one of his stand-up, as well as his directorial works - films "Carmina y revienta" and "Carmina y am?n".Now let's look at him in the role of an actor!???	

(*continued*)

Table 1. (*continued*)

Class ID	Description	Sample post	Note
1	Hero as a rank	Khamzat Chimaev commented on Artur Beterbiev's victory over Dmitry Bivol: "It is with great pleasure that I congratulate our brother, the undisputed absolute world heavyweight boxing champion Artur Beterbiev on the well-deserved and brilliant victory he has just won. 21 fights, 21 wins and 20 knockouts with just one decision - it's just something incredible. Arthur became the only absolute champion in the history of boxing in this weight category. It took him a very long time to get to this point, he went through a huge number of heavy big-name opponents, huge difficulties, which only he himself and a narrow circle of his relatives know about, including the Head of the Chechen Republic, Hero of Russia Ramzan Akhmatovich Kadyrov, who has been providing him with invaluable help and support throughout his life and career. Arthur is an example for us both in life and in sport, who with his diligence and perseverance achieved such a historic result, making us all happy. I congratulate you Arthur, you have done what was expected and hoped for from you".	Hero of Russia, Hero of the Soviet Union, Hero of Socialist Labor...

(*continued*)

Table 1. (*continued*)

Class ID	Description	Sample post	Note
2	The hero in the context of military conflict	Alexey Kukhtin, a Severodvinsk resident, our reliable friend and faithful comrade, was heroically killed in the course of military operations. We express our sincere condolences to the family and friends of Alexey! Eternal memory and glory to the Hero! To be a warrior - to live forever????! Farewell to Alexey will be held on October 22 at 11.00, at the address: 42 Tortseva St. (DOF).	
3	The hero in a personal context	Our hero in the kitchen and in life: Daddy Alexei Solovskoy	
4	undefined	Our Hero	Definition of a word in an unclassifiable context

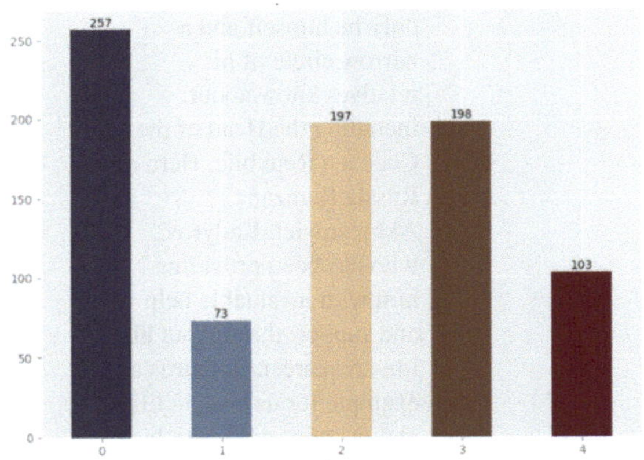

Fig. 2. Class distribution before balancing.

The classes are not mutually exclusive so one option for improvement is multileveling. Most posts can belong to more than one class. In the first iteration, 2 and 3 classes are absorbed into 0 classes. This is due to the availability of a sample of fiction descriptions.

Color coding is based on the HCV color representation model. It describes how colors combine to create the visible spectrum. It has three components: hue, chroma, and luminance value. We propose to evaluate information objects of social influence with the help of the GIC system (Fig. 3).

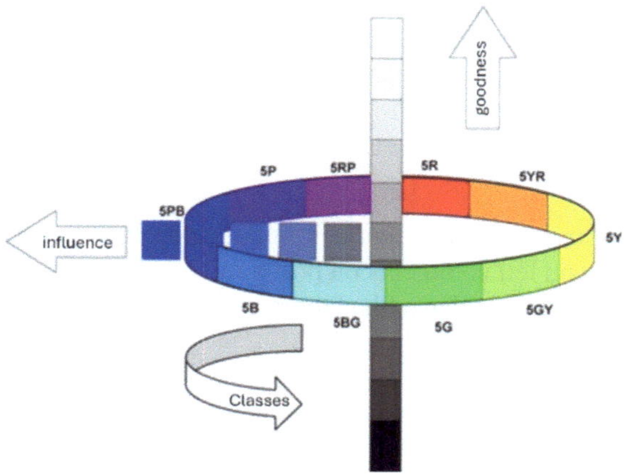

Fig. 3. GIC system.

C (classes) – set of classes, determines the class membership, the intermediate hue covers the cases of multi-class classification and is calculated using a probability vector. For correct color mapping, it is important to set the sequence of classes so that similar classes with large intervals are close to each other on the color wheel.
I (Influence) this value is a reflection of the confidence score in the class prediction.
G (goodness value) - saturation - dark/light. The level of lightness is determined by comparison with an achromatic scale from black (0) to white (10). It is defined by the tonal response where white is a positive value and black is a negative value.

We will represent a social network posts as an undirected weighted graph.

$$G = (V, E) \tag{1}$$

Where V is the set of nodes representing posts in the social network.

E is the set of edges representing the links between posts based on user interactions.

Each node $v_i \in V$ corresponds to a post in a social network. The size of a node (its weight) reflects the popularity of the post, measured by the number of comments.

We propose the following formula for the weight of a node reflecting the number of comments:

$$w_i = log 1 + C_i \tag{2}$$

C is the number of comments on the post v_i (v_i is a post with unique identifier i).

Log - log$_i$arithmic function (e.g. natural logarithm). Using the logarithm allows us to smooth out the differences between posts with very high and moderate numbers of comments, making the graph visualization more informative. We add 1 to avoid a logarithm of zero if a post has no comments.

The edges of the graph connect pairs of posts between which there is user interaction. The weight of an edge reflects the strength of that connection, based on the number of shared users who have liked or commented on both posts.

$e_{ij} \in E$ is an edge connecting posts v_i and v_j. An edge exists if at least one user interacted with both posts.

We propose the following formulas for the weight of an edge:

$$W_{ij} = *U_i \cap U_j + *L_i \cap L_j \quad (3)$$

U_i - the multitude of users who left comments on the post v_i.
U_j - the set of users who left comments on post v_j.
L_i - the multitude of users who have liked the post v_i.
L_j - multiple users who have liked the post v_j.
$|X|$ - power of set X (number of elements in the set).
Is a weighting factor that determines the importance of common commenters?

– Weight coefficient determining the significance of common "likers".

The values of α and β can be adjusted to give more weight to comments or likes depending on the specific goals of the analysis. For example, if comments are considered a more meaningful indicator of interaction, $\alpha > \beta$ can be set. These values can be optimized based on the data and the goals of the study.

An alternative (normalized) formula for the weight of an edge (W_{ij}):

$$W_{ij} = *(|U_i \cap U_j|/|U_i U_j|) + *(|L_i \cap L_j|/|L_i L_j|) \quad (4)$$

U_i - the multitude of users who left comments on the post v_i.
U_j - the set of users who left comments on post v_j.
L_i - the multitude of users who have liked the post v_i.
L_j - multiple users who have liked the post v_j.
$|X|$ - power of set X (number of elements in the set).
Is a weighting factor that determines the importance of common commenters?

– Weight coefficient determining the significance of common "likers".

In this version, we normalize the number of total users by dividing it by the total number of users who commented or liked at least one of the two posts. This allows us to compare links between posts that have different total number of interactions. This normalization avoids bias in favor of posts with more interactions.

The proposed approach has the following advantages: it takes into account heterogeneous types of interactions (comments, likes) through weighting, providing flexibility through customizable coefficients α and β, which adapts the model to the specifics of the social network and analysis tasks. Interpretability is achieved through a clear interpretation of the weights of nodes and edges, which facilitates the understanding of the graph structure.

3 Results

After training the DeBert model, we obtained the results presented in Fig. 4.

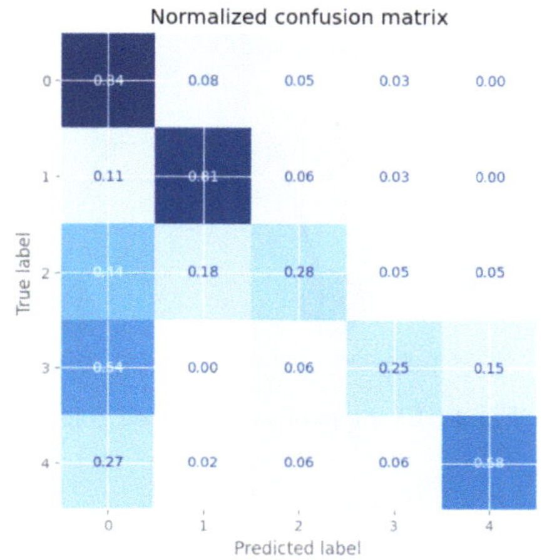

Fig. 4. Model result without dataset modification.

By modifying the dataset with updating the elements with more specific classes, we were able to improve the quality of detection and obtained the results shown in Fig. 5.

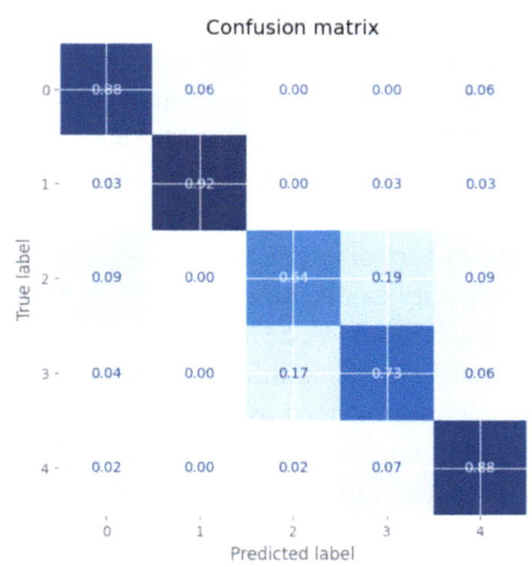

Fig. 5. Model result with dataset modification.

Figure 6 illustrates the interaction graph of information objects linked through publications on a single web page (group page). The graph visualization uses color-coding to differentiate nodes representing different information objects.

Fig. 6. Example of a constructed graph.

4 Discussion

Future research directions, implies several key directions. First, there is a need to focus on improving methods for semantic analysis of text posts to better identify context and hidden links between posts. This includes developing algorithms capable of capturing nuances of language, accounting for sarcasm, irony, and understanding complex contextual dependencies. In particular, methods that combine traditional natural language processing (NLP) approaches with machine learning techniques to detect subtle semantic differences and determine the thematic affiliation of posts should be investigated.

Second, it is critical to consider the temporal aspect when constructing interaction graphs. Models should be developed that can dynamically adapt the weights of links between graph nodes as a function of user interaction time. This may include the use of fading functions that decrease the weights of old links and increase the weights of new links, reflecting the current relevance of the interaction. Alternatively, recurrent models capable of remembering the history of interactions between users and adapting the graph structure according to this history can be used. In addition, methods for visualizing the temporal dynamics of the graph should be developed to allow users to track the changing links between posts and identify trends in information dissemination.

Third, new clustering algorithms need to be investigated that focus on identifying communities of posts that are connected not only by explicit but also by hidden semantic links. Instead of traditional algorithms such as Louvain or Girvan-Newman, methods based on spectral clustering or density clustering should be explored to identify clusters of complex shape and structure. It is important to consider that post communities can be dynamic and their composition may change over time. Therefore, it is necessary to develop algorithms that can track the evolution of communities and adapt the graph structure according to these changes.

Fourth, graph centrality metrics need to be investigated in depth to identify the most important posts and users with the greatest influence on information dissemination. In addition to traditional metrics such as degree centrality, betweenness centrality and eigenvector centrality, new metrics that take into account the semantic context of posts and temporal characteristics of interactions should be explored. For example, metrics that take into account not only the number of links of a post, but also the quality of those links and the time elapsed since the last interaction could be used. Methods should be developed to allow visualization of centrality metrics in the graph, for example, by changing the size or color of nodes based on their importance.

Finally, methods need to be developed to assess the quality of interaction graphs built from semantic analysis of text messages. This includes developing metrics that measure the graph's consistency with real-world data, as well as assessing its ability to detect misinformation and disruptive actors. It is important to keep in mind that the quality of the graph can depend on many factors, including the quality of the semantic analysis of the text, the parameters of the clustering algorithms, and the choice of centrality metrics. Therefore, systematic experiments should be conducted to determine the optimal parameters for constructing interaction graphs.

5 Conclusion

In conclusion, the presented research proposed an innovative approach to social media analysis that integrates graph models, color coding principles, and semantic analysis of text messages in order to detect and prevent the glorification of unlawful leaders and the dissemination of destructive content. Unlike traditional methods that focus primarily on structural graph analysis, the proposed method integrates contextual language understanding through neural networks capable of distinguishing nuances in the use of keywords such as "hero." Color coding based on the HCV (hue, chroma, value) model allows visualizing the level of confidence in classifying content as disruptive or heroic, where hue (hue) defines class, saturation (chroma) reflects confidence in classification, and luminance (value) indicates the tone of the message. The graph model, in turn, is formed based on users' interactions with posts, where the weight of a node reflects the popularity of the post (number of comments), and links between nodes are formed based on the commonality of users interacting with different posts. This approach not only identifies potentially harmful content, but also visualizes its distribution and influence in the network, making it possible to develop more effective strategies to counter misinformation and manipulation. Future research should focus on further improving semantic analysis, taking into account the temporal dynamics of interactions and integrating with other data modalities to improve the accuracy and reliability of the proposed method.

The study was funded by the Russian Science Foundation (RNF) grant (project No. 25-21-20125).

References

1. Burgess, M., Canright, G., Engø-Monsen, K.: A graph-theoretical model of computer security. Int. J. Inf. Secur. **3**, 70–85 (2004). https://doi.org/10.1007/s10207-004-0044-x
2. Chen, Y., et al.: A knowledge graph-based consistency detection method for network security policies. Appl. Sci. **14**(18), 8415 (2024). https://doi.org/10.3390/app14188415

3. Shevchuk, R., Martsenyuk, V., Melnyk, A.: Research hotspots and trend analysis of social media security on CiteSpace knowledge graph. 2024 14th International Conference on Advanced Computer Information Technologies (ACIT), Ceske Budejovice, Czech Republic. **2024**, 543–551. https://doi.org/10.1109/ACIT62333.2024.10712528
4. Sönmez, F.Ö.: Classifying common vulnerabilities and exposures database using text mining and graph theoretical analysis. In: Maleh, Y., Shojafar, M., Alazab, M., Baddi, Y. (eds.) Machine Intelligence and Big Data Analytics for Cybersecurity Applications. Studies in Computational Intelligence, vol. 919. Springer, Cham (2021). https://doi.org/10.1007/978-3-030-57024-8_14
5. Mendsaikhan, O., Hasegawa, H., Yamaguchi, Y., Shimada, H.: Quantifying the significance and relevance of cyber-security text through textual similarity and cyber-security knowledge graph. IEEE Access. **8**, 177041–177052 (2020). https://doi.org/10.1109/ACCESS.2020.3027321. keywords: {Computer security;Data mining;Task analysis;Autonomous systems;Tagging;Computational modeling;Cyber-security knowledge graph;cyber threat;text analysis;textual similarity}
6. Lagraa, S., Husák, M., Seba, H., et al.: A review on graph-based approaches for network security monitoring and botnet detection. Int. J. Inf. Secur. **23**, 119–140 (2024). https://doi.org/10.1007/s10207-023-00742-7
7. Gao, Y., Wang, X., He, X., et al.: Rumor detection with self-supervised learning on texts and social graph. Front. Comput. Sci. **17**, 174611 (2023). https://doi.org/10.1007/s11704-022-1531-9
8. Guo, Z., et al.: Mixed graph neural network-based fake news detection for sustainable vehicular social networks. IEEE Trans. Intell. Transp. Syst. **24**(12), 15486–15498 (2022)
9. Li, S., Yang, J., Liang, G., Li, T., Zhao, K.: SybilFlyover: heterogeneous graph-based fake account detection model on social networks. Knowl.-Based Syst. **258**, 110038 (2022)
10. Kondamudi, M.R., Sahoo, S.R., Chouhan, L., Yadav, N.: A comprehensive survey of fake news in social networks: attributes, features, and detection approaches. J. King Saud Univ. –Comput. Inf. Sci. **35**(6), 101571 (2023)
11. Baziyad, H., Kayvanfar, V., Toloo, M.: A data envelopment analysis model for opinion leaders' identification in social networks. Comput. Ind. Eng. **190**, 110010 (2024)
12. Shebaro, M., de Moura, L.N., Tešić, J.: Improving association discovery through multiview analysis of social networks. Soc. Netw. Anal. Min. **14**(1), 42 (2024)
13. Lee, Y., Alostad, H., Davulcu, H.: Quantifying variations in controversial discussions within Kuwaiti social networks. Big Data. Cogn. Comput. **8**(6), 60 (2024)
14. Rashid, Y., Bhat, J.I.: Topological to deep learning era for identifying influencers in online social networks: a systematic review. Multimed. Tools Appl. **83**(5), 14671–14714 (2024)
15. Vinodha, D., Jenefa, J., Anita, E.A.M., Babenko, M.: Theoretical framework for Blockchain secured predictive maintenance learning model using digital twin. In: Lapina, M., Raza, Z., Tchernykh, A., Sajid, M., Zolotarev, V., Babenko, M. (eds.) AISMA-2024: International Workshop on Advanced Information Security Management and Applications AISMA 2024. Lecture Notes in Networks and Systems, vol. 863. Springer, Cham (2024)
16. Mary Anita, E.A., Jenefa, J., Vinodha, D., Lapina, M.: Advancements in Sybil attack detection: a comprehensive survey of machine learning-based approaches in wireless sensor networks. In: Lapina, M., Raza, Z., Tchernykh, A., Sajid, M., Zolotarev, V., Babenko, M. (eds.) AISMA-2024: International Workshop on Advanced Information Security Management and Applications AISMA 2024. Lecture Notes in Networks and Systems, vol. 863. Springer, Cham (2024)
17. Qasim, M., Sajid, M., Lapina, M., Babenko, M.: A Generalized Bi-Objective Scheduling Algorithm for Batch-of-Tasks on Heterogeneous Computing Systems. Adv. Syst. Sci. Appl. (2024). https://doi.org/10.25728/assa.2024.2024.02.1430. EID: 2-s2.0-85199979370, Part of ISSN: 10786236, https://ijassa.ipu.ru/index.php/ijassa/article/view/1430

System for Integrating Social Network Analysis and Information Security in the Context of Countering Radicalism

Marina Rudenko[1(✉)], Milada Krapivina[1], Dmitrii Kuhta[1], Andrei Rudenko[1], and Yulia Andrusenko[2]

[1] V.I. Venadsky Crimean Federal University, Simferopol, Russia
maridigit@gmail.com
[2] North Caucasus Federal University, Stavropol 355017, Russia

Abstract. The spread of destructive and radical content on social media is a growing threat to national and global security. Existing countermeasures often focus only on single aspects (text or network structure) and prove ineffective against veiled propaganda and dynamic information dissemination. This paper proposes the development of a comprehensive system for social media content evaluation that integrates multimodal analysis, social network analytics (SNA) techniques and temporal dynamics. The proposed system analyzes the visual content of posts using neural networks (YOLO), evaluates textual content for relevance to target topics using natural language processing and classification techniques (e.g., BERT), represents social relationships as a graph for influence analysis, and introduces a freshness metric to account for content relevance. The results of each component analysis are integrated into a single final post score, based on which the post is classified into relevance/influence. The key metrics and methods applied at each stage are discussed, and an example of how the system works in the context of identifying potentially radical content is demonstrated. The presented system provides more accurate and timely identification of destructive publications and understanding of their distribution patterns, which is critical for effectively countering radicalization in the digital space. The study was funded by the Russian Science Foundation (RNF) grant (project No. 25-21-20125 "Development of mathematical models and a system for assessing content for extremist content based on text analysis and object detection in images and videos").

Keywords: Social network analysis · Countering radicalism · Content evaluation · Graph analysis · Image analysis · Text analysis · Neural networks · Temporal analysis · Information security · Destructive content

1 Relevance

The growth of extremist and terrorist activity in the digital space, especially in social networks and messengers, poses a serious threat to security both nationally and globally. According to a study by the National Center for Information Counteraction to Terrorism

M. Lapina et al. (Eds.): AISMA 2025, LNNS 1456, pp. 397–412, 2026.
https://doi.org/10.1007/978-3-032-07275-7_36

and Extremism in the Educational Environment and the Internet (2021–2024), in recent years there has been a significant increase in illegal content on the Runet, including propaganda of terrorism, extremism and violence, with Telegram and YouTube becoming the main distribution platforms. For example, in 2024, 90% of drug sales and 40% of hate speech content will be distributed via Telegram, highlighting the vulnerability of today's digital platforms to attackers.

Statistics on extremist offenses (2018–2022) show an increase in the number of crimes related to discrediting the aut horities, rehabilitation of Nazism and dissemination of extremist materials. This confirms the need to develop tools capable of promptly identifying and blocking destructive content. Furthermore, Demir and Çona (2024) show that lone far-right terrorists are actively using manifestos and online platforms to justify violence and recruit supporters, forming global radicalization networks.

The integration of social network analysis (SNA) and information security technologies is becoming critical to counter these threats. Modern systems should not only detect banned content, but also analyze structural links between users, identify key propagators of ideologies and predict potential radicalization risks. This approach will allow not only to promptly block illegal materials, but also to stop the formation of destructive communities at an early stage.

Thus, the development of a comprehensive system combining SNA methods and cybersecurity tools responds to the challenges identified in the attached materials: the growth of extremism in the digital environment, the transnational nature of threats, and the need for preventive measures. This makes the study relevant both for improving legislative mechanisms and for practical application in the sphere of national security.

2 Literary Review

The growth of extremist and terrorist activity in the digital space requires the development of comprehensive systems capable of effectively identifying, analyzing and countering the spread of radical ideologies. Modern threats are manifested not only in the spread of destructive content, but also in the formation of stable online communities, identification of key figures and coordination of actions. In this regard, the integration of social network analysis (SNA) methods and information security technologies becomes an urgent task. This review analyzes the provided scientific publications in terms of their contribution and applicability to the creation of such an integrated system.

Much of the radical content is in the nature of disinformation or propaganda aimed at manipulating public consciousness, inciting hatred and recruiting supporters. Reviews by Kondamudi et al. (2023) and Mishra et al. (2022) provide a fundamental understanding of the problem of fake news and disinformation in social media. They describe the attributes, characteristics, and taxonomy of malicious content and analyze various machine learning and deep learning techniques to detect it. These reviews are important because they lay the groundwork for understanding the types of content that a counter-radicalization system will have to deal with and introduce basic approaches for its automatic ML/DL-based detection. At the same time, they show that radical content requires more specific analysis beyond the simple definition of "fake".

Understanding the structure of radical online communities and identifying key actors are critical tasks. The work of Smith et al. (2021) proposes a comprehensive approach to automatically detect influential actors in disinformation networks by integrating natural language processing, machine learning, graph analytics, and network causal inference to quantify the impact of actors. This framework is directly applicable to the task of countering radicalism, as it allows not only content detection, but also identifying the most effective disseminators of radical ideas and mapping the structure of the communities in which they operate. The results on detecting influential actors that elude traditional metrics emphasize the need for advanced SNA techniques. A study by Khanday et al. (2023) directly addresses the detection of advocacy communities and their "nuclear" actors using a hybrid approach based on community detection across and within boundaries. This study proposes a specific BCDA method for detecting organized groups spreading propaganda and identifying key individuals within them, which is a direct analogy to detecting radical cells and their leaders in social media.

Analyzing the content itself, beyond its online distribution, requires powerful techniques. Patel and Sutrakar (2025) propose the use of graph neural networks (GNNs) combined with advanced textual analysis to detect fake news, relying solely on textual data. Their ATA-GNN approach, which integrates topic modeling and word clustering into the GNN architecture, demonstrates an improved understanding of text semantics. This is particularly relevant for analyzing radical content, which often contains specific vocabulary, narratives, and veiled meanings. The ability to analyze text in the context of network structure presented by GNN allows for deeper insights into radical messages. The work of Koloski et al. (2022) extends the power of content analysis by proposing the use of Knowledge Graphs-based representations to classify fake news. Integrating structured knowledge about radical ideologies, groups or events (represented in KGs) with textual analysis can significantly improve the accuracy of radical content detection, allowing the system to understand not only words but also the concepts and connections behind them. A review by Lakzaei et al. (2024) also confirms the promise of GNNs for misinformation detection by analyzing different approaches based on graph neural networks.

To effectively counter radicalization in a dynamic social media environment, real-time analysis and the ability to detect incipient threats are essential. A review by Aljabri et al. (2023) details machine learning-based methods for detecting bots in social media. Bots play a significant role in amplifying radical propaganda, creating artificial consensus, and conducting coordinated attacks. Incorporating the bot detection techniques described in this review into an integrated counter-radicalism system will not only filter noise and detect true human radicalization, but also neutralize artificial amplification of destructive content. The work of Jain et al. (2014) presents a scalable CrowdCast framework for analyzing and predicting viral propagation of events in real-time based on streaming data, considering the influence of actors. This approach to monitoring propagation and predicting "virality" is directly applicable to tracking how quickly radical narratives or recruitment messages gain traction. Chen and Neill's (2015) study focuses on detecting human rights violation events using the NPHGS method on heterogeneous social media graphs. Although the context is different, the proposed method for detecting emerging patterns and anomalous clusters in heterogeneous networks is ideal for detecting early signs of radicalization, action planning, or formation of new radical groups, which can manifest as statistical anomalies in online activity and content. Snyder et al.

(2020) add an important dimension by proposing interactive learning to identify relevant tweets in real time. In the face of a noisy data stream and ever-changing specifics of radical content, the ability of security experts to adjust classification and retrain the model in real-time (as implemented in their SMART 2.0 system) is critical to keep the system up-to-date and accurate. The interactive learning proposed by Snyder et al. (2020) also helps with noise control by allowing the system to better distinguish targeted radical content from unrelated data.

Despite significant progress in the development of social network analysis and information security techniques, there are a number of significant challenges to their effective application in the context of countering radicalization.

One of the key challenges is adapting general methods for detecting malicious content to the specifics of radical narratives. As the reviews by Kondamudi et al. (2023) and Mishra et al. (2022) have shown, most studies focus on general approaches to detect fake news. However, as Demir and Çona's (2024) study points out in the context of manifestos of far-right groups, radical materials often use complex, veiled vocabulary and specific ideological constructs that require deep semantic analysis beyond standard text processing techniques (Patel & Sutrakar, 2025) and even the use of knowledge graphs (Koloski et al., 2022) without specific adaptation.

A second important challenge is the reliable and scalable processing of real-time data. While there are frameworks aimed at analyzing streaming data, such as CrowdCast (Jain et al., 2014) or event detection methods (Chen & Neill, 2015), many of them have been tested on historical data or do not demonstrate sufficient robustness when dealing with the huge, high-speed social media data streams characteristic of active radical content or bot activity (Aljabri et al., 2023). Effective detection of incipient threats and rapid response require systems that can perform complex analysis (including network structure according to Smith et al., 2021 and Khanday et al., 2023) with minimal latency.

A third, equally important challenge relates to integrating the results of technical analysis with legislative and law enforcement mechanisms. Detected data on radical communities, actors or content must not only be accurate, but also presented in a format that meets legal standards for use in litigation or administrative blocking. This requires transparency of methods, auditability and documentation of all stages of detection and analysis, which is challenging for the "black boxes" of today's ML models.

Integrating the aforementioned areas of research - the ability to analyze message content, identify the structure of networks and the role of actors within them, do so in real time to detect incipient threats, and use interactive approaches to improve accuracy while countering automated dissemination – is key to creating an effective system to counter radicalization.

3 Methods

Computer vision techniques are to analyze the visual content of social media posts. The main tool is a convolutional neural network YOLO (You Only Look Once), which is used to detect objects in images, such as weapons, symbols of extremist organizations or portraits of key figures.

YOLO (You Only Look Once) is a revolutionary object detection algorithm that transforms the detection problem into a regression problem using a single convolutional neural network for real-world operation. Due to its speed, accuracy and simplicity, YOLO has found applications in autonomous driving, surveillance and medical imaging systems. Unlike two-stage detectors such as R-CNN, YOLO processes the image in a single pass, which significantly speeds up the process (Soni & Rai, 2024)

Since the introduction of YOLO in 2016, the algorithm has undergone many improvements. Versions from YOLOv1 to YOLOv8 show progress in speed and accuracy by optimizing the network structure, learning methods, and data processing strategies (Liang, 2024). The YOLO architecture is based on partitioning the image into a grid, where each cell predicts bounding boxes and class probabilities. A non-maximum suppression (NMS) method is applied to eliminate redundant predictions, which improves the accuracy of object localization.

Comparison of different versions of the model is presented according to AD-YOLOv5: An object detection approach for key parts of sika deer based on deep learning in Fig. 1 (Xiong et al., 2024).

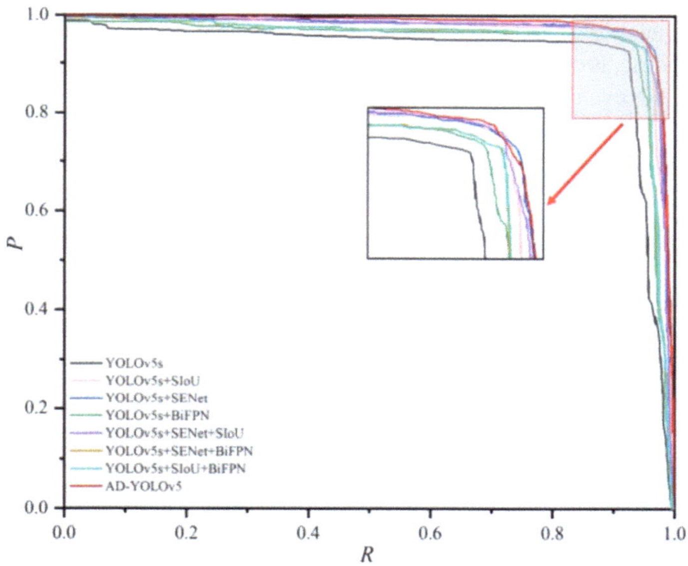

Fig. 1. Coparison of different versions of YOLO.

Key variants of YOLO include YOLOv3 with multi-scale prediction, YOLOv4 with advanced augmentation methods, YOLOv5 with compact architecture, and YOLOv8 optimized for real-world tasks (Kolhatkar et al., 2023). Each version addresses specific problems such as small object detection or the balance between speed and accuracy.

Based on the detection results, metrics are calculated, including the area of the object, its location relative to the center of the image, the number of detected objects, and the confidence level of the neural network in the classification correctness. For example, a large object with high confidences located in the center of the frame is interpreted as a significant sign of radical content. These quantitative indicators are aggregated into an intermediate score that reflects the relevance of the image to the target topic.

Text analysis combines traditional and modern approaches. In the preprocessing stage, lemmatization and stemming are applied to reduce words to a normalized form, which improves the accuracy of the subsequent analysis. TF-IDF, cosine similarity and Jaccard's index metrics are used to evaluate the consistency of the text with radical narratives by comparing textual data with reference samples. For example, high density of key terms combined with semantic proximity to patterns of extremist rhetoric increases the interim score of the textual component. These methods are augmented by neural network models, such as BERT, capable of capturing contextual nuances and hidden meanings.

The theoretical framework of the study includes a systematic literature review of existing works in the fields of social network analysis, natural language processing, and information security. The authors critically evaluate the limitations of previous approaches, such as insufficient adaptation of general methods to the specifics of radical content, and on this basis formulate requirements for a new system. The system approach manifests itself in the development of an architecture that integrates heterogeneous methods - from computer vision to graph analytics - into a single data processing pipeline. For example, temporal analysis, which takes into account the relevance of posts through a freshness metric, is combined with a graph-based representation of social interactions, allowing for the evaluation of not only the content but also the dynamics of content distribution. The logic of the system is based on the sequential combination of interim assessments of each component, providing a comprehensive view of potential threats.

4 General Structure of the System

The system of comprehensive content assessment in social networks is designed for automated analysis, monitoring and classification of online publications. Its main task is to identify and evaluate content by the degree of its compliance with certain target topics, in particular, destructive or radical manifestations, as well as to determine the potential impact of this content on users and its distribution in the social network. The system uses a multidimensional approach, analyzing not only the textual and visual content of posts, but also their social relevance and relevance over time.

The evaluation system implements a sequential analysis algorithm for each post (Fig. 2):

1. Data collection and initial data processing. Extracting posts from the social network, including text, images, user interaction data (comments, likes, reposts) and time of publication. Basic preprocessing (e.g., cleansing text of special characters).
2. Analyzing visual content. Processing images attached to a post to detect objects and extract their characteristics. Calculation of an intermediate score reflecting the relevance of the image to the target topic.
3. Text content analysis. Assessing the semantic relevance of the post text to the target topic using natural language processing and classification methods. Calculation of the interim text relevance score.
4. Basic graph representation of a social network. Generating or updating a graph of posts and users to represent the structure of interactions. Calculation of basic metrics of nodes and edges reflecting the popularity of posts and connections between them.

5. Integration of temporal analysis. Calculation of the post relevance metric based on the time of publication (Freshness Score) and using this metric to correct social influence indicators, generate an intermediate social influence score (Score_Post), and analyze the dynamics of distribution.
6. Metrics Integration. Combining intermediate scores from the image, text, and social influence/relevance analysis stages into a single final post score (Score_Final).
7. Post Classification. Categorizes the post into one of the predefined categories ("Low Compliance/Influence", "Medium Compliance/Influence", "High Compliance/Influence") based on the Score_Final value.

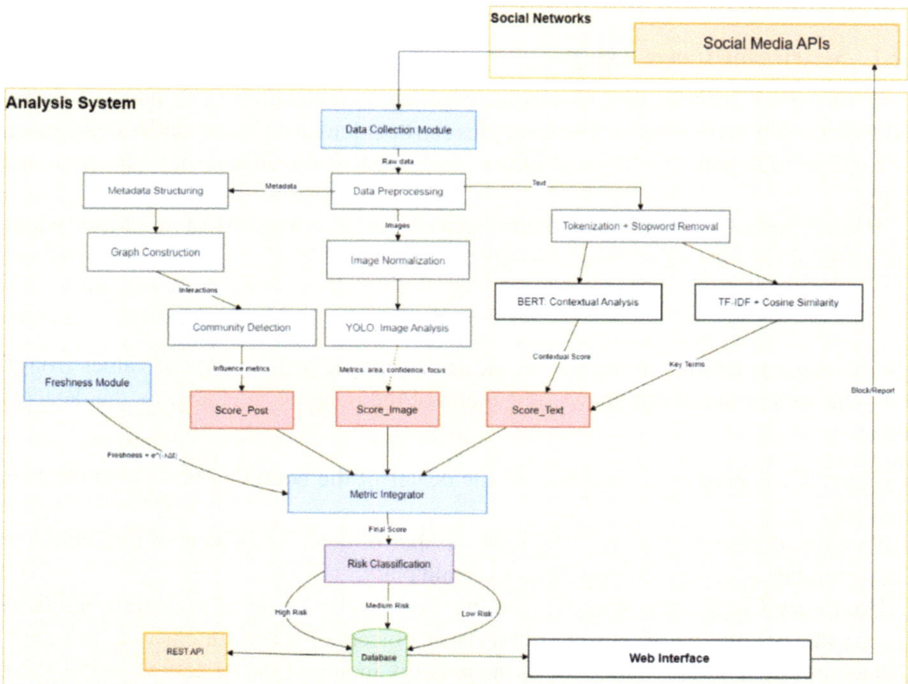

Fig. 2. General structure of the system.

4.1 Visual Content Analysis

Images are analyzed using state-of-the-art convolutional neural networks designed for object detection, we use the YOLO (You Only Look Once) architecture. As a result of image processing by the network, a detection vector containing the following information is generated for each detected object:

$$(n, x, y, w, h, \text{confidence}),$$

where:

n - number of the object class to which the detection is assigned (e.g., "weapon", "symbolism", "flag", "portrait of a certain person", etc.). - the set of classes is determined by the specifics of the target topic).
x and y - relative coordinates of the center of the detected object within the image. The values are in the range from 0 to 1, where (0, 0) corresponds to the upper left corner of the image, and (1) - to the lower right corner.
w and h - relative width and height of the bounding box around the detected object. The values are also in the range from 0 to 1.
confidence - the coefficient of confidence of the neural network in the correctness of the performed detection and classification of this object. The value is in the range from 0 to 1.

4.1.1 Metrics for Image Analysis

Based on the detection vectors, metrics are calculated to characterize the detected objects and their location in the image. These metrics can be calculated for each individual object, for a group of objects of the same class, or aggregated for all detected objects in the image.

•***Object Area (Area):*** The geometric area of the bounding box of an object, which is the product of relative width and height.

$$\text{Area} = w^* h.$$

The value is relative to the size of the entire image, with a range of values from 0 to 1. The area of the object shows how much of the image is occupied by the detected object.

- ***Object focus center:*** Coordinates of the center of the bounding box (x, y) for each object.
- ***Number of objects (Count)***: The total number of detected objects in the image, or the number of objects belonging to a certain class.
- ***Maximum Confidence (Cmax):*** The maximum value of the confidence coefficient (confidence) among all detected objects or objects of a certain class in an image. Indicates the highest confidence of the network in any of the detections in the given image. The range of values is from 0 to 1.
- ***Mean Confident (Cavg):*** The arithmetic mean of the confidence (confidence) values for all detected objects or objects of a particular class. Reflects the average confidence of the network in its detections in the image. The range of values is from 0 to 1.
- ***Maximum Object Area (Smax):*** The maximum Area value among all detected objects or objects of a certain class. Shows the size of the largest detected object in the image. The range of values is from 0 to 1.
- ***Average object area (Savg):*** The arithmetic mean of the Area values for all detected objects or objects of a particular class. Reflects the average size of detected objects. The range of values is from 0 to 1.
- ***Coverage Area (Scom):*** The sum of the areas (Area) of all detected objects (both of the same class and of all found objects). Since objects may overlap each other, the Area coverage can take values greater than 1, which indicates a high "heapiness" of objects in the image. A value $> = 0$.

- ***Image focus center (X, Y):*** The midpoint of the centers of the bounding box centers of all detected objects or objects of a certain class. Calculated as the arithmetic mean of their relative coordinates. If the coordinates of the center of focus are close to (0.5, 0.5), it means that objects are grouped in the center of the image.
- ***Abscissa dispersion (Dx)***: The mean square of the deviation of the relative abscissas (x) of the centers of objects from the center of focus of the image X. Indicates the dispersion of objects horizontally. When only one object is detected, the dispersion is 0. Value $> = 0$.
- ***Dispersion of ordinates (Dy):*** The mean square of the deviation of the relative ordinates (y) of the centers of objects from the center of focus of the image Y. Indicates the dispersion of objects vertically. When only one object is detected, the dispersion is 0. Value $> = 0$.
- ***Distance of the focus center from the center of the image (Dist_Focus):*** The Euclidean distance between the calculated center of focus (X, Y) and the center of the whole image (0.5, 0.5). A value close to 0 means that objects are clustered in the center part of the image. The range of values is from 0 to about 0.707 (for the center in the corner).

4.1.2 Intermediate Score of the Image Component (Score_Image)

Metrics calculated based on object detection are aggregated to produce a single numerical score for the visual content of a post (Score_Image). A high Score_Image value should correlate with a high probability that the image contains objects relevant to the target topic, located in a meaningful way (e.g., centered, large, with high detection confidence). The score is a number between 0 and 1 and is calculated as a weighted sum of the most informative scaled metrics:

$$\begin{aligned}\text{Score\_Image} &= \text{w\_ic1} * \text{Cmax} + \text{w\_ic2} * \text{Cavg} + \text{w\_is1} * \text{Smax} + \text{w\_is2} * \text{Savg} \\ &+ \text{w\_if} * \text{scaled}(1 - \text{Dist\_Focus}) + \text{w\_isc} * \text{scaled}(\text{Scom}) + \text{w\_icn} * \text{s}\end{aligned}$$

where w_ic1, ..., w_icn are weighting coefficients determining the relative importance of each metric for the final image evaluation (the sum of weighting coefficients is equal to 1). scaled() is a function that converts the value of a metric to the range [0, 1]. For example, scaled(Dist_Focus) can be max(0, 1 - Dist_Focus / 0.707) to account for the importance of center location, scaled(Scom) can be min(1, Scom / Scom_max) to account for coverage density, and scaled(Count) can be min(1, Count / Count_max) to account for the number of objects, where Scom_max and Count_max are the expected maximum values determined from the data.

4.2 Analyzing Text Content

Analyzing a text involves the following steps:

1. Preparation of benchmark data. Collection of a representative corpus of documents related to the target topic. Extraction of a set of key terms from the reference documents and/or generation of a reference term distribution.

2. Post text preprocessing. The text goes through stages of cleaning and normalization: tokenization, delete stop words and lemmatization. Tokenization -dividing text into separate lexical units (words, tokens). Delete stop words - deletes frequently occurring but unimportant words (prepositions, conjunctions, etc.). Lemmatization (or stemming) - the reduction of words to their normal form (lemma).

Metrics Calculation or Classification Based on the preprocessed text, an evaluation of its conformance to a benchmark is performed. This can be realized in several ways:

Based on similarity metrics to the benchmark:

- **TF-IDF (Term Frequency-Inverse Document Frequency):** Estimates the importance of each term in the post text relative to a reference collection of documents. It is calculated as the product of the term frequency in a document (TF) by the inverse document frequency (IDF). TF-IDF(t,d) = TF(t,d) × IDF(t).

  - TF(t,d) = (Number of occurrences of term t in document d) / (Total number of terms in document d).
  - IDF(t) = log((Total number of documents in the reference collection) / (Number of documents in the collection containing the term t)). In the presented works there is a variant of the formula IDF(t) = log(DF(t) + 1 N), which requires clarification, but the basic principle of IDF is based on the logarithm of the ratio of the total number of documents to the number of documents containing the term.

- **Keywords Density (KD)**: The specific weight of key terms of the target topic in the post text, expressed as a percentage or as a proportion of the total number of terms. KD = (Number of key terms of the topic in the text) (Total number of terms in the text) × 100%
- **Cosine Similarity (cos(θ)):** A measure of similarity between the feature vector of a post text (e.g., TF-IDF vector) and the feature vector of a reference topic. Calculated as the cosine of the angle between vectors in a multidimensional space. The value ranges from 0 to 1. cos(θ) = (A - B) / (||A|| * ||B||).
- **Jaccard Index (J):** A measure of similarity between the set of unique terms in the post text and the set of unique key terms of the reference topic. It is calculated as the ratio of the power of intersection of sets to the power of their union. The value ranges from 0 to 1. J(A,B) = |A ∩ B| / |A∪ B|.
- **KL-divergence (Kullback-Leibler Divergence, DKL(P||Q)):** A measure of the discrepancy between the probability distribution of terms in the post text (P) and the reference distribution of topic terms (Q). The smaller the DKL value, the closer the distribution of text terms is to the benchmark. The value > = 0. DKL(P||Q) = Σ(P(t) * log(P(t) / Q(t))), where the summation is over all unique terms

Based on these similarity metrics, an intermediate topicality score can be calculated:

$$\text{Score\_Text\_Similarity} = \text{w\_ts1}^* \cos(\theta) + \text{w\_ts2}^* \text{scaled(KD)} + \text{w\_ts3}^* \text{J} + \text{w\_ts4}^* \text{scaled}(1 - \text{DKL}),$$

where w_ts... - are weight coefficients, scaled() are functions for scaling metrics into the range [0, 1].

4.2.1 Based on Text Classification by Neural Network:

- Using a pre-trained deep learning model (e.g., based on a BERT architecture such as BertForSequenceClassification) to classify the post's text into a predefined set of classes. These classes reflect different contexts or shades of meaning relevant to the target topic (e.g., different uses of the word "hero" as described in your article: hero of a work, hero as a title, hero in a military context, hero in a personal context).
- The model outputs the probabilities of text belonging to each class and the predicted class with the corresponding confidence coefficient (confidences).
- An intermediate assessment of text classification can be based on the value of confidences for one or more classes that have been identified as targeted (e.g., directly related to radicalization).

4.2.2 The Intermediate Score of the Text Component (Score_Text)

The final intermediate text score (Score_Text) is a number between 0 and 1. In the context of a system that uses context-based text classification, Score_Text can be directly based on the confidences of the neural network model for the target class(es). A high Score_Text value indicates a high confidence of the system that the textual content of the post is relevant to the target topic (e.g., explicitly uses radical rhetoric or glorifies related concepts).

$$\text{Score\_Text} = \text{Confidence\_in\_Target\_Classes},$$

where Confidence_in_Target_Classes is the corresponding confidence from the classification result (e.g., the confidence of the predicted class if that class is the target, or the aggregate value).

4.3 Basic Graph Representation of a Social Network

In this basic step, the temporal characteristics of the interactions are not directly factored into the weights, but the graph structure will be used for more sophisticated analyses later on.

The social network or part of it related to the analyzed posts and users is represented as an undirected weighted graph.

$$G = (V, E)$$

We propose the following formula for the weight of a node reflecting the number of comments:

$$w_i = log(1 + C_i)$$

C_i is the number of comments on post v_i (v_i is a post with unique identifier i).

Formula for the weight of an edge (W_{ij}):

$$W_{ij} = \alpha * (|U_i \cap U_j|/|U_i \cup U_j|) + \beta * (|L_i \cap L_j|/|L_i \cup L_j)$$

U_i - The many users who left comments on the post v_i.
U_j - set of users who left comments on the post v_j.
L_i - lots of users who liked this post v_i.
L_j - A few users who liked this post v_j.
$|X|$ - is the power of the set X (the number of elements in the set).
α A weighting factor that determines the importance of common commenters.
β - a weighting coefficient that determines the significance of total likes.

We normalize the total number of users by dividing it by the total number of users who commented or liked at least one of the two posts. This allows us to compare links between posts with different total number of interactions.

4.3.1 Integration of Temporal Analysis

To take into account the dynamics of content distribution and identify the most relevant publications, the system integrates the analysis of the temporal characteristics of posts. This is especially important for monitoring rapidly changing information flows and promptly responding to the spread of destructive content.

The freshness metric evaluates the relevance of a post based on the time that has passed since it was published. A post published recently is assigned a higher freshness value, which fades exponentially over time.

$$\text{Freshness Score} = e - \lambda \cdot \Delta t,$$

where:

Δt- time elapsed from the moment of post publication to the current moment of analysis. Measured in selected units (hours, days).
λ is the attenuation coefficient (($\lambda > 0$)). This parameter determines the rate at which content decreases in relevance over time. A larger value of (λ) means that the content becomes obsolete quickly, while a smaller value means that it remains relevant longer. The coefficient (λ) can be chosen empirically or adapted depending on the subject area (e.g., political news may require more (λ) than discussions of fiction).

The value of the freshness metric ranges from $e^{-\infty}$ (very old post) to $e^0 = 1$ (post published right now). Thus, Freshness Score takes values from the range (0, 1].

Calculation of Time-Adjusted Node Weights and Interim Social Impact Assessment The freshness metric is used to modify the base weight of a node derived from the number of comments. This allows the relevance of a post to be taken into account when assessing its social influence and relevance in the graph. The time-adjusted weight of a node (W_{total}) can be calculated by multiplying the base weight by the freshness metric.

This time-adjusted node weight W_{total}reflects the social influence of the post given its relevance. Based on this weight, an intermediate score of the social influence component (Score_Post) is generated, scaled in the range from 0 to 1.

Intermediate Score of the Social Influence Component (Score_Post) Score_Post is a numerical value between 0 and 1. It is based on the time-adjusted node weight W_{total}. A scaling function is applied to obtain a value in the range [0, 1]:

$$\text{Score\_Post} = \text{scaled(W\_total)}$$

where scaled() is a function that translates the value of $W_{(total)}$ (which $> = 0$) to the range [0, 1].

4.4 Metrics Integration (Final Evaluation)

Interim assessments from each component are combined to produce a comprehensive assessment of the post, reflecting a comprehensive analysis of its content, social impact and relevance.

Score_Final Wimage = · Score_Image+Wtext· Score_Text+Wpost· Score_Post, where:

- Score_Image - intermediate score of image analysis (from 0 to 1).
- Score_Text - intermediate score of text analysis (from 0 to 1).
- Score_Post - intermediate social impact score with relevance (from 0 to 1).
- Wimage, *Wtext,* Wpost - weighting coefficients determining the relative importance of each component for the final score.

Weight values are customized depending on the specifics of the target topic.

If any of the analysis components cannot be applied (e.g., the post has no image), the corresponding intermediate score is taken as 0 and the weights of the remaining components are normalized so that their sum remains 1.

4.5 Classification of the Post

The final Score_Final is used to categorize a post based on its relevance to the target topic and potential impact. A Score_Final value between 0 and 1 is compared to predefined thresholds to categorize the post.

The following approximate post classification scale is proposed:

- **Low Relevance/Influence:** Score_Final$<$ Threshold_Low. Posts in this category have a low probability of being relevant to the target topic or having minimal social influence.
- **Average relevance/influence:** Threshold_Low $\leq$ Score_Final$<$ Threshold_High. Posts may contain some indications of targeted topics, have moderate social influence, or their content is ambiguous.
- **High Relevance/Influence:** Score_Final $\geq$ Threshold_High. Posts are highly likely to be relevant to the target topic, have meaningful social impact, and are prioritized for further consideration or countermeasures.

The thresholds (Threshold_Low, Threshold_High) are defined and tuned based on data analysis and task-specific requirements (e.g., in the area of radical content detection, these thresholds can be quite conservative to minimize false positives).

5 Directions for Further Development of the System

The proposed system provides a robust framework for content analysis in social media. Further research and development can focus on the following key areas.

Improved semantic analysis. Develop and apply better natural language processing models that can detect subtle nuances of meaning, sarcasm, irony, innuendo, and euphemisms that are often used in disruptive content to circumvent moderation. Integrate ontologies and knowledge graphs to improve contextual understanding.
Advanced temporal analysis. Further development of methods that take dynamics into account. This includes adaptive tuning of the fading coefficient, application of Temporal Graph Neural Networks to model the evolution of links and information propagation over time, and correlation analysis of bursts of content activity with external events.
Optimization of metrics and thresholds integration. Application of machine learning methods for automatic adjustment of weighting coefficients in the final score formula and determination of optimal thresholds for classification based on labeled data.
Scalability and performance. Ensuring efficient system performance on large volumes of social network data, which requires optimization of algorithms and use of distributed computing resources.

6 Conclusion

The presented work is devoted to the development of a comprehensive system for analyzing and evaluating content in social networks in the context of countering radical manifestations. The relevance of the study is due to the significant growth of destructive content, the complexity of its identification in a veiled form and the need for a prompt response to the rapid spread in the digital environment.

The article substantiates the need to integrate different analytical methods. A review of existing approaches in the field of disinformation detection, SNA for identifying influential actors and communities, and content analysis methods was conducted, showing that the task of countering radicalism requires their synergistic application and specific adaptation.

An original system structure is proposed that integrates the analysis of visual and textual content of a post with its position in the social network graph and temporal relevance. For the first time, within the proposed architecture, the content freshness metric is explicitly integrated with the graph representation to adjust the node weight and generate a social influence score with relevance. Metrics for each of the analysis components (images, text, social influence/time) are developed and a mechanism for integrating them into a single final post score is proposed, which allows for the classification of content according to its potential danger and influence. The use of the GIC color-coding concept to visualize the results on the graph provides a powerful tool for operational monitoring and decision-making by analysts.

The presented system is a step towards creating better tools to counteract destructive content by providing not just the fact of its presence, but also an assessment of its potential for influence and relevance. Further research is focused on improving semantic analysis, developing temporal graph analysis techniques, and optimizing the integration of all system components for maximum efficiency and scalability. The implementation of such systems is critical for improving information security and reducing radicalization risks in today's digital society.

References

Kondamudi, M.R., et al.: A comprehensive survey of fake news in social networks: Attributes, features, and detection approaches. J. King Saud Univ.-Comput. Inform. Sci. **35**(6), 101571 (2023)

Mishra, S., Shukla, P., Agarwal, R.: Analyzing machine learning enabled fake news detection techniques for diversified datasets. Wireless Commun. Mobile Comput. **1**, 1575365 (2022)

Smith, S.T., et al.: Automatic detection of influential actors in disinformation networks. Proc. Natl. Acad. Sci. **118**(4), e2011216118 (2021). https://doi.org/10.1073/pnas.2011216118

Khanday, A., Wani, M., Rabani, S., Khan, Q.R.: Hybrid approach for detecting propagandistic community and Core node on social networks. Sustainability. **15**(1249), 10.3390/su15021249 (2023)

Patel, A., Sutrakar, V.K.: Advanced text analytics--graph neural network for fake news detection in social media arXiv preprint arXiv:2502.16157 (2025)

Koloski, B., et al.: Knowledge graph informed fake news classification via heterogeneous representation ensembles. Neurocomputing. **208**, 496 (2022)

Lakzaei, B., Haghir Chehreghani, M., Bagheri, A.: Disinformation detection using graph neural networks: a survey. Artif. Intell. Rev. **57**, 52 (2024). https://doi.org/10.1007/s10462-024-10702-9

Aljabri, M., et al.: Machine learning-based social media bot detection: a comprehensive literature review. Social Network Anal. Min. **13**, 12023 (2023)

Jain, P., Manweiler, J., Acharya, A., Roy Choudhury, R.: Scalable social analytics for live viral event prediction. Proc. Int. AAAI Conf. Web Soc. Media. **8**(1), 226–235 (2014). https://doi.org/10.1609/icwsm.v8i1.14504

Chen, F., Neill, D.B.: Human rights event detection from heterogeneous social media graphs. Big Data. **3**(1), 34–40 (2015)

Snyder, L.S., Lin, Y.S., Karimzadeh, M., Goldwasser, D., Ebert, D.S.: Interactive learning for identifying relevant tweets to support real-time situational awareness. IEEE Trans. Vis. Comput. Graph. **26**(1), 558–568 (2020). https://doi.org/10.1109/TVCG.2019.2934614. Epub 2019 Aug 20

Kolhatkar, P., Padule, D., Salunkhe, P.: A review of YOLO object detection model in forensic evidence analysis. Int. J. Sci. Res. Eng. Manage. **7**(10), 1–11 (2023). https://doi.org/10.55041/ijsrem26408

Liang, J.: A review of the development of YOLO object detection algorithm. Appl. Comput. Eng. **71**(1), 39–46 (2024). https://doi.org/10.54254/2755-2721/71/20241642

Soni, A., Rai, A.: YOLO for medical object detection (2018-2024). In: 2024 IEEE 3rd International Conference on Electrical Power and Energy Systems (ICEPES), pp. 1–7 (2024). https://doi.org/10.1109/icepes60647.2024.10653506

Xiong, H., Xiao, Y., Zhao, H., et al.: AD-YOLOv5: an object detection approach for key parts of sika deer based on deep learning. Comput. Electron. Agric. **217**, 108610 (2024). https://doi.org/10.1016/j.compag.2024.108610

Zolotarev, V., Arkhipova, A., Lapina, M., Prakasha, G.S., Movzalevskaya, V.: Assessing player interaction for a social networking cooperative educational game. In: Lapina, M., Prakasha, G.S., Grigoriev, S. (eds.) SLET 2022: International Conference on Innovative Approaches to the Application of Digital Technologies in Education and Research. SLET 2022 Lecture Notes in Networks and Systems, pp. 144–125. Springer, Cham (2024)

Lapina, M., Dudun, G., Kotlyarov, D., Rjevskaya, N., Subramanian, S.J.: Analysis of an existing method for detecting adversarial attacks on deep neural networks. In: Alikhanov, A., Tchernykh, A., Babenko, M., Samoylenko, I. (eds.) Current Problems of Applied Mathematics and Computer Systems. CPAMCS 2023 Lecture Notes in Networks and Systems, vol. 1044. Springer, Cham (2024). https://doi.org/10.1007/978-3-031-64010-0_29

Designing an Architecture for a Confidential Data Analysis System in a Fog Computing Network for Low-Power Devices

Egor Shiriaev(✉), Nikolay Kucherov , Ekaterina Bezuglova, and Andrey Gladkov

North-Caucasus Federal University, 1, Pushkin Street, Stavropol 355017, Russian Federation
egor.shiriaev@reaneling.ru, bezuglovakaterina@mail.ru, Agladkov@ncfu.ru

Abstract. Smart solutions are becoming increasingly popular, this is caused by both the increase in the quality and quantity of capabilities of Internet of Things devices, and the development of methods and approaches in organizing distributed networks, in addition, the demand for such solutions is also growing. When a solution such as a Smart City is fully developed, a large amount of confidential data or data of critical importance will be transmitted and processed in its network; if an intruder gains access to this data, or is able to change or destroy it, this can lead to various kinds of problems. In this paper, we investigate the fog computing network and its security, and it is established that fully homomorphic encryption schemes can be used to ensure the best data privacy. As a result, the architecture of a confidential data analysis system in a fog computing network for low-power devices based on the TFHE scheme was obtained, for this purpose, the quantization of values was investigated, as well as the logistic regression algorithm for confidential data processing.

Keywords: fog computing · data analysis · fully homomorphic encryption · logistic regression · TFHE · quantization

1 Introduction

Solutions such as smart cities (SC) [1], smart homes (SH) and other smart solutions (SS) cannot be effectively built on the basis of centralized computing systems [2,3]. SS refers to a network consisting of many low-power devices that exchange data and transmit it to user devices, these can be both low-power devices such as mobile phones, tablets, etc., and more computationally powerful devices such as desktop computers, laptops, etc. In any case, such a network has a large data flow in real time. If we consider such a network as centralized,

M. Lapina et al. (Eds.): AISMA 2025, LNNS 1456, pp. 413–424, 2026.
https://doi.org/10.1007/978-3-032-07275-7_37

then when building the infrastructure, we can encounter such problems as: low data transfer speed, introduction of overloaded communication channels, high cost of the network with many backup data transmission channels that can be idle, as well as solving complex problems of mass service systems. Decentralized systems allow to level out these disadvantages. If we consider cloud computing (CC) [4], then in the case of SS, when building a computing system, they are actually centralized, since communication channels with cloud nodes can also be overloaded. To solve this problem, edge and fog computing paradigms are used. Edge computing (EC) implies a computing network built exclusively on edge devices (low-power devices) [5], while fog computing (FC) is built as an edge device together with an intermediate computing node called a fog node [6]. FC is a more attractive paradigm because the intermediate computing node (FN) allows for more efficient management of the computing load in the network. Therefore, building SC and SH based on FC will allow for a more flexible system with the ability to scale and more efficient load balancing.

When considering the security of FC, several shortcomings can be noticed. The communication channel in FC is open due to the requirement of low data transmission latency, since the edge devices transmit values in real time, in addition, if FN can be protected from an intruder, then the edge devices have either weak protection or no protection at all. When analyzing SC and SH data, either sensitive or confidential data may be used, and an attacker may attempt to access and modify the data. Thus, the task is to ensure the confidentiality of data in the FC network for data analysis. To accomplish this task, it is necessary to build an architecture of a confidential data analysis system in a fog computing network for low-power devices. For this, it is necessary to develop the following:

1. Structures of a computing system taking into account the method of ensuring confidentiality;
2. Data transfer models;
3. Privacy-preserving data analysis algorithms.

As a method of ensuring confidentiality, it is possible to use fully homomorphic encryption (FHE) schemes. Which allows performing mathematical operations on encrypted data (ciphertexts). Thus, in this paper, a study is conducted on the development of an architecture for confidential data analysis based on FHE in FC.

The work is organized as follows: Sect. 2 discusses fog computing and the specifics of building computing systems; Sect. 3 discusses homomorphic encryption methods; Sect. 4 examines privacy-preserving data processing; Sect. 5 presents the developed architecture; and the final section presents the conclusions reached during the work.

2 Fog Computing

Let's take a closer look at FC. FC is a horizontal distributed computing paradigm that was born in 2011 and formed in 2012 as a fork of CC [7]. The emergence of FC was predetermined by the need for an intermediate level of information handling between CC and EC. Since CC lacked a level of data collection and analysis at the user level, and in the case of EC there was only a level of data exchange between devices, where the end point is the user himself. Thus, FC can be considered as a link between CC and EC, and as a hybrid CC-EC system. In essence, the representation of FC depends on the architecture, size and tasks that are set for FC. In this study, FC is considered as a single structure unrelated to both EC and CC. Let us consider the application of FC in various areas. In [8] the authors analyze the application of FC in data processing areas such as augmented reality and real-time video analytics, content delivery and caching, and mobile big data processing. The authors of [9] consider the application of FC in a broader sense and focus on several areas. For example, in healthcare [10–13], FC is often used to organize interactions between personnel. As was said above, in the organization of smart systems [14,15]. In addition, the authors indicate that FC can be used to organize transport systems, such as VANET networks [16–18].

We will analyze FC security in more detail, in the works [19] FC security issues and threats analyzed. The work highlights issues such as denial of access attacks, data leaks, insecure APIs, insiders, vulnerabilities in individual applications, and user account hacks. Various methods are used to counter such threats; asymmetric ciphers are often used to counter internal threats [20], while secret sharing schemes are used to counter internal threats [21,22]. However, both in work [19] and, for example, in works [23–25] the use of FHE methods for organizing FC security is proposed.

Thus, FC allows to deploy a productive horizontal network based on low-power devices. The use of FN makes it possible to organize extensive SC networks, and the distribution of FN can be both geographical, by physical placement of devices, and logical, by network topology.

3 Fully Homomorphic Encryption

‘ FHE is currently a family of cryptographic schemes that began in 2009, when Gentry described the first FHE scheme in his dissertation, the underlying cryptographic apparatus in this case being cryptographic lattices [26,27]. The basic idea of FHE is that these schemes support homomorphic addition and homomorphic multiplication over ciphertexts:

$$c_1 * c_2 = D(E(m_1, pk) * E(m_2, pk), sk) = m_1 * m_2 \tag{1}$$

where $*$ – is a homomorphic operation, m_i – is a plaintext, c_i – is a ciphertext, $D()$ – is a decryption operation, $E()$ – is an encryption operation, pk – is a public key, sk – is a private key. The homomorphism of operations is based on the properties of lattices, so that when performing operations in the domain of the lattice basis pk the computations are homomorphic.

Despite the fact that this option had extremely low productivity, further research by Gentry himself and other researchers allowed to increase the speed of data processing in FHE schemes and expand their scope of application. Currently, FHE schemes support processing of three types of data: logical, integer and real with fixed precision, this became possible due to the expansion of the arsenal of cryptographic methods, techniques and primitives used to build schemes. In addition to cryptographic lattices, FHE schemes also use learning with error, learning with error in the ring and various numerical methods. In addition, some schemes use the residue number system (RNS) to speed up arithmetic [28].

The main drawback of FHE is the computational complexity of encryption schemes. This is caused by the fact that for almost any FHE scheme, the number of homomorphic additions supported is unlimited, but each encryption key supports a limited number of multiplication operations. This situation arises because when performing homomorphic operations, the error introduced into the data also grows, at some point the error value exceeds the value of the main module and the data can no longer be deciphered. The number of allowed multiplications is determined by the multiplicative depth of the scheme, which depends on the set of parameters, i.e. on the size of the modules. To increase the multiplicative depth, it is necessary to increase the parameter sizes, which increases the computational complexity of operations in FHE, in addition, too "large" encryption parameters can make the scheme "insecure". To combat this problem, the so-called bootstrapping was developed, which allows increasing the number of allowed multiplications when recalculating the key, but each such operation increases the computational cost of subsequent operations. Let's consider the most popular FHE schemes and divide them by the type of values processed.

3.1 BFV Scheme

For integer arithmetic, consider the Brakerski/Fan-Vercauteren (BFV) scheme. [29]. BFV uses cryptographic lattices and ring-based learning to organize its cryptography. The multiplicative depth is determined based on the level L, the defining value is B, then the main module of the circuit, defining the module $q \approx B^{2^L}$ in this case, the BFV scheme supports three types of keys: the public key pk, which is used for encryption, the secret key sk which is used for decryption, and the public evaluation key ek, which is used in homomorphic operations. The arithmetic of this scheme was accelerated by using RNS [28].

The additional key ek, although it allows to increase the security of data during processing, in contrast to schemes where one key is responsible for both encryption and homomorphic operations, however, in this case there is an accumulation of internal errors, which neutralizes the advantage of the scheme in terms of accuracy, or, when maintaining it at the required level, performance. This drawback is especially acute when processing non-integer values, where the fractional part is written as a separate ciphertext; such data processing increases the number of homomorphic operations used and requires additional multiplicative depth. If we consider the results of the work [30], we can see how the accuracy of neural network recognition decreases when using the BFV scheme. Thus, this scheme has significant drawbacks that do not allow for effective use in the FC network.

3.2 CKKS Scheme

There is also a scheme that supports the processing of real numbers. The scheme is named after the surnames of its authors Cheon, Kim, Kim, and Song (CKKS), this name has stuck with the scheme, although the authors initially called it Homomorphic Encryption for Arithmetic of Approximate Numbers (HEAAN) [31]. This scheme implements the arithmetic of the scheme based on numerical methods, representing ciphertexts in the field of complex numbers. The cryptosystem itself is built on learning with an error in the ring. Approximate methods allow the scheme to accumulate an error at a lower rate than, for example, the BFV scheme. In this case, the specified parameters of the scheme, similar to BFV - this is a set of large prime numbers (modules), allow for accuracy up to a certain decimal place. In addition, this scheme allows for calculating homomorphic polynomial functions.

The scheme has gained quite a lot of popularity in studies devoted to methods of machine learning that preserve confidentiality. If we consider its disadvantages, we can highlight the following. The scheme requires a special representation of the plaintext in the form of a polynomial, so the message must first be encoded in a polynomial of a certain degree, after which it is possible to encrypt it. The error accumulates exponentially, which is why the need for the bootstrapping operation is higher than in other schemes, as are the requirements for the multiplicative depth. In addition, when setting incorrect parameters, the accuracy of calculations may suffer, and since there are currently no standardized criteria for setting encryption parameters, the optimal parameters for a particular task can only be obtained empirically. In terms of FC, this scheme has a significant drawback, since a large amount of memory is required to process encrypted data, which also depends on the specified parameters.

3.3 TFHE Scheme

Consider a scheme that implements homomorphic arithmetic for Boolean algebra, namely Fast Fully Homomorphic Encryption over the Torus (TFHE) [32]. This scheme has several distinctive features. In terms of security, instead of the usual Gaussian white noise, the scheme generates it based on the mathematical expectation and the standard deviation of the probability distribution in the torus regions, which also allows for learning with an error in the torus. On the other hand, this scheme, unlike the level FHE, such as BFV and CKKS (multiplicative depth level), implements"real" fully homomorphic encryption based on the homomorphic Cmux valve, unlike the NAND Gentry valve. In this scheme, during encryption, a LUT table is built, by rotating which the required homomorphic operation can be obtained by the Cmux valve.

For example, the developers of the library [33] presented several comparison options, but it is worth noting that the appropriate version of the homomorphic comparison operation can currently only be determined empirically. The following can be highlighted as disadvantages of the scheme. With large data volumes, the LUT sizes increase, which also increases the memory consumption, in addition, the number of torus areas used increases, which can lead to calculation errors when crossing areas. However, in the FC area, these disadvantages can be leveled out, since on each computing node the data is processed separately and distributed, which allows both to control the memory consumption and the occupied torus spaces during the key switching operation.

Thus, the TFHE scheme will be applied for further studies.

4 Organizing Privacy-Preserving Computing

To develop the architecture of the data analysis system in the fog network, we first consider the implementation of homomorphic operations required for its implementation. As an example, we consider logistic regression. Logistic regression is a statistical model that is used to predict the probability of some event, which allows us to effectively organize such systems as SC and SH. In addition, this data analysis method satisfies the limitations of FHE schemes in the sense that it does not require operations such as division, since the logistic function can be represented as an approximate polynomial, and discrete logarithm methods can be used to calculate the logarithm.

To obtain a forecast from the variable x, it is necessary to take the logistic function of the product of x and the parameters θ. The logistic function has the following form

$$\sigma(z) = \frac{1}{1+e^{-z}}. \tag{2}$$

However, in FHE this operation is currently difficult to perform. This problem was solved by the authors of [34], where an approximate logistic function was obtained based on a third-degree polynomial, which has the following form:

$$-0.004 \cdot z^3 + 0.197 \cdot z + 0.5. \tag{3}$$

The computation of this polynomial requires five multiplications and two additions, which is quite optimal for use in both FHE in general and in the TFHE scheme.

To apply this function in practice, it is necessary to train the model, training is performed on the basis of the loss function:

$$J(\theta) = -\frac{1}{m}\sum_{i=1}^{m}\left[y^{(i)}\log\left(h_\theta\left(x^{(i)}\right)\right) + \left(1-y^{(i)}\right)\log\left(1-h_\theta\left(x^{(i)}\right)\right)\right], \tag{4}$$

where h_θ – is the error, y – is the target value.

To calculate (4), it is also necessary to prepare the data, since normalized data are often used, which are within the range $0, 1$.

To do this, you can use the following method. Since you need to get integer values, you can convert the significant bits of numbers from the fractional part of the number to the integer. To do this, you need to raise the number to the power of two, i.e.

Example 1: Quantizing the Number of Methods 2^N.

Let's take a number 0.324145 and quantize it 2^8.

Binary representation of a number

$0.324145 = 0.01010010111110110010101010101110001010010111001110011$

Binary representation $2^8 = 256 = 001001010110$

Then the quantized value 1010010

Now let's take the number 0.35432 and immediately quantize it. We get 1011010

Now let's calculate their sum $1010010 + 1011010 = 10101100$

To dequantize, you need to divide the number by 2^8

Then the value has the form $\frac{10101100}{001001010110} = 0.101011$

We convert the number into 10 format and get 0.671875.

Let's run a check

$0.324145 + 0.35432 = 0,678465$, thus we got an accuracy of 2 decimal places.

End of **Example**

Of course, the obtained accuracy is quite low. However, if we take larger powers, such as 16, 32, etc., we will get higher accuracy, which is also optimal for FHE, since the encryption keys support the processing of large numbers.

Then the logistic regression learning algorithm can be displayed as follows.

Algorithm 1. Training Logistic Regression in TFHE

Input: $x = \{n, m\}, y = \{m\}, \alpha, N$
Output: θ

---

1. $\theta = random\,\{m\}$
2. **For** i **from** 1 **to** N
2.1. $z = X \cdot \theta$
2.2. $predictions = -0.004 \cdot z^3 + 0.197 \cdot z + 0.5$
2.3. $errors = predictions\ y$
2.4. **For** j **from** 0 **to** m:
2.4.1 **For** k **from** 0 **to** j: 2.4.1.1. $temp = errors \cdot X$
2.4.2. $gradient = \left(\frac{1}{N}\right) \cdot temp$
2.4.3 $theta_j = theta_j - \alpha \cdot gradient$
3. **return** θ

Now let's consider the architecture of the data analysis system in the fog device network.

5 Architecture of a Data Analysis System in a Fog Computing Networktitle

The main idea of the architecture is distributed data analysis in FN. It is based on the FC organization scheme, which looks like this (Fig. 1).

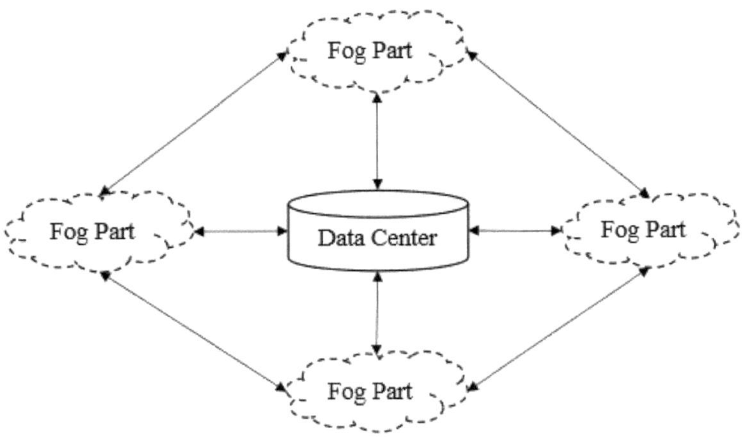

Fig. 1. Structural diagram of fog computing network

Thus, the fog parts are connected both to each other and to the data center (DC). In this case, DC centrally collects processed data from FN into a common storage. Fog parts can exchange data both with DC and with other fog parts. A fog part means several FNs united by one task, such as organizing traffic lights in SC. Data is transmitted exclusively in encrypted form using the TFHE scheme. Let's consider DC in more detail (Fig. 2).

In general, DC is a data storage system, a TFHE key generator, a TFHE block, and a translation block. The key generation module is responsible for generating a public and secret key, with the secret key sent only to the TFHE block, and the public key to the translation block and to the TFHE block. Data coming from the translation block can be sent both encrypted and decrypted, but data from the data storage system comes only encrypted, which is necessary to ensure confidentiality. Now let's consider the model of the fog part (Fig. 3).

In general, it consists of a data translation unit, a fog node, an encryption unit, and an IoT device. It is assumed that the encryption unit and the IoT device in physical form will be one device, or physically connected via a secure communication channel. Data coming from IoT devices will be sent encrypted to the FN, where the values will be quantized according to formula 4, and then logistic regression will be performed according to algorithm No. 1.

This is how the architecture of a confidential data analysis system in a fog computing network for low-power devices is organized.

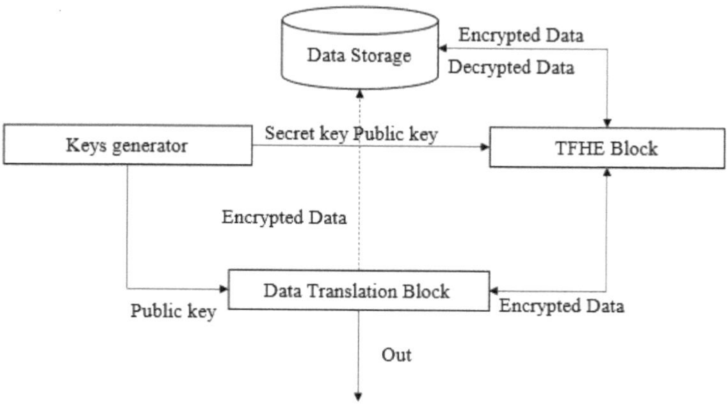

Fig. 2. Data center structure

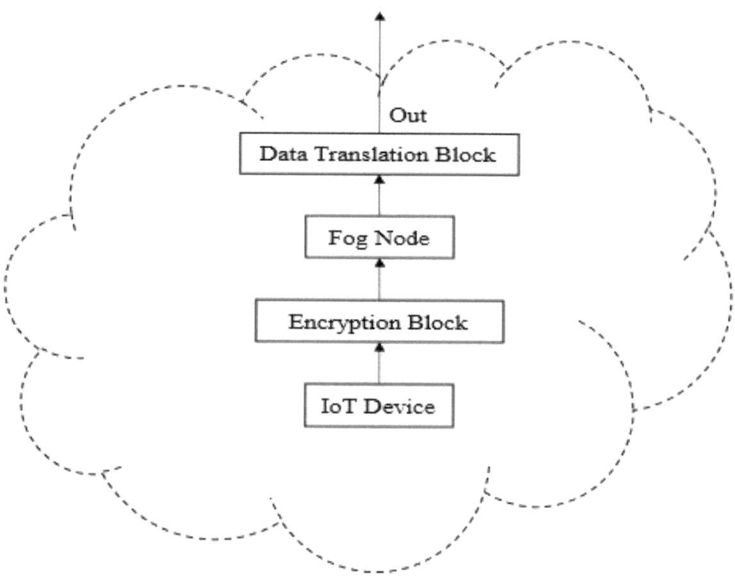

Fig. 3. Structure of the fog part

6 Conclusion

In this paper, a study of FC and its security was conducted, it was found that FC has risks associated with both external and internal threats, in order to mitigate the threats under consideration, it is proposed to build FC security based on the FHE TFHE scheme. In the context of using the FC-TFHE system, an architecture of a confidential data analysis system in a fog computing network for low-power devices was developed, presented on the basis of calculating logistic regression for its calculation in encrypted form, algorithms using homomorphic operations, an approximate logistic function, and also a number quantization method were presented.

In the future, it is planned to expand support for the mathematical apparatus of data analysis, conduct a more detailed study of quantization methods, and conduct a practical study of the architecture being developed.

Acknowledgment. The research was supported by the Russian Science Foundation Grant No. 22-71-10046, https://rscf.ru/en/project/22-71-10046/.

References

1. Deakin, M., Al Waer, H.: From intelligent to smart cities. Intell. Build. Int. **3**, 140–152 (2011). https://doi.org/10.1080/17508975.2011.586671
2. Cardone, G., Bellavista, P., Corradi, A., Foschini, L.: Effective collaborative monitoring in smart cities: converging MANET and WSN for fast data collection. In: Proceedings of ITU Kaleidoscope 2011: The Fully Networked Human? - Innovations for Future Networks and Services (K-2011), pp. 1–8 (2011)
3. Lobo, P., Acharya, S., D'Souza, R.O.: Quality of service for manet based smart cities. 5, 6
4. Foster, I., Zhao, Y., Raicu, I., Lu, S.: Cloud computing and grid computing 360-degree compared. In: 2008 Grid Computing Environments Workshop, pp. 1–10. Ieee (2008)
5. Dolui, K., Datta, S.K.: Comparison of edge computing implementations: Fog computing, cloudlet and mobile edge computing. In: 2017 Global Internet of Things Summit (GIoTS), pp. 1–6. IEEE (2017)
6. Bonomi, F., Milito, R., Zhu, J., Addepalli, S.: Fog computing and its role in the internet of things. In: Proceedings of the First Edition of the MCC Workshop on Mobile Cloud Computing, pp. 13–16. ACM, Helsinki Finland (2012). https://doi.org/10.1145/2342509.2342513
7. Hsu, T.-C., Yang, H., Chung, Y.-C., Hsu, C.-H.: A Creative IoT agriculture platform for cloud fog computing. Sustain. Comput. Inform. Syst. **28**, 100285 (2020)
8. Yi, S., Li, C., Li, Q.: A survey of fog computing: concepts, applications and issues. In: Proceedings of the 2015 Workshop on Mobile Big Data, pp. 37–42 (2015)
9. Hu, P., Dhelim, S., Ning, H., Qiu, T.: Survey on fog computing: architecture, key technologies, applications and open issues. J. Netw. Comput. Appl. **98**, 27–42 (2017)
10. Kraemer, F.A., Braten, A.E., Tamkittikhun, N., Palma, D.: Fog computing in healthcare–a review and discussion. IEEE Access. **5**, 9206–9222 (2017)
11. Kumari, A., Tanwar, S., Tyagi, S., Kumar, N.: Fog computing for Healthcare 4.0 environment: opportunities and challenges. Comput. Electr. Eng. **72**, 1–13 (2018)
12. Pareek, K., Tiwari, P.K., Bhatnagar, V.: Fog computing in healthcare: A review. In: IOP Conference Series: Materials Science and Engineering, p. 012025. IOP Publishing (2021)
13. Mutlag, A.A., Abd Ghani, M.K., Arunkumar, N. al, Mohammed, M.A., Mohd, O.: Enabling technologies for fog computing in healthcare IoT systems. Future Gener. Comput. Syst. **90**, 62–78 (2019)

14. Perera, C., Qin, Y., Estrella, J.C., Reiff-Marganiec, S., Vasilakos, A.V.: Fog computing for sustainable smart cities: A survey. ACM Comput. Surv. (CSUR). **50**, 1–43 (2017)
15. Songhorabadi, M., Rahimi, M., MoghadamFarid, A., Kashani, M.H.: Fog computing approaches in IoT-enabled smart cities. J. Netw. Comput. Appl. **211**, 103557 (2023)
16. Ullah, A., Yaqoob, S., Imran, M., Ning, H.: Emergency message dissemination schemes based on congestion avoidance in VANET and vehicular FoG computing. IEEE Access. **7**, 1570–1585 (2018)
17. Han, M., Liu, S., Ma, S., Wan, A.: Anonymous-authentication scheme based on fog computing for VANET. PLoS ONE **15**, e0228319 (2020)
18. Peixoto, M.L.M., et al.: A traffic data clustering framework based on fog computing for VANETs. Vehicul. Commun. **31**, 100370 (2021)
19. Shiriaev, E., Ermakova, T., Bezuglova, E., Lapina, M.A., Babenko, M.: Reliablity and Security for Fog Computing Systems. Information **15**, 317 (2024)
20. Rivest, R.L., Shamir, A., Adleman, L.: A method for obtaining digital signatures and public-key cryptosystems. Commun. ACM. 21, 120–126 (1978). https://doi.org/10.1145/359340.359342
21. Beimel, A.: Secret-Sharing Schemes: a Survey. In: Chee, Y.M., et al., (eds.) Coding and Cryptology, pp. 11–46. Springer, Berlin, Heidelberg (2011). https://doi.org/10.1007/978-3-642-20901-7_2
22. Ogundoyin, S.O., Kamil, I.A.: Secure and privacy-preserving D2D communication in fog computing services. Comput. Netw. **210**, 108942 (2022)
23. Murugesan, A., Saminathan, B., Al-Turjman, F., Kumar, R.L.: Analysis on homomorphic technique for data security in fog computing. Trans. Emerg. Tel. Tech. 32, e3990 (2021). https://doi.org/10.1002/ett.3990
24. Gupta, S., Garg, R., Gupta, N., Alnumay, W.S., Ghosh, U., Sharma, P.K.: Energy-efficient dynamic homomorphic security scheme for fog computing in IoT networks. J. Inf. Secur. Appl. **58**, 102768 (2021)
25. Amuthan, A., Sendhil, R.: Hybrid GSW and DM based fully homomorphic encryption scheme for handling false data injection attacks under privacy preserving data aggregation in fog computing. J. Ambient. Intell. Humaniz. Comput. **11**(11), 5217–5231 (2020). https://doi.org/10.1007/s12652-020-01849-8
26. Gentry, C.: A fully homomorphic encryption scheme. Stanford university (2009)
27. Gentry, C.: Fully homomorphic encryption using ideal lattices. In: Proceedings of the Forty-first Annual ACM Symposium on Theory of Computing, pp. 169–178. ACM, Bethesda MD USA (2009). https://doi.org/10.1145/1536414.1536440
28. Al Badawi, A., Polyakov, Y., Aung, K.M.M., Veeravalli, B., Rohloff, K.: Implementation and Performance Evaluation of RNS Variants of the BFV Homomorphic Encryption Scheme. IEEE Trans. Emerg. Top. Comput. **9**, 941–956 (2021). https://doi.org/10.1109/TETC.2019.2902799
29. Brakerski, Z.: Fully homomorphic encryption without modulus switching from classical GapSVP. In: Safavi-Naini, R., Canetti, R. (eds.) CRYPTO 2012. LNCS, vol. 7417, pp. 868–886. Springer, Heidelberg (2012). https://doi.org/10.1007/978-3-642-32009-5_50
30. Falcetta, A., Roveri, M.: Privacy-preserving deep learning with homomorphic encryption: an introduction. IEEE Comput. Intell. Mag. **17**, 14–25 (2022)
31. Cheon, J.H., Kim, A., Kim, M., Song, Y.: Homomorphic encryption for arithmetic of approximate numbers. In: International conference on the theory and application of cryptology and information security, pp. 409–437. Springer (2017)

32. Chillotti, I., Gama, N., Georgieva, M., Izabachène, M.: TFHE: fast fully homomorphic encryption over the torus. J. Cryptol. **33**, 34–91 (2020)
33. Zama: Concrete ML: a Privacy-Preserving Machine Learning Library using Fully Homomorphic Encryption for Data Scientists, (2022)
34. Chen, H., et al.: Logistic regression over encrypted data from fully homomorphic encryption, (2018). https://eprint.iacr.org/2018/462

UK-IDS-Machine Learning Based Intrusion Detection System for Unknown Attack Detection

T. Sowmya[1], E. A. Mary Anita[1](✉), and Maria Lapina[2]

[1] Christ University, Bangalore, India
maryanita.ea@christuniversity.in
[2] North Caucasus Federal University, Stavropol, Russia

Abstract. Computer networks have become the major focus for attackers. Hence intrusion detection system plays a significant role in detecting attacks. Many researchers have already focused on the domain of cyber security by developing an efficient framework. However, developing an efficient IDS is still a challenging task because of its effectiveness in determining novel attacks. Hence in the current study, a machine learning based IDS called UK-IDS is proposed by incorporating OC-SVM and a basic SVM model. The aim of the proposed system is to achieve high accuracy and F1 score by detecting novel attacks. The OC-SVM approach identifies the novel attacks by collaborating the clustering and thresholding mechanism. The basic SVM model is to distinguish the type of attack. The experimental study reveals that UK-IDS framework shows good performance in terms of accuracy and F1 score.

Keywords: One Class SVM(OC-SVM) · Intrusion Detection System(IDS) · Unknown -Intrusion Detection System(UK-IDS) · Artificial Intelligence (AI)

1 Introduction

The drastic growth of computer networks has influenced many cyber attackers to destroy the core of the network. The development of an efficient Intrusion Detection System is the one solution to safeguard the network [1]. There are basically two types of Intrusion Detection Systems: Signature based and Anomaly based IDS [2]. Signature based IDS detects the known attacks and on the other hand, anomaly based IDS detects the unknown attacks. All the traditional IDS didn't give promising results. With the emergence of artificial intelligence, researchers started integrating machine learning and deep learning to provide security to the network. Many machine learning based IDS have been proposed and ML has become the core component that can detect and classify the attack by its outstanding learning capability [3]. Although the AI based IDS showed good results in comparison with traditional IDS, they are disadvantaged by poor performance for unknown attacks. This challenge may affect the capability of AI in the domain of intrusion detection system.

In [4] highlighted the importance of detecting unknown attacks. Nowadays attackers are launching new attacks, hence an efficient and reliable mechanism is needed to detect

M. Lapina et al. (Eds.): AISMA 2025, LNNS 1456, pp. 425–433, 2026.
https://doi.org/10.1007/978-3-032-07275-7_38

these novel attacks. Therefore, to provide an efficient IDS in the field of cyber security, a two-layer Reliable SVM model called UK-IDS is proposed, and the model can detect all the unknown threats and can protect the network efficiently. The entire process is divided into two stages. In the first stage, a one class SVM is used which can find whether it is a known attack or not. If it is a known attack, in the second stage normal SVM algorithm is used and can categorize the data into attack and normal. For evaluating the effectiveness of the proposed UK-IDS, nine attacks are considered from UNSWNB15 dataset. The aim of the proposed system is to detect the unknown attacks and to categorize the known attacks effectively. One class SVM(OC-SVM) together with basic SVM increases the overall effectiveness in the domain of cyber security. The overall performance is evaluated based on the UNSWNB15 dataset. The goals of the proposed system are as follows.

Develop a Reliable Machine Learning based IDS that can detect unknown attacks efficiently.

Detect and classify the known attacks from UNSWNB15 with high accuracy.

The remainder of the paper is structured as follows: a detailed review of the work is illustrated in the second section. In the third section, the working of the proposed methodology is covered and in the fourth section, a detailed experimental study is performed with proper results and discussion. Finally, the conclusion of the proposed methodology and future scopes are presented.

2 Literature Review

In this section existing studies related to machine learning based intrusion detection systems are depicted in detail. Researchers have already focused on machine learning based IDS and achieved remarkable results. In that supervised learning achieved more attention. In [5] the attacks on the KDD CUP dataset and HTTP large dataset attained good efficiency. Here attribute selection and attribute extraction are used as an intermediate step. The work uses One class SVM methodology for the detection of attacks and achieved good performance on Affinity Propagation based exemplar extraction.

In [6] proposed an intrusion detection system on NSL-KDD dataset and SVM based approach is used to detect the anomalies. The approach achieved an accuracy of 99.7%. The major challenge in the proposed system is that resource consumption and performance degradation can happen in a few other datasets. In [7] provides an intrusion detection system based on the Autoencoder and Isolation Forest called Auto-IF for the fog environment. The approach focuses only on the binary classification by differentiating between normal and attack data. In comparison with many cutting edge intrusion detection techniques, the Auto-IF mechanism provides outstanding performance and achieves a higher accuracy of 95.14%.

In [8] a host based IDS was proposed using deep learning. The model was able to detect application level attacks with optimal accuracy [9]. uses the NSL-KDD dataset for the detection of attacks. A combination of various ML based models is used for this approach to gain high accuracy. The model achieved a slightly high accuracy of 86.8% [10]. proposed a ML based approach for the determination of Zero day attacks and achieves good accuracy in comparison with other state of art studies related with this domain.

3 Proposed Methodology

3.1 Data Collection

For evaluation purposes, the UNSWNB15 dataset [11] is utilized which consists of 9 attacks namely Fuzzers, Analysis, Backdoors, DoS, Exploits, Generic, Reconnaissance, Shellcode, and Worms. The Argus and Bro-IDS tools are developed to extract the 49 features with the attack class information.

3.2 Data Preprocessing

Data preprocessing is a crucial step in the proposed IDS to gain more accuracy and reduce the overfitting. In the UNSWNB15 dataset, the preprocessing method removes the null value columns and performs processing to the column names by removing the unwanted space. The next step is to scale the features by means of the standard scaler method to maintain a consistent value. Here in the current study Z-score normalization is used which significantly enhances the accuracy. The final step is to perform label encoding, which assigns numerical values to the attack categories. For example 0 for the Fuzzers,1 for the Analysis and so on.

3.3 Random Under Sampling

The UNSWNB15 dataset is a huge one with 49 features. Hence it is very important to balance the dataset before the classification. To balance the data between majority and minority classes, the Random Under sampling method is utilized and this sampling method randomly deletes the majority classes. For example, in the UNSWNB15 dataset, the count for worm attacks is very lower than all the other attacks. Hence to balance the dataset the Random Undersampling method was utilized and the method was able to maintain balance with respect to majority and minority classes.

3.4 Feature Selection Using Genetic Algorithm

Feature selection plays a major role in improving the accuracy and to reduce the overfitting [12]. Within our experimental work, Genetic Algorithm based feature selection is used to select the best features. The huge dimensional dataset is a big issue in machine learning which in turn reduces the accuracy and the resource consumption also will be more. The optimal number of features is selected using fitness, selection, cross over, and mutation operations. To select the best features in our current work, the fitness value is assigned for each feature. The selection operator selects the features with the highest fitness value to recombine for the next generation. The next step is to recombine with the next generation using cross over operation. Finally, the mutation operator generates the best features by randomly changing the value of some features in the offspring.

In the current study, 13 best features are selected for the next stage of classification. The following Table 1 shows the 13 best features for the further evaluation of attacks.

Table 1. Selected Features.

FEATURES	Proto, sttl, state, ct_dst_sport_ltm, srcip, dstip, dsport, ct_src_ ltm, sbytes, ct_state_ttl, trans_depth, ‘synack, dintpkt

3.5 Machine Learning Based Intrusion Detection System

Most of the systems [13] use machine learning models to detect the abnormalities in the data. In this section, a two layer intrusion detection system is proposed. The two layers are comprised of One class SVM and basic SVM for the detection of attacks. To get the overall frame the two models are merged together. The basic idea behind the use of two layer strategy is to reduce resource consumption by identifying the unknown attack at the beginning itself. Further to that in the first layer itself, novel attacks can be determined. The train test ratio considered in this study is 60: 40. Most of the methods use machine learning models to detect the abnormalities in the data.

(a) One Class SVM (OC-SVM) layer

The proposed work is applied firstly on OC-SVM is used to detect whether it is a known attack or not. Unlike traditional SVM which helps in detecting multiple attacks, One class SVM trains on data with a single class. Here in our methodology One Class SVM detects whether it is known attack or not. In this study One class SVM is actually employed for determining anomaly detection or outlier detection. In our proposed approach to provide an extra layer of security, OC-SVM is trained with the whole data from UNSWNB15 including normal and attack. Hence the model finds the best fit line that encapsulates only the target class and removes the data outside the best fit line. Here unsupervised learning strategy is utilized for the detection of novel attacks by determining the similarities based on the decision from its boundary. In this stage, unknown data are further stored for future analysis. If the data is identified as known, it is passed to the next level of SVM to determine which family the attack belongs to. The main motive behind the use of One class SVM is outlier detection which in turn determines the novel attacks also.

(b) Determination of Unknown attack (Novel Attack)

The unknown instances from OC-SVM are evaluated in detail in this step. Here the data retained from the previous step is further evaluated using clustering. In our current study K-Means clustering approach is used and different clusters of data points are accumulated based on the similarity metrics. The clusters are further evaluated to determine whether attack or not. Inspired by [14] thresholding approach is considered for the determination of unknown attacks. Data points with a positive probability ≥ 0.5 are considered as attack data and the rest of the data is considered as normal data. For determining the unknown attack, a random data point ‘R’ from each cluster is considered, and the threshold score is determined for the R. If the value of R is considered as a attack data, the remaining data in the cluster is also identified as attack. Hence resource is also managed properly by determining only one data point in each cluster instead of considering each and every data point.

(c) Basic SVM layer

After collecting the known data from the previous layer, basic SVM is utilized to classify the data into normal and other multiple attacks. Here basic SVM uses suitable decision boundaries to separate the normal data and different families of attacks. Finally, the Performance is evaluated in terms of accuracy and F1 score.

(d) Training and Testing Phase

The training and testing phase is considered separately for each layer in our current study. In the first layer of OC-SVM, the entire model is trained with all the classes including normal and attack from UNSWN15. Here for the training purpose, 60% of the data is only considered. The remaining 40% is considered for the testing purposes. Hence during the testing phase normal data and attack data are used for the determination. OC-SVM will determine the known data and unknown data. Further analysis of unknown data is determined by means of clustering and thresholding mechanisms. Like the previous layer, in the basic SVM model also 60: 40 train test ratio is also considered. The known data are further classified into normal and attack classes. Here multi class classification approach is used.

Algorithm for UK-IDS Model

Input: UNSWNB15 dataset with attack and normal data
Output: UK-IDS model
Initial Setup
  Train the system using 60 % of the UNSWNB15 dataset
  Test the system using 40% of the UNSWNB15 dataset
 For each data point in the dataset ***do***
   use **OC-SVM** to determine the unknown attack
  If each data point is known ***then***
   Classify the data point into normal and family of attacks using the Basic SVM model
  end
  else
   Store the unknown data
 end
end
 # Handling Unknown data
Cluster the unknown data based on K-means clustering : cluster 1,cluster 2,cluster3….clusterN
Apply Thresholding to determine the Unknown attack
For each set of clusters ***do***
 Randomly pick any data point in each cluster and name it 'R'
 If the positive probability of R>=0.5 ***then***
  Consider the corresponding cluster as an attack
 end
 else
  Consider the corresponding cluster as an attack
 end
end

The algorithm describes the working mechanism for the detection of unknown attacks and the overall framework is depicted in the Fig. 1.

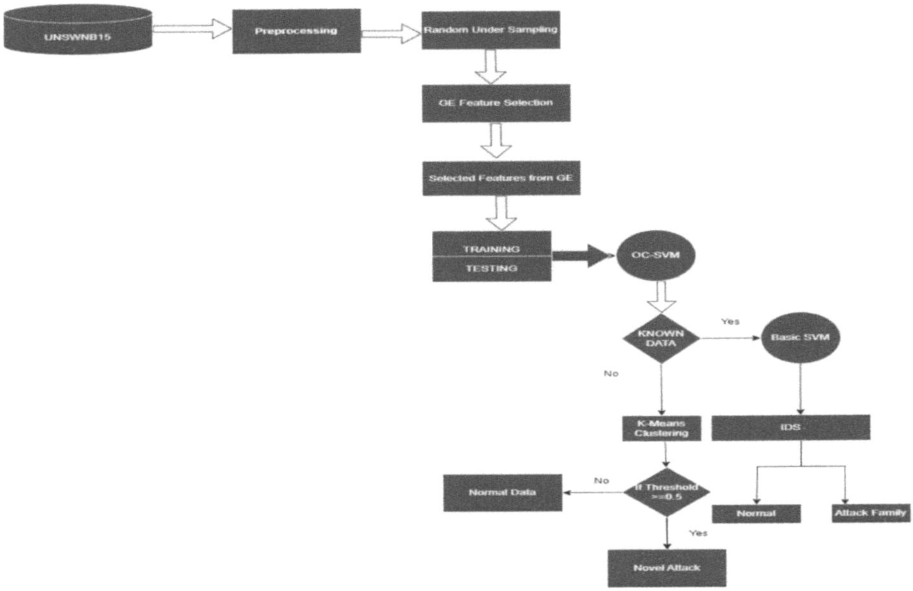

Fig. 1. UK-IDS Framework.

4 Results and Discussion

Our experimental study is performed on an 8GB laptop using python for the execution purpose. Scikit toolkit is also utilized to determine the performance. The evaluation of the framework is done in terms of accuracy, and F1 score. The results of each level are depicted in detail.

4.1 Unknown Data Detection Using OC-SVM

For the classification of known and unknown data after applying feature selection is depicted in a detailed way.

After applying the OC-SVM layer, the accuracy obtained is about 93.8% and the corresponding performance metrics are outlined in Fig. 2. Although the accuracy and F1 score yield satisfactory results, it helps to work adds some advantage for the classification task.

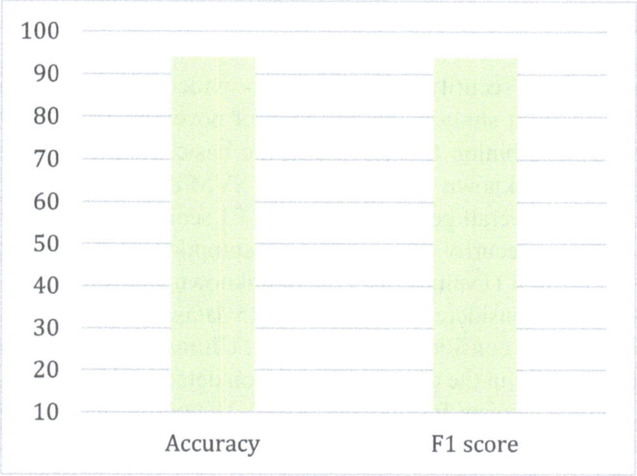

Fig. 2. Performance metrics for Unknown attack Detection.

4.2 Attack Classification

For the multi class classification, the best result of 98.2% accuracy is obtained for the UNSWNB15 dataset. The results show good performance in terms of accuracy and F1 score because of the combined effect of OC-SVM and Basic SVM model.

The results obtained from Figs. 2 and 3 show that the current approach is well suited for the UNSWNB15 dataset. The proposed framework forms a base model for future researchers to determine novel attacks. The current study can be extended by evaluating the unknown attack in detail. The current study lacks a detailed evaluation of unknown attacks.

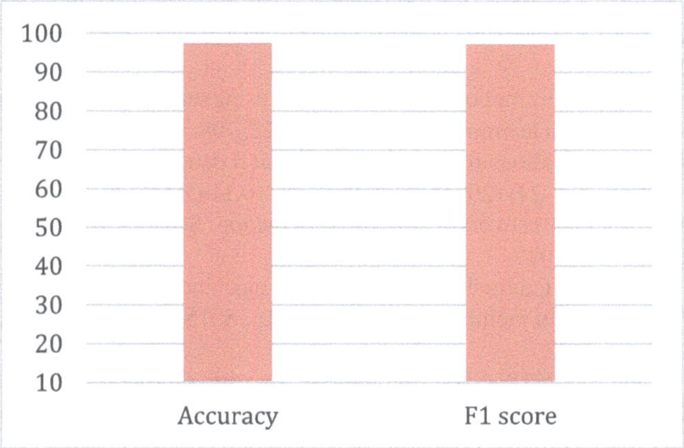

Fig. 3. Performance metrics for multiclassification.

5 Conclusion

In this study, a two layer security mechanism is provided for the detection of attacks. The novelty in the current study is the detection of novel attacks. The ML based IDS model was built by combining OC-SVM and the basic SVM model. The OC-SVM mainly focuses on the unknown attack and basic SVM determines the type of attack. The model achieved an overall good accuracy and F1 score. The main advantage of the proposed IDS system is security and resource consumption. However, there are certain limitations the model didn't evaluate the type of unknown attack in detail. In addition to that, the proposed study considered the UNSWNB15 dataset only for evaluation purpose; other datasets can also be considered in the future. Ultimately, the framework provides a significant advancement in the domain of intrusion detection systems by providing an efficient protection mechanism for network assets. The performance rates make it well suited for identifying unknown attacks and known attacks in a better way.

References

1. Liao, H.J., Lin, C.H.R., Lin, Y.C., Tung, K.Y.: Intrusion detection system: a comprehensive review. J. Netw. Comput. Appl. **36**(1), 16–24 (2013)
2. Alazzam, H., Sharieh, A., Sabri, K.E.: A feature selection algorithm for intrusion detection system based on pigeon inspired optimizer. Expert Syst. Appl. **148**, 113249 (2020)
3. Mishra, P., Varadharajan, V., Tupakula, U., Pilli, E.S.: A detailed investigation and analysis of using machine learning techniques for intrusion detection. IEEE Commun Surv Tutor. **21**(1), 686–728 (2018)
4. Masdari, M., Khezri, H.: A survey and taxonomy of the fuzzy signature-based intrusion detection systems. Appl. Soft Comput. **92**, 106301 (2020)
5. Wang, W., Zhang, X., Pitsilis, G.: Abstracting audit data for lightweight intrusion detection. In: Proceedings of the 6th International Conference on Information Systems Security (Iciss'10), pp. 201–215. Springer-Verlag, Berlin, Heidelberg (2010)
6. Bagaa, M., Taleb, T., Bernabe, J.B., Skarmeta, A.: A machine learning security framework for Iot systems. IEEE Access. **8**, 114066–114077 (2020). https://doi.org/10.1109/ACCESS.2020.2996214
7. Yan, W.: Detecting gas turbine combustor anomalies using semi-supervised anomaly detection with deep representation learning. Cogn. Comput. **12**, 398–411 (2020)
8. Ring IV, J.H., et al.: Methods for host-based intrusion detectionwith deep learning. Digital Threats: Res. Pract. **2**(1–29) (2021). https://doi.org/10.1145/3461462
9. Tang, Y.: Deep stacking network for intrusion detection. Sensors. **22**(1), 25 (2022). https://doi.org/10.3390/s22010025
10. Albulayhi, K., et al.: IoT intrusion detection using machine learning with a novel high performing feature selection method. Appl. Sci. **12**(10), 5015 (2022). https://doi.org/10.3390/app12105015
11. Moustafa, N., Slay, J.: UNSW-NB15: a comprehensive data set for network intrusion detection systems (UNSW-NB15 network data set). In: 2015 Military Communications and Information Systems Conference (Milcis), Canberra, ACT, Australia, pp. 1–6 (2015). https://doi.org/10.1109/MilCIS.2015.7348942
12. Fernando, C.A., Thomas, R., Mary Anita, E. .A., Lapina, M.: A feature selection study on the Bot-IoT dataset using ensemble classification techniques. In: 2024 IEEE International Conference on Contemporary Computing and Communications (Inc4), Bangalore, India, vol. 2024, pp. 1–9. https://doi.org/10.1109/InC460750.2024.10649035

13. Lapina, M., Dudun, G., Kotlyarov, D., Rjevskaya, N., Subramanian, S.J.: Analysis of an existing method for detecting adversarial attacks on deep neural networks. In: Alikhanov, A., Tchernykh, A., Babenko, M., Samoylenko, I. (eds.) Current Problems of Applied Mathematics and Computer Systems. CPAMCS 2023 Lecture Notes in Networks and Systems, vol. 1044. Springer, Cham (2024). https://doi.org/10.1007/978-3-031-64010-0_29
14. Sheng, V.S., Ling, C.X.: Thresholding for making classifiers cost-sensitive. In: Proceedings of the 21st National Conference on AI, pp. 476–481 (2006)
15. Zolotarev, V., Arkhipova, A., Lapina, M., Prakasha, G.S., Movzalevskaya, V.: Assessing player interaction for a social networking cooperative educational game. In: Lapina, M., Prakasha, G.S., Grigoriev, S. (eds.) SLET 2022: International Conference on Innovative Approaches to the Application of Digital Technologies in Education and Research. SLET 2022 Lecture Notes in Networks and Systems, pp. 144–125. Springer, Cham (2024)

Development of an Information Security Risk Assessment Model Through the Calculation of the Vulnerability Danger Factor

Maria Utenkova[1], Andrey Komkov[1], Elena Maksimova[1](✉), and Oksana Mezentseva[2]

[1] MIREA – Russian Technological University, Moscow 119048, Russia
maksimova@mirea.ru
[2] North Caucasus Federal University, Stavropol 355017, Russia

Abstract. The relevance and necessity of research in the field of information security risk assessment are justified against the backdrop of the trend towards digitalization and as part of the national projects "Digital Economy" and its successor "Data Economy and Digital Transformation of the State". The existing standards for information security risk management have been analyzed. A description has been given of the classic statistical risk assessment model and a dynamic model that takes into account a parameter responsible for the intensity of destructive effects accumulation. The disadvantages of the classic risk assessment model have been identified, which relate to problems with expert assessments that affect the quality of the assessment: the lack of accumulated statistics over the years or lack of qualified experts. An analysis has been conducted of the open standard for vulnerability danger quantification, CVSS. A risk assessment model has been developed based on an evaluation of the risk associated with CVSS vulnerabilities. The possibility of defining risk as a combination of two linguistic variables has also been considered. The model is described using linguistic fuzzy logic, and the levels of danger for each parameter in the revised model are determined. Additionally, the option of modifying the model to include the cost of addressing the vulnerability has been explored. Prospects for further development of this approach are discussed.

Keywords: Risk Assessment · Information Security · CVSS

1 Introduction

Digitalization has been one of the most observed trends for many years. Its relevance is noted at the highest levels of government and is implemented within the framework of the national projects "Digital Economy" and its logical successor "Data Economy and Digital Transformation of the State" [2, 6]. In the structure of this project, special attention is paid to the federal project "Cybersecurity Infrastructure", which defines some goals: ensuring the security of the Russian segment of the Internet, countering Internet fraud and combating personal data leaks.

M. Lapina et al. (Eds.): AISMA 2025, LNNS 1456, pp. 434–441, 2026.
https://doi.org/10.1007/978-3-032-07275-7_39

This highlights the urgency of the issue of information protection both at the state level and at the level of private information infrastructures [9, 15].

Information systems are becoming more complex in response to the need to process an increasing volume and variety of data, as well as process automation. In this regard, the number of threats is increasing, the implementation of which will disrupt the correct operation of systems, which will inevitably lead to losses: for example, material or reputational. Information security risk assessment is necessary for the proper allocation of resources, which makes it possible to increase the effectiveness of information security systems and prevent potential losses. The results of the risk assessment can be used to predict information security events, allowing for a better response to incidents [7, 8].

Special attention in the processes of optimizing information flows at significant CII objects should be paid to the dynamics of their behavior and threats at the software level in different types of object relationships [16, 17].

2 Existing Methods for Assessing Information Security Risks

The problem of assessing information security risks has existed for a long time. To date, there are already quite a lot of scientific papers and international standards that offer various approaches and methodologies to information security risk management [3].

Recognized as a classic approach in the Russian Federation, the approach to assessing information security risks is described in [4]. According to the standard, the quantitative assessment of information security risk is the sum of the products of the cost of possible damage from the implementation of the vulnerability i D_i by the probability of its implementation p_i (1):

$$R = \sum_i D_i * p_i. \tag{1}$$

However, this approach does not take into account the possibility of a change in the value of risk over time [10]. That is, the risk assessment is relevant only at the time of the calculations. To solve this problem a third time-dependent parameter is added to the risk assessment formula, namely, the intensity of the accumulation of destructive effects of vulnerability i on the system $\lambda_i(t)$ [1] (2):

$$R = \sum_i D_i * p_i * \lambda_i(t). \tag{2}$$

Within the framework of these approaches, in most cases decisions on the choice of parameter values are made on the basis of an expert assessment. For example, the parameter p_i is determined based on both general statistics on the exploitation of this vulnerability and the decision of experts on the possibility of exploiting this vulnerability within a specific information infrastructure. At the same time, the result is a compromise between the opinions of each of the experts. If there are no statistics developed over the years or there are not enough qualified experts, the assessment obtained may not correspond to the actual situation.

3 Proposed Solution

In order to improve the transparency and objectivity of information security risk assessments, this study proposes replacing the assessment of the probability of vulnerability implementation with an assessment of the danger associated with the vulnerability CVSS (Common Vulnerability Scoring System) [4, 11, 14].

CVSS is an open standard for quantifying vulnerability risks based on a set of parameters that can vary from version to version. The most recent version released is 4.0, which includes the following basic metrics:

- Exploitability metrics: attack vector, attack complexity, attack requirements, privileges required, user interaction.
- Vulnerable system impact metrics: confidentiality, integrity, availability.
- Subsequent system impact metrics: confidentiality, integrity, availability.

In addition to the basic metrics, this standard also includes additional metrics that change over time or depend on the specific context. These metrics are assigned a qualitative value, and then, according to the algorithm described in the CVSS specification, a quantitative risk assessment of the vulnerability is obtained.

Previous versions of CVS are still relevant and are being actively used. In the FSTEC Information Security Threat Database, vulnerabilities that have been released over the past year are usually rated CVSS 2.0 and CVSS 3.0, but have not yet been rated 4.0 [13]. The choice of the standard version is up to the experts of each organization. This is because the range of acceptable values remains the same for all versions of the CVSS. However, it should be noted that the outcome of evaluating the same vulnerability may vary depending on the version used [12].

The CVSS 3.0 vulnerability risk score is a real number $V \in [0, 10]$. In order to maintain consistency with the classical approach, this parameter is normalized within the proposed framework by dividing it by 10 and converting it to a range of [0, 1]. As a result, the following formula for assessing the risk of information security based on the cost of damage, D_i, and the risk of vulnerability, V_i (3):

$$R = \sum_i D_i * \frac{V_i}{10}. \tag{3}$$

To account for the variability in risk over time, it is possible to consider the intensity of the accumulation of destructive effects on the system, if necessary $\lambda_i(t)$ (4):

$$R = \sum_i D_i * \frac{V_i}{10} * \lambda_i(t). \tag{4}$$

The methods described above may not always be suitable for accurately assessing information security risks, since this number does not take into account the relationship between the potential damage from a threat and the actual damage it can cause to an organization. Therefore, in some cases it may be more rational to consider risk as a combination of two linguistic fuzzy variables (5):

$$R = \langle D_i, V_i \rangle, \tag{5}$$

where:

D = "Damage", for which:

universal set $U_D = [0, 10]$,
term-set T(D) = {"missing", "low", "medium", "high", "critical"},
semantic rule M for determining the correspondence of each fuzzy variable $\tilde{D}$ to the fuzzy subset $M(\tilde{D})$ of the universal set U_D is presented in Table 1.

V = "CVSS", for which:

universal set UV = [0, 10],
term-set T(V) = {"missing", "low", "medium", "high", "critical"},
semantic rule M for determining the correspondence of each fuzzy variable $\tilde{V}$ to the fuzzy subset $M(\tilde{V})$ of the universal set UD is presented in Table 2.

The meaning of each linguistic variable V is as follows:

"Missing" - there is no danger.
"Low" - a violation of confidentiality, integrity, and accessibility may have limited adverse consequences for an organization or individual.
"Medium" - a violation of confidentiality, integrity, and accessibility is likely to have a negative impact on an organization or individual.
"High" - a violation of confidentiality, integrity, and accessibility is likely to have a significant negative impact on an organization or individual.
"Critical" - a violation of confidentiality, integrity, and accessibility is likely to have a critical negative impact on an organization or individual.

Table 1. The correspondence between the values of linguistic and fuzzy variables D.

Values of linguistic variables	Values of fuzzy variables
missing	0.0
low	0.0–4.0
medium	4.0–7.0
high	7.0–9.0
critical	9.0–10.0

Table 2. The correspondence between the values of linguistic and fuzzy variables V.

Values of linguistic variables	Values of fuzzy variables
missing	0.0–2.0
low	2.0–4.0
medium	4.0–6.0
high	6.0–8.0
critical	8.0–10.0

4 Discussion

In CVSS, a qualitative assessment is compared to a quantitative one based on the specifications. At the same time, each organization has the opportunity to independently interpret the scale of damage.

A graphical representation of information security risk as a vector with two linguistic variables is shown in the Fig. 1.

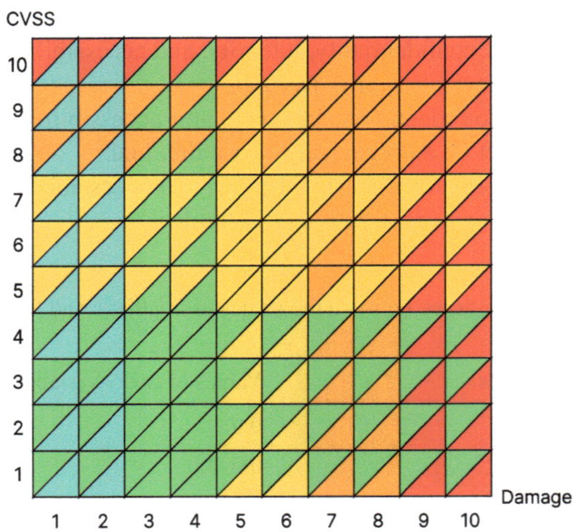

Fig. 1. The ratio of the danger of vulnerability (CVSS) and potential damage in case of its implementation.

In this case, each color represents the corresponding name of the linguistic variable: green represents "low" and red represents "critical". The square, which is obtained by combining the two triangles (CVSS and damage), represents the risk itself.

The presented method will help determine the most relevant information security risks to eliminate. The organization's experts can independently determine the procedure for closing vulnerabilities to reduce the risk of information security in accordance with the needs of a particular organization. Within the framework of this study, the following order of priority for risk management is proposed:

1. Risk with a critical or high CVSS and damage score.
2. Risk with critical or high damage score, but with a CVSS score below high.
3. Risk with critical or high CVSS score, but with damage score below high.
4. Risk with medium damage score and another CVSS score.
5. Risk with medium CVSS score and another damage score.
6. Risk with other CVSS and damage scores.

However, this approach does not take into account important parameter: the cost of fix vulnerability. If the damage caused by the implementation of a vulnerability is less than the cost of fixing it, it is not always rational to fix this vulnerability. In order to take this metric into account, a new linguistic variable C = "Cost" is created with parameters similar to the variable D = "Damage", which is added to the vector R (see Fig. 2) (6):

$$R = \langle D_i, V_i, C_i \rangle \tag{6}$$

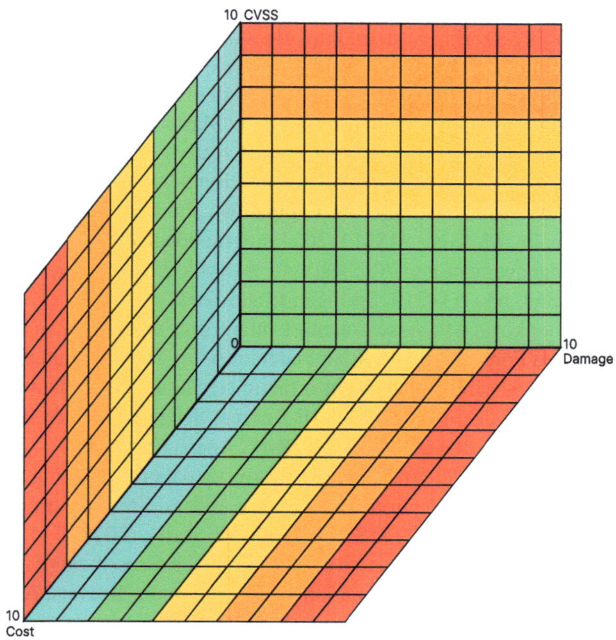

Fig. 2. The vulnerability closure priority identification cube.

5 Conclusion

The proposed information security risk assessment model enables us to solve the challenge of efficiently allocating resources to ensure an effective system for countering these threats. By further exploring and delving deeper into this topic, we can achieve better threat assessment outcomes. Due to its flexibility, the model can be integrated into the workflows of information security teams in organizations to streamline and optimize their operations. Given the transparency of the final results, the model can serve as a tool to educate non-specialists about the potential damage caused by exploiting these threats, as well as demonstrating the benefits of preventing them. This promotes better collaboration between information security professionals and other members of the organization.

References

1. Maksimova, E.: Models and methods for assessing the information security of a subject of critical information infrastructure under destructive impacts of infrastructural genesis: specialty 23.60.00: dissertation for the degree of Doctor of Technical Sciences (2022)
2. National project Data Economy and Digital Transformation of the State. http://government.ru/rugovclassifier/923/about/. Accessed 20 Feb 2025.
3. ISO/IEC 27005:2022. Information security, cybersecurity and privacy protection — Guidance on managing information security risks
4. GOST R ISO/IEC 27005: Information technology. Methods and means of ensuring security. Inform. Secur. Risk Manag. (2010)
5. Common Vulnerability Scoring System. https://www.first.org/cvss/. Accessed 20 Feb 2025
6. Yanushkina, Y., Komkov, A.: Analysis of foreign experience in ensuring the stability of the national currency. Collection of scientific papers digitalization of science and education: prospects for creating a unified knowledge system. In: Materials of the International Scientific and Practical Events of the Society of Science and Creativity, pp. 304–311., Kazan (2024)
7. Utenkova, M., Maksimova, E.: Forecasting the development of information security events in the context of information warfare. In: Lapina, M., Raza, Z., Tchernykh, A., Sajid, M., Zolotarev, V., Babenko, M. (eds.) AISMA-2024: International Workshop on Advanced Information Security Management and Applications. AISMA 2024 Lecture Notes in Networks and Systems, vol. 863, pp. 311–319. Springer, Cham (2024)
8. Maksimova, E., Utenkova, M.: Assessment of interobject interaction in the infrastructural system of the Armed Forces of the Russian Federation. In: The state and prospects of development of modern science in the field of "Information technologies in the Armed Forces of the Russian Federation": Collection of the III All-Russian Scientific and Technical Conference, pp. 33–36. Military innovation technopolis "ERA", Anapa (2024)
9. Maksimova, E., Lapina, M., Lapin, V.: Synthesis of models for ensuring information security of subjects of critical information infrastructure under destructive influences. In: CEUR Workshop Proceedings, pp. 108–117. Stavropol, Krasnoyarsk (2022)
10. Maksimova, E., Sadovnikova, N.: Proactive modeling in the assessment of the structural functionality of the subject of critical information infrastructure. In: Kravets, A.G., Shcherbakov, M., Parygin, D., Groumpos, P.P. (eds.) Creativity in Intelligent Technologies and Data Science. CIT&DS 2021. Communications in Computer and Information Science, vol. 1448. Springer, Cham (2021)
11. Baranov, L., Ivanova, N., Mikhalevich, I.: Modeling and assessment of safety risks of intelligent water transport systems. Autom. Transport. **11**(1), 7–15 (2025)
12. Wunder, J., Kurtz, A., Eichenmüller, C., Gassmann, F., Benenson, Z.: Shedding light on CVSS scoring inconsistencies: a user-centric study on evaluating widespread security vulnerabilities. In: 2024 IEEE Symposium on Security and Privacy (SP), pp. 1102–1121. San Francisco, CA, USA (2024)
13. FSTEC Information Security Threat Database. https://bdu.fstec.ru/vul. Accessed 20 Feb 2025

14. Vovik, A., Larin, A.: An approach to formalizing the assessment of security threats using the information method of fuzzy modeling. H&ES Res. **15**(3), 30–37 (2023). https://doi.org/10.36724/2409-5419-2023-15-3-30-37
15. Andita, R., Faizan, A.: Systematic literature review on information security risk management in public service organizations. J. Teknik Informatika (Jutif). **5**(1), 89–96 (2024). https://doi.org/10.52436/1.jutif.2024.5.1.1223
16. Vinodha, D., Jenefa, J., Anita, E.A.M., Babenko, M.: Theoretical framework for blockchain secured predictive maintenance learning model using digital twin. In: Lapina, M., Raza, Z., Tchernykh, A., Sajid, M., Zolotarev, V., Babenko, M. (eds.) AISMA-2024: International Workshop on Advanced Information Security Management and Applications. Lecture Notes in Networks and Systems, vol. 863. Springer, Cham (2024)
17. Mary Anita, E.A., Jenefa, J., Vinodha, D., Lapina, M.: Advancements in Sybil attack detection: a comprehensive survey of machine learning-based approaches in wireless sensor networks. In: Lapina, M., Raza, Z., Tchernykh, A., Sajid, M., Zolotarev, V., Babenko, M. (eds.) AISMA-2024: International Workshop on Advanced Information Security Management and Applications. AISMA 2024 Lecture Notes in Networks and Systems, vol. 863. Springer, Cham (2024)

Solving High-Performance Computing Problems Using Distributed Neural Networks with Numerical Methods

Nikolay Vershkov(✉), Mikhail Babenko, Vladislav Lutsenko, and Nataliya Kuchukova

North-Caucasus Federal University, Stavropol, Russia
vernick61@yandex.ru

Abstract. This paper presents a study on distributed artificial neural networks implemented using wavelet transform-based modular architectures. The research compares the performance of monolithic, vertically partitioned, and horizontally partitioned artificial neural network configurations, with particular focus on computational efficiency and recognition accuracy. Experimental results demonstrate that horizontally partitioned artificial neural networks employing Haar wavelet transforms (2 × 2 kernel) achieve comparable recognition accuracy to monolithic networks (within 1% difference) while significantly reducing processing time. The four-module configuration shows particular promise, with average training time of 0.0754 s per cycle and inference time of 0.0393 s.

Keywords: orthogonal transformations · modular artificial neural networks · neural network optimization · wavelet transformations

1 Introduction

Exploring the computational capabilities of Artificial Neural Networks (ANNs), A.N. Gorban highlights the critical role of approximation methods when representing multivariable functions through single-variable functions [1]. By generalizing the Stone-Weierstrass theorem, Gorban demonstrated that ANNs can approximate any continuous multivariable function $f(x_1, x_2, \ldots, x_n)$ with arbitrary accuracy, given the presence of arbitrary nonlinearity. This approach provides an alternative to Kolmogorov's method of exact representation of multivariable functions using functions of fewer variables.

In his work, A.I. Galushkin [2] defines the objective of neuromathematics as "the development of highly parallelizable algorithms for both well-defined and complex or ill-defined problems." Initially, parallelism in ANNs was conceptualized as the simultaneous processing of input data by all neurons within a layer. However, the increasing volume of data, particularly in linguistic tasks solved via ANNs, demands substantial computational resources. Consequently, parallelism must be interpreted more broadly: partitioning input data into segments, processing them independently (i.e., in parallel) across multiple processors or devices, and subsequently aggregating the results [3].

M. Lapina et al. (Eds.): AISMA 2025, LNNS 1456, pp. 442–450, 2026.
https://doi.org/10.1007/978-3-032-07275-7_40

Thus, to address the challenge of building high-performance computing systems for large-scale data processing using ANNs, it is essential to parallelize the input data processing workflow. This involves decomposing the data into modules, solving subproblems in parallel across distributed modules, and integrating the results.

2 Theoretical Approach to Modular Neural Network Design

We begin by generalizing A.N. Gorban's extension of the Stone theorem, proven in [1], for arbitrary nonlinearities.

Theorem 1. Let $E \subseteq C(X)$ be a closed linear subspace of $C(X)$, where $1 \in E$, the functions in E separate points in X, and E is closed under a unary nonlinear operation $f \in C(R)$. Then, $E = C(X)$.

From Theorem 1, it follows that any nonlinearity possesses universal approximation properties. That is, using a polynomial p of degree $m > 1$ and first-degree polynomials, along with superposition, addition, and scalar multiplication operations, one can generate all elements of the space $\boldsymbol{C}[x_1, x_2, \ldots, x_n]$.

Thus, if there exist neural networks $S_1, S_2, \ldots, S_k$ computing functions $y_1, y_2, \ldots, y_K$ of the input vector x, then the operator $f(w_0 + w_1 y_1 + \ldots + w_k y_k)$ can approximate the result with arbitrary accuracy. According to Theorem 1, the set of functions computable by neural networks with a given continuous nonlinear activation function is dense in the space of continuous functions of the input signals [1].

The input x and output y can be represented as sets of numerically defined functions (i.e., tabular or grid functions). The fastest way to parallelize the input processing in ANNs is to partition x_i into subsets. However, this approach faces a critical issue: input features x_i often have uneven weights in classification or approximation tasks.

Shannon's information measure, defined as $H = -\sum_i p_i \log(p_i)$, cannot always be applied in ANNs because not all tabulated x_i values contribute equally. Splitting input features x_i by isolating those with the highest decision weights may inadvertently amplify the influence of less significant features, thereby degrading the approximation quality of the ANN modules.

To reduce input dimensionality and extract low-dimensional features, convolutional layers in ANNs are employed. These layers perform a convolution operation between the input x_i and a kernel w_k:

$$y_i = \sum_{k=0}^{l} x_i^k w_k. \tag{1}$$

The output will consist of p arrays $\{y\}$, representing feature sets of dimensionality n/d, where d is the kernel size of the layer, provided the convolution stride matches the kernel dimension. However, this approach does not fully solve the problem of parallelizing neural network input processing, since the reduced-dimensionality features are still processed jointly.

One promising approach for analyzing, identifying, and controlling non-stationary systems involves projecting the system onto a finite-dimensional subspace and solving the resulting approximation problem. Methods implementing this idea are known as projection-based approximation techniques.

An approximate solution to the original problem can be expressed as an expansion of xx over a set of basis functions $\{u_i(t)\}_{i=1}^{k}$ (the approximation basis):

$$x_k = \sum_{i=1}^{k} a_i u_i(t). \tag{2}$$

Basis functions in (2) may be defined analytically, algorithmically, or numerically. Among the first to propose basis decomposition for feature extraction were N. Ahmed and K. R. Rao [5], who emphasized optimizing input signal structure via orthogonal transforms rather than classifier architecture.

The discretization of continuous functions is governed by the Whittaker-Kotelnikov-Shannon theorem [6], which states that the sampling rate must exceed twice the signal's maximum frequency. This relates closely to decimation–reducing sample counts after low-pass filtering. The expansion coefficients a_i in (2) represent projections of the function onto orthogonal basis functions over $[0, T]$.

For closed orthogonal systems [7], approximation accuracy can be evaluated using Parseval's equality:

$$B = \sum_{i=0}^{\infty} x_I^2 = \sum_{n=0}^{\infty} a_n^2.$$

The relative integral approximation accuracy is then:

$$\gamma = \frac{\sum_{n=0}^{k} a_n^2}{B} = \frac{a_0^2}{B} + \frac{a_1^2}{B} + \cdots + \frac{a_n^2}{B} = \frac{B_1}{B} + \frac{B_2}{B} + \cdots + \frac{B_d}{B}. \tag{3}$$

The Fourier transform is a cornerstone of information theory. For discrete ANN inputs, the Discrete Fourier Transform (DFT) applies:

$$Ax(t) = F(\omega) = \sum_{i=0}^{n} x_i e^{-\frac{2\pi j \omega i}{n}}, \tag{4}$$

where $x \in L^1(R)$, $\omega = 2\pi f$ is angular frequency, and t is time. For square-integrable functions, the Fourier operator $A : L^2(R) \to L^2(R)$ is well-defined.

It is known from the properties of the Fourier transform that the higher the degree of smoothness of the function $x(t)$, the faster its Fourier transform $F(\omega)$ decreases. Consequently, the partial sums from series (4) will be decreasing. Then the integral accuracy of the approximation (3) will not decrease significantly when the last terms of the series are discarded. The Fourier transform is an effective tool for the study of functions, but has a significant disadvantage: having a high degree of frequency localisation, it has a very weak temporal (spatial) localisation.

If instead of Fourier-like transforms we use the wavelet transform, we can get a gain in the number of significant features without additional research. In general, wavelets are a system of functions of the following form

$$\varphi_{a,b}(x) = \sqrt{2^a}\varphi(2^a x - b), \tag{5}$$

If V_a is the space spanned by (5), nested subspaces [8] $V_0 \subset V_1 \subset \ldots \subset V_a$ emerge, each with an orthonormal basis $\{\varphi_{i,\,b}(x)\}$. This allows hierarchical approximation via orthogonal projection:

$$P_a : L^2(R) \to V_a, P_a(f) = \sum_{b \in Z} \left(f, \varphi_{a,b}\right)\varphi_{a,b}(x).$$

Projections P_i yield increasingly precise approximations. For ANNs, input vectors $\{f_i(x_j)\}$ in $L^2(R)$ can be projected onto $V_0 \subset V_1 \subset \ldots \subset V_a$, where each P_i refines the input representation. Training on P_0 provides a coarse but computationally efficient approximation.

The well-known Haar wavelet [5, 8, 9] allows us to divide the space $L^2(R)$ into 2 subspaces V_0 and V_1. Due to this, it is possible to solve the problem by parallelisation of the input signal and separate processing of subspaces V_0 and V_1 by a smaller ANN. From expression (5) it is clear that the wavelet spectrum has a two-dimensional nature, so, despite the reduction in the size of the ANN modules, training should be carried out for all levels of the wavelet spectrum. This raises the question of constructing parallel modular ANNs.

3 Implementation of Parallel High-Performance Neural Network

The practical implementation of modular parallel neural networks was carried out using the PyTorch library [11] on a personal computer running the Simply Linux operating system with an AMD Athlon X2 370K FM2 processor and 4 GB of RAM. The goal of the experiment was to compare the time and accuracy characteristics of the obtained models with a monolithic neural network.

The distributed package included in PyTorch [11] allows parallelizing computations across processes and machine clusters. It uses message passing semantics, enabling each process to communicate data to any other process. Unlike the multiprocessing package in Python, these processes use different communication backends and can run on different machines. Parallel computing support is only available under Linux-like operating systems. When using the “gloo” backend, the entire computing environment is divided into processes with the ability to set the required number of threads in each process. That is, each process represents a virtual computing environment where a specified segment of code is executed.

The following approach was used to divide the neural network into processes:

1. The first layer(s) perform wavelet transformation by embedding a filter from a filter bank obtained using the PyWavelets library [12] into the convolutional layer kernel. The experiment was conducted using the Haar wavelet with a 2×2 filter size. The wavelet transform is performed in a separate module.
2. The result, consisting of two or more sequences with dimensionality 2 (or more) times smaller than the original signal, is fed to 2 (or more) neural networks with smaller dimensionality than the monolithic one. Each neural network runs in its own process with 16 threads by default. Data transfer to the neural network inputs and their synchronization are handled by a data server running in a separate process.

3. The processing results from the modular neural networks are transmitted to a merging layer, which also runs in a separate process. During training, a version of the optimizer that supports training of distributed neural networks is used.

To compare the obtained results, a monolithic neural network was initially trained and applied in a separate process. The monolithic neural network represents a fully connected four-layer network trained on the MNIST dataset for 10 epochs. Figure 1 shows the graphs of relative error change and recognition accuracy.

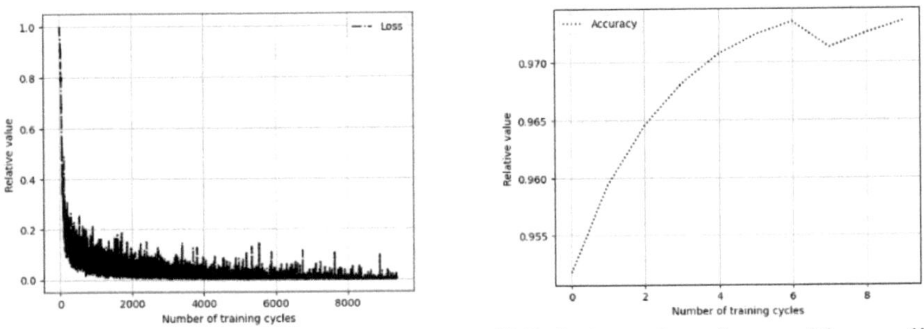

a) Relative value of the error b) Relative value of recognition quality

Fig. 1. Variation of relative error value and recognition quality of monolithic ANN in the process of training. (a) Relative value of the error, (b) Relative value of recognition quality

Table 1 presents the timing characteristics of a monolithic neural network trained and tested on the test portion of the dataset. For subsequent comparisons with the monolithic neural network, we will focus on the average execution time of one training and inference cycle, i.e., the time required to obtain test results.

Table 1. Timing characteristics of the monolithic neural network.

	Maximum	Minimum	Average
Training	0.5358	0.1276	0.2021
Inference	0.1586	0.0188	0.0645

Currently, the sequential (vertical) partitioning approach is widely used for constructing modular neural networks [10]. In this method, one or more layers are placed on a separate machine (computing node) and connected via communication channels. A significant drawback of this approach is the high communication overhead.

In the experiment, the neural network was divided into two modules, each containing two layers. Figure 2 illustrates the training process of the vertically partitioned neural network managed by the data server.

Comparing Figs. 1 and 2, it can be observed that after the same number of cycles, the training quality remained virtually unchanged. This result was expected since the

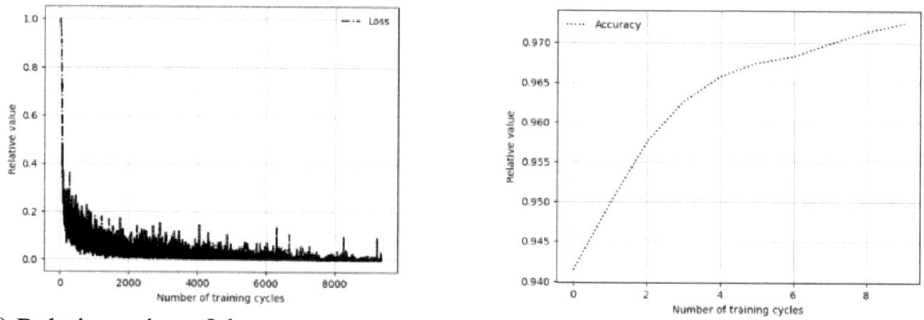

a) Relative value of the error

b) Relative value of recognition quality

Fig. 2. Change in relative error value and recognition accuracy of the vertically partitioned neural network during training. (a) Relative value of the error, (b) Relative value of recognition quality

architecture of the vertically partitioned neural network is completely identical to the monolithic neural network.

When comparing the timing characteristics of the monolithic neural network and the vertically partitioned neural network, we can conclude that the values in Table 2 are slightly higher due to data transfer between modules.

Table 2. Timing characteristics of the vertically partitioned neural network.

	Maximum	Minimum	Average
Training	0.9405	0.1247	0.2212
Inference	0.1237	0.0206	0.0657

When constructing a horizontally partitioned neural network, architectural differences must be considered: two additional layers have been added to the neural network. The first one is a convolutional layer with a fixed kernel that performs wavelet transformation, while the second (final) layer combines the results from the modules. Thus, the horizontally partitioned neural network consists of 4 modules connected by communication channels.

The training process of the horizontally partitioned modular neural network is shown in Fig. 3. From Fig. 3, we can see that during the selected training period, the horizontally partitioned modular neural network achieves practically the same recognition accuracy as both the monolithic and vertically partitioned modular neural networks. However, the timing characteristics presented in Table 3 show significant differences compared to the first two experiments.

Comparison of Tables 1 and 3 shows that despite having two additional modules and time costs for data transfer, the average training time per cycle for the horizontally partitioned modular neural network is slightly lower than for the monolithic one, while the average inference time is nearly 1.5 times shorter.

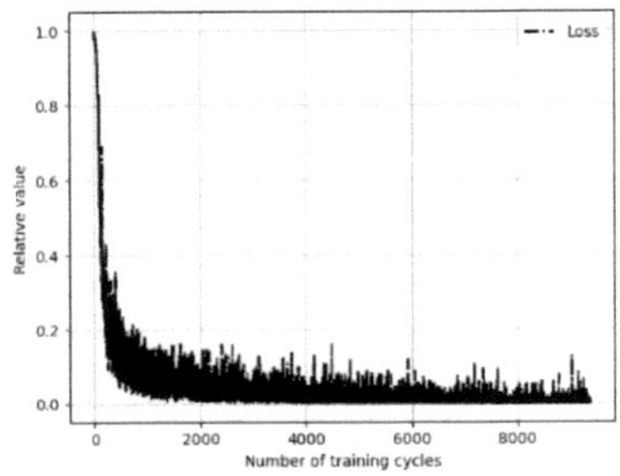

a) Relative value of the error

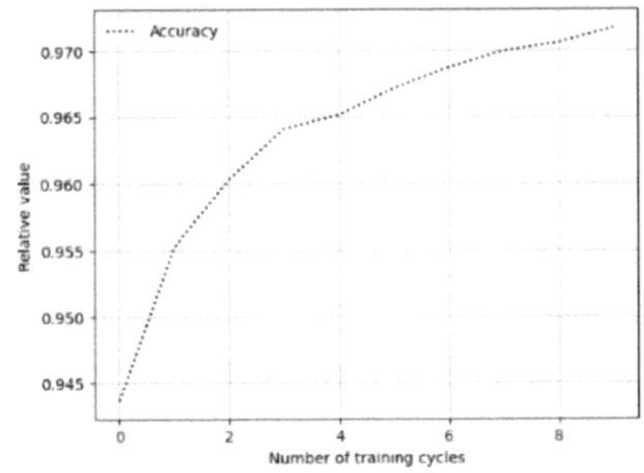

b) Relative value of recognition quality

Fig. 3. Change in relative error value and recognition accuracy of the horizontally partitioned neural network during training. (a) Relative value of the error, (b) Relative value of recognition quality.

Table 3. Timing characteristics of the horizontally partitioned neural network.

	Maximum	Minimum	Average
Training	1.4162	0.0649	0.1198
Inference	0.1269	0.0204	0.0403

The final experiment with training the neural network at the second level of the wavelet spectrum was designed to evaluate time costs when increasing the number of modules to four. In this setup, the input sequence passes through the first layer three times: first being split into 2 modules, each of which is then split again into 2 modules. As a result, the first module outputs 4 input sequences that are 4 times shorter than the original. The accuracy characteristics of the 4-module neural network are shown in Fig. 4.

Figure 4b demonstrates that the training quality at the second level is somewhat worse than that of the monolithic neural network, though only marginally - by less than 1%. The experimental results showed that increasing the number of training epochs doesn't improve quality - the 97% threshold wasn't surpassed even when increasing epochs to 20. This is quite explainable: depending on the quality and features of the original image, recognition quality decreases as module sizes reduce. However, by training modular neural networks at different levels of the wavelet spectrum, one can select the level that provides the required quality given known computational node capabilities. Moreover, if a significant number of machines are available, training can proceed in parallel across multiple wavelet spectrum levels.

The timing characteristics of the 4-module neural network are presented in Table 4.

From Table 4, it can be seen that the average time spent on one training and inference cycle of the 4-module neural network has become even shorter. This means that when training multiple neural networks in parallel at different wavelet spectrum levels, the total training time will not exceed the training time at the first level.

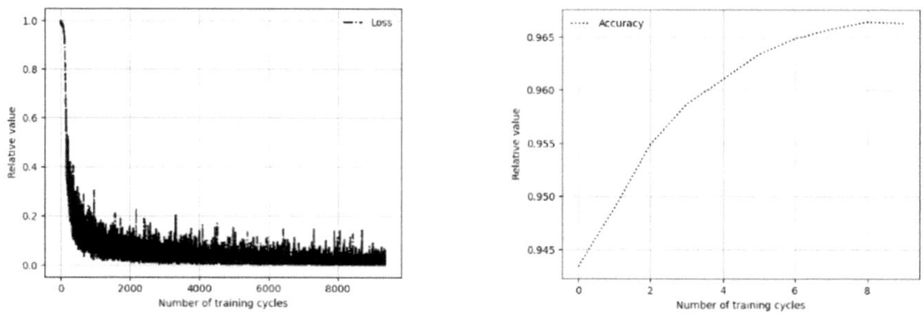

a) Relative value of the error b) Relative value of recognition quality

Fig. 4. Change in relative error value and recognition accuracy of the 4-module horizontally partitioned neural network during training. (a) Relative value of the error, (b) Relative value of recognition quality

Table 4. Timing characteristics of the 4-module horizontally partitioned neural network.

	Maximum	Minimum	Average
Training	1.6123	0.051	0.0754
Inference	0.1021	0.0271	0.0393

4 Conclusion

The conducted experiments confirmed the correctness of the theoretical conclusions that using wavelet transformation to divide the neural network into several lower-bit modules allows solving, at minimum, the following serious tasks:

1. In the absence of a powerful centralized computing node, the task of training and applying the neural network can be solved by a group of lower-power computing nodes. Such a task arises when managing a group of objects using neural networks, for example, unmanned aerial vehicles, cars, and other similar objects. Creating distributed neural networks will allow utilizing unused resources of object controllers, obtaining neural networks with high image recognition quality in the absence of connection to high-power computing resources. This approach will enable, when using the Internet of Things, the placement of neural networks not in cloud data centers, but on fog and edge computing nodes, which, in turn, increases the confidentiality of user information.
2. Very large volumes of information requiring processing by neural networks, accumulated in a number of sciences: astronomy, chemistry, biology, etc., require very significant amounts of computing resources, high processing costs, and long queues for access to cloud data centers. Moreover, a large amount of information has a distributed structure and requires lengthy preparation and repeated transmission over high-speed communication channels. The proposed approach, through distributed

processing, can significantly reduce information preparation time, as well as lower the requirements for data center computing resources and reduce processing costs.

Thus, using projection methods, the task of creating distributed high-performance artificial neural networks that do not require high-power centralized computing resources can be solved.

Funding. The research was supported by the Russian Science Foundation Grant No. 24-21-00149, https://rscf.ru/en/project/24-21-00149/

References

1. Gorban, A.N.: Generalised approximation theorem and computational capabilities of neural networks. Siberian J. Comput. Math. **1**, 11–24 (1998)
2. Galushkin, A.I., Sudarikov, V.A., Shabanov, E.V.: Neuromathematics: methods for solving problems on neurocomputers. Math. Model. **3**(8), 93–111 (1991)
3. Boreskov, A.V., et al.: Parallel computing on GPU. CUDA architecture and software model. (2015) (In Russian)
4. Lecun, J.: How the Machine Learns: The Revolution in Neural Networks and Deep Learning. Alpina Publisher (2023)
5. Ahmed, N., Rao, K.R.: Orthogonal Transforms for Digital Signal Processing. Svjaz' publ, Moscow (1980)
6. Jerry, A.J.: Shannon's reference theory, its various applications and generalisations. Rev. Tiier. **65**(11), 53–89 (1977)
7. Dedus, F.F., Kulikova, L.I., Pankratov, A.N., Tetuev, R.K.: Classical orthogonal bases in problems of analytical description and processing of information signals. FCMIC OF MSU. (2004)
8. Smolencev, N.K.: Fundamentals of Wavelet Theory Wavelets in MATLAB. DMK Press (2019)
9. Haar, A.: Zur theorie der orthogonalen funktionensysteme. Georg-AugustUniversitat, Gottingen (1909)
10. Ushakov, Y.A., Polezhaev, P.N., Shukhman, A.E., Ushakova, M.V.: Vertical distribution of neural network between mobile device and cloud infrastructure services. Mod. Inf. Technol. IT Educ. **14**(4), 903–910 (2018)
11. PyTorch. Source: https://pytorch.org/
12. PyWavelets. Source: https://pypi.org/project/PyWavelets/
13. Qiao Y. The Mnist Database of handwritten digits. (2007). Source: http://www.gavo.t.utokyo.ac.jp/qiao/database.html

Modern Methods for Anomaly Detection in Enterprise System Logs: Algorithms, Implementations, and Practical Case Studies

Sergey Yarushev[1] and Aleksandr Anurov[2(✉)]

[1] Federal Research Center "Computer Science and Control" of the Russian Academy of Sciences, Moscow, Russia
[2] Plekhanov Russian University of Economics, Moscow, Russia
Anurov.AO@rea.ru

Abstract. System logs are indispensable for maintaining the reliability and security of modern IT infrastructures, yet they generate vast volumes of data in which anomalies—ranging from performance degradations to security breaches—can be easily overlooked. This review article offers a focused survey of practical statistical methods for log anomaly detection, examining both univariate and multivariate approaches. We begin with classic thresholding techniques such as Z-score and interquartile range (IQR), illustrating their implementation in tools like Prometheus, Zabbix, Splunk, and ELK Stack. To address non-stationarity and seasonal trends, we discuss adaptive strategies based on sliding windows and exponentially weighted moving averages (EWMA), as found in Grafana and Datadog. We then explore multivariate analysis via Mahalanobis distance, highlighting its use in SIEM platforms (IBM QRadar, LogRhythm) and AIOps solutions (Moogsoft). Each method is complemented by real-world use cases and historical context, tracing origins from Shewhart control charts and Tukey's boxplot to contemporary DevOps practices. By emphasizing transparency, ease of integration, and limitations, this article equips practitioners with a robust first-line toolkit for automated anomaly monitoring. The surveyed statistical techniques serve as the foundation for more advanced machine-learning and hybrid detection systems, which will be addressed in subsequent sections.

Keywords: Anomaly detection · System logs · Statistical methods · Z-score · Interquartile range · EWMA · Mahalanobis distance · Monitoring tools · SIEM · AIOps

1 Introduction

System logs are rich repositories of events and status messages generated by operating systems, applications, and network devices. Detecting anomalies in these logs—such as unusual patterns, errors, or malicious activity—is critical for maintaining security and reliability. An anomaly (or outlier) in log data can indicate either a software defect or an attempted security breach. Unlike supervised tasks, anomalies are typically rare

M. Lapina et al. (Eds.): AISMA 2025, LNNS 1456, pp. 451–461, 2026.
https://doi.org/10.1007/978-3-032-07275-7_41

and unlabeled in advance, so detection often relies on unsupervised algorithms that model "normal" log behavior and flag deviations. This paper provides a comprehensive survey of anomaly-detection methods for system logs: statistical approaches, clustering algorithms, tree-based models, neural-network techniques, and hybrid methods. We also cover popular Python libraries (e.g., Scikit-Learn, PyOD, TensorFlow/Keras) and discuss commonly used benchmark datasets (CICIDS2017, UNSW-NB15, DARPA'99, etc.).

2 Practical Statistical Methods for Log Anomaly Detection

2.1 Threshold-Based Detection Using Z-Score

Using historical data for a metric (for example, service response time or CPU utilization), compute the mean (μ) and standard deviation (σ) [1]. Any event that falls outside $\pm 3\sigma$ is flagged as anomalous (see Fig. 1).

- **Prometheus** implements this via Alertmanager.
- **Zabbix** lets you set data-item thresholds based on statistical parameters.

In one large online retailer, Z-Score alerts were the first to reveal a "progressive decline" in API throughput following an evening deployment. The services automatically opened an incident before users even noticed any degradation.

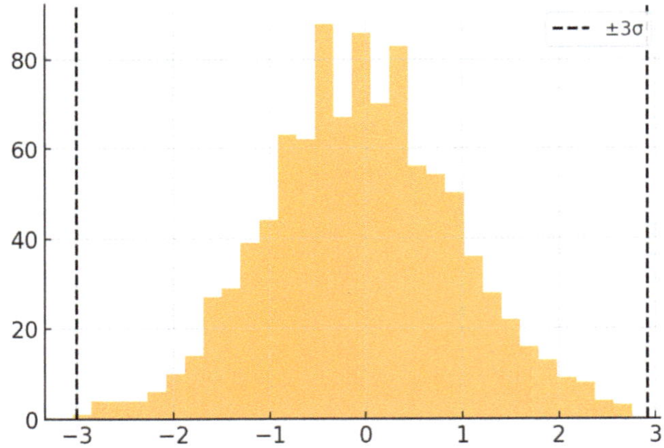

Fig. 1. Histogram of a metric with $\pm 3\sigma$ (Z-Score) thresholds.

2.2 Interquartile Range (IQR) and Boxplot

Divide the metric into quartiles; values lying outside the IQR boundaries are treated as outliers [1, 8].

- **Splunk's** built-in boxplot visualization makes it easy to spot anomalous spikes in error rates.

- In **Elasticsearch Watcher**, quartile calculations can be scripted using Painless.
- During a microservice version upgrade, an IQR-based check instantly pinpointed a single HTTP server exhibiting unusually high peak response times—averting a widespread outage.

2.3 Sliding Window and EWMA

Sliding Window
Recompute statistics (μ, σ) over the most recent N data points, allowing quick adaptation to changing trends [2].

EWMA (Exponentially Weighted Moving Average)
Applies more weight to recent observations while smoothing out small fluctuations. Where it's used:

- **Grafana** offers moving-average functions and alerting based on sliding-window statistics.
- **Datadog** incorporates EWMA into its built-in anomaly-detection algorithms.

When monitoring Kubernetes clusters, EWMA alerts have predicted sustained memory growth several hours before an OOM event—allowing time to rebalance workloads (Fig. 2).

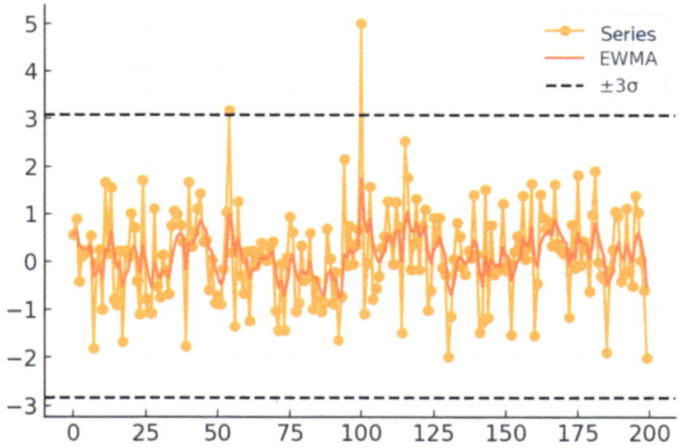

Fig. 2. EWMA chart with adaptive ±3σ thresholds.

2.4 Multivariate Analysis: Mahalanobis Distance

For a feature vector (e.g., CPU, memory, I/O), compute the Mahalanobis distance to the center of the "normal" distribution and compare it against a percentile threshold (typically 97–99%). See Fig. 3.

SIEM platforms like **IBM QRadar** and **LogRhythm** use multivariate statistics to detect sophisticated attacks [3, 8].

AIOps solutions (e.g., **Moogsoft**) combine multiple metrics into feature vectors and trim out anomalies using Mahalanobis distance.

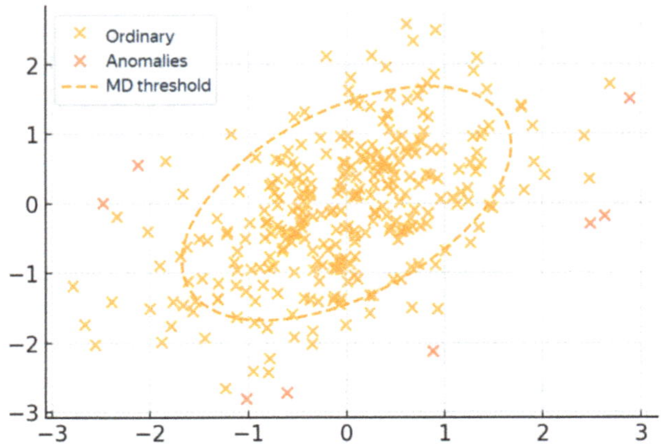

Fig. 3. Mahalanobis threshold contour on two-dimensional data.

2.5 Historical Roots of Statistical Anomaly Detection

Statistical techniques for anomaly detection predate the IT-monitoring era, originating in engineering and quality control:

- **1924**: Walter Shewhart developed control charts, laying the groundwork for Statistical Process Control—an idea that survives today in Z-Score and EWMA checks in monitoring tools.
- **1936**: Prasanta Chandra Mahalanobis introduced his eponymous distance measure, accounting for feature correlations—a foundation for modern multivariate analysis in SIEM and AIOps.
- **1959**: Robert R. Roberts formalized EWMA control for sensitive detection of slow process drifts.
- **1977**: John Tukey popularized the boxplot and IQR, providing a rapid visualization tool for distributions and outliers—now ubiquitous in analytics platforms.

The shift from physical manufacturing to digital services has only broadened the application of these ideas: what once assured product quality on the assembly line now helps "produce" reliable IT services and ensure fault tolerance in distributed systems. The statistical advances of the 1920s–1970s still form the backbone of today's "first line" of anomaly detection in logs, remaining indispensable for DevOps and SRE teams alike.

3 Anomaly Detection Based on Clustering

Clustering algorithms group similar log entries or feature vectors, and anything that doesn't fit well into any cluster can be treated as anomalous. One popular method is DBSCAN (Density-Based Spatial Clustering of Applications with Noise). DBSCAN builds clusters around dense regions of points and labels points in sparse regions as "noise" (outliers). Unlike k-means, DBSCAN doesn't require you to specify the number of clusters in advance and automatically flags outliers as noise. This is especially handy when normal log events naturally form tight groupings (e.g., common message patterns) in some feature space, and anomalous entries "stick out" from these clusters (see Fig. 4).

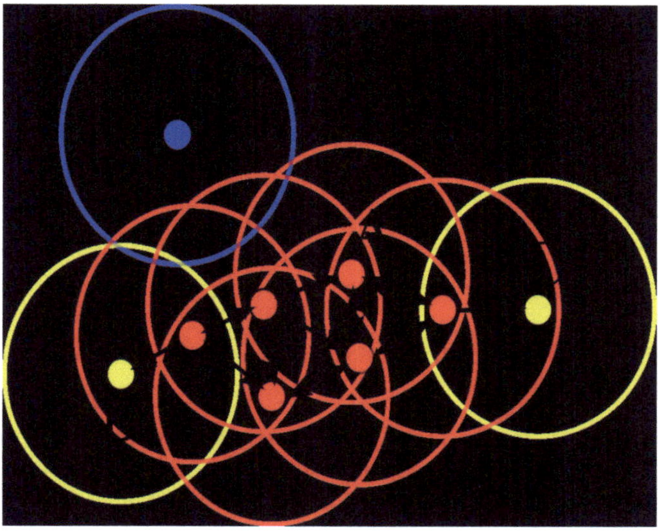

Fig. 4. Illustration of DBSCAN in action.

In a logging context, you'd first transform log messages into feature vectors (for example, via hashing or embeddings) and then run DBSCAN to group similar events. Any record that doesn't fall into a dense cluster of known events gets marked as an anomaly [1, 6].

Other clustering approaches include k-means—where after clustering, points with large distances from their nearest centroid can be treated as anomalies—and hierarchical clustering, where very small leaf clusters may represent rare events. However, k-means is less ideal for anomaly detection since it forces every point into some cluster and lacks the notion of "noise." DBSCAN, by explicitly labeling noise points, is generally better suited for uncovering outliers.

4 Tree-Based Methods and Isolation Algorithms

Tree-based anomaly detection methods primarily use ensembles of decision trees to "isolate" outliers. A prototypical example is the Isolation Forest—a model specifically designed for anomaly detection. Isolation Forest randomly partitions the feature space using tree splits; anomalies, being rare and distinct from the bulk of the data, tend to be "isolated" (that is, end up in singleton leaves) after fewer random splits compared to normal points. The model comprises an ensemble of such random trees. An anomaly score for each point is calculated as the average path length through the trees required to isolate it: the shorter the path, the quicker the point was separated from the rest, indicating an anomaly. This unsupervised method runs in linear time and scales well to high-dimensional data, making it well suited for large volumes of logs.

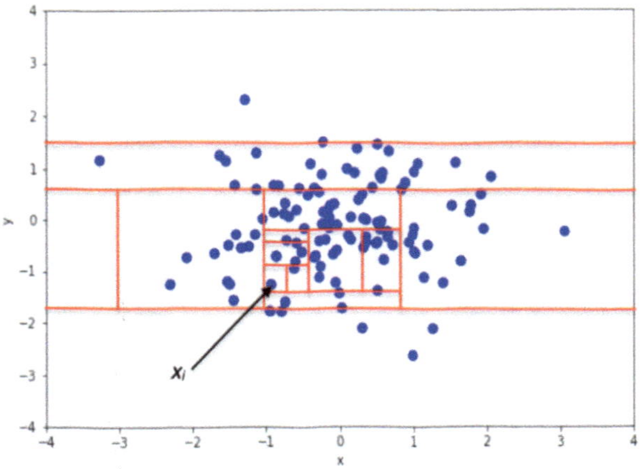

Fig. 5. Illustration of the Isolation Forest concept in two dimensions.

As shown in Fig. 5, the red rectangles represent random splits that isolate point i among the blue points. Here, i is a normal point (not an anomaly), requiring many splits to isolate. In contrast, an anomalous point (located far from the others) is isolated with far fewer splits. Isolation Forest leverages this principle: anomalies yield shorter isolation paths than normal instances.

Unlike density-based methods, decision-tree approaches hinge on "random isolation," which makes them effective and robust even with complex data distributions.

Another tree-based strategy involves training ensembles of decision trees (for example, Random Forest) in a supervised setting when labeled examples of "anomalous" and "normal" records are available. However, such labels are often scarce. In the unsupervised scenario, beyond the basic Isolation Forest, there are extended variants (e.g., Extended Isolation Forest) and other ensemble algorithms. Historically, single decision trees were also used for outlier detection—building a tree that maximally grouped normal objects and treating early splits as indicators of anomalies—but the randomized ensemble of Isolation Forest has largely superseded this approach.

5 Neural-Network Methods (Deep Learning Approaches)

With the advent of deep learning, neural-network methods have taken center stage in log-based anomaly detection. They're able to capture complex patterns and dependencies that simpler algorithms often miss. A common neural approach is the autoencoder—a network trained to compress data into a lower-dimensional representation and then reconstruct it. The idea is that an autoencoder learns to accurately recreate "normal" log data, but when you feed it an anomalous example, the reconstruction error spikes (because that pattern wasn't part of its training). By setting a threshold on reconstruction error, you can flag anomalies.

The encoder–decoder network maps an input feature vector from a log entry into a compact latent space, then tries to reconstruct the original (Fig. 6). Having learned to reproduce the most frequent patterns, the autoencoder "fails" on anomalies. Variants

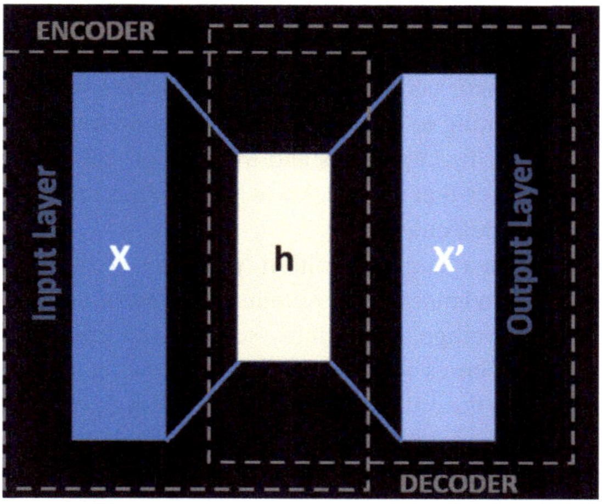

Fig. 6. Autoencoder architecture for anomaly detection

include denoising autoencoders (to strip out noise and learn more robust features) and sequential or LSTM autoencoders when you treat the problem as sequence modeling (e.g., spotting abnormal chains of messages rather than isolated entries) [4, 9].

Other neural approaches for log anomaly detection include:

- Recurrent Neural Networks (RNNs), especially LSTM-based models, trained on event sequences to detect deviations in temporal patterns (widely used for time-series analysis and system traces).
- Generative Adversarial Networks (GANs), where a generator crafts "normal" log entries and a discriminator learns to tell real from fake; anomalies show up when the discriminator deems an entry inconsistent with the learned norm.
- Graph Neural Networks, suitable for structured logs (when logs form event- or call-graphs), detecting vertices or subgraphs that don't fit the expected structure.

Deep-learning methods typically require large volumes of training data (mostly normal behavior) and substantial compute resources. Deploying them can be more involved, but they excel on high-dimensional log data (for example, embedding entire log lines or parameter vectors). One must also consider explainability—unlike simple rules or even Isolation Forest (which can surface feature importances), autoencoders and LSTMs remain "black boxes." Nonetheless, their power to capture subtle structures makes them cutting-edge for tasks like spotting anomalies in system-call traces or unusual message combinations that precede failures [11].

6 Hybrid Approaches

No single anomaly-detection technique works perfectly across all scenarios, so hybrid methods strive to combine the strengths of multiple approaches. In the context of log anomaly detection, a "hybrid" can mean:

- Combining Detection Methods
  A system might first apply simple rule-based or threshold checks for known issues (for example, specific error codes or metrics that exceed set limits), then augment those with machine learning to catch new, previously unseen anomalies. This blend of "expert" detectors for fixed patterns and ML models for unknown cases reduces false positives and ensures that critical, known anomalies aren't missed.
- Ensemble of Anomaly Detectors
  Instead of relying on a single algorithm, you run several (statistical, clustering-based, tree-based, etc.) and merge their outputs. The ensemble can vote on whether an entry is anomalous or average the anomaly scores from each method. For instance, you might average the scores from an Isolation Forest and a deep autoencoder, or train a meta-classifier on the outputs of all detectors. Ensembles generally yield more stable results across diverse data, though they add complexity.
- Feature and Model Hybridization
  Here, the output of one method feeds into another. A classic example is first parsing logs and extracting message templates—turning unstructured logs into structured events—and then feeding those into an anomaly-detection model. Another variant is clustering logs to identify common patterns, then training a supervised classifier on those clusters to prioritize suspicious groups. This two-stage pipeline first reduces noise and dimensionality, then makes a more precise anomaly decision.

Pros of Hybrid Methods

- Higher accuracy and reliability by leveraging multiple techniques.
- Combining expert rules with ML ensures known errors are caught while still detecting novel issues.

Cons

- Increased development and maintenance complexity: multiple components must be configured and kept in sync.
- Potential conflicts between detectors (for example, when a threshold rule and an ML model flag different events).

In practice, many production log-anomaly pipelines are hybrid: simple checks catch the obvious problems, and more advanced models scrutinize the remainder to uncover subtle or unexpected patterns.

7 Datasets for Evaluating Anomaly-Detection Methods

To assess anomaly-detection techniques on logs or network data, researchers frequently rely on public benchmark datasets. Each of the following contains labeled "normal" and "anomalous" instances, making them ideal for head-to-head algorithm comparisons:

- **CICIDS2017 (Intrusion Detection Evaluation Dataset)**
  Created by the Canadian Institute for Cybersecurity, CICIDS2017 includes both benign network traffic and a variety of real-world 2017 attack scenarios. It provides raw packet captures (PCAP) plus extracted flow features in CSV format, with each record labeled as "normal" or with its specific attack type. Designed to reflect modern traffic mixes and attack techniques (covering DoS/DDoS, brute-force logins, web attacks, intrusions, botnets, etc.), it addresses many shortcomings of earlier datasets. With millions of records and over 80 features, it's become the de facto standard for testing network anomaly-detection systems and IDS.
- **UNSW-NB15**
  Published in 2015 by the Australian Centre for Cyber Security, this dataset contains about 2.54 million connection records, each described by 49 features and labeled as "normal" or one of nine attack categories (fuzzers, reconnaissance, backdoors, DoS, exploits, generic, worms, and others). It blends real traffic with synthetic attack injections, and comes pre-split into training and test sets. Compared to the older KDD Cup '99, UNSW-NB15 captures up-to-date attack vectors and a more realistic baseline of normal behavior. It's widely used for evaluating both anomaly-based and signature-based detection methods.
- **DARPA 1998/1999 & KDD Cup 1999**
  Among the earliest benchmarks for IDS and anomaly detection. DARPA '99 simulated the U.S. Air Force network under multiple attack scenarios, producing weeks of traffic and audit logs for training/testing. The KDD Cup '99 dataset is a refined extract of those logs: roughly 5 million records (with a 10% subset of ~494 000), each with 41 numeric features and labeled "normal" or one of 37 attack types (DOS, probe, R2L, U2R categories). Despite well-known issues (duplicate records, outdated attacks), KDD '99 remains a standard comparison; its successor NSL-KDD addresses many of its flaws.

7.1 Additional Log-Centric Datasets

- HDFS Log Anomaly Dataset – Hadoop filesystem logs annotated by success vs. failure of operations.
- BGL (Blue Gene/L) Log Dataset – Supercomputer logs tagged with error events.
- HTTP Attack Injection Logs (e.g., NASA HTTP Dataset) – for benchmarking web-IDS.
- UNM & ADFA-LD – Linux system-call traces labeled for anomaly detection.

When benchmarking, it's crucial to use the same dataset and metrics (ROC curves, F1-score, etc.). Most of these collections come with predefined train/test splits and evaluation guidelines. For instance, CICIDS2017 and UNSW-NB15 are commonly used to report detection rates versus false-alarm rates. Researchers should account for each dataset's quirks—KDD '99 is almost entirely categorical features encoded as numbers, whereas CICIDS2017 mixes continuous and discrete attributes—and apply appropriate preprocessing (normalization, encoding) before clustering or distance-based methods. Despite their imperfections, these public datasets are indispensable for validating new anomaly-detection techniques and comparing them against established solutions.

8 Conclusion and Future Directions

Anomaly detection in system logs is a cornerstone capability for modern IT operations, security, and reliability. In this survey, we've covered a broad spectrum of techniques: from simple statistical threshold checks that catch gross outliers to sophisticated neural-network models that learn the system's normal behavior. Each category—statistical, clustering-based, tree-based, neural, and hybrid—has its own strengths. Simple methods execute quickly, are transparent, and deploy easily, but they may miss subtle anomalies or struggle to adapt to evolving conditions. Advanced approaches detect complex deviations but often demand large datasets, careful tuning, and tend to be "black boxes."

In practice, a layered strategy is most common: start with straightforward checks for known issues, then hand off to more complex models to catch the "unknown unknowns." The choice of method depends on your context—for instance, tree-based and clustering algorithms working on numeric flow features may excel at detecting network attacks, whereas sequence-aware neural models are better suited for spotting anomalies in application log message sequences.

Key avenues for further research and real-world development include:

- Online and Streaming Anomaly Detection. Logs arrive continuously, so it's crucial to use methods that update incrementally (online learning) rather than retraining from scratch.
- Deep Learning and AI Innovations. The rise of transformers and large language models (LLMs) opens new possibilities: researchers are already exploring LLM-based log parsing and semantic-driven anomaly detection.
- Integration with Monitoring and Alerting Systems. Production deployments need intuitive dashboards and "smart" alerts: prioritizing anomalies, batching repeated signals, and reducing alert fatigue.
- Explainability and Forensics. In security contexts, clear explanations are essential: analysts must understand why a record was flagged. Techniques like SHAP values for tree models or attention-visualization in neural networks can link anomalies to event sequences for incident investigation.
- Standardized Benchmarks and Reproducibility. As interest grows, the community is building standardized datasets and frameworks for comparable experiments (for example, ADBench in the PyOD ecosystem).

Detecting anomalies in system logs remains a challenging yet fascinating problem at the intersection of systems engineering, data science, and AI. The methods outlined here form a practical toolkit—from a simple outlier-detection script to multi-stage pipelines powering large-scale systems. By understanding each approach's assumptions and trade-offs, engineers and researchers can pick the optimal solution for their needs. Continued algorithmic advances combined with careful engineering will yield more reliable, "smarter" monitoring systems—ones that learn normal behavior in complex environments and sound the alarm only when genuinely unusual events occur. While future automation and AI will only enhance log analysis, the fundamental principles—define "normal," spot the "difference," and do so accurately, efficiently, and pragmatically—will remain timeless.

References

1. Chandola, V., Banerjee, A., Kumar, V.: Anomaly detection: a survey. ACM Comput. Surv. 41(3), 15:1–15:58 (2009)
2. Patcha, A., Park, J.-M.: An overview of anomaly detection techniques: existing solutions and latest technological trends. Comput. Netw. **51**(12), 3448–3470 (2007)
3. Ahmed, M., Mahmood, A.N., Hu, J.: A survey of network anomaly detection techniques. J. Netw. Comput. Appl. **60**(1), 19–31 (2016)
4. Du, M., Li, F., Zheng, G., Srikumar, V.: DeepLog: anomaly detection and diagnosis from system logs through deep learning. In: Proceedings of the 24th ACM Conference on Computer and Communications Security, pp. 1285–1298. ACM, New York (2017)
5. Liu, F.T., Ting, K.M., Zhou, Z.-H.: Isolation forest. In: Giannotti, F., et al. (eds.) Proceedings of the Eighth IEEE International Conference on Data Mining, pp. 413–422. IEEE, Pisa (2008)
6. Breunig, M.M., Kriegel, H.-P., Ng, R.T., Sander, J.: LOF: identifying density-based local outliers. In: Proceedings of the 2000 ACM SIGMOD International Conference on Management of Data, pp. 93–104. ACM, Dallas (2000)
7. Xu, W., Huang, L., Fox, A., Patterson, D., Jordan, M.I.: Detecting large-scale system problems by mining console logs. In: Proceedings of the 22nd ACM Symposium on Operating Systems Principles, vol. 43, pp. 117–132. ACM, Stevenson, WA (2009)
8. Aggarwal, C.C.: Outlier Analysis, 2nd edn. Springer, Cham (2017)
9. Chalapathy, R., Chawla, S.: Deep learning for anomaly detection: a survey. arXiv:1901.03407 (2019)
10. Li, Z., Shi, J., van Leeuwen, M.: Graph neural networks based log anomaly detection and explanation. arXiv:2307.00527 (2023)
11. He, S., et al.: A survey on automated log analysis for reliability engineering. arXiv:2009.07237 (2020)
12. Yang, Z., Harris, I.G.: LogLLaMA: transformer-based log anomaly detection with LLaMA. arXiv:2503.14849 (2025)
13. Lee, Y., Kim, J., Kang, P.: LAnoBERT: system log anomaly detection based on BERT masked language model. arXiv:2111.09564 (2021)
14. Guan, W., Cao, J., Qian, S., Gao, J., Ouyang, C.: LogLLM: log-based anomaly detection using large language models. arXiv:2411.08561 (2024)
15. LNCS Homepage, http://www.springer.com/lncs. Accessed 18 May 2025

Algebraic Generalization of Wilson's Theorem and Its Applications in Cryptographic and Intelligent Systems

Vladimir N. Yevstifeyev[1], Yelizaveta S. Vitulyova[2,3](✉), Dina B. Shaltykova[2], and Ibragim E. Suleimenov[2]

[1] Institute "KazNIPIEnergoprom" JSC, 050004 Almaty, Kazakhstan
[2] Satbayev University, 050013 Almaty, Kazakhstan
lizavita@list.ru
[3] "Institute of Digital Engineering and Technology" JSC, Almaty, Kazakhstan

Abstract. The generalized Wilson theorem we present, based on an algebraic delta function that enables the explicit formulation of operations over finite fields. Matters to be discussed include sums of multinomial coefficients and particularly how classical Wilson's theorem arose as a particular instance of the result de-rived. The authors established that there exist applications of our approach methods in practically every domain of information security, from secure classification implementation within artificial intelligence systems up to error correcting code design. This mathematical model will enhance robustness in AI systems toward input perturbation when there is limited trust regarding the execution environment. The work is interdisciplinary in nature, combining tools from number theory, coding theory, and algebraic logic, thus offering new perspectives for applications in cryptography, digital testing, and the development of secure computational architectures.

Keywords: Multinomial Coefficients · Number Theory · Galois Fields · Wilson's Theorem · Error-Correcting Codes · Interdisciplinary Collaboration

1 Introduction

The results obtained in classical number theory are increasingly used in modern information technology. Important examples in this regard are the use of prime numbers in cryptography [1–3]. In particular, Mersenne primes are used to generate public keys or protect against factorization attacks [4, 5].

One of the important results obtained in classical number theory and of interest to modern information technology are analogues [6, 7] and generalizations [8] of Wilson's theorem [9]. This theorem states that if p is a prime number, then the expression (p-1)! + 1 is divisible by p. This theorem can also be proven by means of Galois field theory [6].

The factorial is one of the main functions used in combinatorics. Along with this function, binomial and multinomial coefficients [10, 11] are widely used in combinatorics and its applications. Accordingly, it is permissible to consider the possibilities of

M. Lapina et al. (Eds.): AISMA 2025, LNNS 1456, pp. 462–470, 2026.
https://doi.org/10.1007/978-3-032-07275-7_42

proving analogues of Wilson's theorem as applied to such coefficients. One of the variants of proof of the analogue of Wilson's theorem for binomial coefficients was given in Ref. [12], it is also based on the use of the connection between abstract algebra and number theory.

The purpose of this work is to prove the possibility of studying the properties of multinomial coefficients using the algebraic delta function defined in Ref. [13, 14]. This approach is also a special case of the application of abstract algebra to obtain results in the field of number theory. Further development of such methods is of interest, including for the improvement of processor technologies used in UAV systems, which contributes to reducing their cost [15, 16], a matter that is becoming increasingly relevant due to the growing use of UAVs for various purposes [17, 18]. The developed models can be adapted for systems implementing stable transmission of permutations over noisy channels, as proposed in [19].

As an illustration, the study also discusses certain prospects for applying the obtained results to improve psychological testing methodologies (and, potentially, sociological testing in the future). This issue is particularly relevant with regard to the so-called projective techniques, the legitimacy of which remains a subject of ongoing debate in the literature [20, 21]. The relevance of this matter is further underscored by the increasing integration of modern digital technologies into psychological research, which is becoming ever more widespread [22, 23]. This approach offers additional opportunities for application in the field of cyber-security, particularly in the development of secure communication protocols within distributed systems based on artificial intelligence. For example, the evolution of mobile communication systems towards 5G networks highlights stringent requirements for reliability, security, and efficient data processing, which can benefit from algebraic approaches such as those presented here [24].

2 Background

The paper uses a method of reducing multi-valued logic operations, specified in the form of truth tables, to explicit algebraic expressions [13, 14]. This method is based on the use of an algebraic delta function, which is defined as follows.

Any element of the Galois field $GF(p)$ satisfies the equality

$$x^{p-1} = 1. \tag{1}$$

Therefore, the function

$$\delta_i(x) = 1 - (x - x_i)^{p-1}, \tag{2}$$

where x_i is a fixed element of the field $GF(p)$, have the following property [12].

$$\delta_i(x) = \begin{cases} 1, x = x_i \\ 0, x \neq x_i \end{cases}. \tag{3}$$

Formula (2) serves as the foundation for proving the main theorem of the present study.

3 Results

3.1 Algebraic Function for Identifying a Given Number of Errors

For clarity, we will proceed from the application of Formula (2) to error-correcting coding methods, which are currently in widespread use [23, 25]. The applicability of Formula (2) to such methods is justified by the following considerations [26, 27].

In accordance with the noise-resistant coding technique, in the original digital sequence

$$A = (0, 1, 1, 0, 0, \ldots, 0, 1, 1, 0) \quad (4)$$

additionally entered characters associated by a certain rule with the characters contained in the original sequence.

$$A \to A^{+} = \left(0, 1, 1, 0, 0, \ldots, 0, 1, 1, 0, a_1^{+}, .., a_n^{+}\right) \quad (5)$$

This redundant information allows to reconstruct the source code in the same way as redundant information contained in messages in natural languages allows to reconstruct their meaning from the context.

The noise-resistant coding technique also allows the following interpretation.

The set of all sequences of the form is decomposed into subsets (for simplicity, the case of extremal codes is considered [28]), which can be denoted as $\{A^{+} + \delta A\}$. This notation assumes that each subset contains the sequence A^{+} and some of its neighborhoods. Error correction can therefore be considered as a homomorphism

$$\left\{A^{+} + \delta A\right\} \to A^{+} \quad (6)$$

which associates the subset $\{A^{+} + \delta A\}$ with its element A^{+}.

It should be noted that the morphism defined in Eq. (6) is applicable not only to error correction in code sequences. Equally important is its potential use in various classification tasks, which arise, for example, in the context of psychological testing. In such cases, a set of responses to test items can be represented as a code sequence (in the case of "Yes" or "No" answers, this sequence is binary, of course). One can also define a set of reference sequences, each corresponding to a specific psychological type to be diagnosed. The comparison between an individual respondent's answer sequence and the reference sequences (i.e., the solution of the classification problem) is then performed according to the minimum code distance criterion.

The morphism expressed by Eq. (6) can be represented in an explicit algebraic form using Formula (2).

Such a formulation of the classification problem, expressed through algebraic structures over finite fields, can be regarded as a formalization of the machine learning mechanism within the context of secure computation. This is particularly relevant for the development of robust AI systems that require a trusted correspondence be-tween input and reference data under untrusted environments or in the presence of data tampering threats.

Based on it, it is permissible to define the identification function of a given number of errors K as

$$Q_K = 1 - \left(\sum\nolimits_1^{p-1} q_i - K\right)^{p-1} \tag{7}$$

where

$$q_i = (x_i - y_i)^{p-1};\ i = 1, 2, \ldots, N \tag{8}$$

The variables x_i correspond to the resulting code sequence, while the variables y_i correspond to the reference sequence used for comparison. The purpose of function (7) is to identify the situation in which the code distance (Hamming distance) between these two sequences is exactly equal to K.

It is assumed that the calculations in formula (7) are performed in the Galois field $GF(p)$. The validity of formula (7) follows from the following considerations. By virtue of (1), the value (8) will be strictly equal to one for any deviation of the value x_i from y_i. Consequently, the sum of such values $\sum_1^{p-1} q_i$ is exactly equal to the number of "errors", i.e. the number of values x_i deviating from y_i. The difference on the right side of formula (7) will be equal to one in the only case when the number of "errors" is equal to K. In all other cases, by virtue of (2), it will be equal to zero.

Let us emphasize that both positive and negative numbers can be used to represent the elements of the Galois field, as was clearly demonstrated in [12]. Therefore, it can be assumed that the elements of any field $GF(p)$, $p \neq 2$, are the elements -1,0,1. This allows using the algebraic delta function directly in form (2).

Obviously, a special case of formula (7) is a function identifying the absence of errors.

$$Q_0 = 1 - \left(\sum\nolimits_1^{p-1} q_i\right)^{p-1} \tag{9}$$

It is equal to one only when there are no errors. In all other cases it becomes zero. It is this function, as follows from what follows, that is convenient to use for studying the properties of multinomial coefficients.

The sum of function (9) and functions (7) up to and including the function indexed by K_0 allows for the identification of cases in which the code distance between a given sequence and the reference sequence does not exceed K_0. This, in turn, provides an algebraic representation of the morphism defined in Eq. (6).

However, the practical implementation of this morphism, particularly in the context of enhancing psychological testing methodologies, lies beyond the scope of the present study. Instead, the primary interest here concerns the structural form of function (7), which also serves to clarify the proof of the theorem presented below.

3.2 Analogue of Wilson's Theorem for Sums of Multinomial Coefficients

The identity holds

$$(a_1 + a_2 + \ldots a_m)^n = \sum\nolimits_{k_1+k_2+\cdots+k_m=n} C^n_{\vec{k}} a_1^{k_1} a_2^{k_2} \ldots a_m^{k_m} \tag{10}$$

where the multinomial coefficients $C^{n}_{\overrightarrow{k}}$ are given by the expressions

$$C^{n}_{\overrightarrow{k}} = \binom{n}{k_1, k_2, \ldots, k_m} = \frac{n!}{k_1! k_2! \ldots k_m!}, \tag{11}$$

We define the coefficients β_r as

$$\beta^{n}_{r} = \sum\nolimits_{k_1+k_2+\cdots+k_l=r} \binom{n}{k_1, k_2, \ldots, k_l}, \ k_i \neq 0 \tag{12}$$

The meaning of these coefficients is as follows. β^n_r is the sum of all multinomial coefficients corresponding to terms that contain non-zero powers of the first r variables on the left-hand side of formula (10), i.e. variables from a_1 to a_r. For example, if the expression $(a_1 + a_2 + a_3 + a_4)^4$ is considered, then β^4_2 is the sum of multinomial coefficients corresponding to terms of the form $a_1 a_2^3$, $a_1^2 a_2^2$ and $a_1^3 a_2$.

The following theorem holds.

If p is a prime number, then one of the following relations is true

$$\beta^{p-1}_{s} - 1 \equiv 0 \ (\text{mod}\ p) \tag{13}$$

$$\beta^{p-1}_{s} + 1 \equiv 0 \ (\text{mod}\ p) \tag{14}$$

Proof.

Directly from (1) follows

$$q^{n}_{i} = q_i \tag{15}$$

Let us emphasize that here and below the calculations are carried out in the sense of the field $GF(p)$. Let us consider the expression

$$\xi^{p-1}_{0} = \left(\sum\nolimits_{1}^{p-1} q_i\right)^{p-1} \tag{16}$$

From relations (10), (12) and (15) it follows that the function ξ^{m}_{0} can be represented as a series

$$\xi^{p-1}_{0} = \beta^{p-1}_{1} \sigma_1\left(\{q_i\}\rceil^{i=p-1}_{i=1}\right) + \beta^{p-1}_{2} \sigma_2\left(\{q_i\}\rceil^{i=p-1}_{i=1}\right) + \cdots + \beta^{p-1}_{p-1} \sigma_{p-1}\left(\{q_i\}\rceil^{i=p-1}_{i=1}\right) \tag{17}$$

where σ_s are symmetric functions

$$\sigma_s\left(\{q_i\}^{i=p-1}_{i=1}\right) = \sum\nolimits_{i_1<i_2<\ldots<i_s} q_{i_1} q_{i_2} \cdots q_{i_s} \tag{18}$$

For illustration, let us consider the special case of the coefficients β^4_2 and β^4_3. From the formula

$$(q_1 + q_2 + q_3 + q_4)^4 = \beta^4_1 \sigma_1 + \beta^4_2 \sigma_2 + \beta^4_3 \sigma_3 + \beta^4_4 \sigma_4 \tag{19}$$

It follows that

$$\beta_2^4 = \frac{4!}{1!3!} + \frac{4!}{2!2!} + \frac{4!}{3!1!} = 14 \equiv -1 \pmod 5 \tag{20}$$

$$\beta_3^4 = \frac{4!}{1!1!2!} + \frac{4!}{1!2!1!} + \frac{4!}{2!1!1!} = 36 \equiv 1 \pmod 5 \tag{21}$$

Multinomial coefficients in formula (20) correspond to terms of the form $q_1 q_2^3$, $q_1^2 q_2^2$ and $q_1^3 q_2$, and in formula (21) to terms of the form $q_1 q_2 q_3^2$, $q_1 q_2^2 q_3$ and $q_1^2 q_2 q_3$.

Further, the function on the left-hand side of formula (19), as follows from (18), is linear in any of the variables q_i.

By construction, function (9) must vanish when $q_i = 1$ for any i and must be equal to zero if all $q_i = 0$. Consequently,

$$Q_0 = 1 - \left(\sum\nolimits_1^{p-1} q_i\right)^{p-1} = 1 - \sum\nolimits_{i=1}^{i=p-1} \beta_i^{p-1} \sigma_i \prod\nolimits_{i=1}^{i=p-1} (1 - q_i) \tag{22}$$

Let us emphasize once again that formula (22) is written in the sense of the field $GF(p)$.

Further, the formula

$$\sum\nolimits_{r=0}^{r=p-1} t^r \sigma_r\left(\{q_i\}_{i=1}^{i=p-1}\right) = \prod\nolimits_{s=0}^{s=p-1} (1 + t q_s) \tag{23}$$

is valid, including in calculations in the sense of Galois fields. Comparing expressions (22) and (23), we obtain that in calculations in the field $GF(p)$ we have

$$\beta_i^{p-1} = \pm 1 \tag{24}$$

Indeed, to obtain the right side of formula (22), it is sufficient to set $t = -1$. The entry (24) when moving to modulo calculations is equivalent to entries (13) and/or (14). The theorem is proven.

4 Discussion

The connection between the proven theorem and Wilson's theorem follows directly from formula (10), applied to the case where $m = n = p - 1$. In this case, $\beta_{p-1}^{p-1} = (p-1)!$ takes place. The result obtained in the work means that $(p-1)! \pm 1 \equiv 0 \ (mod\ p)$, which coincides with Wilson's theorem, i.e. it is a special case of the theorem under consideration.

Further, the authors are aware that there are various options for proving both Wilson's theorem itself and its analogues, including those related to multinomial coefficients. The history of Wilson's theorem itself shows that it can be proven in various ways, and all of them are of some interest [29, 30]. With regard to the proposed method of proving the theorem formulated above, the point of view [31] is of interest, according to which research in the field of finite fields actually serves as a tool for interdisciplinary cooperation between number theory, computer science, coding theory and cryptography. The results of this study indicate that this list may also include tasks related to classification problems arising in the humanities, such as methods of psychological or sociological testing. Indeed, tasks of this kind, as follows from the calculations presented in this work, can be solved by algebraic means. The effectiveness of such an approach, of course, remains debatable, but this does not cancel the conclusion that attempts to develop algebraic approaches to solving the problem of classifications. In any case, such an approach allows us to obtain, for example, non-trivial proofs of analogues of Wilson's theorem. The approach proposed in this study can be integrated into the framework of se-cure machine learning, error analysis, and adaptive transmission protocols, aligning with the objectives of developing reliable and interpretable AI models.

5 Conclusions

Thus, the analysis of the pattern recognition problem using the error number identification function allows us to propose a non-trivial version of the proof of the theorem on the sums β_i^m multinomial coefficients related to the same set of variables in the expansion of the multinomial $(a_1 + a_2 + \ldots a_m)^n$. According to this theorem, if $m = p - 1$, then one of the equalities $\beta_s^{p-1} - 1 \equiv 0 \pmod p$ or $\beta_s^{p-1} + 1 \equiv 0 \pmod p$ is true, and Wilson's theorem is a special case of this result.

Funding. This research is funded by the Committee of Science of the Ministry of Science and Higher Education of the Republic of Kazakhstan (Grant No. BR24992908 «Support system for agricultural crop production optimization via remote monitoring and artificial intelligence methods (Agroscope)»).

References

1. Islam, M.A., Islam, M.A., Islam, N., Shabnam, B.: A modified and secured RSA public key cryptosystem based on "n" prime numbers. J. Comput. Commun. **6**(3), 78 (2018)
2. Bajard, J.C., Duquesne, S.: Montgomery-friendly primes and applications to cryptography. J. Cryptogr. Eng. **11**(4), 399–415 (2021)
3. Baikenov, A., Faure, E., Tynymbayev, S., Abramkina, O., Skutskyi, A.: Research of the effectiveness of frame synchronization methods for creating markerless telecommunication systems for exchanging short packets. In: Proceedings of the 9th International Conference on Digital Technologies in Education, Science and Industry (DTESI 2024), CEUR Workshop Proceedings, Almaty, Kazakhstan, vol. 2024,
4. Aggarwal, D., Joux, A., Prakash, A., Santha, M.: A new public-key cryptosystem via Mersenne numbers. In: Shacham, H., Boldyreva, A. (eds.) CRYPTO 2018, LNCS, vol. 10993, pp. 459–482. Springer, Cham (2018)
5. Raghunandan, K.R., Aithal, G., Shetty, S.: Comparative analysis of encryption and decryption techniques using Mersenne prime numbers and phony modulus to avoid factorization attack of RSA. In: The 2019 International Conference on Advanced Mechatronic Systems (ICAMechS 2019), pp. 152–157. IEEE (2019)
6. Bluher, A.W.: New Wilson-like theorems arising from Dickson polynomials. Finite Fields Appl. **72**, 101819 (2021)

7. Levaillant, C.: Wilson's theorem modulo P2 derived from Faulhaber polynomials. arXiv preprint arXiv:1912.06652 (2019)
8. Aebi, C., Cairns, G.: Generalizations of Wilson's Theorem for double-, hyper-sub- and superfactorials. Am. Math. Mon. **122**(5), 433–443 (2015)
9. Connes, A.: Around Wilson's theorem. J. Number Theory. **194**, 1–7 (2019)
10. Athanasiadis, C.A.: Binomial Eulerian polynomials for colored permutations. J. Comb. Theory Ser. A. **173**, 105214 (2020)
11. Balasubramanian, K.: Computational multinomial combinatorics for colorings of 5D-hypercubes for all irreducible representations and applications. J. Math. Chem. **57**, 655–689 (2019)
12. Suleimenov, I.E., Matrassulova, D.K.: Using the relationship between the theory of algebraic fields and number theory for developing promising methods of digital signal processing. In: 2022 3rd Asia Conf. Comput. Commun. (ACCC), pp. 121–125. IEEE (2022)
13. Suleimenov, I.E., Vitulyova, Y.S., Kabdushev, S.B., Bakirov, A.S.: Improving the efficiency of using multivalued logic tools. Sci. Rep. **13**(1), 1108 (2023)
14. Suleimenov, I.E., Vitulyova, Y.S., Kabdushev, S.B., Bakirov, A.S.: Improving the efficiency of using multivalued logic tools: application of algebraic rings. Sci. Rep. **13**(1), 22021 (2023)
15. Kadyrzhan, A., Kadyrzhan, K., Bakirov, A., Suleimenov, I.: Prospects for the use of quasi-Mersen numbers in the design of parallel-serial processors. Appl. Sci. **15**, 741 (2025). https://doi.org/10.3390/app15020741
16. Suleimenov, I., et al.: Advanced applications of polymer hydrogels in electronics and signal processing. Gels. **10**, 715 (2024). https://doi.org/10.3390/gels10110715
17. Mukhamediev, R.I., et al.: Coverage path planning optimization of heterogeneous UAVs group for precision agriculture. IEEE Access. **11**, 5789–5803 (2023)
18. Nesha, M.K., et al.: An assessment of data sources, data quality and changes in national forest monitoring capacities in the Global Forest Resources Assessment 2005–2020. Environ. Res. Lett. **16**, 054029 (2021)
19. Faure, E., Baikenov, A., Skutskyi, A., Faure, D., Abramkina, O.: Algorithms for reliable permutation transmission protocols in noisy communication channels. CEUR Workshop Proc. **3826**, 40–49 (2024)
20. Suleimenov, I., et al.: Empirical validation of the use of projective techniques in psychological testing using Galois fields. Front. Appl. Math. Stat. **10**, 1455500 (2024)
21. Ibragim, S., Mariya, K., Yelizaveta, V.: Projective psychological tests from the point of view of neural network theory of society. J. Ecohumanism. **3**(4), 2603–2609 (2024)
22. Matz, S.C., Netzer, O.: Using big data as a window into consumers' psychology. Curr. Opin. Behav. Sci. **18**, 7–12 (2017)
23. Dyriv, A., et al.: The user's psychological state identification based on big data analysis for person's electronic diary. In: 2021 IEEE 16th International Conference on Computer Sciences and Information Technologies (CSIT), vol. 1, pp. 101–112. IEEE (2021)
24. Ancans, G., Stafecka, A., Bobrovs, V., Ancans, A., Caiko, J.: Analysis of characteristics and requirements for 5G mobile communication systems. Latvian J. Phys. Tech. Sci. **54**(2), 41–50 (2017)
25. Zou, J.Y., Sun, M.X., Liu, K.H., Wu, Q.Q.: The design of dynamic ensemble selection strategy for the error-correcting output codes family. Inf. Sci. **571**, 1–23 (2021)
26. Huang, L., Zhang, H., Li, R., Ge, Y., Wang, J.: AI coding: Learning to construct error correction codes. IEEE Trans. Commun. **68**(1), 26–39 (2019)
27. Bakirov, A.S., Suleimenov, I.E.: On the possibility of implementing artificial intelligence systems based on error-correcting code algorithms. J. Theor. Appl. Inf. Technol. **99**(1), 83–99 (2021)

28. Suleimenov, I.E., Gabrielyan, O.A., Bakirov, A.S.: Initial study of general theory of complex systems: physical basis and philosophical understanding. Bull. Electr. Eng. Inform. **14**(1), 774–789 (2025)
29. Vanstone, S.A., Van Oorschot, P.C.: An Introduction to Error Correcting Codes with Applications, vol. 71. Springer, New York (2013)
30. Ore, O.: Number Theory and Its History. Courier Corporation, New York (1988)
31. Shparlinski, I.: Finite Fields: Theory and Computation. Springer, New York (2013)

Modeling Intruder Behavior in Information Systems: Methodology Using Graphs with Non-standard Reachability

Boris Zakhir[1] (✉) and Anna Oleynikova[2,3]

[1] Federal State Autonomous Educational Institution of Higher Education Bauman Moscow State Technical University, Moscow, Russia
bzakhir@gmail.com
[2] Telecom Integration LLC, Moscow, Russia
[3] Reshetnev Siberian State University of Science and Technology, Krasnoyarsk, Russia

Abstract. In this study, a novel methodology is developed to refine information security analysis by leveraging graph theory to explore attack pathways, with a focus on non-standard reachability. This research critiques the narrow focus of traditional analyses on immediate attack vectors and targets, proposing instead a dynamic and integrated approach for assessing the impacts of cyber-attacks on information systems. By establishing a novel framework for the in-depth analysis of intruder behaviors using graph-based path analysis, the study aims to equip security professionals with a comprehensive tool for understanding the full scope of an attack's implications. The methodology is designed to facilitate a holistic examination of potential vulnerabilities and to bolster system resilience against cyber threats. Findings from the study suggest that a more expansive perspective on information system responses to cyber-attacks can significantly elevate the effectiveness of security measures and incident response strategies. By contributing an innovative analytical tool to the cybersecurity field, this work offers a more nuanced strategy for comprehending and counteracting cyber risks, potentially guiding the development of more robust security protocols and frameworks.

Keywords: Directed graphs · cyberattack · mathematical model · prediction · resilience

1 Introduction

The task of ensuring the information security of information systems under attack is considered fragmentarily in modern approaches to analysis. Researchers' attention is focused on the attack itself, its structure, and the system elements that are attacked. This may lead to the oversight of certain aspects of the attack and its impact on the elements of the information system (and related systems) during the analysis, which, in turn, could lead to additional losses in the event of serious incidents.

Furthermore, an interesting issue in attack modeling is the development of the idea of continuous incident response [1], which suggests that the assessment of the impact on

M. Lapina et al. (Eds.): AISMA 2025, LNNS 1456, pp. 471–481, 2026.
https://doi.org/10.1007/978-3-032-07275-7_43

the objects of the information system (and the assessment of the state of the information system itself) will dynamically change.

The purpose of this study is to develop an approach to the structured description of the information system as an object of information security.

The main scientific task of the work is to create a method for researching the behavior of an intruder in the system by analyzing paths on a graph with non-standard reachability.

Consequently, it is necessary to move towards understanding the integrity of the impact on the information system during an attack, considering all possible consequences and changes to the system itself as an object of impact.

2 Review of Existing Works

An information system, according to the definition provided by ISO/IEC 2382:2015, is a system intended for storing, searching, and processing information. Accordingly, damage to it would be considered a disruption of processes aimed at achieving the goals for which this system is designed. The article [2] examines the structure of the information system as an object of management and defines the concept of an attack on its information security. In the existing work, we will use the term "object of impact," which better reflects the subject of this study. In article [2], an IS is presented as a combination of modules that make it up and functions/programs it performs, which, as a generalized concept, suits the task of researching the behavior of an intruder in this system. The behavior of the intruder itself can be represented from two positions - from the perspective of the intruder and from the perspective of the protected IS. In article [2], an attack is called an information-technical impact and is defined as an event that changes the state of one or more characteristics of the model. The change in the system's state as a result of an attack can be understood as a breach of some security component of a module or program, for example, a denial-of-service event on a traffic encryption server. This representation of an attack is useful for calculating its consequences and the extent of damage, but to compile a detailed picture of the perpetrator's actions, a different approach is required. To begin, it is necessary to define the actions of the perpetrator.

From the perspective of regulatory documents, the definition of an attack highlights the composition of its objectives and/or consequences for the protected information system, that is, it is given from the perspective of the protected IS. For example, according to GOST R ISO/IEC 27000-2021, an attack is called an attempt to destroy, disclose, alter, block, steal, gain unauthorized access to an asset, or use it without authorization [3]. Or, as stated in NIST SP 800–12, it is any type of malicious activity aimed at collecting, disrupting operation, denying access, degrading the quality, or destroying the resources of the information system or the information itself [4]. In these definitions, certain actions are characterized as an attack exclusively by their objectives and consequences, but not by their composition and structure. To examine the attack process in more detail, let's turn to the concept of the Cyber Kill Chain [5, 6], which represents an attack as an ordered sequence of specific steps. It is assumed that the attacker first selects suitable targets (reconnaissance), prepares the necessary tools (weaponization), and delivers them to the targeted information system (delivery). Then, these tools allow, by exploiting system or application vulnerabilities (exploitation), to establish a backdoor or trojan (installation), create a command and control channel for malware (command and control, C&C), and, finally, act according to the objectives of the attack (actions). However, this model has often been criticized for its limited applicability, for example, focusing on malware and not adequately addressing the mechanics of phishing attacks. Therefore, researcher Paul Pols introduced the concept of a unified kill chain, consisting of 18 tactical steps, or phases [7]. The tactics describe the establishment of a foothold by the intruder in the targeted system, its propagation through the network, and actions to achieve the objectives of the attack. The concept uses confidentiality, integrity, and availability as security properties that are subject to compromise.

Additionally, the MITRE ATT&CK framework is widely recognized — it is a knowledge base for modeling attacks, based on the kill chain concept [8]. ATT&CK is a matrix composed of columns and rows. Its tactics (columns) can be seen as phases of a multi-stage attack, while the techniques (rows) represent specific steps of an attack. Some techniques may not appear in a single multi-stage attack and may not follow the same order. Recently, the MITRE D3FEND initiative has been developed, which highlights countermeasures that can be applied against some of the ATT&CK techniques.

The concept of attack graphs, proposed by Sandia Laboratories researchers in 1998, represents attacks on information systems as an ordered sequence of steps, with nodes symbolizing potential attack states and edges reflecting state transitions due to attacker actions or victim's unintended actions [9]. These graphs are generated from attack patterns, system configurations, and attacker profiles, enabling a tailored analysis of network vulnerabilities by considering factors like the attacker's capabilities and the specific network setup. Also in [10] a unified attack graph model was introduced, that aids in the early identification of potential attack paths and weak security points, enabling administrators to preemptively counteract cyber threats through strategic defense measures.

3 Formalization of the System and Its Components

The analysis of the attack process on an information system in this study is based on the scientific sources from the field of information security and graph theory described above. The attack graph, as a directed weighted subgraph, will be presented, unlike existing approaches, as part of the unified graph of the information system. This will make the strategic and comprehensive assessment of the attack's impact on the target system more effective.

Let us denote the weighted directed multigraph G:

$$G = (V, E, I, J) \quad (1)$$

Here, V represents the set of nodes of the graph; E represents the set of edges of the graph, where each edge $e \in E$ is represented by a triplet (u, v, k), indicating that this is the k-th edge leading from node u to node v, and $u, v \in V$; $I : V \rightarrow R$ - the function assigning weights to the nodes of the graph; $J : E \rightarrow R$ - the function assigning weights to the edges of the graph.

3.1 Nodes

In the graph, vertices will be any computing devices (CDs) involved in network interactions with other system elements. It is important to note that the defining entity of a CD is its distinct file system, not the network card. Let's consider several examples.

- A device that uses multiple network cards, and thus has several IP addresses, will still be represented as a single node on the graph, and the existing channels of network interaction of these IP addresses will be aggregated in this node;
- A device on which multiple virtual machines are deployed will not be represented as a single node on the graph; instead, each virtual machine will be a separate node.

To determine the parameters of the vertex weight function V, we propose to refer to information security practices, from which conditions for non-standard reachability will be formulated for the formation of paths on graph G. For further convenience in calculating reachability and conditions for the spread of the attack across the information network, we suggest aggregating the characteristics of vertices using weight coefficients. We will introduce attractiveness and accessibility ratings for a node to an attacker, which collectively will determine the importance of the node to the attacker in terms of attack (and criticality for security from the IS perspective). These indicators will help assess the strategic value of each node in the network and identify which of them are most likely to become targets for the attacker.

Let's define the importance of a vertex $I(V)$ as the product of the attractiveness rating and the accessibility rating:

$$I(V) = A(V) * D(V) \tag{2}$$

This formula considers that an attacker is interested in nodes that are not only attractive in terms of further development of the attack but are also sufficiently accessible as a new step in spreading across the network.

The attractiveness rating $A(V)$ is defined as follows:

$$A(V) = k_1 * S_a(V) + k_2 * S_l(V) + k_3 * S_b(V, d) + k_4 * S_s(V) - k_5 * S_h(V) - k_6 * S_n \tag{3}$$

where:

$S_a(V) \in [0, 1]$ — the importance rating of the asset represented as node V;
$S_l(V) \in \mathbb{N}$ — the number of logon sessions of high-privileged users on node V;
$S_b(V, d) \in [0, 1]$ — the relative contribution of the node's importance in the broadcast network domain d in case the node V belongs to this domain;
$S_s(V) \in \{0, 1\}$— the presence of weak security policies on the node V;
$S_h(V) \in \{0, 1\}$ — the presence of host-based information protection tools on the node V;
$S_n \in \{0, 1\}$ — the presence of network-based information protection tools;
$k_1, k_2, k_3, k_4, k_5, k_6 \in [0, 1]$ — weight coefficients reflecting the importance of each factor for the overall attractiveness.

The accessibility rating $D(V)$ is defined as follows:

$$D(V) = m_1 \cdot (1 - S_h(V)) + m_2 \cdot (1 - S_n) + m_3 \cdot S_x(V) \quad (4)$$

where:

$m_1, m_2, m_3 \in [0, 1]$ — weight coefficients that reflect the importance of each factor for the overall accessibility.
$S_x(V) \in \{0, 1\}$ — the presence of vulnerabilities on the node, where $X(V)$ – is a function mapping the vulnerabilities present on the node. Let's define this function. Let $\mathcal{X}$ — be the set of all vulnerabilities identified in the information system. Then, for each vertex $V_i \in V$ of the graph G, the function $X(V_i)$ is defined as:

$$X(V_i) = \{x_1, x_2, \ldots, x_n\} \subseteq \mathcal{X} \quad (5)$$

where each element $x_j \in X(V_i)$ represents a vulnerability associated with the vertex V_i.

The remaining variables have been described above.

To maintain the completeness of the study, let's expound on the concept of a security policy. According to [10], an information security policy is the aggregate of directives, regulations, and rules that prescribe how an organization manages, protects, and distributes information. In this article, the security policy is interpreted in a narrower sense, as a set of organizational rules and measures that minimize the likelihood of a given set of threats being realized. And, based on this definition, we denote a weak security policy as one that does not fully cover the given set of threats.

Let's examine in more detail the methodology for calculating the relative importance of node V in the broadcast domain. Let d be the set of nodes belonging to the broadcast domain d, и $I'(V)$ — be the overall importance of node V, calculated without considering its membership in the broadcast domain.

For broadcast domain d we will calculate the total importance of all nodes in this domain, without considering their membership in the broadcast domain:

$$d = \{V_1, V_2, \ldots V_q\}, |d| = q \quad (6)$$

$$I'(d) = \sum_{i \in [1,q]} I'(V_i) = \sum_{i \in [1,q]} A'(V_i) * D(V_i) = \sum_{i \in [1,q]} (k_1 * S_a(V_i) + k_2 * S_l(V_i) + k_4 * S_s(V_i) - k_5 * S_h(V_i) - k_6 * S_n) * D(V_i) \quad (7)$$

The degree of importance of node V membership in broadcast domain d, denoted as $S_b(V)$, can be expressed as the ratio of the overall importance of node V to the total importance of all nodes within the domain:

$$S_b(V, d) = \frac{I'(V)}{I'(d)} \quad (8)$$

This approach allows for the consideration not only of the individual characteristics of a node but also its role within the entire broadcast domain. Nodes with high overall importance in a domain with high total importance can potentially serve as critical points for the spread of attacks in the network. Thus, attention is focused not only on the node's ability to propagate broadcast traffic but also on the significance of this node in the context of potential attacks on other nodes within the domain.

Representing the importance of a node for an attack or its criticality for defense as an aggregated value $I(V)$ enables a comprehensive consideration of the diverse characteristics of assets in the information system, their importance, level of protection, and compliance with regulatory requirements.

3.2 Edges

For each edge e_{uv} in E, we introduce the following notations:

- $P(e_{uv})$ denotes the network interaction protocol of edge e_{uv}, including the protocol version (e.g., HTTP/2, SSHv2).
- $X(e_{uv})$ represents the set of vulnerabilities present at the terminal node of edge e_{uv} and relevant to the network interaction protocol.

Then, the weight of the edge e_{uv} is determined as:

$$J(e_{uv}) = \alpha * WP(P(e_{uv})) + \beta * \sum\nolimits_{x \in X(e_{uv})} WU(x, P(e_{uv})) \tag{9}$$

where:

$WP(p)$ is the weight assigned to the network interaction protocol p based on its prevalence in attack techniques;

$WU(x,p) = \begin{cases} cvss(x), \textit{ if } x \textit{ exploits } p \\ 0, \textit{otherwise} \end{cases}$ the weight (criticality) of vulnerability x from the set of vulnerabilities $X(v)$, exploiting protocol p, found on the terminal node v of edge e_{uv};

α and β are coefficients that allow for balancing the contribution of the protocol and vulnerabilities to the overall weight of the edge.

This formulation allows for a comprehensive assessment of the weight of each edge in the attack graph, considering both the characteristics of network interaction and specific vulnerabilities associated with protocols used for communication between nodes.

3.3 Attack

Based on the formulated scientific task of the work, as well as the analysis of existing studies, let us define an attack on the information system as an ordered set of the intruder's impacts on the elements of the information system. We will establish two types of impacts: those not leading to propagation to other nodes, which we will call the set of actions $\mathcal{R}$, and those leading to propagation (i.e., related to TA0008), which we will call the set of actions $\mathcal{R}'$. The intruder's actions will be defined by the technique of the MITRE ATT&CK matrix and the set of graph vertices – that is, the assets affected by this action. We will denote the set of techniques as T, and an individual technique as t.

$$\begin{cases} \mathcal{R} = \{r_1, r_2, r_3, \ldots, r_g\}, |\mathcal{R}| = g \\ \mathcal{R}' = \{r'_1, r'_2, r'_3, \ldots, r'_l\}, |\mathcal{R}'| = l \\ t \in T \\ r = \{t, v\} \in \mathcal{R}, v \in V \\ r' = \{t, \{u, v\}\} \in \mathcal{R}', u, v \in V \end{cases} \tag{10}$$

It should be noted that the actions under consideration invariably require some time for implementation. Information about the time required to implement various techniques will be accumulated in the Security Orchestration, Automation, and Response (SOAR) system for incidents of information security, individually for each protected information system. Thus, let us record the attack $\mathbb{A}$ as an ordered set of actions r_i^a, and define the function τ_r, as a mapping of actions into the space of real numbers, indicating the time required for the implementation of these actions. It is important to note that an attack is formed from actions already carried out by the intruder. It is also established that an attack begins with a specific action from the set $\mathcal{R}$, and consequently, from a specific node.

$$\begin{cases} \mathbb{A} = \left\{r_1^a, r_2^a, \ldots, r_y^a\right\}, |\mathbb{A}| = y \\ r^a \in \mathcal{R} \cup \mathcal{R}' \\ \tau_r(r) \to \mathbb{R} \end{cases} \tag{11}$$

We define the attack path M as a subgraph of the graph G:

$$M = (V_M, E_M, I, J) \subset G \tag{12}$$

The vertices V_M and edges E_M of the attack path are composed of vertices involved in the actions that make up the attack $\mathbb{A}$.

$$V_M = \left\{v | \exists r_a = (t, v) \in \mathcal{R} \; or \; r_a = (t, \{u, v\}) \in \mathcal{R}'\right\} \tag{13}$$

$$E_M = \left\{(u, v) | \exists r_a = (t, \{u, v\}) \in \mathcal{R}'\right\} \tag{14}$$

The weight functions are the same as for graph G.

According to the definition of an attack, the subgraph M has one source vertex, referred to as the start of the attack.

4 Introduction of Rules for Non-standard Reachability

Having defined the initial state of the information system – a weighted directed multigraph $G = (V, E, I, J)$, specifying the characteristics of its vertices and edges, and introducing the term of an attack as a subgraph in the mathematical interpretation of the considered information system, we can proceed to describe the process of modeling the further behavior of an intruder based on the listed information. Let's call the further behavior of the intruder the hypothetical attack path M_h of attack $\mathbb{A}$. To define the principles for constructing these paths, let's accept the following postulates:

1. Only those attacks that are capable of inflicting damage on the system are considered.
2. Only attacks aimed at damaging critical nodes of the system can inflict damage (i.e., attacks whose terminal vertices of routes are critical vertices of the system's graph G).
3. We assume critical vertices of the system to be known. In practice, it is indeed possible to identify them thanks to Business Continuity Management (BCM) systems.

Considering all these hypotheses, we will construct the hypothetical attack path only towards the critical nodes of the system. Let's define the construction of the hypothetical attack path as the construction of paths on the weighted directed multigraph G, which will be carried out taking into account the so-called conditions of non-standard reachability. Let's examine in more detail the process of constructing paths that make up the hypothetical route M_h of attack $\mathbb{A}$..

The start of these paths are the vertices M, adjacent to the vertices of graph G not included in M. The end of the paths are vertices belonging to the set C of critically significant nodes for the protected system.

To ensure that a path on the graph, symbolizing to some extent the route of the intruder's propagation on the graph, reflects the real conditions of this process, we will use the theory of non-standard reachability on graphs, as discussed in articles [11, 12, 14]. The author of this scientific concept, Y. M. Yerusaliemsky, writes [13] that in relation to a graph, the concept of non-standard reachability implies that not all possible paths on the graph are permissible, but only those that satisfy certain additional conditions. This approach provides the opportunity to consider a multitude of important factors affecting the accessibility of various network nodes to the attacker during the attack development.

To articulate these principles in relation to an attack on an information system represented as a graph, let's use the following practical principles of information security. Let's denote the intruder by the symbol $\mathcal{N}$.

1. Presence of vulnerabilities on the node. $\mathcal{N}$ cannot utilize any tactics or techniques in the absence of necessary vulnerabilities for their implementation - cannot pave their way;
2. Presence of logon sessions of high-privileged users on the node. Such a circumstance significantly increases the priority of $\mathcal{N}$ transition to this vertex;
3. Node's presence in a broadcast network domain containing other nodes. This condition is determined by the device's connection to a switch that has connections from other nodes. This also increases the priority of $\mathcal{N}$ transition to the node;
4. Presence of weak local or domain security policies on the node increases the priority of $\mathcal{N}$ transition to the node (essentially, increases the probability of node compromise);
5. Presence of host-based information protection tools (HSI) on the node. Affects the ability of the intruder to conduct malicious activity on the host, leading to the impossibility of implementing certain techniques and tactics by $\mathcal{N}$;
6. Presence of network-based information protection tools (NSI) in the system. The influence is similar to that of host-based information protection tools, but applies to the entire set of graph vertices;
7. Time constraints. By measuring the time after the attack begins, it can be said which actions $\mathcal{N}$ could have carried out, and which take too much time. Thus, the reachability of vertices by this principle depends on the time elapsed since the attack began. The more time has passed, the more actions $\mathcal{N}$ has available (including those related to spreading through the protected system), and the more paths $\mathcal{N}$ can pave.

Let's mathematically formulate the principles for constructing path components of the hypothetical attack route M_h.

Presence of vulnerabilities on the node necessary for implementing actions:

$$U(V_i) = \begin{cases} 1, \text{if vertex } V_i \text{ has vulnerabilities} \\ 0, \textit{otherwise} \end{cases} \quad (15)$$

Presence of logon sessions of high-privileged users:

$$LSP(V_i) = \begin{cases} 1, \text{if vertex } V_i \text{ has sesssions} \\ 0, \textit{otherwise} \end{cases} \quad (16)$$

Node's presence in a broadcast network domain:

$$B(V_i) = \begin{cases} 1, \text{if vertex } V_i \text{ s in a broadcast network domain} \\ 0, \textit{otherwise} \end{cases} \quad (17)$$

Presence of weak local or domain security policies (DSP):

$$WSP(V_i) = \begin{cases} 1, \text{if vertex } V_i \text{ has weak local or DSP} \\ 0, \textit{otherwise} \end{cases} \quad (18)$$

Presence of HSI tools:

$$HSI(V_i) = \begin{cases} 1, \text{if vertex } V_i \text{ has HSI} \\ 0, \textit{otherwise} \end{cases} \quad (19)$$

Presence of NSI tools:

$$NSI = \begin{cases} 1, \text{if the system has NSI} \\ 0, \textit{otherwise} \end{cases} \quad (20)$$

Time constraints:

$$\mathrm{T}_r = \tau - \tau_0 \quad (21)$$

where τ — is the current time, τ_0 — is the start time of the attack, and T_r denotes the time elapsed since the start of the attack at time τ.

For constructing the hypothetical attack route M_h on the weighted directed multigraph G, considering non-standard reachability conditions:

$$M_h(V_i) = \{V_j | V_i \rightarrow V_j, U(V_j) == 1, LSP(V_j) + B(V_j) + WSP(V_j) + HSI(V_j) + NSI > 0, \mathrm{T}_r \geq \Delta\tau_{ij}\} \quad (22)$$

where $\Delta\tau_{ij}$ is the time required to perform the lateral movement action from the vertex V_i to the V_j, considering all the above conditions.

Based on the entire mathematical framework presented above, there is ample scope for further refinement of methods for applying the model of calculating paths on a weighted directed multigraph with non-standard reachability. What is outlined in this article serves as a starting point for subsequent research. Specifically, for instance, it can be noted that based on the number of possible attack paths and their likelihood, conclusions can be drawn about the danger level of this attack to the protected information system.

5 Practical Perspectives of the Model

Mathematical modeling of an intruder's action path is determined based on the organization's policies or charter, which identifies critically important nodes (information systems or processes) of the enterprise. Narrowing the scope of objects involved in modeling allows prioritizing the chains of investigated incidents based on their actual significant impact in the company's context, i.e., focusing on and investigating primarily those attacks that will lead to significant losses for the organization.

The level of losses for each specific organization when a significant business unit (or production unit) is disabled is assessed during the impact analysis procedure in the process of business continuity evaluation. Consequences types can be not necessarily quantitative (such as financial losses) but also qualitative (e.g., regulatory-legal, social, environmental). Assessing the level of losses and categorizing assets as critically significant (for example, in the context of Critical Information Infrastructure within the framework of Federal Law No. 187 [14]) provides the first limitation of the set of attacker paths of interest to us. As a result, we prioritize work on attack chains so that those attacks leading to the greatest damage to the company are investigated first, and the level of response to them occurs in a more "operational mode".

Possible paths of the attack route are also characterized by data obtained from asset management and vulnerability management processes. Through these tools, the graph is enriched with context on the presence of network connectivity between assets; access through specific ports and protocols; the presence of tools allowing lateral movement from one host to another; and the presence of vulnerabilities that can be exploited by an attacker to escalate privileges, gain command line access (or code execution), and further develop the attack with a transition to the next graph node.

Compliance checking systems against best practices enrich the graph with data on proper host configuration settings and the node's level of cyber resilience, which also affects the possibility of further advancement of the attacker through additional defensive/offensive strength coefficients.

Incident management systems provide a retrospective history of similar situations involving the analyzed infrastructure objects, as an additional danger metric - the probability of an attack passing through this path.

As a result of accounting for permissible development paths of the situation, the organization achieves an improvement in the quality of the process of investigating and responding to information security incidents by reducing the time to access response variations. Variations are achieved thanks to additional analytics that consider hidden connections and statistical calculations based on big data from various sources.

Besides the obvious value of improving the effectiveness of the investigation and response process to incidents, the information security specialist receives a more precise analysis and practically valuable, updated threat model.

References

1. Oleynikova, A., Zolotarev, V.: The concept of information security management based on the cycle of continuous detection and response to information security incident. Izvestia YFU. Tech. Sci. **235**(5), 66–81 (2023) [In Russian]

2. Derbin, A., Zakhir, B.: Mathematical model for assessing the consequences of information impacts on a sociotechnical system in cyberspace in the interests of ensuring its information security. Inf. Wars. **66**(2), 82–87 (2023) [In Russian]
3. GOST R ISO/IEC 27000-2021. Information Technology-Security Techniques-Information Security Management Systems-Overview and Vocabulary, 2021 [In Russian]
4. NIST. Guide to Industrial Control Systems (ICS) Security. Special Publication 800-12
5. Yadav, T., Rao, A.: Technical aspects of cyber kill chain. In: Abawajy, J.H., Mukherjea, S., Thampi, S.M., Ruiz-Martínez, A. (eds.) Security in Computing and Communications, pp. 438–452. Springer International Publishing (2015)
6. Hutchins, E., Cloppert, M., Amin, R.: Intelligence-Driven Computer Network Defense Informed by Analysis of Adversary Campaigns and Intrusion Kill Chains. Lockheed Martin Corporation (2024)
7. Pols, P.: The unified kill chain, from https://www.unifiedkillchain.com/assets/The-Unified-Kill-Chain.pdf (retr. Feb. 15, 2024)
8. Strom, B., et al.: MITRE ATT&CK®: Design and philosophy, from https://attack.mitre.org/docs/ATTACK_Design_and_Philosophy_March_2020.pdf (retr. Feb. 15, 2024)
9. Phillips, C., Swiler, L.: A graph-based system for network vulnerability analysis. Proceedings of Workshop New Security Paradigms, 71–79 (1998)
10. Sadlek, L., Čeleda, P., Tovarnak, D.: Identification of Attack Paths Using Kill Chain and Attack Graphs, IEEE/IFIP Network Operations and Management Symposium (NOMS) (2022)
11. Basangova, E., Yerusaliemsky, Y.: Mixed reachability on partially oriented graphs. In: Deposited in VINITI vol. 5892-82 (1982) [In Russian]
12. Basangova, E., Yerusaliemsky, Y.: Various types of mixed reachability. Algebra and discrete mathematics, pp. 70–75, publ. in Kalmyk State University, Elista, 2015. [In Russian]
13. Yerusaliemsky, Y., Skorokhodov, V., Kuzminova, M., Petrosyan, A.: Graphs with Non-standard Reachability: Problems, Applications. Southern Federal University, Rostov-on-Don (2009) [In Russian]
14. Antonova, V., Zakhir, B., Kuznetsov, N.: Modeling of graphs with different types of reachability in python. J. Commun. Technol. Electron. **64**, 1464–1472 (2019)
15. Russian Federation, Federal Law No. 187-FZ of July 26. On the Security of Critical Information Infrastructure (2017). [In Russian]

SFOA-Driven Defense: Leveraging SFOA for Optimal Feature Selection in Network Intrusion Detection Systems

Mohammad Zayeem and Mohammad Sajid(✉)

Department of Computer Science, Aligarh Muslim University, Aligarh 202002, India
sajid.cst@gmail.com

Abstract. Intrusion detection systems (IDS) are critical for safeguarding networked environments against evolving cyber threats. This paper presents an innovative approach for feature selection in IDS using the Superb Fairy-wren Optimization Algorithm (SFOA). The proposed method effectively balances exploration and exploitation through three unique behavioral phases, i.e., growth, breeding, and feeding, and predator avoidance behaviors of fairy-wren birds. Using UNSW-NB15, the performance of SFOA is evaluated against four established optimization algorithms: Cheetah Optimizer (CO), Harris Hawk Optimization (HHO), Whale Optimization Algorithm (WOA), and Parrot Optimization (PO), in terms of accuracy and selecting optimal feature subsets. Experimental results demonstrate that SFOA achieves the best classification accuracy and faster convergence while reducing the feature dimensionality. By balancing exploration and exploitation, SFOA enhances IDS efficiency, ensuring high detection rates and low false alarms for robust cybersecurity.

Keywords: Intrusion Detection System · Feature Selection · Metaheuristic Algorithms · Random Forest · Decision Tree · SVM · UNSW-NB15

1 Introduction

In recent years, the proliferation of interconnected devices and networks has led to an exponential growth in cybersecurity threats, making intrusion detection systems (IDS) increasingly vital for securing digital infrastructure. Intrusion detection systems are critical components in cybersecurity frameworks, designed to identify and respond to unauthorized access attempts and malicious activities within computer networks [1]. However, the effectiveness of these systems depends significantly on their ability to distinguish between normal and anomalous behavior patterns amidst vast volumes of network traffic data containing numerous features. Feature selection has emerged as a crucial preprocessing step in enhancing the performance of intrusion detection systems by identifying the most informative subset of features while eliminating redundant and irrelevant ones [2]. The challenge lies in selecting an optimal feature subset that maximizes detection accuracy while minimizing computational complexity. Traditional

M. Lapina et al. (Eds.): AISMA 2025, LNNS 1456, pp. 482–496, 2026.
https://doi.org/10.1007/978-3-032-07275-7_44

feature selection methods often struggle with high-dimensional data and can become trapped in local optima, leading to suboptimal solutions [1]. Figure 1 demonstrates the overview of the IDS framework.

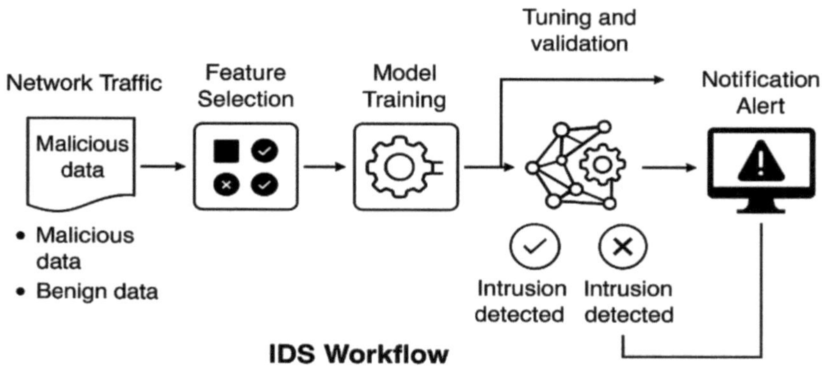

Fig. 1. Intrusion detection system (IDS) framework overview.

Metaheuristic optimization algorithms, inspired by natural phenomena and biological systems, have shown promising results in addressing complex optimization problems across various domains [3]. These algorithms provide efficient strategies for exploring large search spaces and finding near-optimal solutions without exhaustive enumeration. In the context of intrusion detection, metaheuristic approaches can significantly improve classification accuracy while reducing computational overhead through effective feature selection [1]. Recent advancements in nature-inspired computing have introduced several innovative optimization algorithms that simulate various biological behaviors and natural processes. The Harris Hawks Optimization (HHO) algorithm mimics the cooperative hunting behavior of Harris' hawks, employing a surprise pounce strategy to efficiently explore the search space [4, 5]. Similarly, the Whale Optimization Algorithm (WOA) simulates the hunting mechanism of humpback whales using bubble-net feeding behavior, though it faces challenges related to convergence speed and local optima avoidance [6]. The Parrot Optimizer (PO), a relatively recent addition to the metaheuristic family, draws inspiration from the behaviors of trained Pyrrhura Molinae parrots, incorporating specialized foraging behaviors and Lévy flight strategies to balance exploration and exploitation [7]. The Cheetah Optimizer, while less documented in the literature, attempts to harness the hunting efficiency and speed of cheetahs to navigate complex optimization landscapes [8]. Among these nature-inspired techniques, the Superb Fairy Wren Optimization Algorithm (SFOA) stands out for its unique approach based on the life habits of the Superb fairy-wren bird species. SFOA incorporates three distinct behavioral phases-young bird growth, breeding and feeding, and predator avoidance-providing a comprehensive framework for balancing global exploration and local exploitation in optimization tasks [9].

This paper presents an innovative approach for feature selection in IDS using the Superb Fairy-wren Optimization Algorithm (SFOA). Using UNSW-NB15, the performance of SFOA is evaluated against four established optimization algorithms: CO, HHO, WOA, and PO, in terms of accuracy and selecting optimal feature subsets. The experimental evaluation demonstrates that the proposed SFOA-based approach achieves superior classification accuracy and convergence speed while effectively reducing the dimensionality of the feature space. The main contributions of this paper include:

1. Implementation and evaluation of the Superb Fairy Wren Optimization Algorithm for optimal feature subset selection
2. A comprehensive comparison of five nature-inspired metaheuristic algorithms for feature selection in intrusion detection systems
3. Empirical validation of the proposed approach using standard benchmark datasets for intrusion detection
4. Analysis of convergence behavior and classification performance across different optimization techniques

The paper is organized as follows: Section 2 reviews related work in feature selection and metaheuristic optimization for intrusion detection. Section 3 presents the problem formulation, whereas in Sect. 4, the proposed algorithm is described. Section 5 presents implementation details, experimental evaluation, convergence behavior, and accuracy analysis. Section 6 presents the conclusion and ideas for future research directions.

2 Existing Research

Metaheuristic algorithms for feature selection have been thoroughly researched in the context of intrusion detection systems. This section presents a summary of key contributions for IDS using metaheuristic algorithms. These algorithms encounter difficulties with high-dimensional datasets, which frequently include superfluous and irrelevant features. To address these issues, researchers have suggested a range of feature selection methods. Filter, wrapper, and hybrid approaches.

Wang et al. proposed an enhanced WOA with a chaotic search strategy for feature selection in intrusion detection systems, demonstrating improved detection accuracy compared to the standard WOA [10]. Their approach incorporated chaos theory to enhance the algorithm's population initialization and search capability, addressing the premature convergence issue commonly observed in metaheuristic techniques.

The application of PSO for feature selection has been investigated by Pourbahrami, who introduced ISPSO-GLOBAL, an improved seeding PSO method that enhances the exploration capability of traditional PSO [2]. The method utilizes chaos theory for population initialization and employs a novel formula to automatically determine the optimal number of selected features, showing superior performance compared to conventional approaches.

HHO has emerged as a competitive algorithm for optimization problems due to its efficient balance between the exploration and exploitation phases. Heidari et al. demonstrated the effectiveness of HHO in solving various benchmark problems and engineering applications, highlighting its potential for feature selection tasks in classification problems [4]. The cooperative hunting behavior modelled in HHO provides a robust framework for navigating complex search spaces encountered in high-dimensional feature selection problems.

In IoT and industrial systems, metaheuristic optimization has gained traction for enhancing intrusion detection capabilities. Recent research has shown that metaheuristic algorithms can effectively optimize machine learning model parameters, including feature selection, improving accuracy, and reducing false-positive rates in attack detection systems [1]. This is particularly valuable in critical infrastructure such as automotive controller area networks (CAN), where security breaches can have severe consequences.

The recently introduced Parrot Optimizer (PO) represents a novel approach to optimization problems, drawing inspiration from the behaviors of trained parrots. Lian et al. demonstrated the application of PO to various optimization challenges, highlighting its efficiency in balancing exploration and exploitation through specialized foraging behaviors and Lévy flight strategies [7]. The algorithm's boundary control mechanisms and fitness tracking capabilities make it suitable for feature selection tasks.

Several researchers have conducted comparative studies of metaheuristic algorithms for feature selection. These studies evaluate algorithms based on classification accuracy, convergence speed, and computational efficiency. For instance, Gutierrez et al. investigated random initialization strategies in PSO for high-dimensional search spaces, highlighting the importance of effective initialization in achieving optimal results [2]. The application of chaos theory to enhance metaheuristic algorithms has gained attention in recent years. Chaotic maps can improve optimization algorithms' initialization and parameter adaptation, leading to better exploration of the search space and reduced likelihood of premature convergence [11]. This approach has been successfully applied to various algorithms, including PSO and WOA, resulting in improved performance for the feature selection task.

In the specific context of intrusion detection systems, hybrid approaches that combine multiple metaheuristic algorithms or integrate metaheuristics with machine learning techniques have shown promising results. These hybrid models leverage the strengths of different approaches to overcome limitations in individual algorithms, resulting in more robust feature selection and improved classification performance [12].

While significant progress has been made in applying metaheuristic optimization to feature selection for intrusion detection, there remains a need for comprehensive comparative studies that evaluate the performance of newer algorithms, such as SFOA, against established techniques. This research addresses this gap by thoroughly analyzing five nature-inspired optimization algorithms in the context of intrusion detection feature selection.

3 Problem Formulation

The feature selection problem in intrusion detection systems can be formulated as a multi-objective optimization problem where we aim to simultaneously minimize the number of selected features and the classification error rate. Let $X = [x_1, x_2, x_3, \ldots\ldots,$ $x_d]$ represent the complete set of d features in the network traffic data. Our objective is to find an optimal subset S ⊆ X such that:

$$Minimze\, f(S) = \alpha.E(S) + (1 - \alpha).\frac{|\, S \,|}{d} \tag{1}$$

$$Subject\ to\ |S| \geq 1$$

$$x_i \in \{0, 1\} \forall i \in \{1, 2, \ldots., d\} \tag{2}$$

$$\alpha \in [0, 1] \tag{3}$$

Where:

- $E(S)$ represents the Classification Error Rate using feature subset S
- $|S|$ represents the Cardinality of the selected feature subset.
- α represents the Weighting parameter balancing accuracy and dimensionality reduction.

This formulation addresses the dual challenges of maintaining detection accuracy while reducing computational complexity (features), particularly critical for real-time intrusion detection system [13]. The binary constraint x_i indicates a feature's presence (1) or absence (0) in the selected subset.

4 Proposed Algorithm

This section explains the proposed algorithm along with Machine learning algorithms.

4.1 Superb Fairy-Wren Optimization Algorithm (SFOA)

Superb Fairy Wren Optimization Algorithm (SFOA) has shown promising results in numerical optimization, engineering design, and high-dimensional feature selection classification problems [9]. By simulating three natural behaviors of the Superb fairy-wren-young bird growth, breeding, feeding, and predator avoidance, SFOA establishes a comprehensive mathematical model that efficiently navigates different areas of the search space, demonstrating high efficiency and robustness when compared against both classical and novel optimization algorithms.

4.1.1 Biological Inspiration

The SFOA draws inspiration from three distinct behavioral phases of Malurus cyaneus:

1. **Growth Stage**: Juvenile birds develop survival skills through social learning
2. **Breeding & Feeding**: Cooperative breeding with strategic food gathering
3. **Predator Avoidance**: Complex alarm system with rapid evasion patterns

These biological mechanisms are translated into mathematical models for efficient search space exploration and exploitation [9].

4.1.2 Algorithm Framework

(a) Initialization Phase

In the initial phase, the population of N solutions are initialized using chaotic mapping:

$$X_{i,j} = lb_j + Chaos(0, 1).\left(ub_j - lb_j\right) \tag{4}$$

Where $Chaos(0, 1)$ generates chaotic values using the logistic map equation.

(b) Growth Stage Model

In the growth stage model, juvenile development is modeled through progressive solution refinement:

$$X_{new} = X_i + \gamma.(X_{best} - X_{rand}) \tag{5}$$

Where γ represents the learning rate derived from the bird's growth curve.

(c) Breeding & Feeding Dynamics

In breeding & feeding dynamics, the cooperative feeding behaviour is captured through a special updating mechanism:

$$X_{updated} = \begin{cases} X_{best} - A.D \; if \; p < 0.5 \\ D'.e^{bl}.\cos(2\pi l) + X_{best} \; otherwise \end{cases} \tag{6}$$

Where $D = \mid C.X_{best} - X_i \mid$ and $l \in [-1, 1]$ controls spiral tightness.

(d) Predator Avoidance Strategy

Incorporates Lévy flight patterns for escaping local optima:

$$X_{new} = X_i + \beta.Levy(\lambda).X_i - X_{best} \tag{7}$$

The Lévy distribution parameter λ adapts dynamically based on the population diversity.

4.2 Machine Learning Algorithm

In this study, three widely used machine learning algorithms were employed to assess the effectiveness of the selected feature subsets for intrusion detection:

4.2.1 Decision Tree (J48)

The decision tree classifier constructs a tree-based model that splits the data based on feature values to maximize information gain using the entropy criterion [14]. This approach is valued for its interpretability and ability to handle numerical and categorical data. In intrusion detection, decision trees can reveal important decision rules for distinguishing between normal and malicious network behaviour.

4.2.2 Random Forest

The random forest algorithm, an ensemble method, creates multiple decision trees and combines their predictions to improve accuracy. By using randomness in feature selection and data sampling, random forests reduce overfitting and effectively handle high-dimensional datasets [15]. It makes the random forests well-suited for intrusion detection tasks for high-dimensional datasets consisting of irrelevant or redundant features.

4.2.3 Support Vector Machine (SVM)

The linear support vector machine obtains the optimal hyperplane that separates classes in the feature space. SVMs are well-known for their effectiveness in high-dimensional datasets and their ability to maximize the margin between classes, which is crucial for robust intrusion detection. Moreover, SVMs exhibit lower sensitivity to outliers and generalize well, especially when the number of features significantly exceeds the number of samples [16].

The evaluation employs three distinct classifiers—Decision Tree (J48) as a baseline interpretable model, Random Forest for ensemble-based learning, and Support Vector Machine (SVM) for margin-based classification— to rigorously assess the effectiveness of selected feature subsets across diverse machine learning paradigms [17].

4.3 Fitness Function

The solution vector $X = [x_1, x_2, x_3, \ldots\ldots, x_d]$ is converting into the binary vector $S = [s_1, s_2, s_3, \ldots\ldots, s_d]$ using the given equation.

$$s_i = \begin{cases} 1 \; if \; x_i > \theta \\ 0 \; otherwise \end{cases} \tag{8}$$

Where θ represents the threshold value.

The feature selection process in this research is formulated as a weighted-sum optimization problem, balancing two primary objectives: minimizing the classification error and the number of selected features. The fitness of a candidate feature subset is computed as a weighted sum of these objectives:

$$Fitness = \alpha \times Error\ Rate + (1 - \alpha) \times Feature\ Ratio \tag{9}$$

The Error Rate is the proportion of misclassified samples when the selected feature subset is used to train and validate a classifier. It directly reflects the predictive performance of the chosen features. The Feature Ratio is the ratio of selected features to the total number of available features. It encourages the selection of smaller, more informative feature subsets, which can reduce computational cost and improve generalization.

The parameter α controls the trade-off between classification accuracy and feature reduction. By varying α(e. g., 0, 0.3, 0.5, 0.7, and 1),the optimization process can prioritize minimizing the number of features, maximizing classification accuracy, or finding a balance between the two. This approach is a typical wrapper-based feature selection strategy, where the performance of each candidate subset is evaluated using machine learning models, and the optimization algorithm searches for the subset that offers the best overall fitness [18].

This weighted-sum fitness computation is essential for intrusion detection systems, as it ensures that the resulting models are both accurate and efficient, capable of operating effectively in real-world, high-dimensional network environments.

5 Experimental Evaluation

This section presents the experimental setup, results, and observations to evaluate the performance of SFOA against Cheetah Optimizer (CO), Harris Hawk Optimization (HHO), Whale Optimization Algorithm (WOA), and Parrot Optimization (PO), in terms of accuracy and selecting optimal features using the UNSW-NB15 dataset.

5.1 Implementation Details

The experimental framework was developed using Python 3.12.0 and the scikit-learn library for machine learning and evaluation. The **UNSW-NB15** dataset was used as the primary benchmark, with categorical features encoded and all features standardized before feature selection and classification. The metaheuristic algorithms, including the Superb Fairy-wren Optimization Algorithm (SFOA), were implemented as modular components, allowing for direct comparison under identical conditions.

The experiments were conducted on a macOS operating system. The setup included an Apple M3 chip, 8-core CPU, 10-core GPU, installed with 16GB unified memory and 1TB SSD, ensuring reproducibility and fair resource allocation. For each metaheuristic algorithm, the same random seed was used to ensure consistency across runs. Feature selection was performed with a population size and iteration count as specified in Table 1. After feature selection, the reduced feature sets were evaluated using three classifiers: Decision Tree (J48), Random Forest, and Support Vector Machine (SVM), with 5-fold cross-validation to ensure robust performance estimates. The fitness function incorporated classification error and the proportion of selected features, with the trade-off controlled by varying the alpha parameter.

Table 1. Input Parameter for Metaheuristic Algorithms.

Parameter Name	Values	Description
Population Size (N)	10	Number of candidate solutions
Iterations (T)	50	Maximum number of iterations
Threshold value (θ)	0.5	Threshold Value for Feature Selection
Alpha (α)	0, 0.3, 0.5, 0.7, 1	Trade-off between accuracy and feature reduction
Classifiers	J48, Random Forest, SVM	Used for evaluating selected feature subsets
Cross-validation	5-fold	For a robust performance estimation
Dataset	UNSW-NB15	Benchmark for intrusion detection

For a fair comparison, the same parameter settings were used for all metaheuristic algorithms.

5.1.1 Convergence Behavior

The convergence behavior of SFOA was analyzed by computing the fitness value, as per Eq. (1), over 50 iterations for different values of weighting factor. The convergence curves in Fig. 2 depict that SFOA efficiently navigates the search space, achieving rapid and stable convergence. Typically, lower alpha values, which prioritize feature reduction, converge more quickly, whereas higher alpha values, which focus on accuracy, may require more iterations to reach optimal fitness. These findings indicate that SFOA is stable, robust and effective, consistently identifying high-quality feature subsets irrespective of the value of weighting factor.

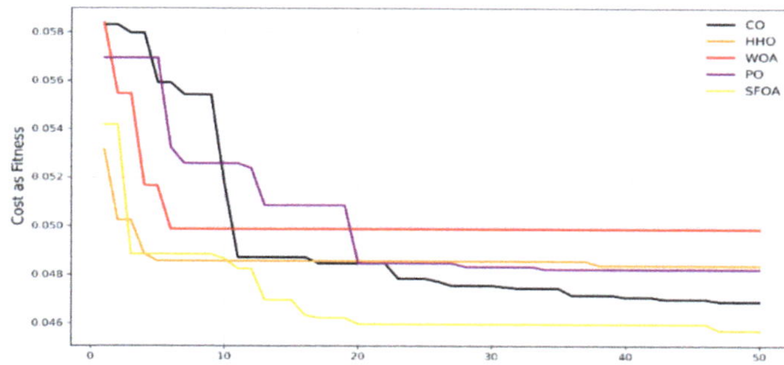

Convergence Behaviour for $\alpha = 0.0$

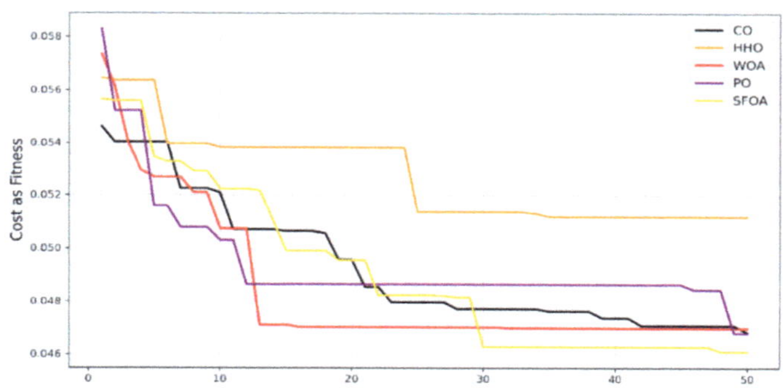

Convergence Behaviour for $\alpha = 0.3$

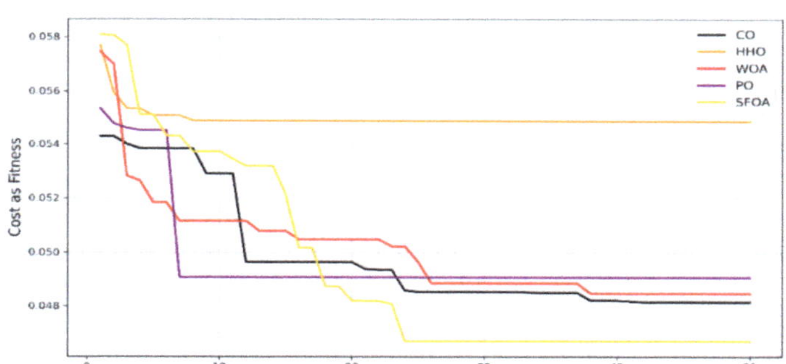

Convergence Behaviour for $\alpha = 0.5$

Fig. 2. Convergence Behaviour of Metaheuristics Algorithms.

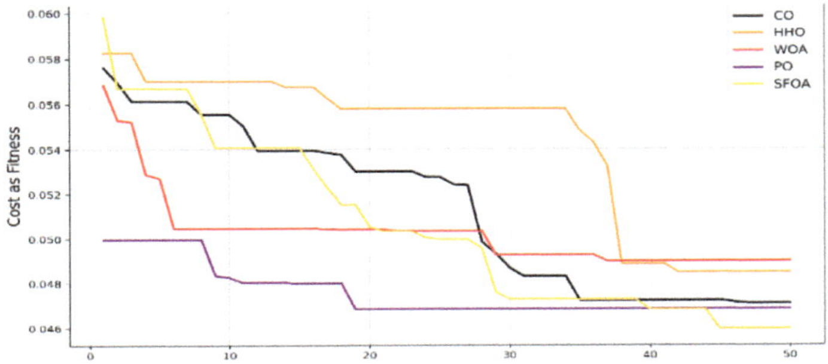

Convergence Behaviour for $\alpha = 0.7$

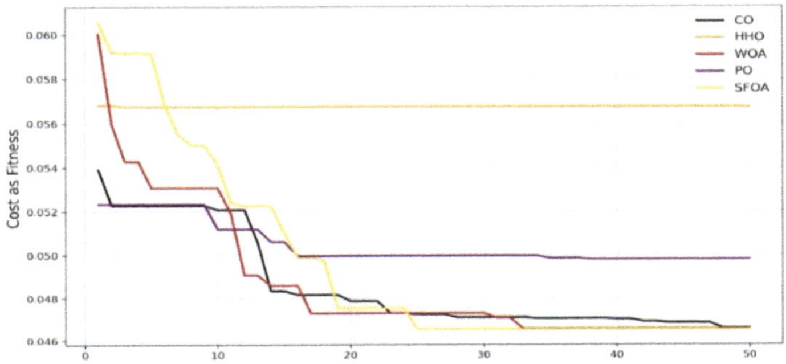

Convergence Behaviour for $\alpha = 1.0$

Fig. 2. *(continued)*

5.1.2 Accuracy Analysis Across

The performance of the Superb Fairy-wren Optimization Algorithm (SFOA) was systematically evaluated across a range of alpha values (0, 0.3, 0.5, 0.7, and 1) to investigate the trade-off between feature reduction and classification accuracy. As depicted in Fig. 3 below, the results demonstrate that SFOA consistently achieves high classification accuracy for all tested alpha values. Notably, increasing the alpha value, which emphasizes classification accuracy in the fitness function, leads to a corresponding improvement in detection performance.

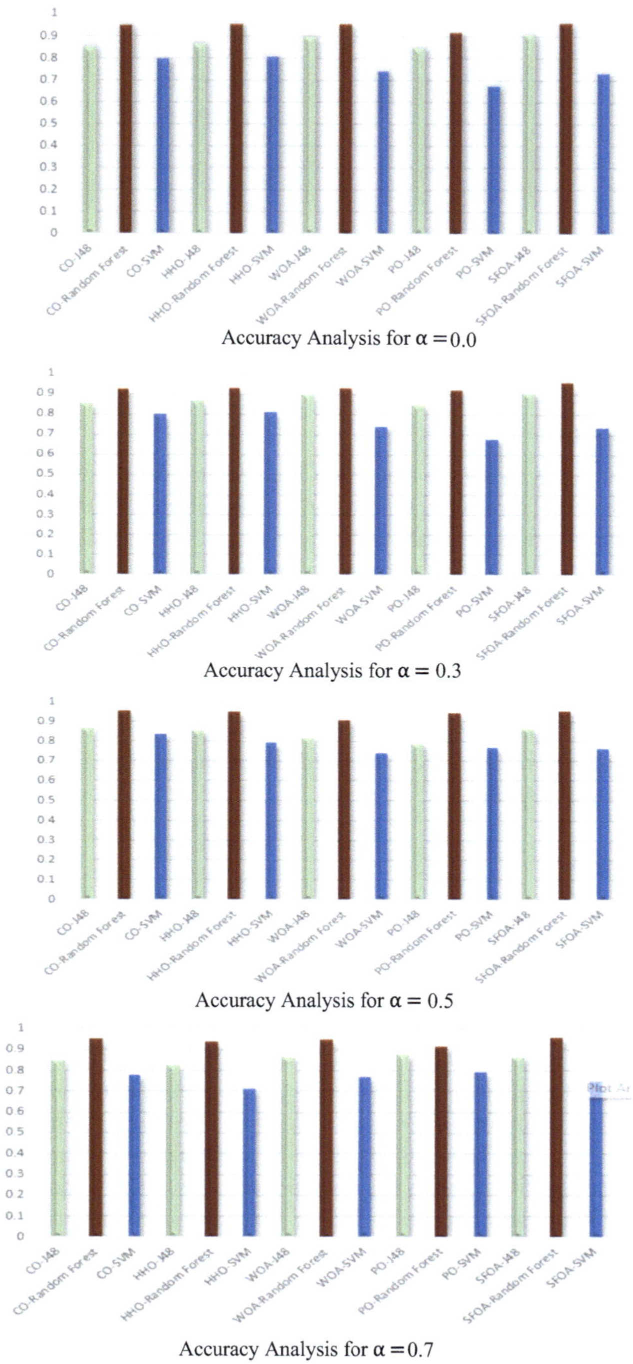

Fig. 3. Accuracy Analysis of Metaheuristics Algorithms.

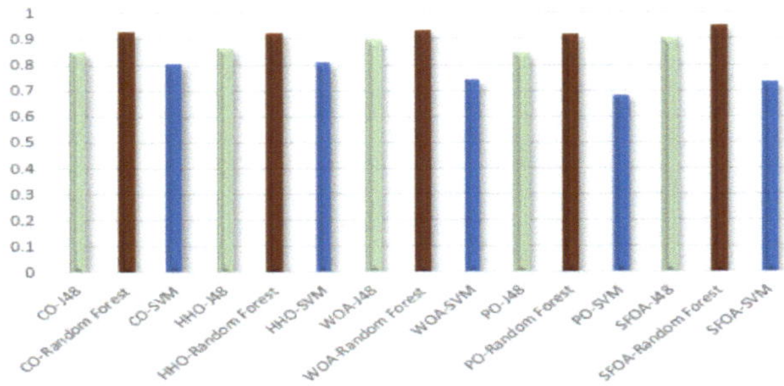

Accuracy Analysis for $\alpha = 1.0$

Fig. 3. *(continued)*

5.1.3 Observations

The experimental analysis shows that when it comes to feature selection for intrusion detection, the Superb Fairy-wren Optimisation Algorithm (SFOA) continuously performs better than alternative metaheuristic techniques. Across all tested values of the trade-off parameter alpha, SFOA achieves robust and high classification accuracy. The convergence curves for each alpha value show that SFOA rapidly and steadily reduces the fitness value, indicating efficient search space exploration and exploitation. These results confirm that SFOA is efficient and reliable, effectively identifying high-quality feature subsets regardless of the specific balance between accuracy and feature reduction. This adaptability makes SFOA a strong candidate for real-world intrusion detection systems, where operational requirements may vary, and detection performance and computational efficiency are crucial [19–21].

6 Conclusion

This work proposes an innovative approach for feature selection in intrusion detection system (IDS) using the Superb Fairy-wren Optimization Algorithm (SFOA). The SFOA demonstrates the best performance in feature selection for intrusion detection systems, achieving higher classification accuracy and faster convergence compared to existing metaheuristic approaches. By balancing exploration and exploitation through its biologically inspired phases, SFOA effectively reduces feature dimensionality while maintaining robust detection capabilities, making it a promising solution for real-time cybersecurity applications. The adaptability of its alpha parameter further allows practitioners to tailor the algorithm to specific operational needs, whether prioritizing compact feature subsets or maximizing accuracy. Future research endeavours will investigate integrating SFOA with reinforcement learning models for more complex intrusion scenarios and explore its dynamic adaptation in real-time network environments. Using reinforcement learning, the algorithm can be extended to handle large-scale, distributed datasets to further enhance its practical applicability.

Acknowledgments. I (Mohammad Sajid) would like to thank the authorities of the Inter University Centre for Astronomy and Astrophysics, Pune, India, for providing research facilities.

References

1. Dakic, P., Zivkovic, M., Jovanovic, L., et al.: Intrusion detection using metaheuristic optimization within IoT/IIoT systems and software of autonomous vehicles. Sci. Rep. **14**(1), 22884 (2024). https://doi.org/10.1038/S41598-024-73932-5
2. Pourbahrami, S.: Improving PSO global method for feature selection according to iterations global search and chaotic theory. arXiv:1811.08701 (2018). https://doi.org/10.48550/arXiv.1811.08701
3. Li, G., Zhang, T., Tsai, C.Y., et al.: Review of the metaheuristic algorithms in applications: visual analysis based on bibliometrics. Expert Syst. Appl. **255**(Part D, 124857) (2024). https://doi.org/10.1016/j.eswa.2024.124857
4. Heidari, A.A., et al.: Harris hawks optimization: algorithm and applications. Futur. Gener. Comput. Syst. **97**, 849–872 (2019). https://doi.org/10.1016/J.FUTURE.2019.02.028
5. Qasim, M., Sajid, M.: An efficient IoT task scheduling algorithm in cloud environment using modified firefly algorithm. Int. J. Inform. Technol. (Singapore). **17**(1), 179–188 (2024). https://doi.org/10.1007/S41870-024-01758-5/FIG.S/5
6. Mirjalili, S., Lewis, A.: The whale optimization algorithm. Adv. Eng. Softw. **95**, 51–67 (2016). https://doi.org/10.1016/J.ADVENGSOFT.2016.01.008
7. Lian, J., et al.: Parrot optimizer: algorithm and applications to medical problems. Comput. Biol. Med. **172**, 108064 (2024). https://doi.org/10.1016/J.COMPBIOMED.2024.108064
8. Akbari, M.A., Zare, M., Azizipanah-abarghooee, R., Mirjalili, S., Deriche, M.: The cheetah optimizer: a nature-inspired metaheuristic algorithm for large-scale optimization problems. Sci. Rep. **12**(1), 1–20 (2022). https://doi.org/10.1038/S41598-022-14338-Z
9. Jia, H., Zhou, X., Zhang, J., Mirjalili, S.: Superb fairy-wren optimization algorithm: a novel metaheuristic algorithm for solving feature selection problems. Clust. Comput. **28**(4), 246 (2025). https://doi.org/10.1007/S10586-024-04901-W
10. Tang, J., Wang, L.: A whale optimization algorithm based on atom-like structure differential evolution for solving engineering design problems. Sci. Rep. **14**(1), 1–27 (2024). https://doi.org/10.1038/S41598-023-51135-8
11. Jawed, M.S., Sajid, M.: COBLAH: a chaotic OBL initialized hybrid algebraic-heuristic algorithm for optimal S-box construction. Comput. Stand Interfaces. **91**, 103890 (2025). https://doi.org/10.1016/J.CSI.2024.103890
12. Sajid, M., Mittal, H., Pare, S., Prasad, M.: Routing and scheduling optimization for UAV assisted delivery system: a hybrid approach. Appl. Soft. Comput. **126**, 109225 (2022). https://doi.org/10.1016/J.ASOC.2022.109225
13. Alazzam, H., Sharieh, A., Sabri, K.E.: A feature selection algorithm for intrusion detection system based on pigeon inspired optimizer. Expert Syst. Appl. **148**, 113249 (2020). https://doi.org/10.1016/J.ESWA.2020.113249
14. Panigrahi, R., Borah, S.: Rank allocation to J48 group of decision tree classifiers using binary and multiclass intrusion detection datasets. Procedia Comput. Sci. **132**, 323–332 (2018). https://doi.org/10.1016/J.PROCS.2018.05.186
15. Hasan, M.A.M., et al.: Support vector machine and random forest modeling for intrusion detection system (IDS). J. Intell. Learn. Syst. Appl. **6**(1), 45–52 (2014). https://doi.org/10.4236/JILSA.2014.61005
16. Mohammadi, M., et al.: A comprehensive survey and taxonomy of the SVM-based intrusion detection systems. J. Netw. Comput. Appl. **178**, 102983 (2021). https://doi.org/10.1016/J.JNCA.2021.102983

17. Al-Manaseer, H., et al.: A novel big data classification technique for healthcare application using support vector machine, random forest and J48. In: Classification Applications with Deep Learning and Machine Learning Technologies, Studies in Computational Intelligence, vol. 1071, pp. 205–215. Springer International Publishing (2023). https://doi.org/10.1007/978-3-031-17576-3_9
18. Varzaneh, Z.A., Hosseini, S.: An improved equilibrium optimization algorithm for feature selection problem in network intrusion detection. Sci. Rep. **14**(1), 1–14 (2024). https://doi.org/10.1038/S41598-024-67488-7
19. Fernando, C.A., Thomas, R., Mary Anita, E.A., Lapina, M.: A feature selection study on the bot-IoT dataset using ensemble classification techniques. In: 2024 IEEE International Conference on Contemporary Computing and Communications (InC4), Bangalore, India, pp. 1–9 (2024). https://doi.org/10.1109/InC460750.2024.10649035
20. Baleev, M., Shevchenko, A., Basan, E., Lapina, M., Elashmawi, W.H.: Detection of anomalous activity in wireless communication channels. In: Raza, Z., Babenko, M., Sajid, M., Lapina, M., Zolotarev, V. (eds.) AISMA-2023: International Workshop on Advanced Information Security Management and Applications. AISMA 2023 Lecture Notes in Networks and Systems, vol. 1207. Springer, Cham (2024). https://doi.org/10.1007/978-3-031-77229-0_3
21. Lapina, M., Dudun, G., Kotlyarov, D., Rjevskaya, N., Subramanian, S.J.: Analysis of an existing method for detecting adversarial attacks on deep neural networks. In: Alikhanov, A., Tchernykh, A., Babenko, M., Samoylenko, I. (eds.) Current Problems of Applied Mathematics and Computer Systems. CPAMCS 2023 Lecture Notes in Networks and Systems, vol. 1044, pp. 316–329. Springer, Cham (2024). https://doi.org/10.1007/978-3-031-64010-0_29

Provision of Security of Wireless Information Transmission Systems in View of Correlation Properties of Code Sequences

Alexander Zhuk(✉), Kirill Stogniy, and Elsi Midaev

North Caucasus Federal University, Stavropol 355017, Russia
alekszhuk@ncfu.ru

Abstract. The purpose of the work is to study ensembles of orthogonal multiphase code sequences obtained by computing the eigenvectors of Hermitian matrices with different values of arguments satisfying the required correlation characteristics. One of the properties of Hermitian matrix eigenvectors is their orthogonality, which allows to arrange them in orthogonal ensembles suitable for use in code-separated information transmission systems. Being one of the main characteristics of the code sequence, determining the stability and reliability of information transmission are autocorrelation and cross-correlation functions. Concurrently, the use of Hermitian matrices will enable the existence of a vast array of orthogonal ensembles that possess the requisite correlation properties It is known that the number of orthogonal structures used in code-division signal communication systems directly affects the structural secrecy of the entire data transmission system. The novelty of the work consists in studying the influence of arguments of coefficients of the basic Hermitian matrix on the values of side peaks of correlation functions of coded sequences obtained from its eigenvectors. The result consists in obtaining the ranges of admissible values that can be set by the arguments of the Hermitian matrix coefficients so that the side peaks of the correlation functions of coded sequences satisfy the necessary conditions.

Keywords: Hermitian matrices · Complex numbers · Orthogonal matrices · Code sequences · Structural secrecy · Information security · Communication systems · Eigenvectors · Autocorrelation function · Cross-correlation function

1 Actuality

To ensure information exchange in information transmission systems with code-based channel separation, two types of sequences are used: pseudorandom codes of maximum length based on shift registers intended for scrambling and short orthogonal codes representing ensembles of binary orthogonal code sequences intended for channel separation with subsequent spectrum expansion.

In [1–5] the necessity to increase the number (arsenal) of used ensembles of orthogonal code sequences (EOCS) in order to increase the structural secrecy of information transmission systems with code division of channels was proved. Structural secrecy of

M. Lapina et al. (Eds.): AISMA 2025, LNNS 1456, pp. 497–509, 2026.
https://doi.org/10.1007/978-3-032-07275-7_45

the data transmission system directly affects the level of system security, complicates the realization of detection and measurement of signal parameters in order to change them. A substantial quantity of used ensembles facilitates frequent modification of signal shapes, thereby enhancing resistance to radio-technical intelligence measures. However, an increase in the number of EOCS can lead to the fact that not all of them will meet the specified requirements for correlation characteristics. In view of the above, we can conclude that it is necessary to search for a way to obtain an increased number of EOCS, satisfying the requirements for correlation characteristics.

As demonstrated in works [6–11], it is possible to obtain an increased number of EOCPs by considering the eigenvectors of symmetric matrices. A salient property of symmetric matrices is that the eigenvectors associated with distinct eigenvalues are orthogonal to each other. The eigenvectors of a symmetric matrix obtained in this manner can be arranged in an orthonormalized basis.

The use of Hermitian matrices, which are a generalized case of symmetric matrices and, as a consequence, possess the property of orthogonality of their eigenvectors, allows to model ensembles of orthogonal multiphase code sequences (EOMCS), with greater, in comparison with symmetric matrices, variability, due to the increase in the number of degrees of freedom of the simulated sequences, due to the presence of a complex component in them. The increased degrees of freedom will allow generating more orthogonal structures, which in turn will lead to an increase in the structural secrecy of the system.

One of the main characteristics that determine the possibility and efficiency of using code sequences in the information transmission system are their autocorrelation and cross-correlation functions. This is due to the fact that the reception of code sequences is carried out with the help of matched filters (correlators), and the normalized filter response can be defined using the convolution integral [17].

$$\mathrm{R}_{jk}(\tau) = \frac{1}{E}\int_{-\infty}^{\infty} u_j(t)u_k(t+\tau)dt, \tag{1}$$

where E is the energy of signals (code sequences) $u_j(t)$ and $u_k(t)$, $u_j(t)$ is the signal (code sequence) at the input of the filter matched with the signal (code sequence) $u_k(t)$, τ is the time shift between $u_j(t)$ and $u_k(t)$.

Under the condition $j = k$ and $\tau = 0$

$$R_j(0) = R_{j\ \max} = \frac{1}{E}\int_{-\infty}^{\infty} u^2(t)dt = 1.$$

Which determines the maximum value of the signal level at the output of the matched filter at the moment of matching.

The right part of expression (1) defines the relationship between signals (code sequences) $u_j(t)$ and $u_k(t)$ at some time shift τ. Therefore, the index $R_{jk}(\tau)$ is called the correlation function characterizing the degree of similarity of signals (code sequences) $u_j(t)$ and $u_k(t)$ among themselves.

Autocorrelation and cross-correlation functions of signals (code sequences) are distinguished.

The autocorrelation function (ACF), of a signal (code sequences) $u(t)$ with finite energy, is a quantitative integral characteristic of the signal form, of revealing in the signal the nature and parameters of the mutual temporal relationship of samples, which always takes place for periodic signals, as well as the interval and degree of dependence of the values of samples at current moments of time on the prehistory of the current moment. The ACF can be defined by (1) under the condition that $j = k$, i.e. for the case when under the integral expression there is one signal and its copy shifted by the time interval τ at the analysis interval equal to the sequence period. Therefore, we will denote the autocorrelation function hereafter as $R_j(\tau)$.

The cross-correlation function (CCF) of two different signals (code sequences) describes the degree of similarity of their shape and their mutual location relative to each other. For two different signals (code sequences) $u_j(t)$ and $u_k(t)$, the CCF can be determined by (1), considering them on the analysis interval equal to the period of the sequences. Since the cross-correlation function is defined for two signals, hereinafter we will denote with double index $R_{jk}(\tau)$.

In [18] the concept of a system (ensemble) of signals, which is understood as a set of signals united by a single rule of their construction. In the author's opinion, the concept of system (ensemble) is fully applicable to a set of code sequences obtained by a single rule.

Since in the systems of information transmission with code division of channels with increased structural secrecy it is possible to use ensembles of orthogonal code sequences modeled by eigenvectors of Hermite matrices, in addition to solving the problem of their synthesis in an increased number, it is necessary that they have ACF and CCF satisfying the given requirements.

The purpose of this work is experimental modeling of possible variants of EOMCS on the basis of eigenvectors of Hermitian matrices of the 4th order at different values of arguments of diagonal coefficients of Hermitian matrices and estimation of their compliance with the required autocorrelation and cross-correlation characteristics.

2 Methodology

To obtain the EOMCS as an orthogonal matrix A_n based on the Hermitian matrix H_n where n is the order of the matrix, it is necessary to compute the eigenvectors of the matrix H_n for each eigenvalue $\lambda = \{\lambda_1, \lambda_2, \ldots\lambda_n\}$ [12–14].

An eigenvector of the matrix H is a nonzero vector x such that

$$H \cdot x = \lambda \cdot x, \tag{2}$$

where λ is the eigenvalue of matrix H.

The eigenvalues of the matrix H are defined as the roots of its characteristic equation

$$\det(H - \lambda I) = 0, \tag{3}$$

where I is the identity matrix.

Obtained on the basis of Hermitian matrices eigenvectors of the form $x = \{x_1, x_2, \ldots, x_n\}$ can be represented as an orthonormalized basis

$$H = \begin{bmatrix} a_{11} & a_{12} & \cdots & a_{1n} \\ a_{21} & a_{22} & \cdots & a_{2n} \\ \vdots & \vdots & \ddots & \vdots \\ a_{n1} & a_{n2} & \cdots & a_{nn} \end{bmatrix} \Rightarrow \lambda = \begin{bmatrix} \lambda_1 \\ \lambda_2 \\ \vdots \\ \lambda_n \end{bmatrix} \Rightarrow A = \begin{bmatrix} x_{11} & x_{12} & \cdots & x_{1n} \\ x_{21} & x_{22} & \cdots & x_{2n} \\ \vdots & \vdots & \ddots & \vdots \\ x_{n1} & x_{n2} & \cdots & x_{nn} \end{bmatrix}. \quad (4)$$

Consider a basic Hermitian matrix of order 4 with superdiagonal and subdiagonal filled by three diagonal coefficients as follows [14, 15].

$$H_4 = \begin{bmatrix} 0 & a_1 & 0 & 0 \\ \overline{a_1} & 0 & a_2 & 0 \\ 0 & \overline{a_2} & 0 & a_3 \\ 0 & 0 & \overline{a_3} & 0 \end{bmatrix}, \quad (5)$$

where $a_1 = A_1 e^{-i\varphi_1}$ means that the element is complex-conjugate to the element $\overline{a_1} = A_1 e^{i\varphi_1}$.

Based on the eigenvalues of the matrix (5), it is possible to determine the eigenvectors using one of the well-known methods [11–13].

Experimentally, the authors of the paper have found that the values of the modules of the diagonal coefficients of the Hermitian matrix of the form (5), allowing to obtain an orthonormalized basis with two-level modules of the eigenvectors (for an orthonormalized basis with values 0.5 and −0.5), are $a_1 = a_3 = 100$, $a_2 = 0.01$, and the matrix of the modules of the orthogonal eigenvectors itself will have the following form.

$$A_4 = \begin{bmatrix} 0,5 & 0,5 & 0,5 & 0,5 \\ 0,5 & 0,5 & -0,5 & -0,5 \\ 0,5 & -0,5 & -0,5 & 0,5 \\ 0,5 & -0,5 & 0,5 & -0,5 \end{bmatrix}.$$

To obtain EOMCS with variable phases of code sequence elements, it is shown in [7–9, 15] that this is possible by changing the arguments of complex numbers of diagonal coefficients of the basic Hermitian matrix of the form (5). In this case, different variants of the arguments of the diagonal coefficients yield different EOMCS with different correlation properties.

An important property of a coded sequence that ensures reliable synchronization during data transmission is its “good” correlation properties, which means that its ACF and CCF should have “small” maximum side peaks, i.e., they should tend to zero [17, 18]. However, it is impossible to obtain zero peaks if the coding sequence has nonzero elements, because at time shifts in the process of calculating the ACF and CCF only the last shift for the coding sequence will always give a nonzero result [19, 29].

To establish the correlation properties of EOMCS obtained on the basis of Hermitian matrices of order 4, the authors enumerated all possible arguments φ_i, $i = 1, \ldots, n$ for all i-th coefficients of the basic Hermitian matrix of the form (5).

To find out how changing the argument in one of the three coefficients of the basic Hermitian matrix affects the correlation functions of the EOMCS, we enumerate the arguments in the range $\varphi_i = [0, 2\pi]$ with step $\Delta\varphi$ in one degree increments of each individual coefficient with fixing all other arguments of the diagonal HM coefficients and calculating the ACF and CCF side peaks of the obtained EOMCS.

The plots of the dependences of the ACF and CCF side peaks of the EOMCS obtained on the basis of the Hermitian matrix of the form (5) on the value of the argument of its diagonal coefficient $a_1 = 100 \cdot e^{i\varphi_1}$, $arg(a_1) = \varphi_1 \in [0, 2\pi]$, with constant values of the coefficients $a_2 = 0.01$, $a_3 = 100$, are presented in Figs. 1 and 2, respectively.

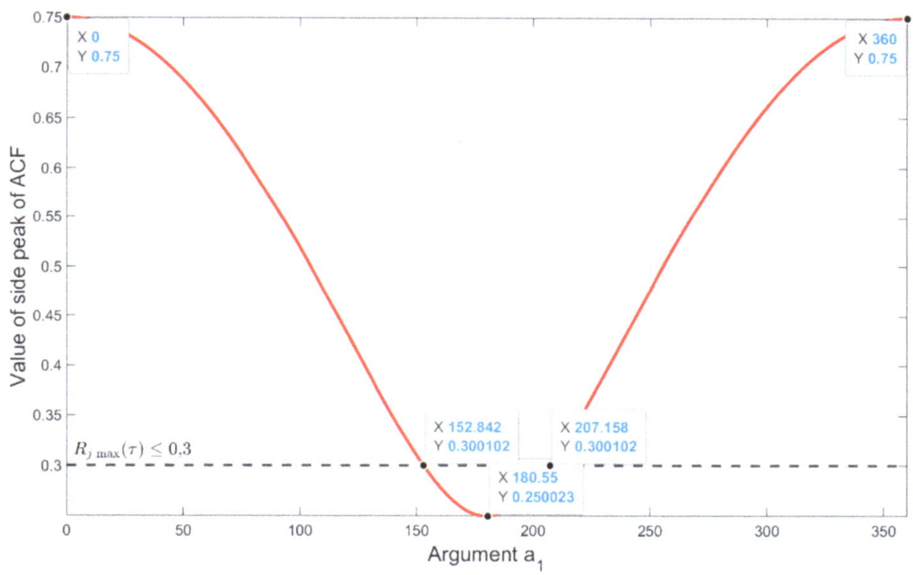

Fig. 1. Dependence of side peaks of ACF EOMCS on the value of the argument of the diagonal HM coefficient a_1.

It follows from Fig. 1 that at the value of the argument $arg(a_1) = \pi$, the minimum value of the ACF side peaks of EOMCS equal to $R_j(\tau) = 0.25$ is provided, and at the value of the $arg(a_1) = 0$; 2π, the maximum value of the ACF side peaks of the EOMCS equal to $R_j(\tau) = 0.75$ is provided.

Figure 2 shows that at the value of the $\arg(a_1) = \frac{\pi}{2}; \frac{3\pi}{2}$, the minimum value of the side peaks of the CCF of EOMCS equal to $R_{j,k}(\tau) = 0.56$ is provided, and at the value of the $arg(a_1) = \pi$; 2π, the maximum value of the side peaks of the CCF of EOMCS equal to $R_{j,k}(\tau) = 0.75$ is provided.

The plot of the dependences of the ACF side peaks of the EOMCS obtained on the basis of the Hermitian matrix of the form (5) on the value of the argument of its diagonal coefficient $a_2 = 0.01 \cdot e^{i\varphi_2}$, $arg(a_2) = \varphi_2 \in [0, 2\pi]$, with constant values of the coefficients $a_1 = 100$, $a_3 = 100$, are presented in Fig. 3.

Figure 3 shows that at the value of the arg $(a_2) = \frac{\pi}{2}; \frac{3\pi}{2}$, the minimum value of the side peaks of the ACF of EOMCS equal to $R_{j,k}(\tau) = 0.56$ is provided, and at the value of

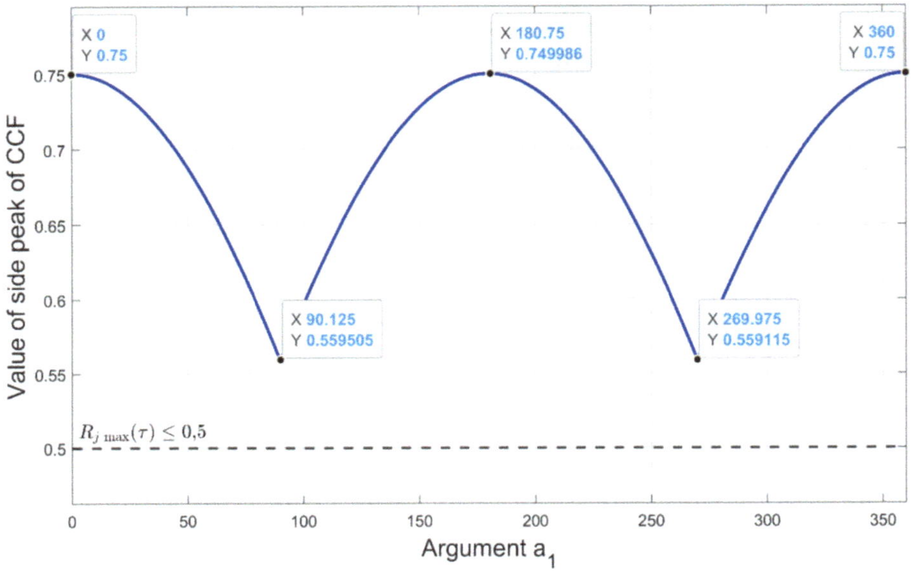

Fig. 2. Dependence of side peaks of CCF EOMCS on the value of the argument of the diagonal HM coefficient a_1.

the arg $(a_2) = \pi$; 2π, the maximum value of the side peaks of the ACF of EOMCS equal to $R_{j,\,k}(\tau) = 0.75$ is provided.

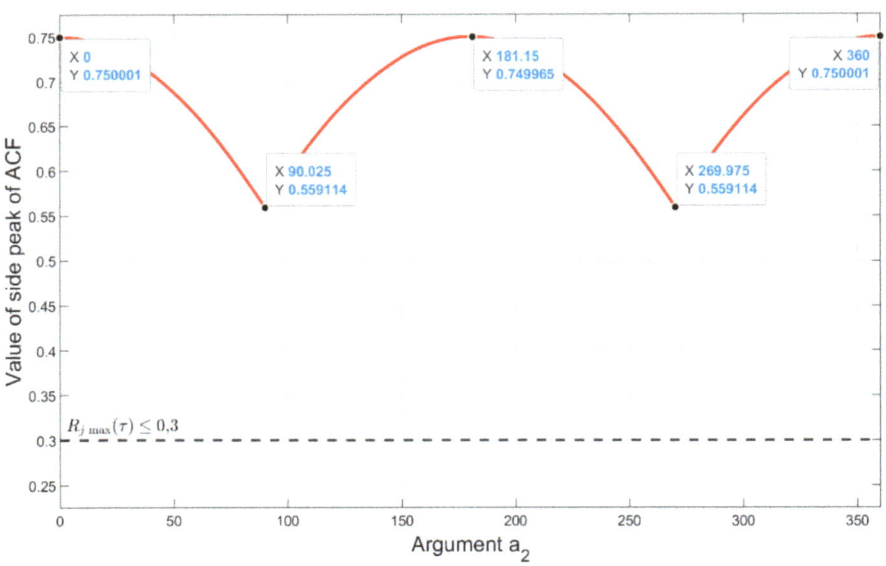

Fig. 3. Dependence of side peaks of ACF EOMCS on the value of the argument of the diagonal HM coefficient a_2.

The dependence of the side peak of the CCF of EOMCS on the change of the argument of the diagonal HM coefficient a_2 is similar to the dependence of the CCF of EOMCS at the change of the argument a_1 (see Fig. 2).

The dependences of the side peak of ACF and CCF of EOMCS on the change of the argument of the diagonal HM coefficient a_3 at constant values of the coefficients a_1 and a_3 coincide with the dependences of the side peak of ACF and CCF of EOMCS at the change of the argument of the diagonal HM coefficient a_1 (see Figs. 1 and 2).

The plots of the dependences of the ACF and CCF side peaks of the EOMCS obtained on the basis of the Hermitian matrix of the form (5) on the values of the arguments of its diagonal coefficients $a_1 = 100 \cdot e^{i\varphi_1}$, $arg(a_1) = \varphi_1 \in [0, 2\pi]$ and $a_2 = 0.01 \cdot e^{i\varphi_2}$, $arg(a_2) = \varphi_2 \in [0, 2\pi]$ with constant value of the coefficient $a_3 = 100$, are presented in Figs. 4 and 5, respectively.

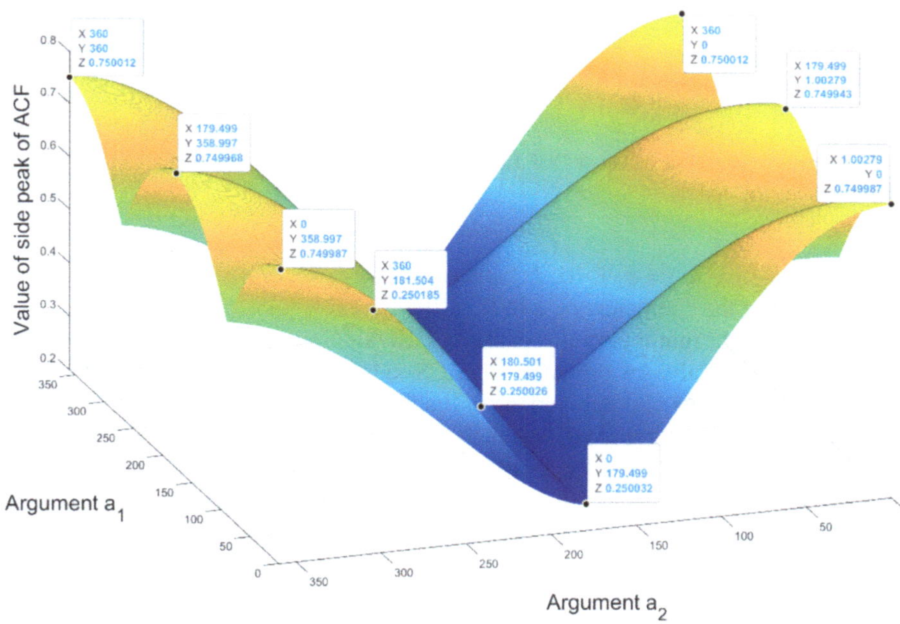

Fig. 4. Dependence of side peaks of ACF EOMCS on the values of the arguments of the diagonal HM coefficients a_1 and a_2.

The results of research of the values of the ACF and CCF side peaks of EOMCS at constant value of the coefficient $a_1 = 100$ from the change of the arguments of the diagonal coefficients of HM a_2 and a_3, are identical to the results of the research of the dependence of the values of the ACF and CCF side peaks of EOMCS at a constant value of the coefficient a_3 from the change of the arguments of the diagonal coefficients of HM a_1 and a_2 (see Figs. 4 and 5).

Figure 6 shows a plot of the dependence of the ACF of the EOMCS side peak on the arguments of the diagonal HM coefficients of the form (6) a_1 and a_3 with coefficients

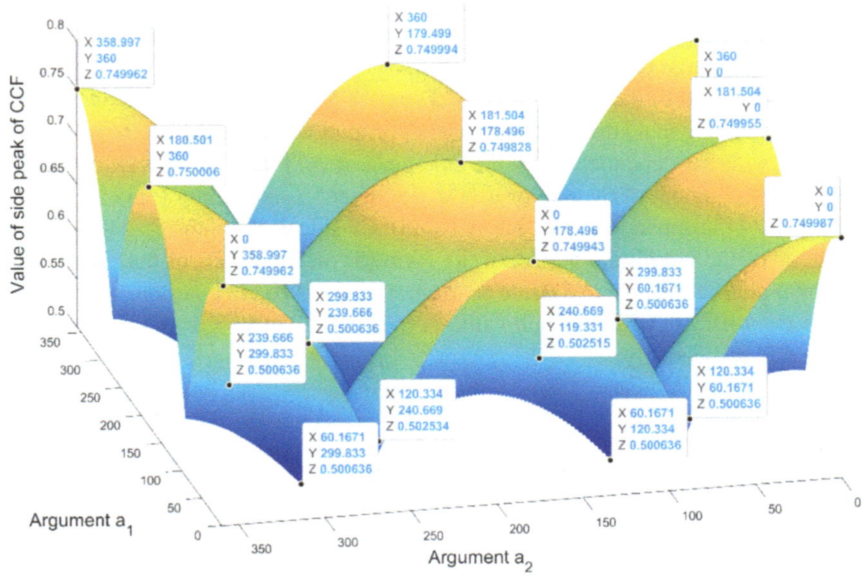

Fig. 5. Dependence of side peaks of CCF EOMCS on the values of the arguments of the diagonal HM coefficients a_1 and a_2.

$a_1 = 100 \cdot e^{i\varphi_1}$, $arg(a_1) = \varphi_1 \in [0, 2\pi]$ and $a_3 = 100 \cdot e^{i\varphi_3}$, $arg(a_3) = \varphi_3 \in [0, 2\pi]$ at a constant value of the coefficient $a_2 = 0.01$.

It follows that the minimum value of the ACF EOMCS side peaks equal to $R_j(\tau) = 0.25$ and the maximum value of the ACF EOMCS side peaks equal to $R_j(\tau) = 0.75$ are observed at the variants of combinations of the values of the arguments of the diagonal HM coefficients of the form (5) a_1 and a_3 (see Fig. 6).

The results of research the values of the side peaks of the CCF of the EOMCS and their analysis also shows that the dependence of the side peaks of the CCF on arguments a_1 and a_3 is identical to the dependence of the side peaks of the CCF on arguments a_1 and a_2 (see Fig. 4).

Visual representation of the dependence of the maximum side peak of the correlation functions on the arguments of all 3 coefficients of the basic Hermitian matrix is difficult due to the increase in dimensionality. However, the analysis shows that the addition of the dependence on the third coefficient leads only to a shift of the already showed plots (Figs. 4, 5 and 6) with respect to the X and Y axes.

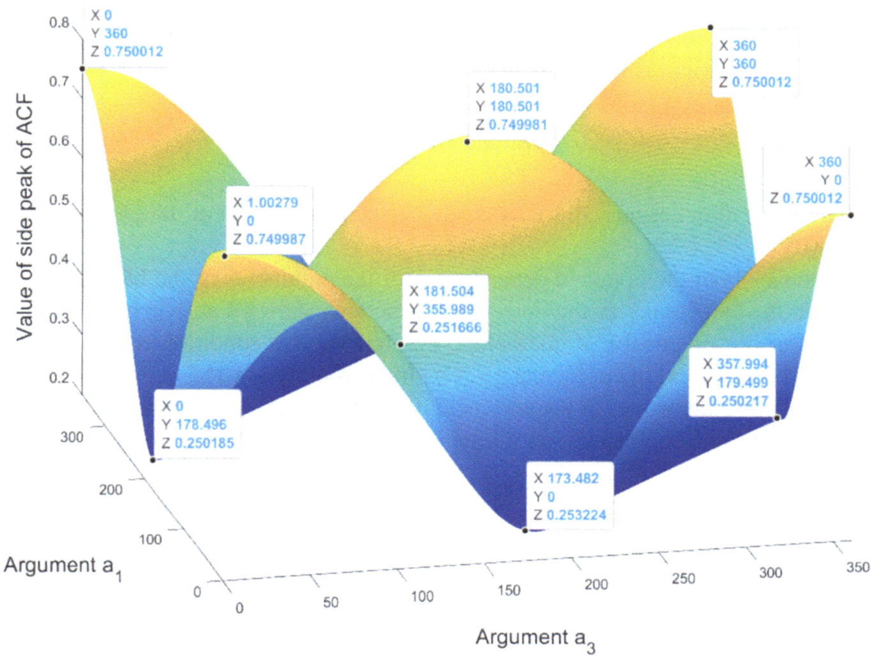

Fig. 6. Dependence of side peaks of ACF EOMCS on the values of the arguments of the diagonal HM coefficients a_1 and a_3.

3 Estimation of Compliance of Correlation Properties of Ensembles of Orthogonal Multiphase Code Sequences with Necessary Conditions

Summarizing the above and taking into account the necessity of using EOMCS with small side peaks of correlation functions, it is necessary:

1. Determine the effect of side peaks of ACF on the probability of correct detection of signals.
2. Determine the effect of side peaks of CCF on the probability of error in desynchronization.
3. Find the condition of smallness of the side peaks of ACF and CCF.

These problems were solved by Varakin L. E. in [16–18]. He revealed two conditions of smallness of side peaks of correlation functions. The condition of smallness of ACF side peaks will be called the first "Varakin condition". The condition of smallness of the side peaks of the ACF will be called the second "Varakin condition".

The first Varakin condition states that the influence of the maximum side peak of the ACF on the probability of correct detection of signals can be neglected if the value of the maximum side peak does not exceed 0.3

$$R_{j\ \max}(\tau) \leq 0.3.$$

The second Varakin condition states that the influence of the maximum side peak of the CCF on the error probability in the case of desynchronization can be neglected if the value of the maximum side peak is not greater than

$$R_{jk\ \max}(\tau) \leq \frac{1}{\sqrt{N}},$$

where N is the digit capacity of EOMCS. At $N = 4$ the side peak of the CCF should not exceed 0.5.

The analysis of the obtained dependences of the arguments of the basic Hermitian matrix on the side peaks of the correlation functions shows that, when changing the argument of one diagonal HM coefficient, the first Varakin condition is fulfilled at the values of the argument of coefficient a_1 or a_3 in the range from 152 to 207 degrees, and is not fulfilled at any values of the argument of coefficient a_2 second Varakin condition, for a change of one coefficient is not satisfied at any values of the arguments of any of the three coefficients.

By varying arguments a_1 and a_3, first Varakin condition is satisfied for the values of the arguments shown in Fig. 7.

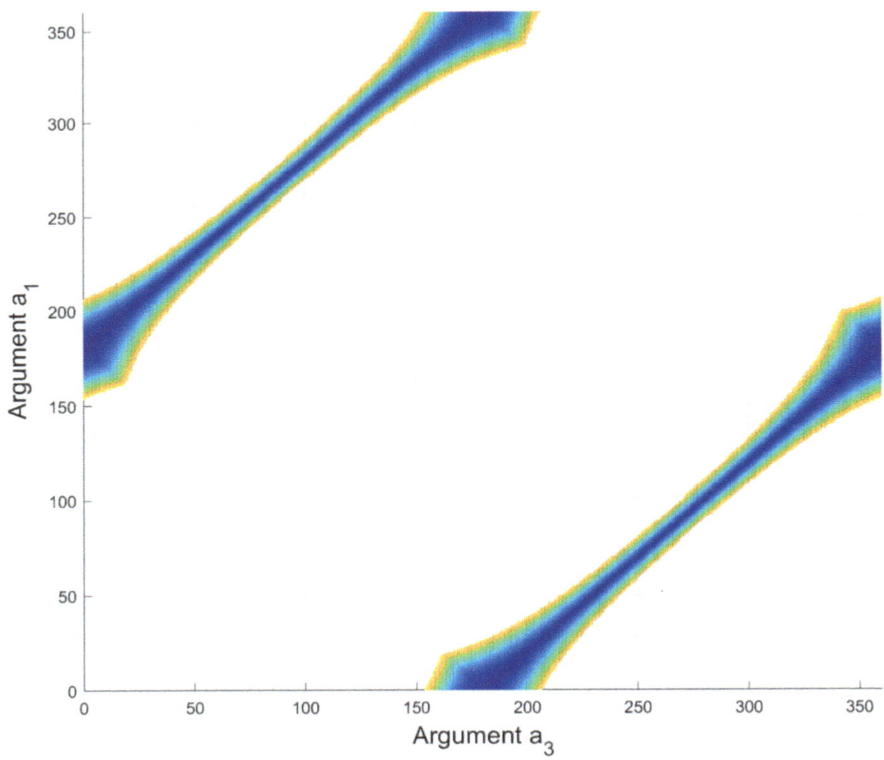

Fig. 7. The value of arguments a_1 and a_3 at which first Varakin condition is satisfied.

The number of EOMCS with ACF side peaks satisfying the first Varakin condition relative to all possible EOMCS obtained by changing arguments a_1 and a_3 is 10.09%.

Second Varakin condition, when arguments a_1 and a_3 are changed, is satisfied only at 8 specific values of arguments a_1 and a_3, which are

$$\begin{gathered}\varphi_1 = 60^\circ \& \varphi_3 = 120^\circ;\ \varphi_1 = 60^\circ \& \varphi_3 = 300^\circ;\\ \varphi_1 = 120^\circ \& \varphi_3 = 60^\circ;\ \varphi_1 = 120^\circ \& \varphi_3 = 240^\circ;\\ \varphi_1 = 240^\circ \& \varphi_3 = 120^\circ;\ \varphi_1 = 240^\circ \& \varphi_3 = 300^\circ;\\ \varphi_1 = 300^\circ \& \varphi_3 = 60^\circ;\ \varphi_1 = 300^\circ \& \varphi_3 = 240^\circ,\end{gathered}$$

where φ_1 and φ_3 is the arguments of coefficients a_1 and a_3, respectively.

As a consequence, the relative number of EOMCS satisfying the second Varakin condition is negligible.

By changing arguments a_1 and a_2, first Varakin condition is satisfied for the values of the arguments shown in Fig. 8.

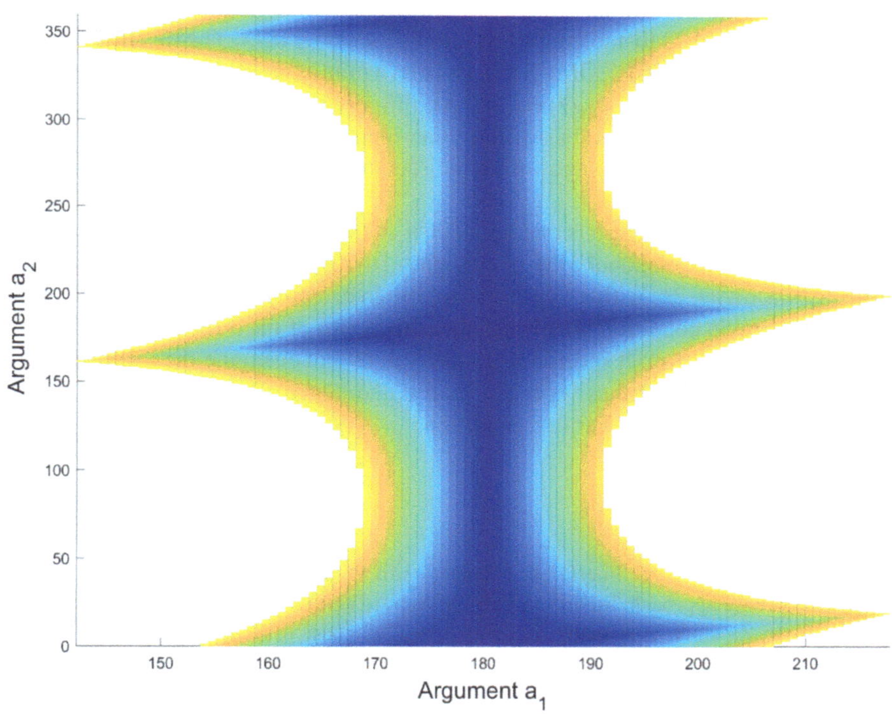

Fig. 8. The value of arguments a_1 and a_2 at which first Varakin condition is satisfied.

The number of EOMCS with ACF side peaks satisfying the first Varakin condition relative to all possible EOMCS obtained by changing arguments a_1 and a_2 is 10.06%.

Second Varakin condition, when arguments a_1 and a_2 are changed, is satisfied at the same values as when arguments a_1 and a_3 are changed.

The analysis of the CCF and CCF side peaks of the EOMCS when the arguments of all three diagonal PM coefficients a_1, a_2 and a_3 are changed has shown that the relative number of CCF and CCF side peaks satisfying the first and second Varakin conditions remains the same as when the arguments of two coefficients out of three are changed. The ranges of arguments at which the EOMCS satisfying the first and second Varakin conditions occur shift linearly relative to the argument of the third coefficient along the X and Y axes.

4 Conclusions

The results of analyzing the dependences of correlation functions of different EOMCS on the values of the arguments of the diagonal coefficients of the basic Hermitian matrix of the 4th order, obtained on the basis of the coefficients $\{a_1, a_2, a_3\}$, which are complex numbers, show that the eigenvectors of this matrix represent the EOMCS model, which has the following correlation characteristics.

The range of variation of the side peaks of the ACF EOMCS ranges from a minimum value of $R_{j\,min}(\tau) = 0.25$ to a maximum value of $R_{j\,max}(\tau) = 0.75$. The range of variation of the side peaks of the CCF EOMCS ranges from a minimum value of $R_{j,\,k\,min}(\tau) = 0.25$ to a maximum value of $R_{j,\,k\,max}(\tau) = 0.25$. At different variants of values of arguments of diagonal coefficients of the basic Hermitian matrix of the 4th order, side peaks of ACF and CCF of EOMCS change within the specified limits from the minimum to the maximum value.

The first Varakin condition is met by 10.06% of the obtained EOMCS, which, when the arguments of all diagonal HM coefficients $\{a_1,\ a_2,\ a_3\}$ are varied in the range $argarg\ (a_i) = \varphi_i \in [0; 2\pi]$ with a step $\Delta\varphi$ of one degree, amounts to 4665600 ensembles.

Second Varakin condition is only satisfied when two of the three coefficients of the Hermitian matrix have strict specific argument values, but changing the argument of the third coefficient will shift these specific values, and hence allow more EOMCS to be obtained that satisfy second Varakin condition.

References

1. Borisov, V., Zinchuk, V.: Interference Protection of Radio Communication Systems. Time-and-probability approach. RadioSoft Publ. Moscow (2008)
2. Borisov, V., Zinchuk, V., Limarev, A., Muhin, N., Nahmanson, G.: Interference Protection of Radio Communication Systems with Expansion of Signals Spectrum by Modulation of the Carrier with Pseudorandom Sequence. RadioSoft Publ, Moscow (2003)
3. Uriadnikov, I., Adzhemov, S.: Ultra-Wide-Band Communication. Theory and Application. Solon-Press Publ, Moscow (2013)
4. Makarenko, S., Ivanov, M., Popov, S.: Noise Immunity of Communication Systems with Pseudorandom Tuning of the Operating Frequency. Svoe izdatel'stvo Publ, Saint-Petersburg (2013)
5. Zhuk, A., Cherniak, Z., Sazonov, V.: On the feasibility of using ensembles of orthogonal signals with varying dimensions in the CDMA system. Infokommunikacionnye Tehnologii. **6**(4), 16–20 (2008)

6. Zhuk, A., Petrenko, V., Kuz'minov, I., Zhuk, E., Luganskaia, L.: Improving information exchange methods in high-speed wireless information networks using new types of discrete sequence ensembles. Mod. Probl. Sci. Educ. **5**, 144–145 (2013)
7. Zhuk, A., Burmistrov, V., Gavrishev, A.: Information transfer system using stochastic orthogonal ensembles of discrete multilevel signals. Modern Inform. Technol. IT-educ. **2**(11), 493–498 (2015)
8. Zhuk, A., Zhuk, E., Troshkov, A.: A method for transmitting information with pseudorandom rearrangement of the signal form for communication systems with code-division multiple access. In: Information Security-2012: Materials of the XII International Scientific and Practical Conference, pp. 346–350. Southern Federal University, Taganrog (2012)
9. Gavrishev, A., et al.: Stochastic orthogonal codes generator. Patent Russia, RU2615322C1 (2017)
10. Kosjakin, S., Moskvitin, I., Smirnov, A.: Method of transmitting information in systems with code division of channels and the device for its implementation. Patent Russia, RU2234191 (2004)
11. Verzhbitskii, V.: Numerical methods (linear algebra and nonlinear equations). Izdatel'skii dom ONIKS 21 vek Publ., Moscow (2005)
12. Golovina, L.: Linear Algebra and some of its Applications. Nauka Publ, Moscow (1971)
13. Demidovich, B., Maron, I.: Fundamentals of Computational Mathematics. Nauka Publ, Moscow (1970)
14. Popenko, V.: Estimation of the discrete signal's spectrum width. Radiotekhnika. **11**, 57–59 (1996)
15. Popenko, V.: Vector Synthesis of Orthogonal Signal Ensembles. Part 2. Stavropol Higher Military Engineering College of Communications, Stavropol (1993)
16. Varakin, L.: Communication Systems with Noise-like Signals. Radio i sviaz' Publ, Moscow (1985)
17. Varakin, L.: Complex Signal Theory. Sovetskoe radio Publ, Moscow (1970)
18. Varakin, L.: Signal System Theory. Sovetskoe radio Publ, Moscow (1978)
19. Golomb, S., Gong, G.: Signal Design for Good Correlation for Wireless Communication, Cryptography, and Radar. University Press, Cambridge (2005) Cambridge University Press
20. Viterbi, A.: CDMA: Principles of Spread Spectrum Communication. Addison-Wesley Pub. Co., Reading, Mass (1995)

Evaluation of Approaches to Modeling Information Risk of Telecommunication Systems Based on Hypergraphs

Vyacheslav Zolotarev[1], Elena Gorodetskaya[2(✉)], and Alexey Kuznetsov[2]

[1] Reshetnev Siberian State University of Science and Technology, 660037 Krasnoyarsk, Russian Federation

[2] Siberian State University of Telecommunications and Information Science, 630008 Novosibirsk, Russian Federation

elena.shmanina-97@yandex.ru

Abstract. The article considers the main features and suitability for solving the task of complex information security risk assessment of various risk management models based on hypergraphs or using hypergraphs to assess individual properties. The purpose of reviewing and comparing the models is to select a modeling tool for the task of multi-level risk assessment of a telecommunication system. The paper provides both an overview of the main models suitable for this task and their characteristics (the elements of the hypergraph approach used, the main estimated and applied in the models estimated system characteristics). The value of the study lies in the selection of useful properties of the models for the subsequent development of complex or hybrid models based on them for the assessment of various characteristics of information risk or a comprehensive indicator of information risk in the transmission of information in a telecommunications system. The obtained results can also be used in the analysis of information security incidents in telecommunication systems and applied (with appropriate refinement) in related areas, such as technological networks and corporate networks of organizations. The considered models can also be the basis for algorithms and protocols used in information protection tasks in the context of information risk assessment.

Keywords: Information Risk · Information Risk Model · Hypergraph · Information Risk Assessment

1 Introduction

Modern telecommunication systems are characterized by a high degree of interconnectivity and complexity, which generates new attack vectors and vulnerabilities. In this regard, hypergraphs, which allow modeling relationships between multiple objects simultaneously through hyperedges, become an effective tool for representing complex interdependencies. Their application allows predicting the development of incidents, optimizing real-time response and providing control over the information security system.

M. Lapina et al. (Eds.): AISMA 2025, LNNS 1456, pp. 510–521, 2026.
https://doi.org/10.1007/978-3-032-07275-7_46

In the course of scientific research the task was set to develop a graph-based mathematical model for risk assessment in information transmission in complexly organized dynamic systems. This model should take into account the type of information being transmitted, be adaptive to the conditions of information transmission and external destructive influences, as well as allow quantitative risk assessment. To represent complex multilevel interconnections between the nodes of the network, it is proposed to use a hypergraph model, where the vertices will represent the objects of the system, and hyperedges - group interconnections of nodes by a set of attributes.

In order to select a model that will act as a basis for achieving the task at hand, it is necessary to analyze existing mathematical models for information security management.

2 Mathematical Models

2.1 A Model Based on Markov Processes

Discrete or semi-continuous Markov models describe the dynamics of the states of a security system, where each state reflects the level of threat (e.g., "no threat", "under suspicion", "attack", "compromise"). Transitions between states are given by the transition probability matrix $P = [p_{ij}]$, where p_{ij} is the probability of transition from state i to state j [1, 14].

Dynamic programming algorithms (e.g., value iteration) are used to optimize the model: a strategy $\pi*$, which for each state selects the action a that minimizes the long-run risk, is computed [2].

Markov processes are combined with hypergraph models:

- The vertices of the hypergraph $H = (V, E)$ correspond to the states S_0, S_1, S_2.
- Hyperedges $e \in E$ link groups of nodes involved in the same attack (e.g. $S_1 \rightarrow S_2$).
- The weights $\omega(e)$ of the hyperedges specify the transition probabilities p_{ij}, strengthening the model for analyzing complex scenarios.

The model allows to:

- Predict the development of incidents.
- Select optimal defense measures in real time.
- Take into account multivariate interactions (via hyperedges) in risk assessment.

2.2 Bayesian Networks

The model is based on a probabilistic graph representation, where nodes represent events (e.g., vulnerabilities, incidents, indicators of abnormal behavior) and arcs reflect dependencies between them. Bayesian networks allow updating the risk representation as new data become available and combining expert knowledge with empirical observations [3, 4].

Bayesian networks can be augmented with hypergraph models:

- The vertices of the hypergraph $H = (V,E)$ correspond to the nodes of the Bayesian network.

- Hyperedges $e \in E$ link groups of nodes involved in complex dependencies (e.g., coordinated attacks).
- The weights *ω(e)* of the hyperedges specify the a priori probabilities or risk levels for the connected nodes [5].

The model allows to:

- The model is dynamically updated, allowing probabilities and risks to be recalculated in real time as new data becomes available.
- Expert knowledge (a priori probabilities) and empirical observations (likelihoods) are combined for more accurate risk assessment.
- Hypergraph structure helps to reveal complex hidden relationships between groups of nodes that are difficult to describe using traditional methods.

2.3 Game Theory Model (Security Games)

In this model, the opposing parties (attacker and defender) are represented as players seeking to maximize their winning functions. The model considers both static and dynamic (multi-step) scenarios, which allows to analyze the adaptability of attacks and responses. Nash equilibrium, the Stackholmer model (leader-sequester) and other game concepts are used to formalize strategies [6].

Game models are combined with hypergraphs:

- Vertices of the hypergraph $H = (V,E)$: nodes of the system (servers, devices).
- Hyperedges $e \in E$: groups of nodes attacked simultaneously (e.g., a coordinated DDoS attack) [7].
- The weights of the hyperedges of $\omega(e)$:
  - For the defender: the cost of defending a group of nodes.
  - For the attacker: the probability of a successful attack on the group.

The model allows to:

- Realize optimal resource allocation by determining which nodes or groups of nodes (hyperedges) require enhanced protection.
- By analyzing the attacker's strategies, the ability to predict vulnerable areas is realized.
- In multi-step games, strategies are updated in real time to account for changes in the system (e.g., new nodes in the hypergraph), which is an adaptation to the dynamics of change.

2.4 Graph models

Graphs are well suited for modeling the topology of telecommunication networks, where nodes V represent devices or servers and edges E represent communication links. Centrality, clustering and substructure detection metrics help to identify critical elements and potential attack vectors [8].

The model allows to:

- Identifying critical nodes:
  Centrality metrics help identify which nodes require enhanced protection (e.g., hubs with $k_i > 5$).
- Attack localization:
  High Q modularity indicates isolated subnets where attacks can be blocked locally.
- Monitoring Optimization:
  Nodes with high mediation $C_b(v_i)$ become prioritized for security sensor placement.

2.5 A Model of Attributed Attack Graphs

This model represents potential attack paths in a system as a graph, where nodes represent security states (e.g., vulnerabilities, compromised nodes) and edges represent possible transitions between them due to exploits. Attributes (probability of attack success, time or resource cost) are mapped to nodes and edges, allowing quantitative risk assessment for different attack scenarios.

The mathematical formalization of the graph is as follows:

$$\mathrm{A} = (V, E, A_v, A_E) \tag{1}$$

where:

- V - vertices corresponding to security states;
- $E \subseteq V x\ V$ - edges reflecting transitions between states;
- A_v - vertex attributes;
- A_E - edge attributes.

This model allows us to estimate the risk for an attack path, given the probability of success, cost and damage, and to solve the problem of selecting the optimal risk minimization strategy, given the existing resource constraints [9].

When extending attributed attack graphs to hypergraphs, hyperedges connect groups of nodes involved in complex attacks, the weights of the hyperedges *w(e)* include aggregate attributes:

$$w(e) = \prod\nolimits_{ve} Risk\ (v) * \prod\nolimits_{ee} p\ (e) \tag{2}$$

The result of the implementation of the attacking attributed graph model is:

- Quantitative risk assessment, by ranking attack paths by *R(π)*, highlighting the most critical scenarios;
- optimization of system protection by selecting nodes for vulnerability remediation based on the *Risk/Cost* ratio;
- Attack prediction, based on analyzing the dynamics of attributes to predict new nodes.

2.6 A Dynamic Model of Trust Graphs

In this model, telecommunication system elements (servers, workstations, IoT devices) are represented by nodes, and trust relationships and security policies are represented by edges. The graph reflects not only static topology, but also dynamic changes due to updates, configuration changes or node behavior.

The mathematical formalization of the graph is as follows:

$$T(t) = (V, E(t), W(t)) \tag{3}$$

where:

- V - nodes of the system;
- $E(t) \subseteq VxV$ are edges reflecting current trust relationships;
- $W(t)$ is a matrix of trust weights, where $Wij\ (t) \in [0;1]$ is the trust level of node v_i to node v_i at time t.

The trust level is updated based on events (e.g., successful authentication, anomaly detection):

$$W_{ij}(t+1) = W_{ij}(t) + \alpha * \Delta_{pos}(t) - \beta * \Delta_{neg}(t), \tag{4}$$

where:

- α, β - learning coefficients ($0 < \alpha, \beta < 1$);
- $\Delta_{pos}(t)$ - a positive event (e.g., correct execution of a request);
- $\Delta_{neg}(t)$ - negative event (e.g., unauthorized access attempt).

When calculating the security policy, thresholds are set for each edge e_{ij}:Θ_{min} - a minimum level of trust to authorize the interaction;

- Θ_{block} - the level at which the interaction is blocked.

The application of a dynamic trust risk model allows us to calculate the risk of system compromise through a specific individual node v_i:

$$R(v_i) = 1 - \prod\nolimits_{jNeighbors(v_i)} W_{ij}(t), \tag{5}$$

where $Neighbors(v_i)$ are the nodes associated with v_i.

When extending trust graphs to hypergraphs, hyperedges connect groups of nodes governed by a common policy (e.g., all IoT devices in the same subnetwork), the weights of the hyperedges $w(e)$ reflect the aggregate trust level of the group [10]:

$$w(e) = \frac{1}{e}\sum\nolimits_{v_i e} W_{ij}(t). \tag{6}$$

The result of implementing the dynamic trust graph model is:

- The model automatically monitors trust levels in real time.
- Prevent attacks by blocking interactions when trust drops below Θ_{block}.
- Model scalability due to trust management for groups of nodes via hyperedges.

2.7 Hypergraph Model of Threat Analysis

The model is based on a directed weighted graph and represents the telecommunication system infrastructure as a hypergraph *H* = *(V, E)*, where:

- The *V* vertices correspond to devices, network nodes, or services (e.g., server, IoT device, API gateway);
- Hyperedges *E* reflect multiple relationships (e.g., cooperative authentication, data sharing among a group of devices).

Mathematical formalization of the hypergraph. The structure of a hypergraph:

Each hyper edge *eE* connects *k* 2 vertices:

$$e = \{v_1, v_2, \ldots, v_k\}, v_i V. \tag{7}$$

- The weight of a vertex *w(v)*:

$$w(v) = \alpha * Criticality(v) + \beta - Vulnerability(v), \tag{8}$$

where *α, β* are the criticality and vulnerability significance coefficients.

- The weight of the hyperedge *w(e)*:

$$w(e) = \gamma * Frequency(e) + \delta * ThreatPotential(e), \tag{9}$$

where *γ, δ* are the weights of interaction frequency and threat potential. The total risk of the system is calculated as:

$$R(H) = \sum\nolimits_{eE(v)} w(e) * \left(\prod\nolimits_{ve} w(v)\right)^{\frac{1}{e}} \tag{10}$$

where *e* is the number of vertices in the hyperedge *e*.

The hypergraph centrality metric is used to identify hidden dependencies:

$$C(v) = \sum\nolimits_{eE(v)} \frac{w(e)}{\sum\nolimits_{ue} w(u)}, \tag{11}$$

where *E(u)* is the set of hyperedges containing vertex *v* [11].

Results of the model:

- Hypercreepers enable detection of coordinated attacks (e.g., DDoS via a group of IoT devices).
- The model optimizes protection by identifying bottle-necks through clustering.
- The temporal dynamics of the hypergraph predicts new threats.

2.8 Hypergraph Model of Attacks

This model extends the traditional weighted oriented graph approach by representing the system as a hypergraph *H* = *(V, E)*, where:

- The V vertices correspond to network components (servers, devices, services) or vulnerabilities.
- Hyperedges E reflect complex relationships, such as the need to exploit multiple vulnerabilities simultaneously for a successful attack [12].

Results of the model:

- Hyperedges enable detection of complex attack scenarios that require coordination.
- The model allows us to optimize protection by ranking hyperedges $R(e)$, highlighting critical vulnerabilities.
- Temporal dynamics helps to predict the evolution of attacks (e.g., propagation through new hyperedges).

2.9 Dynamic Anomaly Model Based on Temporal Hypergraphs

The model is based on a dynamic weighted oriented graph and represents the hypergraph as a dynamic structure $H(t) = (V,E(t))$, where time slices $H(\tau)$ reflect the state of the system at time τ. Changes in the structure of hyperedges (appearance/disappearance, modification of links) allow tracking anomalies, such as sudden spikes in node activity or atypical inter-node interactions. The methods of machine learning, including hypergraph convolutional neural networks, are used to detect patterns that deviate from the norm.

The temporal graph has the following mathematical formalization:

- Vertices V: static (e.g. devices, services)
- The hyperedges $E(t)$: Are dynamically time-varying.
- Time slice $H(\tau)$: the state of the hypergraph at time τ, where $\tau = 1,2, \ldots, \mathrm{T}$.

The anomaly metric at time t is defined through the discrepancy between the current and previous states:

$$A(t) = \sum\nolimits_{eE(t)} w(e,t) - w(e,t-1) * \frac{\sum_{ve} w(v)}{e}, \tag{12}$$

where:

- $w(e,t)$ is the weight of the hyperedge e at time t, reflecting risk or activity.
- $w(v)$ is the criticality of node v.

A loss function that minimizes the normal state prediction error is used to detect anomalies:

$$L = \sum\nolimits_{t=1}^{T} X(t) - \hat{X}(t)_F^2 + \lambda * Reg(\Theta), \tag{13}$$

where:

- $X(t)$ - matrix of node attributes at time t.
- $\hat{X}(t)$ - matrix predicted by the model.
- $Reg(\Theta)$ - regularization of model parameters Θ [13].

Results of the model:

- The model can identify abrupt changes in the structure of the hypergraph, thereby detecting anomalies in real time.
- *Reg(Θ)* regularization minimizes noise from random fluctuations, reducing false alarms
- The model is adaptive, by automatically learning hypergraph convolutions on new data, updating the norm patterns.

Table 1 compares approaches to modeling information risks in telecommunication systems based on hypergraphs.

Table 1. Summarized description of graph-based models

№	Model	Model description	Used parameters	Model capabilities
1	Model based on Markov processes	Dynamics of security system states. Each state reflects the level of threat. Transitions between states are defined by the transition matrix	t - point in time. s(t) - probability of the system being in the state i. Calculation of risk function J. Minimizing long-term risk. Transition matrix without protection.	Predict the development of incidents. Select optimal protection measures in real time. Take into account multivariate interactions when assessing risks.
2	Bayesian networks	Nodes represent random events, edges represent dependencies between them. When new events occur, the event probabilities are updated.	P - event probabilities. Risk assessment through expected losses. L - losses associated with the condition of the node.	The model is dynamically updated, allowing probabilities and risks to be recalculated in real time as new data become available. Expert knowledge (a priori probabilities) and empirical observations (likelihoods) are combined for more accurate risk assessment. The hypergraph structure helps to reveal complex hidden relationships between groups of nodes that are difficult to describe using traditional methods.

(*continued*)

Table 1. (*continued*)

№	Model	Model description	Used parameters	Model capabilities
3	Game Theory Model (Security Games)	Maximizing winning features for the attacker and the defender.	Value and defense are evaluated. Players' strategies are calculated.	The model considers both static and dynamic (multi-step) scenarios, which allows analyzing the adaptability of attacks and responses. The model determines which nodes (groups of nodes) require enhanced protection. Strategies can be analyzed to predict linguistic areas (attack prediction). Strategies are updated in real time, taking into account changes in the system.
4	Graph models	Telecom topology modeling. Networks Nodes - devices or servers Ribs - communication channels	Centrality metrics - calculation of node degree through elements of adjacency matrices. Calculation of node proximity and shortest paths between nodes. Shortest distance between nodes. Paths passing through a node Clustering coefficient by calculating triangles. Modularity is the quality of partitioning a graph into communities.	Identification of isolated groups of nodes for localizing attacks. Estimating the density of subnets vulnerable to local attacks (High Q modularity indicates isolated subnets where attacks can be blocked locally). Identification of priority nodes - nodes with high mediation counter-loops data flows between subnets (become a priority for security sensor placement). Identify the most vulnerable nodes - nodes with high mediation are more likely to be targeted for attacks Identify key data paths.

(*continued*)

Table 1. (*continued*)

№	Model	Model description	Used parameters	Model capabilities
5	A model of attributed attack graphs	It represents potential attack paths in the system as a graph, where nodes represent security states and edges represent possible transitions between them due to exploits.	Probability of attack success, costs, damage. Costs of eliminating the risk of compromise. Defense budget. Set of all possible attack paths. Risk of compromise.	Risk assessment for the attack path. Selecting the optimal defense strategy by minimizing risk with limited resources. Attack path ranking to identify the most critical scenarios. Optimizing the protection system by selecting nodes based on cost/value ratio Attack Prediction.
6	A dynamic model of trust graphs	In this model, the elements of the telecommunication system (servers, workstations, IoT devices) are represented by nodes, and trust relationships and security policies are represented by edges	A matrix of trust weights, where W is the trust level of a node to a node at time t. Learning coefficients, trust level. Risk of system compromise through a node through all nodes associated with the main one.	The model automatically monitors trust levels in real time. Attack prevention by blocking interactions when trust drops below ϴblock. Model scalability through trust management for groups of nodes via hyperedges.
7	Hypergraph model of threat analysis	The model is based on the representation of the telecommunication system infrastructure as a hypergraph Vertices – devices Hyperedges - interconnections (e.g., joint authentication, data exchange between a group of devices).	Criticality and vulnerability significance coefficients. Interaction frequency and threat potential weights total system risk. Hypergraph centrality. Set of hyperedges containing vertices.	Detecting hidden envy. Detecting coordinated attacks. Predicting new threats.
8	Hypergraph model of attacks	This model extends traditional graph-based approaches by representing the system as a hypergraph Vertices - network components or **vulnerabilities** Hyperedges - complex relationships, such as simultaneous exploitation of multiple vulnerabilities.	Vertex - characterized by the level of vulnerability. Rib - the complexity of the attack, depending on the number of vertices in the hyperlattice. Probability of success of the attack. System risk. Weight of the edge - potential damage from the attack.	Hyperedges allow detection of complex attack scenarios that require coordination. The model allows to optimize protection by ranking hyperedges R(e), highlighting critical vulnerabilities. Temporal dynamics helps to predict the evolution of attacks (e.g., propagation through new hyperedges).

(*continued*)

Table 1. *(continued)*

№	Model	Model description	Used parameters	Model capabilities
9	Dynamic anomaly model based on temporal hypergraphs	The model represents the hypergraph as a dynamic structure H(t)=(V,E(t)), where the time slices H(τ) reflect the state of the system at time τ.	Vertexes – devices. Hyperedges - change in time. Time slice - state of the hypergraph at time t. Edge weight - reflects risk or activity. Node criticality. Matrix of node features at time t.	The model allows us to identify abrupt changes in the hypergraph structure, thereby detecting anomalies in real time. Reg(Θ) regularization minimizes noise from random fluctuations, reducing false positives. The model is adaptive, by automatically training hypergraph convolutions on new data, updating the norm patterns.

3 Conclusions

Due to the fact that none of the considered models does not fully meet the requirements set in the course of scientific research to assess the risk of information transfer in complexly organized dynamic systems, for further work it is proposed to develop a modified mathematical model, taking as a basis a dynamic model of anomalies based on temporal graphs (№9). The model is also supposed to use such model parameters as calculation of centrality metric and total system risk from the hypergraph model of threat analysis (№7), calculation of the risk of system compromise through the central node (Dynamic Trust Graph Model - №6), as well as the ability to calculate the proximity of nodes and the shortest distance between them (Graph Model – №4).

The use of hypergraphs in mathematical modeling will allow not only to identify anomalous events characterized by a sharp change in eigenvalues, but also to scale the model for structures with similar types of risks with a different set of nodes. Clustering and community search in the hypergraph will help to detect groups of nodes with high interdependence, which may indicate possible "bottle necks" in the system. The temporal dynamics of the hypergraph, reflecting changes in the structure of relationships between components over time (e.g., the emergence of new hyperedges as a result of fragmentation attacks or system updates), can serve as a tool for incident prediction and real-time risk management.

References

1. Rosenko, A.: Markov models of confidential information security assessment taking into account the impact of internal threats on the automated information system. Sci. Innovat. Technol. **43**, 34–40 (2005)
2. Eryshov, V.G., Ilina, D.V.: Method of the information security monitoring process in information and telecommunication systems based on the application of methods of the theory

of Markov random processes. In: Wave Electronics and its Application in Information and Telecommunication Systems (WECONF), pp. 236–242 (2020)
3. Node Bayesian network. https://www.ibm.com/docs/ru/spss-modeler/saas?topic=models-bayesian-network-node. Accessed 28 Apr 2025
4. Krasyuk, P.P.: Information technology risk management: a review of mathematical models in information security. In: RISK MANAGEMENT - MODERN CHALLENGES: Collection of Scientific Works of Interuniversity Scientific-Practical Conference 2021, pp. 80–90, Moscow (2021)
5. Soloveichik, K., Svetlov, R., Arkin, P.: Bayesov decision making networks in risk management challenges of organization information system. Proceeding of UNECON no.2 (128), pp. 88–99 (2021)
6. Gromov, A.N.: Game models of information security. Dynamics of systems, mechanisms and machines no.4. pp. 71–73 (2014)
7. Erokhin, S.D., Borisenko, B.B., Fadeev, A.S.: Analysis and development of game-theoretic models information security of critical information infrastructure. System of synchronization, formation and processing of signals no.6, pp. 9–17 (2023)
8. Kulikov, I., Zhukova, N.: Practical aspects of using knowledge graphs for telecommunication networks modelling. In: Informatics and automation, pp. 1–5 (2020)
9. Kasyanov, V.N., Kasyanova, E.V.: A Model of Attributed Hierarchical Graphs with Ports for Visualizing Structured Information. NSU, Novosibirsk (2023)
10. Sukara, L.: Probabilistic Graph Models. Principles and Applications. DMK Press, Moscow (2021)
11. Zharikov, I.N.: Multimodal Topic Models Hypergraphs. MIPT, Moscow (2018)
12. Finogeev, A.A., Finogeev, A.G., Nefedova, I.S., Finogeev, E.A., Kamaev, V.A.: Analysis of information risks in distributed monitoring system based on FOG-computing model, https://vestnik.astu.org/en/nauka/article/32478/view. Accessed 28 Apr 2025
13. Kirzhin, D.A., Ilyushin, E.A.: Review and comparative analysis of attack and defense algorithms on graph-based ANN architectures. Int. J. Open Inform. Technol. **12**, 12–22 (2024)
14. Zolotarev, V., Maro, E., Nissenbaum, O., Ishchukova, E.: Markov and Semi-Markov Models of Real-Time Quests in Information Security Education. In: Ural Symposium on Biomedical Engineering, Radioelectronics and Information Technology (USBEREIT), pp. 221–224 (2019)
15. Fernando, C.A., Thomas, R., Mary Anita, E.A., Lapina, M.: A feature selection study on the Bot-IoT dataset using ensemble classification techniques. In: 2024 IEEE International Conference on Contemporary Computing and Communications (InC4), Bangalore, India, pp. 1–9 (2024). https://doi.org/10.1109/InC460750.2024.10649035
16. Baleev, M., Shevchenko, A., Basan, E., Lapina, M., Elashmawi, W.H.: Detection of anomalous activity in wireless communication channels. In: Raza, Z., Babenko, M., Sajid, M., Lapina, M., Zolotarev, V. (eds.) AISMA-2023: International Workshop on Advanced Information Security Management and Applications. AISMA 2023 Lecture Notes in Networks and Systems, vol. 1207. Springer, Cham (2024). https://doi.org/10.1007/978-3-031-77229-0_3
17. Lapina, M., Dudun, G., Kotlyarov, D., Rjevskaya, N., Subramanian, S.J.: Analysis of an existing method for detecting adversarial attacks on deep neural networks. In: Alikhanov, A., Tchernykh, A., Babenko, M., Samoylenko, I. (eds.) Current Problems of Applied Mathematics and Computer Systems. CPAMCS 2023 Lecture Notes in Networks and Systems, vol. 1044, pp. 316–329. Springer, Cham (2024). https://doi.org/10.1007/978-3-031-64010-0_29

Author Index

A
Abakumova, Veronika 255
Abramkina, Olga 1
Akhatov, Akmal Rustamovich 11
Alam, Mahfooz 19, 105
Andrusenko, Yulia 31, 397
Anurov, Aleksandr 451
Arkhipova, Anastasiya 40
Athavale, Vijay Anant 51, 63, 120, 298, 323
Awasekar, Dipali 51
Azarov, Ivan 323

B
Babenko, Mikhail 442
Babenko, Tetiana 1
Bagautdinova, Alina 277
Bezuglova, Ekaterina 63, 413
Bogadurov, Vasiliy 72

C
Chetyrbok, Petr 83

D
Deepak, 19, 105
Dixit, Rashmi Kamlakar 120
Dmitrienko, Alexey 31, 268
Dmitrii, Kuhta 384
Dolzhenkov, Sergey 131
Dorofeeva, Anna 83

E
Elashmawi, Walaa H. 255

F
Fedash, Diana 31
Fernandes, Melsa Magdalene 362

G
Geryugova, Aisanat 156
Gish, Alexander S. 298
Gladkov, Andrey 234, 413
Goncharov, Victor O. 224
Gorlacev, Dmitriy 156
Gorodetskaya, Elena 510
Govorov, Egor 72, 168
Govorova, Svetlana 72, 168
Grobova, Tatyana 362

I
Iakovleva, Anastasia 177
Ivanov, Andrey 186
Ivanova, Irina 286

J
Jenefa, J. 145
Jumanov, Isroil 200, 212

K
Kaladeep Yalagi, Pratibha C. 323
Karpova, Nadezhda E. 224
Karshiev, Khusan 200, 212
Kazak, Anatoliy 83
Komkov, Andrey 434
Krapivina, Milada 384, 397
Kucherov, Nikolay 234, 413
Kuchukova, Nataliya 442
Kuhta, Dmitrii 397
Kushchenko, Alisa 31
Kustov, Nikita 243
Kuznetsov, Alexey 510

L
Lapin, Vitalii 51, 120, 145, 255, 268
Lapina, Maria 19, 93, 277, 286, 425
Lidzhiev, Artem B. 298
Lobo, L. M. R. J. 51, 120
Lukyanov, Dmitry 31
Lukyanova, Yelena 83
Lutsenko, Vladislav 63, 156, 442

M. Lapina et al. (Eds.): AISMA 2025, LNNS 1456, pp. 523–524, 2026.
https://doi.org/10.1007/978-3-032-07275-7

M
Makarov, Igor 310
Maksimova, Elena 131, 277, 434
Mandritsa, Igor 310
Mandritsa, Olga 310
Manish, Kumar Singla 11
Mary Anita, E. A. 93, 145, 234, 362, 425
Mathpati, Shreya Shivkumar 323
Mayorova, Angela 83
Melnikov, Sergey 72, 168
Mezentseva, Oksana 434
Midaev, Elsi 497
Mirny, Nikita 156
Mishra, Divya 19
Modibbo, Umar Muhammad 277
Mustafa, Zeya 335

N
Nazarov, Fayzullo 352
Nechaev, Dmitry 186

O
Oleynikova, Anna 471

P
Peleshenko, Tatyana A. 298
Peleshenko, Victor S. 298
Pereira, Sharlene Anna 362
Petrenko, Vyacheslav 310

Q
Qasim, Mohammad 335, 370

R
Raza, Zahid 286
Rudenko, Andrei 384, 397
Rudenko, Marina 384, 397

S
Sajid, Mohammad 268, 335, 370, 482
Sajini, S. 145
Sayidkulov, Asliddin 352
Shahid, Mohammad 19, 168, 370
Shaltykova, Dina B. 462
Shiriaev, Egor 234, 413
Sokolovskii, Maksim 310
Sowmya, T. 425
Stogniy, Kirill 497
Suleimenov, Ibragim E. 462
Sumegh, Tharewal 200, 212
Syedneva, Daria 83

T
Tokmakova, Marina 255, 268

U
Ulugmurodov, Shokh Abbos Bakhodir ugli 11
Upadhyay, Manoj Kumar 105
Utenkova, Maria 434

V
Vershkov, Nikolay 442
Vinodha, D. 93
Vitulyova, Yelizaveta S. 462

X
Xamidov, Munis 11, 200, 212, 352

Y
Yakubova, Muborak 1
Yarmatov, Sherzodjon 352
Yarushev, Sergey 451
Yevstifeyev, Vladimir N. 462
Yurdanov, Evgeny 286

Z
Zakhir, Boris 471
Zayeem, Mohammad 482
Zgonnikov, Mikhail 156
Zhuk, Alexander 497
Zhukova, Marina 177
Zolotarev, Vyacheslav V. 224
Zolotarev, Vyacheslav 510

MIX
Papier aus verantwortungsvollen Quellen
Paper from responsible sources
FSC® C105338

If you have any concerns about our products,
you can contact us on
ProductSafety@springernature.com

In case Publisher is established outside the EU,
the EU authorized representative is:
Springer Nature Customer Service Center GmbH
Europaplatz 3, 69115 Heidelberg, Germany

Printed by Libri Plureos GmbH
in Hamburg, Germany